Bacterial and Bacteriophage Genetics

Fifth Edition

D1576857

Bacterial and Bacteriophage Genetics

Fifth Edition

Edward A. Birge

With 167 Illustrations

 Springer

Edward A. Birge
School of Life Sciences
Arizona State University
Tempe, AZ 85287-4501
USA

Cover illustration: Micrographs courtesy of Dr. Yuri Lyubchenko and Dr. Luda Shlyakhtenko, Department of Pharmaceutical Sciences, College of Pharmacy, 986025 Nebraska Medical Center, Omaha, Nebraska.

Library of Congress Control Number: 2005923811

ISBN-10: 0-387-23919-7
ISBN-13: 978-0387-23919-4

Printed on acid-free paper.

Printed in the United States of America.   (SPI/MVY)

9 8 7 6 5 4 3 2 1

springeronline.com

For Lori, Anna, and Mara

Preface

It has been exactly 25 years since I completed the first edition of this text, and its evolution is not yet complete. Continued from the previous edition are some boxed questions for students to read and think about before moving on to the next section. The **Thinking Ahead** boxes are intended to prod the students to make predictions based on their existing knowledge before reading the new material. The **Applications** boxes are intended to encourage the students to try their hands at experimental design. The answers to these latter questions are provided in an appendix, usually in the form of a reference to a research paper that addressed that specific question.

For instructors who have used the fourth edition of this book, there is only one major rearrangement of material. The material on DNA and chromosome structure has moved from Chapter 2 to Chapter 1 to allow room for more discussion of analytic techniques. Once again, the final four chapters are shorter and much less dependent on each other, so they can be used in any order or omitted at an instructor's discretion. All chapters contain updated material, but there are significant additions to Chapters 14, 16, and 17. I have tried to convey the concept of the "Age of Phage" as a theme throughout the book.

There is an appendix that provides Internet addresses where current information about maps and organisms is located because the amount of detail now available precludes publishing current maps in a book of this size. A supporting Web site for this book is located at http://lifesciences.asu.edu/bactgen/

Once again, it is a pleasure to acknowledge the support and able assistance of Paula Callaghan, Andrea Macaluso, Francine McNeill, and the other members of the editorial and production staff at Springer Science+Business Media, Inc. I am also grateful to my family for their patience while I was closeted with this manuscript.

Edward A. Birge
Yakima, Washington

Contents

Contents

Contents

11   Conjugation and the *Escherichia coli* Paradigm   351

12   Plasmids and Conjugation Systems Other than F   383

15 Site-Specific Recombination   463

16 Applied Bacterial Genetics   479

Contents

1

Fundamentals of Bacterial and Viral Genetics

Before beginning the formal study of the genetics of bacteria and their viruses, it is important to understand the fundamentals of microbial cell, chromosome, and DNA structures, and the basic viral processes. This chapter briefly reviews cell anatomy as well as genetic terminology and processes with respect to bacteria, yeast, and viruses and attempts to ensure that all students have the necessary background information for subsequent chapters. Detailed coverage of major genetic processes is provided later in this book. However, these summaries are essential for understanding some of the experiments presented in the next few chapters.

Major topics include:

- The cellular similarities and differences between prokaryotic and eukaryotic microorganisms
- Principal structural features of a DNA molecule

- DNA organization into chromosomes
- Types of genetic transfer processes operative in prokaryotic microorganisms
- Possible forms of prokaryotic chromosomes

THINKING AHEAD

What functions must a system of genetic nomenclature fulfill?

Nomenclature

All genetic nomenclature appears arcane when first encountered, but bacterial genetic terminology reads somewhat better than most. Prior to the adoption of rules proposed by Demerec et al. (1966), the nomenclature of bacterial genetics had developed in a somewhat haphazard fashion. The rules were intended to bring the nomenclature more in line with conventions of eukaryotic genetics. Each observable trait is given a three-letter italicized symbol that in many cases is an abbreviation of a mnemonic. For example, a mutation that affects proline biosynthesis is designated *pro*, and a mutation that affects utilization of proline as a carbon source is designated *put*. If the function of a gene is not known, a meaningless tag such as *yyy* is used.

In many cases, discrete genetic loci can be shown to affect the same phenotype. These loci are differentiated by assigning capital letters (e.g., *proA*, *proB*, *proC*). If more than 26 loci are needed, the mnemonic is incremented one letter in the third position, and the series repeats (e.g., large subunit ribosomal protein loci are *rplA*,..., *rplZ*, *rpmA*). As each new mutation is isolated, it is assigned a unique allele number that identifies it in bacterial pedigrees. An example of a complete notation for one mutation is *proA52*. When it is not certain if more than one genetic locus can affect a particular trait or the precise locus is not yet known, the capital letter is replaced by a hyphen (e.g., *pro-106*). Some genetic stock centers that coordinate naming and numbering of mutations are listed in Table 1.1. They are also sources of strains carrying known mutations. The commonly used gene symbols are given in Table 1.2.

The rules in the preceding paragraph concern the **genotype** of an organism (the total catalog of genetic capabilities of an organism, regardless of whether they are presently in use). A convention has developed in the literature

Table 1.1. Some better-known genetic stock centers.

American Type Culture Collection, 10801 University Blvd., Manassas, VA
20110-2209, USA

Bacillus Genetic Stock Center, Department of Biochemistry, The Ohio State
University, 776 Biological Sciences Building, 484 West 12th Avenue,
Columbus, OH 43210, USA

E. coli Genetic Stock Center, Department of Human Genetics, Yale University
School of Medicine, New Haven, CT 06520-8193, USA

S. typhimurium Genetic Stock Center, Department of Biological Sciences,
University of Calgary, Alberta, Canada T2N 1N4

German Collection of Microorganisms and Cell Cultures (DSMZ-Deutsche
Sammlung von Mikroorganismen und Zellkulturen GmbH) Mascheroder
Weg 1b, 38124 Braunschweig, Germany

The *Pseudomonas* Genetic Stock Center, Biotechnology Program, East Carolina
University School of Medicine, Greenville, NC 27858, USA

Note: The Internet addresses of the genetic stock centers are listed in Appendix 2.

Table 1.2. Frequently encountered genotype abbreviations for bacterial
genetics.

Symbol	Phenotype
ace	Acetate utilization
ade	Adenine requirement
ala	Alanine requirement
ara	Arabinose utilization
arg	Arginine requirement
aro	Aromatic amino acid requirement
asn	Asparagine requirement
asp	Aspartic acid requirement
cys	Cysteine requirement
cyt	Cytosine requirement
fla	Flagella biosynthesis
gal	Galactose utilization
glt	Glutamic acid requirement
gln	Glutamine requirement
gly	Glycine requirement

(*continued*)

3

Table 1.2. (*Continued*)

Symbol	Phenotype
gua	Guanine requirement
his	Histidine requirement
hut	Histidine utilization
ile	Isoleucine requirement
ilv	Isoleucine + valine requirement
lac	Lactose utilization
leu	Leucine requirement
lys	Lysine requirement
mal	Maltose utilization
man	Mannose utilization
met	Methionine requirement
mtl	Mannitol utilization
nal	Nalidixic acid sensitivity
phe	Phenylalanine requirement
pho	Alkaline phosphatase activity
pro	Proline requirement
pur	Purine biosynthesis
put	Proline utilization
pyr	Pyrimidine biosynthesis
rec	Recombination proficiency
rha	Rhamnose utilization
rif	Rifampin resistance
rpo	RNA polymerase activity
rpsE	Spectinomycin resistance (ribosomal protein, small subunit)
rpsL	Streptomycin resistance
ser	Serine requirement
spo	Spore formation (*Bacillus*); magic spot production (*E. coli*)
str	Streptomycin resistance = *rpsL*
sup	Suppressor (usually a tRNA)
thi	Thiamine (vitamin B$_1$) requirement
thy	Thymine requirement
ton	Phage T1 resistance
trp	Tryptophan requirement
tsx	Phage T6 resistance
tyr	Tyrosine requirement

Table 1.2. (*Continued*)

Symbol	Phenotype
ura	Uracil requirement
uvr	Ultraviolet radiation sensitivity
val	Valine requirement
xyl	Xylose utilization

Note: Occasionally one or more of these abbreviations will be printed with Δ as a prefix to indicate that the corresponding DNA is missing (deleted) from the chromosome.

that when referring to the phenotype of an organism the same three-letter abbreviation is used, but it is not italicized and the first letter is capitalized. It is possible, therefore, to talk about the "Pro" phenotype of an organism, so that a strain carrying *proA52* would be phenotypically Pro⁻ (i.e., unable to synthesize the amino acid proline).

There is one holdover from the old genetic nomenclature that may still be found occasionally in the contemporary literature and can confuse students. It has to do with suppressor mutations (see Chapter 3). Current nomenclature designates a strain not carrying such a mutation as sup^+ and a strain that does carry the mutation as sup^-. In the old system the terminology was exactly reversed: $sup^+ = su^-$ and $sup^- = su^+$. Sometimes bacterial strains thought to be free of suppressor mutations are described as su^0 or sup^0.

THINKING AHEAD

What are the similarities and differences between prokaryotic and eukaryotic cell structures? How do these differences impact genetic analysis?

Prokaryotic Cells and Eukaryotic Cells

Structure

The key feature that distinguishes prokaryotic organisms from eukaryotic organisms is their lack of an organized, membrane-bounded nucleus. Most prokaryotes do not compartmentalize their genetic material. The DNA is found clumped within the cytoplasm, and RNA synthesis occurs at the cytoplasm–DNA boundary.

Notable exceptions are members of the planctomycetes (Lindsay et al. 2001), which have a single, intracytoplasmic membrane that separates the DNA from the outer paryphoplasm. Prokaryotic cells are also typically smaller than eukaryotic cells, with several noteworthy exceptions. For example, an average, rapidly growing *Escherichia coli* cell is a cylinder about 1×0.5 µm, whereas a typical *Saccharomyces cerevisiae* (baker's yeast) cell is spherical to ovoid and about 3–5 µm in diameter. The approximately 64-fold difference in the cell volume between these two organisms is reflected in their internal cytoplasmic complexity, with *S. cerevisiae* containing the usual intracellular organelles such as mitochondria and endoplasmic reticulum, and prokaryotic cells generally having no real compartmentalization of function. In all the comparisons that follow, only nuclear activity in eukaryotic cells is considered, as modern genetic evidence indicates that mitochondria and chloroplasts are descended from ancestral prokaryotic endosymbionts, most of whose genes have migrated to the nucleus.

A major difference between cell types is found in the way in which cells carry out the processes of cell division and DNA segregation. In both cases, cell volume increases during metabolism until a particular size is attained. At that point, a complex series of events begins that culminates in the production of two daughter cells, each containing an exact copy of the DNA found in the parent cell.

Eukaryotic cells in general divide by a process coupled to mitosis, yielding two equal-sized cells. However, in the case of many of the yeasts and *Saccharomyces* in particular, the process of cell division is called **budding** because each new cell is produced as a small, rapidly enlarging protrusion from the surface of its parent cell. During bud formation, mitosis occurs. Microtubular spindle fibers form and attach to centromeres on the previously duplicated chromosomes. Pairs of chromosomes line up and then separate, moving along the spindle toward the poles of the elongating nucleus. In yeast, the nuclear membrane persists at all times, unlike in animals and plants. The nucleus continues to elongate, one end eventually entering the already large bud. When the nucleus splits in two, cytokinesis occurs to form the new cells.

Prokaryotic cells reproduce without the formal mitotic apparatus, either by budding or more often by binary fission (Fig. 1.1), in which cell mass and volume enlarge exponentially until the cell undergoes cytokinesis to yield two equal-sized daughter cells. The best-studied cell division system in prokaryotic cells is that of *E. coli*, with *Bacillus subtilis* a close second (Errington et al. 2003). Cellular buoyant density (determined by the relative amounts of protein and nucleic acids) remains roughly constant throughout the cycle. This mechanism of cell growth is grossly similar to that in eukaryotic cells, but the process

Figure 1.1. Simple binary fission as cell division. A cell elongates but changes little if at all in width. When the mass of the cell reaches a critical value, fission occurs.

of mitosis is unknown in prokaryotic organisms. There is, however, a tubulin analog in *E. coli* encoded by the *ftsZ* gene with closely related genes in other bacteria. There is no prokaryotic structure physically analogous to a centromere, although functional analogs exist. Therefore, bacterial cells obviously must use other means to ensure proper segregation of their DNA molecules. In sporulating *B. subtilis*, Ben-Yehuda et al. (2003) have shown that the RacA protein serves to anchor the origin of DNA replication to the pole of the cell. Presumably there is some aspect of the origin that serves the function of a centromere.

In the case of *Vibrio cholerae*, which has two chromosomes, Fogel and Waldor (2005) have shown that the respective origins of replication segregate to different specific regions of the developing daughter cells. They suggest that each chromosome has its own, unique segregation mechanism.

APPLICATION BOX

What kind of an experiment could you perform to test the prediction that the origins of replication in a bacterial chromosome move in coordinated ways during the cell cycle?

The prokaryotes are not a uniform group of organisms. In fact, taxonomic studies by Woese and his collaborators (Graham et al. 2000) have shown that there are two taxonomic domains of prokaryotes: true *Bacteria* (also called eubacteria) and the *Archaea* (formerly called archaebacteria). Members of the two domains differ with respect to metabolism, cell wall structure, and normal habitat. Archaea are frequently known as extremophiles because the preferred habitats of the best-studied members are very warm, very salty, or in other ways removed from ecosystems that support normal plant and animal life. Molecular biologists differentiate among members of the two domains by the base sequence of the RNA that is part of the small ribosomal subunit. Several different archaeans, including *Halobacterium salinarum* (formerly *halobium*, an organism that grows only in high salt solutions), *Sulfolobus acidocaldarius*, and *Methanococcus vannielii*, are intensively studied by bacterial geneticists.

Ploidy

In eukaryotic cells the process of mitosis serves to ensure that after cell division each daughter cell has the appropriate chromosome complement. Yeast cells may be either **haploid** (one copy of each chromosome) or **diploid** (two copies of each type of chromosome, one from each parent). Unlike animals, in which diploid cells comprise most of the organism and haploid cells are gametes, the diploid and haploid forms of yeast can each reproduce themselves and are inter-convertible via meiosis (diploid to haploid) or mating (haploid to diploid).

In order for two haploid cells to mate, they must be of different mating types, one designated **a** and the other designated α. The necessity of two mating types makes *S. cerevisiae* a **heterothallic** yeast.

Because bacterial cells lack the ability to undergo mitosis or meiosis, they are fundamentally haploid; otherwise, there would be no way for a cell to guarantee the DNA content of its progeny. In the strict genetic sense, the term haploid means that there is only one copy of each piece of genetic information per cell. Obviously, shortly before any haploid cell divides, the chromosome(s) will have already replicated even though cytokinesis has not yet begun. This sequence of events does not violate the principle of haploidy because the two chromosomes are identical, whereas in a true diploid they are permitted to be from different parents and therefore not identical.

This description of the haploid state is complicated, however, by the fact that many bacteria are capable of growing at a rate such that the **generation time** (the average time interval between cell divisions) is shorter than the length of time required to replicate the entire DNA molecule in a cell (one round of DNA replication). Cells obviate this problem by beginning a second round of DNA replication prior to completion of the first. As the generation time decreases, the time interval between initiations of new rounds of replication also decreases. The net result of these processes is that a rapidly growing bacterial cell actually has multiple copies of most genetic information. Moreover, the amount of genetic information located near the origin of replication is proportionally higher than near the termination site (Fig. 1.2).

An interesting variation on ploidy is present in the highly radiation-resistant *Deinococcus radiodurans*. Even in the stationary growth phase when most bacteria have only one chromosome, *Deinococcus* has four identical chromosomes. The advantage is that if one of the chromosomes is damaged by radiation, the other chromosomes can carry on. Even if all four chromosomes suffer damage,

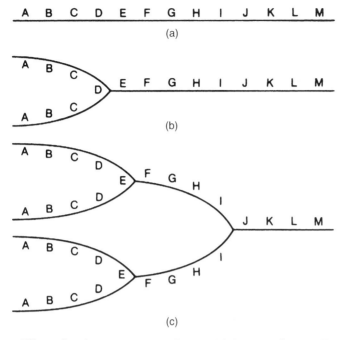

Figure 1.2. Effect of replication on gene dosage. (**a**) A nonreplicating DNA duplex. (**b**) The first round of replication has begun, initiated at the left-hand end of the duplex. (**c**) The second round of replication has begun before the first round of replication has finished. Once again, the initiation occurred at the left-hand end, giving rise to two new replication forks. The same effect would be seen in a cell with a circular chromosome except that the DNA duplex would be longer and would be looped back on itself.

there is still the possibility of assembling one intact chromosome from pieces of the original four (recombination).

Strictly speaking, it is thus not possible to talk about the number of sets of information per bacterial cell (**genomes**) because most of them are incompletely replicated. Instead, the term **genome equivalent**, which refers to the number of nucleotide base pairs contained in one complete genome, is generally used. For example, a cell in which one genome has replicated halfway along the molecule contains 1.5 genome equivalents of DNA. A cell containing several genome equivalents of DNA is nonetheless haploid, because all DNA results from replication of one original molecule. An example is shown in Fig. 1.2, in which there are four copies of gene *A* but only one of *J*. Figure 1.3 demonstrates the relation between the number of genome equivalents and the growth rate for *E. coli*.

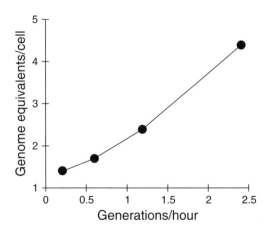

Figure 1.3. Relation between the amount of DNA per *E. coli* cell and the growth rate of the cell. (Data are from Maaløe, O., Kjeldgaard, N.O. [1966]. *The Control of Macromolecular Synthesis*. Reading, MA: Addison-Wesley.)

In the case of a bacterial cell that has received a new piece of DNA via some type of genetic process, it is possible to have two distinctly different sets of genetic information in the same cytoplasm. Such a cell is effectively diploid for that information. However, most DNA transfer processes move only a fraction of the total genome (see Chapters 2 and 8–11), and the resulting cell is only a **merodiploid**, or partial diploid. If the new piece of DNA is a **plasmid** (an independent DNA molecule that is capable of self-replication), the merodiploid state may persist indefinitely. If it is not a plasmid, only one daughter cell at each cell division is a merodiploid, and the lone merodiploid cell is soon lost among the large number of haploid cells unless the extra, nonreplicating DNA confers some selective advantage on the cell possessing it. Eventually, the nonreplicating DNA is degraded and the bases reused, although it may not happen for several generations.

DNA Structure

All known cells contain DNA, and the physical structure of a DNA molecule is generally assumed to be some variant of the double-helical structure proposed by James Watson and Francis Crick. This structure has a built-in chemical polarity due to the position of various substituents on the deoxyribose moiety (Fig. 1.4). Each linear nucleic acid molecule has a 5'-end or 3'-end, depending on the point

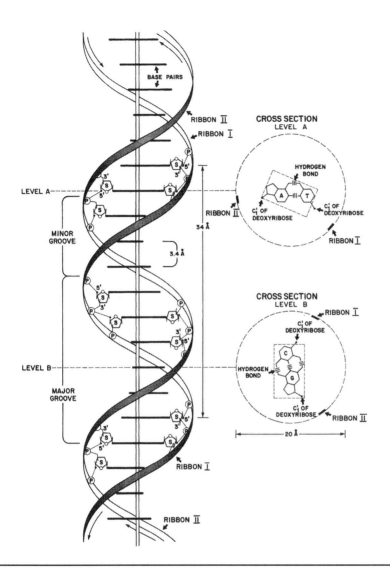

Figure 1.4. Double-helical DNA (B form). At left, the molecule is drawn in side view with the fiber axis indicated by the *vertical rod*. The backbone of the molecule consists of two polynucleotide chains that form right-handed helices. These chains are coiled together in a plectonemic (i.e., intertwined and not freely separable) manner to form a double helix having two grooves, one shallow (minor) and one deep (major), and an overall diameter of 2 nm. Each chain is composed of D-2'-deoxyribose sugar moieties (**S**) linked by phosphate groups (**P**), thus forming 3',5'-phosphodiester bridges and producing a long unbranched polymer. The individual bases are attached to the sugar molecules through β-*N*-glucosyl linkages. The two

(*continued*)

of attachment of the last substituent (phosphate or hydroxyl group) to the pentose ring of the last nucleotide in the chain. In Fig. 1.4, for example, the arrowheads are always at the 3'-hydroxyl end of the chain. Each chain is considered to be a single **strand** of DNA.

In DNA, bases are paired such that a pyrimidine (thymine, cytosine) from one strand is always opposite a purine (adenine, guanine) in the complementary strand. The pairing of bases is stabilized by hydrogen bonds (three for G:C pairs, two for A:T pairs) and it generates a helical structure containing a major (wide) and a minor (narrow) groove (see Fig. 1.4). B-form DNA, which most closely matches the Watson–Crick model, has a right-hand helix and a constant width of about 2 nm. One complete helical turn requires about 10 base pairs (bp) or 3.4 nm. A-form DNA, on the other hand, has a more shallow pitch with 11 bp per helical turn. Unlike B-form DNA whose sugars are horizontal, A-form has its sugars inclined about 30°. Double-strand RNA and RNA•DNA hybrids assume structures more closely resembling A-form than B-form. The helix is not always straight. Certain base sequences tend to cause a curve or bend in the DNA

Figure 1.4. (*Continued*) chains are antiparallel with the 5' to 3' direction proceeding upward for one chain but downward for the other. This 5' to 3' direction is illustrated by the *arrows* at the top and bottom of the diagram. For the sake of clarity, the molecular structure of the sugar–phosphate backbone is shown only over small regions. The two ribbons represent the continuity of the two chains, the shaded regions being closest to the viewer. The hydrogen-bonded base pairs, represented by *horizontal heavy lines*, are planar molecules occupying the central core of the helix (the region indicated in cross section within the *dotted rectangles* at the *right* of the diagram). Only the bases lie in the plane of the cross sections, and thus only the base pairs are drawn, with attachment to the sugar merely being indicated. The position of each ribbon at either of the two cross-sectional levels is indicated. The *broken line* forming a circle indicates the outer edge of the double helix that would be observed when the molecule is viewed end on. An adenine (A)–thymine (T) pair is shown as the pair representative of level A, whereas a guanine (G)–cytosine (C) pair is shown to represent level B. The surface planes of the bases are perpendicular to the vertical axis and are separated from each neighboring base pair by a vertical distance of 0.34 nm. There are 10 bp per complete turn of the helix so that each turn of the helix has a vertical length of 3.4 nm, and each base pair is rotated 30° relative to its nearest neighbors. As a result of this rotation, the successive side views of the base pairs appear as lines of varying lengths depending on the viewing angle. The hydrogen bonding between the bases and the hydrophobic interactions resulting from the parallel "stacking" of the bases stabilizes the helical structure. (From Kelln, R.A., Gear, J.R. [1980]. A diagrammatic illustration of the structure of duplex DNA. *Bioscience* 30: 110–111.)

molecule, and the effect can be accentuated by the binding of specific proteins such as HU (see next section). A positive (+) curve means that the minor groove is opened.

Alexander Rich and Robert Wells and their respective collaborators have shown that an alternative physical state of DNA exists. In this molecule, the same base-pairing rules apply, but the stereochemical configuration of the bases is strictly alternating *anti* and *syn*. The result is a double-strand DNA molecule with a left-hand configuration possessing a single deep groove (Fig. 1.5), a structure designated as Z-DNA. Transitions between the B and Z forms of DNA have been shown to be all or none. Z-DNA is a poor substrate for the repair enzyme DNA polymerase I. Plasmids carrying sequences that should adopt a Z configuration are not fully accessible in vivo to enzymes that use B-form DNA as substrates. A run of 56 G:C base pairs was sufficient to see the effect, while a run of 32 bp was insufficient.

Unlike B-DNA, Z-DNA has sufficient tertiary structure to be a reasonable antigen, and Z-DNA-specific antibodies can be prepared. Such antibodies can be shown to react specifically with DNA isolated from whole cells, which seems to indicate that there are Z-DNA regions present within native DNA molecules (or that an antibody can force DNA into the Z configuration). Certain bands of *Drosophila* chromosomes have also been shown to contain Z-DNA. One suggestion is that Z-DNA regions may serve a regulatory function. Antibody binding studies in *E. coli* have given rise to estimates that there is an average of one Z-DNA segment for every 18 kilobases (kb). Laboratory studies in vitro have shown that purines most easily assume the *syn* configuration necessary for Z-DNA structure. Therefore, most work has been done with synthetic **oligonucleotides** (short chains of bases) that have a strictly alternating sequence of purines and pyrimidines, usually cytosine and guanine. Similar tracts have been observed in natural DNA molecules. However, runs of A:T pairs can also form the Z-DNA configuration, provided that three to five G:C pairs flank them.

Chromosome Structure

Eukaryotes

The genetic material in eukaryotic cells is organized into chromosomes, precisely organized structures of protein and nucleic acid. Proteins incorporated into the chromosome are known as histones and are designated H1, H2A, H2B,

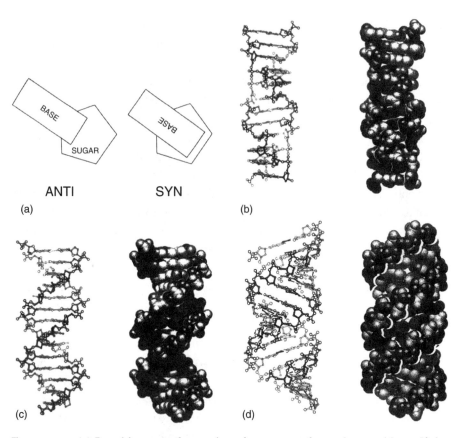

Figure 1.5. (a) Possible stereochemical configurations of a nucleic acid base. If the base is oriented primarily away from the plane of its attached deoxyribose, it is in the *anti* configuration. If the base lies primarily over the plane of the sugar, it is in the *syn* configuration. (b) Side views of a stick-and-ball model and a space-filling model of Z-DNA. The bases are in a strictly alternating *syn* and *anti* pattern, which gives the DNA a zigzag arrangement. The sugar–phosphate backbone is the darker region of the diagram. Only the minor groove is visible, as the major groove is filled with cytosine C_5 and guanine N_7 and C_8 atoms. (c) Side views of a stick-and-ball model and of a space-filling model of B-DNA. (d) Side views of a stick-and-ball model and of a space-filling model of A-DNA. Note the change in major and minor grooves compared to B-form. Both major and minor grooves are visible. (**b–d** from Saenger, W. [1984]. *Principles of Nucleic Acid Structure*. New York: Springer-Verlag.)

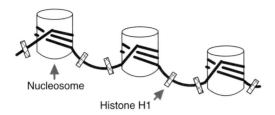

Figure 1.6. Chromosome structure of eukaryotic DNA. Each cylinder of protein is composed of two copies each of histones H2A, H2B, H3, and H4. There are approximately 146 bp of DNA wrapped around the histone core in a left-handed helix to form a nucleosome. In most organisms the DNA between each nucleosome is coated with histone HI (small boxes).

H3, and H4. Two molecules of each protein except H1 join to form an octameric cylinder around which approximately 146 bp of a DNA molecule are wrapped (Fig. 1.6). Individual cylinders of protein with their associated DNA are known as nucleosomes, and in most eukaryotes individual nucleosomes are linked together by a spacer of about 40–60 bp of DNA coated by histone H1. However, histone H1 does not appear in the eukaryotic microorganism *Saccharomyces*, although there is an analogous protein called HHO1 whose loss affects expression of some genes. The internucleosome distance in *Saccharomyces* is reduced to only about 20 bp.

There are 16 physically distinct chromosomes in *S. cerevisiae*, ranging in size from 200,000 to 2200,000 bp (200–2200 kb). The total length of DNA present is about 14,000 kb. Physically, the DNA molecules underlying these chromosomes are linear structures, although the ends of the DNA strands (telomeres) within a single helix seem to be joined by material that is neither protein nor DNA. The telomeres consist of repetitive DNA and are the functional equivalent of double-strand breaks, although repair enzymes do not see them as such (Lydall 2003).

Prokaryotes

Electron microscopists originally thought that bacteria did not have a structure that was truly comparable to that of the chromosome in eukaryotic cells. Indeed, the special term genophore was coined to describe the physical DNA structure of a typical bacterium. However, it is now apparent that there are in fact some striking parallels between a eukaryotic chromosome and a DNA molecule found in a bacterial cell.

John Cairns achieved the first clear visualization of a bacterial DNA molecule, using **autoradiography** (exposure of a film emulsion by the decay of radioactive atoms) to demonstrate the circularity of an intact tritium-labeled *E. coli* DNA molecule (Fig. 1.7). The DNA molecule has a length slightly greater than 1 mm and consists of about 4700 kb that are **covalently closed** (all joined by covalent chemical bonds). The 1000-fold greater contour length of the DNA molecule when compared to the average size of an *E. coli* cell indicates that the DNA must be highly folded within the cytoplasm. David Pettijohn and his coworkers subsequently showed that it was possible to extract intact DNA molecules from bacterial cells and preserve at least some of their folded configuration. These structures have been designated **nucleoids** or **folded chromosomes** and occupy about 20% of the volume of an *E. coli* cell. A typical nucleoid can be seen in Fig. 1.8. Roughly 5% of the volume of a nucleoid is DNA. Immunolabeling studies of sections viewed by electron microscopy have shown that the nucleoid core is basically double-strand DNA with single-strand DNA and proteins at the periphery. The density of DNA in a nucleoid is comparable to that of interphase chromosomes in mammalian cells.

All bacteria are generally assumed to contain a single, circular chromosome. However, there are numerous exceptions (Jumas-Bilak et al. 1998). *Rhodobacter sphaeroides* has two circular chromosomes, chromosome I containing 3.0 million base pairs and chromosome II containing 0.9 million base pairs. Chromosome II codes for some amino acid and vitamin biosynthesis genes (housekeeping genes) and shows no fundamental difference from chromosome I. However, sequencing data suggest that there are 144 **open reading frames** (translatable regions) on chromosome II. Paired chromosomes are also known in *Brucella melitensis* and *Leptospira interrogans*. *Rhizobium meliloti* and *Burkholderia cepacia*, on the other hand, have three chromosomes each. In some cases ribosomal RNA genes are present on two chromosomes.

There are even some examples of bacteria that have linear chromosomes (Casjens 1998). Prominent among these are *Streptomyces coelicolor*, *Streptomyces lividans*, *Rhodococcus fascians*, and members of the genus *Borrelia*. Most unusual is *Agrobacterium tumefaciens* biovar 1, which has both a circular chromosome and a linear chromosome. The immediate question, of course, is what happens to the ends of the linear chromosomes. In eukaryotic cells there are telomeres that protect the end of the DNA. *Borrelia* and *Agrobacterium* use a system in which one DNA strand loops back to become the complementary strand (a hairpin). *Streptomyces* protects

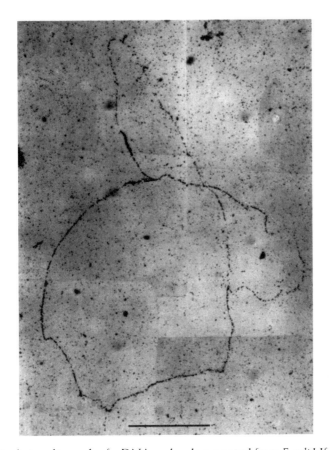

Figure 1.7. Autoradiograph of a DNA molecule extracted from *E. coli* Hfr3000. The DNA was labeled with tritiated thymidine for two generations and then extracted from the cell using the enzyme lysozyme, which attacks the cell wall. A photographic emulsion was overlaid on the DNA and exposed to the radioactive atoms for 2 months. As the tritium decayed, the beta particles that were emitted activated the emulsion in the same fashion as exposure to light. Upon development of the emulsion, silver grains were formed whose positions indicated the location of the original DNA molecule. The scale at the bottom represents a length of 100 μm; the length of the DNA, discounting the replicated portion, is estimated at about 1.1 mm. It should be remembered that the cell from which this DNA molecule was extracted was probably only a few micrometers in length. (From Cairns, J. [1963]. The chromosome of *E. coli. Cold Spring Harbor Symposia on Quantitative Biology* 28: 43–45.)

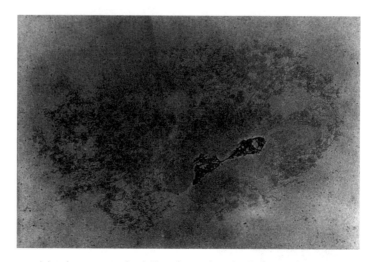

Figure 1.8. Membrane-attached *E. coli* nucleoid. Cells were lysed gently with lysozyme and detergent. The DNA was separated from cellular debris by sedimentation through increasing concentrations of sucrose and then mounted for electron microscopy using a monolayer of cytochrome C molecules on the surface of a formamide solution. The DNA was stained with uranyl acetate and coated with platinum to increase contrast. The remains of the cell envelope can be seen near the center of the photograph. The fine particles surrounding it are probably parts of the cell membrane. Continuous fibers of DNA in various states of supercoiling radiate out from the cell envelope. Short, kinky fibers of single-strand RNA, representing transcription in progress, can be seen along the DNA. (From Delius, H., Worcel, A. [1974]. Electron microscopic visualization of the folded chromosome of *E. coli*. *Journal of Molecular Biology* 82: 107–109.)

its chromosome ends by covalently attached proteins that may have served as primers for DNA replication.

The predominant impression of isolated nucleoids is one of coiled DNA. This complex structure consists of 60% DNA, 30% RNA, and 10% protein. It consists of **superhelical coils** that result from twisting an entire DNA helix over and above the normal helical turns. This is roughly equivalent to taking a two-stranded wire rope, coiling it on the ground, and then gluing the ends together so that the extra turns become an integral part of the structure and give a torsional tension to the structure. On a molecular scale, superhelical turns are added or removed by a group of enzymes known as **topoisomerases**, which act to change only the topology of a DNA molecule and not its base sequence. Superhelicity can be added in a positive sense, in which case the normal helix is coiled tighter, or in a negative sense, in which case the normal helix is coiled more loosely (Fig. 1.9). In nucleoids studied thus far, all supercoiling has been negative supercoiling.

The superhelical density of a nucleoid has been estimated to be about one supertwist for each 15 normal helical turns. Each of the 40–100 superhelical loops constitutes a separate domain. If a **nick** (a broken phosphodiester bond) is introduced into such a loop, only that particular loop relaxes (unwinds). Maintaining the proper superhelical density is important to normal functioning of DNA, and several enzymes are involved in the process. These enzymes are discussed later in connection with DNA replication.

In addition to the chromosomal oddities mentioned earlier, Bendich and Drlica (2000) point out that the line between prokaryotes and eukaryotes is substantially blurred in other ways. For example, the *Myxococcus* (prokaryotic)

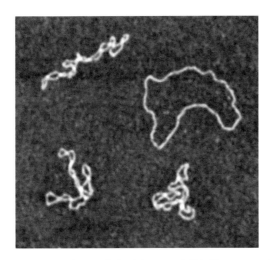

Figure 1.9. Superhelical coiling of double-strand DNA as seen by atomic force microscopy (see Chapter 2). At the *top right* of the figure is a circular DNA duplex formed by one strand that is a circle and one strand that has a nick in it, and therefore, cannot maintain any supercoils. The enzyme DNA ligase can act on this molecule in concert with DNA gyrase to yield a covalently closed, circular molecule that also contains superhelical turns (other three molecules shown). These turns are inserted by rotating the free end of the DNA about the axis of the helix several times prior to the final ligation step. Depending on the direction of rotation, the superhelical turns may be either positive or negative, and the helix is slightly overwound or underwound. Micrograph courtesy of Alexander Lushnikov, University of Nebraska.

genome is larger than the genome of *Pneumocystis* (eukaryotic). Moreover, there is a telomerase-deficient mutant of the yeast *Schizosaccharomyces pombe* that has a circular chromosome like bacteria. Bendich and Drlica propose that a better way to distinguish between the two groups of organisms is to look at their nucleosomal structure.

Archaeal chromosomes from members of the Euryarchaeota (methanogens and extreme halophiles) are associated with true histone proteins, while the supercoiled DNA molecule of a bacterial cell is associated with certain histonelike proteins, basic proteins that lend regular structure to bacterial chromosomes. Predominant among them are proteins designated HU and H-NS.

HU is an abundant protein (2×10^4 to 1×10^5 copies per cell), and closely related proteins are found among cyanobacteria and archaea. When a **nuclease** (an enzyme that cleaves nucleic acids) is added to DNA, it normally makes cuts every 10 bp, corresponding to one turn of the helix (see Fig. 1.4). However, in the presence of purified protein HU, the cuts occur every 8.5 bp, indicating that the DNA helix has become overwound (more tightly coiled). They have also reported that the HU–DNA complex also contains negative supercoils induced by interaction of the two components. The structure described is similar but not identical to that of a eukaryotic nucleosome. However, electron microscopic observations using antibodies against HU have shown that HU is localized at the periphery of the nucleoid in the region of active transcription, ruling out any major nucleosomelike structure for bacterial DNA. Moreover, eukaryotic DNA has a nucleosome associated with each supercoiled loop whereas there are significant numbers of unrestrained supercoils in bacterial nucleoids. The chromosome of *H. salinarum* seems to have both protein-free and protein-associated regions and therefore may occupy a middle ground.

Protein H-NS (histone-like nucleoid structuring protein) has been best studied in *E. coli* and *Salmonella*, but it or a similar protein is common in over 70 Gram-negative bacteria (Trendeng and Bertin 2003). The physical structure of the amino and carboxy termini is known, and the physiologic function of these proteins is becoming clearer (Rimsky, 2004). They appear to have some preference for curved DNA and can act as a zipper to join two strands of DNA together as in the case of a hairpin loop. Binding of H-NS also influences promoter function and therefore gene regulation. In addition to H-NS, there are also IHF (integration host factor), and Fis (factor for inversion stimulation). The higher order structure conferred by these proteins can often be inferred by the pleiotropic phenotypes of deficiency mutants, but this structure does not have a major impact on chromosome condensation (Dorman and Deighan 2003). Examples of the roles played by these proteins are presented in subsequent chapters. One physical model for the bacterial nucleoid is shown in Fig. 1.10.

The archaea also have DNA binding proteins, and some of these have a significant impact on DNA structure because the normal growth temperature for the organism is higher than the temperature at which the hydrogen bonds denature. Reeve (2003) has reviewed the major nucleoid proteins for *Sulfolobus*. These include Sul7d, whose binding stabilizes hydrogen bonds and Alba, whose binding induces negative supercoiling. There are true histones in the Euryarchaeota but not in the Crenarchaeota. The predicted structure of the histones suggests that they form cylinders of protein around which the DNA can wrap.

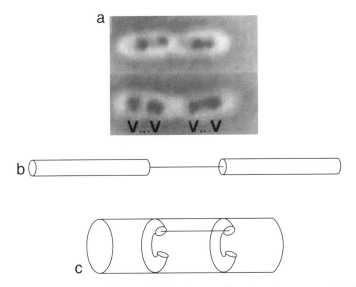

Figure 1.10. A flexible doublet model for bacterial nucleoid structure. (**a**) A sample formaldehyde-fixed slow-growing cell photographed by simultaneous phase-contrast and DAPI (DNA)-fluorescence images. The "V" marks indicate presumptive unit nucleoids. (**b**) Hypothetical unit nucleoid, composed of two subunits joined by a spacer. Position of attachments of the spacer and subunit shapes varied considerably, and an arbitrary selection is made here. (**c**) Unit nucleoid within rapidly grown cell. Nucleoid subunits tended to be radially directed and closely opposed to cell envelope, as drawn here. However, relative orientations between the two subunits varied widely in the cells examined. Reproduced with permission from Zimmerman (2003).

In summary, it seems clear that prokaryotic chromosomes have substantial structure. Although not explicitly considered in many models for genetic processes, it is certain that such structure affects genetic processes in subtle ways.

Selection: An Essential Element of Microbial Genetics

The concept of **selection** is important in understanding microbial genetics. When dealing with microorganisms, the sheer number of individuals in any culture (frequently $\geq 10^8$ cells/ml) precludes any possibility of examining every organism resulting from a genetic manipulation, as can be done for macroscopic organisms such as pea plants or fruit flies. Instead, mixed cultures of cells differing in one or more biochemical pathways are used for genetic experiments. When placed under an appropriate set of nutritional or environmental conditions (the selection), growth of the original cell types is prevented so that only those cells that have acquired a new **phenotype** (one or more new observable characteristics) via DNA transfer are able to grow and divide. Examples of selective agents used to prevent growth of parental cells are antibiotics, bacteriophages, and required growth factors.

Transferred DNA may be in the form of a chromosomal fragment or an intact plasmid. In the case of fragmentary transfer, recombination must occur in order for the new genetic information to replicate. Each recombinant cell ultimately divides many times to produce a **colony**, a macroscopically visible mound of cells on the surface of an agar plate. The usual assumption is that each colony represents one genetic event, even though there is evidence that multiple rounds of recombination may occur.

In eukaryotic genetics, the distance between genetic loci is often expressed in terms of the **recombination frequency**, the total number of recombinants obtained divided by the total number of progeny from the cross. Under selective conditions, however, a geneticist examines only a subset of recombinant progeny, those that survived the selective process. The total number of recombinant progeny of all types is, in fact, unknown. Recombination frequencies in microbial genetics represent the number of recombinant progeny obtained from the selection divided by the number of minority parent cells (assuming that the limiting factor in the transfer process is the number of each type of parent cell). Implicit in this definition is the assumption that the number of recombinants must always be less than or equal to that of the least abundant (minority) parent. Chapter 6 presents a test of this assumption.

The ability to apply stringent selection to a large number of individuals has made microbial genetics a powerful tool for studying rare events. If a particular event is expected to occur one time in 10 million, the classic macroorganismal geneticist has a considerable problem. Yet 10 million *E. coli* cells represent only about 0.1 ml of a typical growing culture. One billion *E. coli* cells, growing at their maximum rate, can easily be accommodated in 10 ml of culture. More-over, their growth rate does not even begin to slow down until there are eight to ten times that many cells in the same culture volume.

However, it is important to bear in mind that selection always introduces a bias into the sample population. Any cell that fails to promptly form a viable recombinant, for any reason, is lost to the sample. For example, if a cell does not display a recombinant phenotype for several generations after the genetic trans-fer and selection is applied too early, the cell is lost. The genetic transfer is not detected even though it did occur. Chapter 3 discusses this problem further. For now it is sufficient to remember that, typically, bacterial genetic experiments examine only those instances in which the entire process of genetic transfer and expression was successful.

Major Genetic Transfer Processes Observed in Microorganisms and Their Viruses

Although each genetic transfer process is discussed at length in subsequent chap-ters, it is advantageous to briefly introduce them now because textbooks can be divided into neat categories, but actual research cannot. As a result, people who are studying the nature of transduction may resort to some conjugation experi-ments and vice versa. Therefore, to provide maximum flexibility in the forth-coming discussion, major features of each process are reviewed in this section.

Genetic Transformation

Genetic transformation was first observed in bacteria and was the first bacterial system for genetic transfer to be discovered. Although it occurs naturally in only certain bacteria, under laboratory conditions it seems to be possible to carry out genetic transformations with any cell type, prokaryotic or eukaryotic. In bacteria the process begins when a bacterial cell (living or dead) releases some DNA into the surrounding medium. This DNA is, of course, vulnerable to degradation but may encounter another bacterial cell before any significant change can occur. The second cell may take up the DNA, transport it across the cell wall and cell membrane, and allow it to recombine with the homologous portion of the resident bacterial chromosome. The resulting recombinant cell is called a **transformant**. In theory, any piece of genetic information may be transferred by this method, although the amount of DNA transferred per event is small, on the order of 10 kb in length. (For a more complete discussion, see Chapter 10.)

Transduction

In transduction, a bacterial virus (**bacteriophage** or **phage**) is involved intimately in the genetic transfer process. Phage infections begin with adsorption of virus particles to specific receptor sites on the host cell surface. The nucleic acid contained inside the viral protein coat is then transferred to the cytoplasm of the bacterial cell, where it becomes metabolically active and undergoes replication and transcription.

Typically there are two possible outcomes of phage infection. During a lytic response, the virus produces structural components of new phage particles, packages its nucleic acid inside them, and then causes the cell to lyse and release progeny phage. During a temperate response, the virus establishes a stable relationship with a host cell in which some phage functions are expressed, but not those that lead to uncontrolled DNA replication or the production and assembly of new particles. Instead, viral DNA is replicated along with host DNA, usually as an integral part of the same molecule, and is transmitted to all progeny cells. Occasionally **lysogens** (cells carrying a temperate phage) undergo a metabolic shift that reactivates the viral DNA. The result is the same as for an initial lytic response. Some phages may give only lytic responses and some only temperate ones; some, however, may give either response, depending on growth conditions.

During the course of a phage infection of a bacterial cell, some or all of the viral DNA inside an individual **virion** (virus particle) may be replaced by bacterial DNA, and this process may occur only rarely or with great frequency. After such an altered phage particle is released into the medium, it may encounter another bacterial cell and attempt to initiate an infection. In so doing, however, it transfers the DNA fragment from the previous host's chromosome. If the newly infected cells are not killed and the DNA fragment can either replicate or recombine, the result is the production of **transductants**.

The amount of DNA transferred by these means varies considerably, but generally is not more than the amount of DNA normally present in a single bacteriophage particle. It may approach 200 kb in length. The actual amount of DNA recombined is significantly less in most cases and, in addition, depends on whether the transduction is generalized or specialized.

During generalized transduction the phage enzyme system that packages viral DNA attaches to the bacterial chromosome and packages some of that DNA instead. The DNA that is packaged is chosen on a more or less random basis, and as a result it is possible for any piece of host genetic information to be transferred. Specialized transduction, on the other hand, involves a temperate phage that has physically integrated its DNA into the bacterial chromosome at a specific site. As mentioned earlier, such an integrated phage may be stable for long periods of time. However, it may reactivate and replicate itself independent of the bacterial chromosome. During the reactivation process, it is possible for a mistake to occur so that some bacterial DNA located adjacent to one end of the viral DNA is also excised from the chromosome instead of the appropriate DNA from the other end of the viral genome. Because the overall size of the excised DNA must be nearly constant, only certain pieces of genetic information can be transferred, and their size depends on the physical nature of the mistake that caused their production. (For further discussion of both types of transduction, see Chapter 9.)

Conjugation

The term **conjugation** can be used in several senses in biology. For example, in yeast the result of conjugation is fusion of haploid cells and formation of a diploid cell type. In a bacterium such as *E. coli*, instead of cell fusion there is unidirectional transfer of DNA from a donor cell (which carries a **conjugative plasmid**) to a recipient cell beginning at a definite point on the DNA molecule and proceeding in a linear fashion. The transferred DNA may be all or part of the plasmid and may include a portion of the host DNA as well. By analogy to other bacterial transfer processes, the recombinant bacteria are called **transconjugants**. The amount of bacterial DNA that can be transferred by conjugation ranges from a few kb to the entire chromosome. (This process is discussed at length in Chapters 11–13.)

Protoplast Fusion

Protoplast fusion has been used successfully for many years with eukaryotic cells. Its use with prokaryotic cells is comparatively rare, but apparently the technique is applicable to most cells. For protoplast fusion to work, the **protoplasts** (cells that have been stripped of their walls) must be prepared by various enzymatic or antibiotic treatments. Fusion of cell membranes is aided by a high concentration of polyethylene glycol. The resulting diploid cell usually segregates haploid offspring, many of which show extensive recombination of parental characters. Formation of stable, noncomplementing diploid *B. subtilis* cells have also been reported. The diploid state can be stable over many generations, as evidenced by successful transformation of parental genes whose phenotype was not present in the diploid donor cell. Successful fusions have been reported with *Actinoplanes*, *Brevibacterium*, *Bacillus*, *Mycobacterium*, *Providencia*, *Staphylococcus*, and *Streptomyces*.

Experimenters often use protoplasts in a simpler way. Protoplasts are good recipients in genetic transformation and readily take up plasmid DNA such as that prepared by genetic engineering technology. This technique avoids many of the competency problems discussed in Chapter 10.

Electroporation

When a high voltage (as much as 2500 V) is passed from a capacitor through a solution containing living cells, significant damage occurs to cell membranes, and many cells die. Among the survivors, however, are cells that developed small holes (pores) in their cell membranes as a result of the brief passage of current. These pores are quickly sealed, but while they are open, solutes can pass into or out of the cytoplasm. What is important to geneticists is that plasmid DNA molecules can also enter a cell if the exterior concentration is sufficiently high. This technique has been very successful with Gram-negative bacteria and somewhat less successful with Gram-positive bacteria.

Bacteriophage Genetic Exchange

Viral genetics can be studied effectively by arranging the virus/cell ratio so that a cell is simultaneously infected by more than one virus particle. Assuming that the two viruses are genetically distinguishable, selection is applied to prevent parent-type phage particles from successfully completing an infection. Under these conditions, only cells in which phages carrying recombinant DNA have been produced yield progeny virus particles. The resulting virions are tested for phenotype, and recombination frequency is calculated in the same manner as for bacteria. (For a more extensive discussion of this subject, see Chapter 6.)

Summary

DNA molecules are the genetic material in all cells, but cell structure in prokaryotes is dramatically different from that in eukaryotes. Prokaryotes lack internal membrane-bounded organelles, including nuclei. Instead they have dense masses of DNA known as nucleoids. Bacterial chromosomes are commonly circular and all are supercoiled to allow such long molecules to fit into such small cells. They have certain histonelike proteins associated with them that impart an unstable, nucleosomelike structure. Some members of the archaea have true histone proteins and are predicted to have DNA packaged like chromatin. Prokaryotic cells are haploid and reproduce by binary fission. Eukaryotic cells are haploid or diploid, depending on the organism and the life-cycle stage; they divide by mitosis or meiosis.

Natural genetic transfer in prokaryotes is accomplished by genetic transformation, transduction, or conjugation. In the laboratory, protoplast fusion has also been shown to be possible. These techniques enable a true study of genetics in bacteria. The nomenclature used to describe genetic traits is similar to that used for eukaryotes and is intended to apply to all bacteria.

Questions for Review and Discussion

1. Suppose a diploid bacterial cell did not inactivate its second chromosome, what would be the consequences with respect to the genetic properties of that cell? What are the advantages to a bacterium of inactivating a second chromosome?
2. Why is it important that allele numbers not be duplicated between laboratories?
3. What do the following genotype symbols indicate: lys-108, lysB99, lysD3, lysB48?
4. List the similarities and differences of a bacterial nucleoid and a true nucleus.

References

General

Bendich, A.J., Drlica, K. (2000). Prokaryotic and eukaryotic chromosomes: What's the difference? *BioEssays* 22: 481–486.

Brock, T.D. (1990). *The Emergence of Bacterial Genetics.* Cold Spring Harbor, NY: Cold Spring Harbor Laboratory Press. (Historical review of the field.)

Demerec, M., Adelberg, E.A., Clark, A.J., Hartman, P.E. (1966). A proposal for a uniform nomenclature in bacterial genetics. *Genetics* 54: 61–76.

Dorman, C.J., Deighan, P. (2003). Regulation of gene expression by histone-like proteins in bacteria. *Current Opinion in Genetics & Development* 13: 179–184.

Errington, J., Daniel, R., Scheffers, D. (2003). Cytokinesis in bacteria. *Microbiology and Molecular Biology Reviews* 67: 52–65.

Higgins, P.N. (ed.)(2004). *The Bacterial Chromosome.* Washington, DC: ASM Press.

Jumas-Bilak, E., Michaux-Charachon, S., Bourg, G., Ramuz, M., Allardet-Servent, A. (1998). Unconventional genomic organization in the alpha subgroup of the proteobacteria. *Journal of Bacteriology* 180: 2749–2755.

Lederberg, J. (1987). Genetic recombination in bacteria: A discovery account. *Annual Review of Genetics* 21: 23–46.

Lydall, D. (2003). Hiding at the ends of yeast chromosomes: Telomeres, nucleases and checkpoint pathways. *Journal of Cell Science* 116: 4057–4065.

Perry, J.J., Staley, J.T., Lory, S. (2005). *Microbial Life.* Sunderland, MA: Sinauer Associates (A general text suitable as a tool for review.)

Reeve, J.N. (2003). Archaeal chromatin and transcription. *Molecular Microbiology* 48: 587–598.

Sherratt, D.J. (2003). Bacterial chromosome dynamics. *Science* 301: 780–785.

Tendeng, C., Bertin, P.N. (2003). H-NS in Gram-negative bacteria: A family of multifaceted proteins. *Trends in Microbiology* 11: 511–518.

Specialized

Akamatsu, T., Sekiguchi, J. (1987). Genetic mapping by means of protoplast fusion in *Bacillus subtilis*. *Molecular and General Genetics* 208: 254–262.

Ben-Yehuda, S., Rudner, D.Z., Losick, R. (2003). RacA, a bacterial protein that anchors chromosomes to the cell poles. *Science* 299: 532–536.

Graham, D.E., Overbeek, R., Olsen, G.J., Woese, C.R. (2000). An archaeal genomic signature. *Proceedings of the National Academy of Sciences of the USA* 97: 3304–3308.

Grandjean, V., Hauck, Y., LeDerout, J., Hirschbein, L. (1996). Noncomplementing diploids from *Bacillus subtilis* protoplast fusion: Relationship between maintenance of chromosomal inactivation and segregation capacity. *Genetics* 144: 871–881.

Lindsay, M.R., Webb, R.I., Strous, M., Jetten, M.S.M., Butler, M.K., Forde, R.J., Fuerst, J.A. (2001). Cell compartmentalisation in planctomycetes: Novel types of structural organisation for the bacterial cell. *Archives of Microbiology* 175: 413–429.

Zimmerman, S.B. (2003). Underlying regularity in the shapes of nucleoids of *Escherichia coli*: Implications for nucleoid organization and partition. *Journal of Structural Biology* 142: 256–265.

Reference Books for Specific Organisms

Hatfull, G.F., Jacobs, Jr., W.R. (eds.) (2000). *Molecular Genetics of Mycobacteria*. Washington, DC: ASM Press.

Neidhardt, F.C. (editor-in-chief) (1999). *Escherichia coli and Salmonella: Cellular and Molecular Biology*, 2nd ed. Washington, DC: ASM Press.

Sonenshein, A.L. (editor-in-chief) (2002). *Bacillus subtilis and its Closest Relatives: From genes to Cells*. Washington, DC: ASM Press.

2

Replication and Analysis of DNA

Genetics is the study of the properties, synthesis, and inheritance of nucleic acids. Most of the topics in this book revolve around deoxyribonucleic acid (DNA), although some of the viruses discussed in later chapters have ribonucleic acid (RNA) as their genetic material. This chapter presents basic information about DNA replication, segregation, and the methods used to study the structure and function of DNA. The material is fundamental for all discussions that follow.

Major topics include:

- Reproduction of DNA molecules and their migration into daughter cells
- Determination of the base sequence of a DNA molecule
- Locating a specific sequence within a mixture of DNA molecules
- Linking of DNA for the creation of new DNA molecules

- DNA cloning
- Computer databases and their contribution to bacterial genetics

DNA Replication

Replication is in vivo synthesis of new DNA molecules and occurs in essentially the same manner in both eukaryotic and prokaryotic cells. It is particularly well studied in *Escherichia coli*, and subsequent discussion focuses on that organism.

As postulated by James Watson and Francis Crick and first demonstrated by Matthew Meselson and Franklin Stahl, DNA replication is a semiconservative process in which new bases are hydrogen-bonded with old ones and then joined covalently to create a new strand. The resulting DNA duplex is thus composed of one new and one old strand. Enzymes that catalyze this process are said to be DNA-dependent DNA polymerases because they copy DNA into more DNA. Conventionally they are referred to simply as DNA polymerases.

However, if a DNA polymerase is presented with a single strand of DNA, it cannot synthesize a complementary strand. This property of DNA polymerase means that it cannot start a new strand de novo but must have some sort of primer to extend. The primer can be a short oligonucleotide, either RNA or DNA, so long as its nucleotide sequence matches a portion of the existing strand. New bases are added to the primer following a 5' to 3' polarity. In other words, the first base added is joined to the 3'-end of the primer via the hydroxyl group on its 5'-carbon and has a free hydroxyl group on its 3'-carbon to which the next base will be attached.

If RNA or protein synthesis is inhibited in a cell, DNA replication slowly stops, and the completed DNA is devoid of any RNA sequences or protein. This observation indicates that macromolecular synthesis is necessary either for DNA synthesis or for its initiation, and the molecules are removed after their function is completed. Experimentally this suggestion can be verified for RNA by isolating newly synthesized DNA and demonstrating that transiently each newly synthesized piece of DNA carries a few RNA bases (three or four in the case of *Bacillus subtilis*) at its 5'-end. The enzymes involved in primer synthesis are discussed later. Several animal viruses, bacteriophage φ29 (see Chapter 7), and certain plasmids are known to use a protein primer instead of RNA.

Because DNA synthesis always occurs in a 5' to 3' direction, it is not possible for both strands of a DNA helix to begin replicating simultaneously

Instead, one strand (the leading strand) must replicate for some distance and open up the helix so that the second or lagging strand can then begin its replication (Fig. 2.1). Note that the lagging strand must be synthesized as a series of discrete fragments (called **Okazaki fragments**, after their discoverer). Each fragment of the lagging strand must have its primer removed and be joined to the others in order to complete the replication process. At the *E. coli* origin of DNA replication (*oriC*), two initiation events occur so that separate clockwise and counterclockwise replication forks form. Each fork has a leading and lagging strand, and together they constitute a bidirectional replication system.

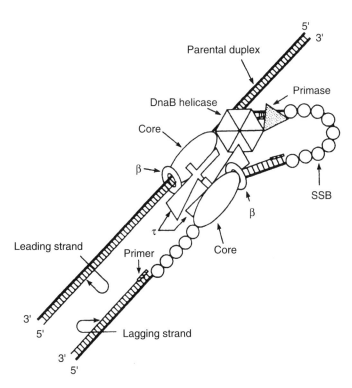

Figure 2.1. One possible arrangement of proteins at the replication fork. In this diagram, the leading- and lagging-strand polymerases are shown in an antiparallel orientation. All single-stranded DNA is shown coated with single-strand binding protein (SSB). The drawing is not to scale. (From Kim, S., Dallmann, H.G., McHenry, C.S., Marians, K.J. [1996]. Coupling of a replicative polymerase and helicase: A τ-DnaB interaction mediates rapid replication fork movement. *Cell* 84: 643–650.)

Most organisms possess a variety of DNA polymerase enzymes. In *E. coli* there are two primary enzymes, designated as DNA polymerase I and DNA polymerase III. DNA polymerase III is responsible for the bulk production of DNA. Its action is rapid, and it can join together up to 60,000 nucleotides per minute. It also has a proofreading capability, a $3' \rightarrow 5'$ exonuclease activity that allows the enzyme to "backspace" if an incorrect base has been inserted. A core enzyme composed of the α, β, and β' subunits with other proteins added forms a holoenzyme complex (Table 2.1).

Table 2.1. Major *E. coli* genetic loci involved in DNA replication.

Genetic Locus in Bacteria (*E. coli*)	Protein Function	Functional Equivalent in Archaea
DnaA	ATP binding protein; initiates replication; binds to origin of replication; also can shut off transcription of itself and other genes	Cdc6/Orc1(origin recognition complex)
DnaB	DNA helicase; unwinds DNA to allow primer synthesis	Minichromosome maintenance (MCM)
DnaC	Part of initiation complex; loading factor for helicase	Cdc6/Orc1
DnaG	DNA primase, not sensitive to rifampin	Primase homolog
DnaN	β Subunit of DNA polymerase III; sliding clamp (dimer)	Proliferating cell nuclear antigen (PCNA)
DnaQ	ε Subunit of DNA polymerase III; fidelity of replication; proofreading exonuclease ($3' \rightarrow 5'$)	
DnaX	δ Subunit of DNA polymerase III	
DnaZ	τ Subunit of DNA polymerase III; connects holoenzyme to primase	
Gyr	DNA gyrase adds and removes superhelical turns	DNA gyrase

Table 2.1. (*Continued*)

Genetic Locus in Bacteria (*E. coli*)	Protein Function	Functional Equivalent in Archaea
Lig (NAD-dependent)	DNA ligase; joins Okazaki fragments	DNA ligase I (ATP-dependent)
NrdAB	Ribonucleotide reductase that synthesizes deoxyribonucleotides from ribonucleotides	
PolA,	Removal of primers, repair enzyme	Flap endonuclease I
RNase H	Removal of primers, repair enzyme	RNase H
PolC	α Subunit of DNA polymerase III; polymerase function	
RpoA	α Subunit of RNA polymerase	
RpoB	β Subunit of RNA polymerase	
RpoD	σ Subunit of RNA polymerase; responsible for normal promoter binding	
Ssb	Single-strand DNA binding protein; facilitates melting of helices and stabilizes single strands	RPA/SSB

The number of proteins involved in DNA replication is far greater than what can be summarized in one table. In *B. subtilis*, 91 different interactions between 69 proteins were identified using the yeast two-hybrid system (see later) (Noirot-Gros et al. 2002). It should be noted that 36% of the proteins they identified had no known function at that time. Two major clusters of proteins are evident, one centered on the polymerase holoenzyme and the other centered on the methyl-accepting chemotaxis protein TlpA.

DNA polymerase I, on the other hand, functions more slowly and seems to be involved primarily in repair activities. It has $5' \rightarrow 3'$ polymerase, $3' \rightarrow 5'$ exonuclease, and $5' \rightarrow 3'$ exonuclease activities, which means that the enzyme can add new bases to the 3'-end of an RNA primer; proofread; or, if it attaches to an Okazaki fragment on the 5'-side of the primer of an adjacent fragment, degrade the primer and replace the RNA bases with DNA bases. The effect is to lengthen fragment 1 and shorten fragment 2. When all of an RNA primer has been removed, the enzyme DNA ligase closes the final nick (missing phosphodiester bond).

Although bacterial chromosomes have only one or a few *oriC* loci, yeast chromosomes have multiple origins, members of a class called ARS (autonomously replicating sequence elements) spaced at approximately 36 kb intervals. Bacterial *oriC* is a highly conserved structure among the Bacteria. Many genes having the same functionality occur in the same map order in different Bacteria, and there is extensive amino acid sequence homology among DnaA proteins. In *E. coli*, DnaA protein associates with ATP and 20–40 of these complexes bind to specific 9 base pair (bp) sites (DnaA boxes) in the *oriC* region (located about 45 kb from the DnaA gene) to create a nucleosomelike structure with DnaA at its core (Fig. 2.2). In *B. subtilis* and many other Bacteria, *oriC* is adjacent to *dnaA*. Additional DnaA boxes flank the *dnaA* gene itself and are scattered throughout the genome. When DnaA is bound to a site located on the noncoding DNA strand, transcription through that region terminates, helping to coordinate expression of some genes with replication. ATP hydrolysis occurs when DnaA is released from DnaA boxes, and DnaA must release ADP and bind new ATP in order to function. The necessity for DnaA protein is one reason why DNA replication stops in the absence of protein synthesis. This process has been reviewed by Messer (2002).

Initiation of DNA replication requires the DNA helix to be opened sufficiently to allow primer synthesis. This action is initiated by DnaA binding and further stimulated by transcription from an adjacent strong promoter (*mioC*) into *oriC*. An A:T-rich region is immediately adjacent to the DnaA boxes and the binding of DnaA to each box bends the DNA by 40°. The effect of the bending is to break the hydrogen bonds in the A:T region and allow the loading of DNA helicase. Binding of protein HU and the action of DNA gyrase are also important. Two complexes composed of DnaB, DnaC, and ATP bind to the partially opened region of DNA, one in each replication direction, leading to hydrolysis of ATP and DnaC release. The helicase (unwinding) action of DnaB allows entry of a primer-synthesizing enzyme and subsequent establishment of two replication forks by two DNA polymerase III enzymes. The complex of replication proteins (Table 2.1) is called a **replisome** and is localized to specific sites within the cell. This localization means that during replication the enzyme complex is stationary, and the DNA moves through it and spreads out into the presumptive daughter cells. See Baker and Bell (1998).

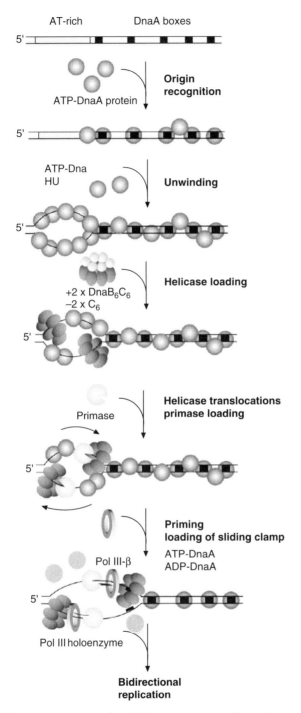

Figure 2.2. The initiation cycle for DNA replication in *E. coli*. DnaA protein binds in a cooperative fashion to open an A:T-rich portion of the DNA. DNA helicase loads followed by primase. Then, the sliding clamp and the DNA polymerase III holoenzyme complex load and the two oppositely oriented replication forks become active. (Reproduced with permission from Messer [2002].)

Premature reinitiation of DNA replication is prevented by hydrolysis of the ATP bound to DnaA and several mechanisms that prevent new binding. These include binding of DnaA protein to sites away from *oriC* as well as a lack of methylation of newly replicated DNA containing the sequence GATC. The *dam* methylase that catalyzes the reaction lags some 10 min behind the replication fork, and the lack of methylation impedes DnaA binding. Other factors preventing premature initiation of new rounds of replication include binding of *oriC* to the cell membrane and the necessity to accumulate additional DnaA–ATP complexes.

While the basic process of replication initiation is similar in the Archaea, there are notable differences as well. For example, instead of a DnaA protein, the Archaea use a protein similar to the proteins used for initiation by Eukaryotes, Orc1 and Cdc6. The Orc1/cdc6 protein binds to Orb sites to initiate DNA replication. Robinson et al. (2004) have shown that, unlike most studied Archaea, the Crenarchaeotan *Sulfolobus solfataricus* has two functional replication origins.

Analysis of RNA molecules transcribed in vitro from *oriC* shows that RNA transcripts may begin at several sites, yet replication of the Bacterial chromosome always begins essentially at the same spot. This specificity is enforced both by the RNase H enzyme, which acts to degrade RNA found in RNA–DNA hybrids, and by topoisomerase I. Presumably, there is an inherent stability in the "true" primer that makes it relatively resistant to the action of RNase H, whereas the abnormal primers are easily degraded. Separate RNA transcripts normally synthesized by DNA primase prime the leading strands of both clockwise-moving and counterclockwise-moving replication forks. In the absence of primase, primers are synthesized by RNA polymerase and are longer.

DNA polymerase III holoenzyme complex attaches to each primer and begins to lay down a new strand of DNA (see Fig. 2.1). A DNA helicase enzyme (ring-shaped *dnaB* hexamer) unwinds the helix ahead of the replication fork, and a single-stranded DNA binding protein (*ssb*) stabilizes the resulting single strands of DNA. Kornberg (reprinted in 2003) proposed that appropriate folding of single strands of DNA would permit a pair of identical DNA polymerase

III oligomers to synthesize both leading and lagging strands in tandem (see Fig. 2.1). The τ subunit of the holoenzyme interacts specifically with DnaB helicase. After the interaction, helicase increases its translocation rate tenfold, and the holoenzyme follows closely behind. The primase shown in Fig. 2.1 must be separately recruited to the complex for each Okazaki fragment synthesized. Delagoutte and von Hippel (2003) have reviewed the extensive involvement of helicases in cellular processes.

DNA polymerase is a highly processive enzyme that does not easily dissociate from its substrate once attached. Mechanically, the processivity results from a sliding clamp that holds the polymerase on the DNA. The sliding clamp for *E. coli* is the β subunit of the holoenzyme, which is dimerized and loaded onto the template by a set of proteins called the clamp loader (γ complex). Clamp loading requires ATP hydrolysis. The DNA template slides through the circular clamp that is firmly attached to the polymerase, thus maintaining a typical high rate of synthesis.

The act of unwinding the DNA duplex introduces extra twists into the unreplicated DNA structure that are removed by topoisomerase I (formerly known as the omega protein). The normal supercoiled state is restored in newly replicated DNA helices by DNA gyrase acting after the replication fork has passed. The amount of superhelicity in the *E. coli* chromosome is thus the result of the relative balance between these two competing enzyme activities. Inactivation experiments have shown that any change in the activity level of either topoisomerase is detrimental to a cell unless offset by a corresponding change in the other enzyme.

Termination of replication in *E. coli* occurs at two loci positioned approximately 180° around the circular genetic map from *oriC*. The two terminator regions are separated by 352 kb and are located roughly at positions 27.6 and 34.6 on the *E. coli* genetic map shown in Fig. 11.3 (also see Fig. 2.3). Each region consists of five unevenly spaced specific 23 bp *ter* sites, each of which inhibits the passage of a replication fork across it in one direction but not in the other. Thus, any given replication fork passes through one terminator region and stops at the next one, generating a short area of overlapping DNA. This mechanism eliminates the necessity of timing the arrival of the two replication forks. Whichever fork arrives first stops and awaits the arrival of the other. The Tus protein interacts specifically with the DnaB helicase to prevent it from unwinding the DNA duplex further (Mulugu et al. 2001). Its gene is located near one termination site and, if that gene is removed, neither terminator region is functional. Termination in *B. subtilis* occurs at a single *terC* locus whose extent is only

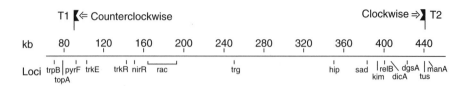

Figure 2.3. Map of the terminus region of the *E. coli* chromosome showing genetic markers and physical coordinates in kb, and TI, T2, and *tus*. Tus is a protein required for function of replication terminators TI and T2.

59 bp and which includes two inverted repeats (identical sequences rotated 180° with respect to each other).

Another major issue for a cell is precisely how the ends of the DNA from the various replication forks are joined together. One potential problem would be the joining of incorrect ends (Fig. 2.4). If this situation were to occur, a circular **concatemer** (a molecule of greater than unit length) would be formed. Another possibility is the formation of a **catenane** (set of interlocking circles). Concatemeric molecules can be reduced to unit length by **recombination** (rearrangement of the phosphodiester bonds linking the bases together so that altered DNA molecules are formed) at *dif* sites located within *terC* in a process catalyzed by XerC, XerD, and FtsK. FtsK is a cell division protein that is required for XerCD function. Therefore, proper resolution of concatemers can only occur during cell division. Mixing experiments have shown that FtsK action is species-specific in *E. coli* and *Hemophilus influenzae* (Yates et al. 2003). Catenanes can be separated through the action of topoisomerase I.

Related to the problem of replication termination is proper segregation of the completed chromosome so that each daughter cell receives one complete copy. This process was originally assumed to be a passive one. Indeed, the segregation model by François Jacob, Sidney Brenner, and F. Cuzin proposed that DNA replication origins were inserted into the membrane and gradually separated from one another due to the random insertion of new membrane material between their attachment points. However, this model is no longer considered tenable.

Fluorescence labeling experiments in *B. subtilis*, in which a fluorescent dye binds to the replication complex, have shown that the replicating DNA is oriented so that the complex lies near the middle of the cell and the respective replication origins are oriented toward the existing poles of the cell (Fig. 2.5), which supports the Monod model. Arguing against this idea is the observation

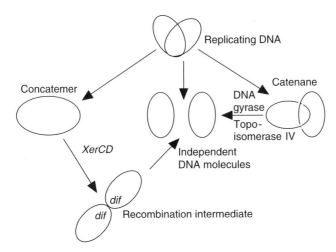

Figure 2.4. Termination of DNA replication. Depending on the way in which the ends of the newly replicated DNA helices are joined together, the product of replication might be one circular concatemer consisting of two complete genomes (*left*), two separate circular molecules (*center*), or two interlocked circular molecules (a catenane). Catenanes can be separated by a combination of topoisomerase IV and DNA gyrase, and concatemers can be turned into monomers by recombination between homologous *dif* regions on the two-component genomes in a reaction catalyzed by the XerCD proteins.

that in *B. subtilis* proper segregation of DNA requires the Spo0J protein and the product of a gene called *smc* (structural maintenance of chromosomes). The Smc protein shows significant sequence similarity to eukaryotic Smc proteins, which are required for chromosome condensation and segregation. The necessity for functional proteins argues for a dynamic and not a passive segregation process. Yamazoe et al. (2005) immunoprecipitated *E. coli* chromosomes using antibody raised against the SeqA protein (sequestration A) that helps regulate initiation of replication by binding to GATC sequences on hemimethylated (recently replicated) DNA. Using cells whose initiation of one replication event was synchronized, they were able to show that newly replicated regions moved to positions 1/4 and 3/4 of the distance across the long axis of the cells in preparation for division. The process is dependent on MukB protein, which is distantly related to the *smc* family. In both *Bacillus* and *Escherichia*, an increase in negative superhelicity (gyrase mutation) will compensate for defects in *smc* or *muk*. There

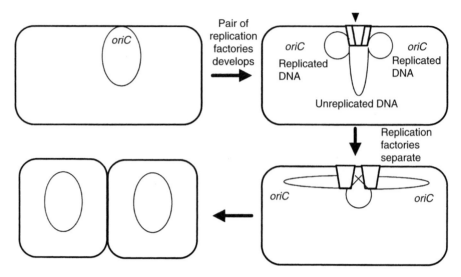

Figure 2.5. Coupling of chromosome replication and segregation in *B. subtilis*. The chromosomal DNA passes through a pair of stationary replication factories and then moves so that the origins of replication are located near the poles of the existing cell. The results of Yamazoe et al. (2005) for *E. coli* suggest that the replication factories actually move to positions in the center of what will be the daughter cells.

is now serious talk, but not definitive proof, that there may in fact be a microtubular mitotic apparatus in the Bacteria. The genetic loci that may be involved are *parA* and *parB*, and their effects have been observed in *Bacillus*, *Caulobacter*, and *Escherichia*.

Linking DNA in Arbitrary Ways

One of the basic tenets of genetics is that indiscriminate exchange of genetic information is disadvantageous to a species. In eukaryotic cells, problems with chromosome pairing during mitosis and meiosis often prevent cells that have acquired foreign chromosomes from surviving and/or producing functional gametes. However, because segregation of the nucleoid in prokaryotic cells requires no such pairing, other strategies must come into play. In particular, many bacterial cells and their viruses use a system of restriction and modification to tag their own DNA and disrupt any foreign DNA that may be present.

Enzymology of DNA Restriction and Modification

As implied by the name, DNA restriction and modification systems actually consist of two enzymatic activities that may be on the same or a different protein molecule. Each enzymatic activity requires a **recognition sequence**, a series of more or less specific bases to which the protein molecule(s) bind(s), but the similarity ends there. A **modification enzyme** makes a chemical alteration, usually adding a methyl group, to a specific site or sites within the recognition sequence. It acts after the replication fork has passed, meaning that its normal substrate is a hemimethylated DNA duplex, one that has one strand methylated and the other one not. A **restriction enzyme**, on the other hand, is stimulated by the presence of an incorrectly modified recognition sequence to act as an **endonuclease** (an enzyme that cuts a phosphodiester bond within a DNA molecule) and cut the DNA one or more times. The effect of this cutting is to prevent or reduce the ability of foreign bacterial DNA to incorporate itself into the bacterial chromosome and to prevent or minimize an infection by an invading viral DNA. Practically speaking, the main effect of restriction enzymes is antiviral because many of them have reduced activity during normal genetic transfer processes. Interestingly, *Pseudomonas aeruginosa* loses its restriction ability when grown for at least five generations at 43°C. While the loss is not permanent, it does persist for about 60 generations at 37°C, making routine genetic transfer much simpler because the source of the new DNA becomes immaterial.

Restriction endonucleases are designated by three italicized letters representing the genus and species names of the producing organism. A fourth letter may be added to indicate the specific strain that produces the enzyme. In cases where an organism produces more than one enzyme, the enzymes are differentiated by Roman numerals. Examples of this nomenclature and the types of base modifications that have been identified are given in Table 2.2.

Restriction enzymes actually fall into four broad classes. Type I enzymes occur only in the enteric bacteria and constitute four families (A, B, C, and D). The best-studied example is family IA, and its major representatives are *Eco*BI and *Eco*KI. The individual enzymes are composed of nonidentical subunits (one binding, two modification, and two restriction) encoded at *hsd* loci and have a high molecular weight (about 300,000 Da). They require ATP, magnesium ions, and *S*-adenosylmethionine (SAM) in order to function properly. The enzyme binds to both the recognition sequence and some neighboring bases in the presence of SAM. If both strands of DNA are methylated, nothing happens. If only one

Table 2.2. Properties of some commonly encountered restriction endonucleases.

Enzyme	Source	DNA Sequence Recognized	No. of Cleavage Sites		
			Phage Lambda	pUC19	pBluescript
Type I					
EcoBI	Escherichia coli B	TGÅA(N$_8$)TGCT	9		0
Type II					
Symmetric					
AluI	Arthrobacter luteus	AG$\overset{*}{C}$T	143	16	17
HaeIII	Hemophilus aegyptius	GG/$\overset{*}{C}$C	149	11	14
MboI (Sau3A1)	Moraxella bovis (Staphylococcus aureus)	/G $\overset{*}{A}$TC	116	15	15
BsmI	Bacillus stearothermophilus NUB36	5'GAATGCN/N 3' 3'CTTAC/GNN5'	46	0	0
BamHI	Bacillus amyloliquefaciens H	G/G $\overset{*}{A}$TCC	5	1	7
EcoRI	Escherichia coli RY13	G/A $\overset{*}{A}$TTC	5	1	1
PstI	Providencia stuartii	$\overset{*}{C}$TGC$\overset{*}{A}$/G	28	1	1
BglI	Bacillus globigii	G$\overset{*}{C}$C(N$_4$)/NGG$\overset{*}{C}$	29	0	2
NotI	Nocardia otidis-caviarum	GC/GG$\overset{*}{C}\overset{*}{C}$GC	0	0	1
Degenerate symmetric					
HaeII	Hemophilus aegyptius	PuG$\overset{*}{C}$GC/Py	48	3	4
Asymmetric					
HgaI	Hemophilus gallinarum	5'GACG$\overset{*}{C}$(N$_5$)/3' 3'CTGC$\overset{*}{G}$(N$^5_{10}$)/5'	102	4	4
Type III					
EcoPI	Escherichia coli (PI lysogen)	5'A G$\overset{*}{A}$CC3' 3'TCTGG5'	49		

Note: The recognition sequences are given as a series of letters beginning at the 5' side in which A, T, C, and G represent the first letters of the corresponding bases, Py represents any pyrimidine, Pu represents any purine, and N represents any nucleotide. The site of methylation (if known) is indicated by an asterisk, and the point of cleavage is indicated by a slash. If only one strand is indicated, the position of the cut in the complementary strand is symmetric. Type II enzymes always cut the DNA at a precise site relative to the recognition sequence, whereas type I and III enzymes do not (see text). Additional information on the properties of restriction enzymes may be obtained from Kessler, C., Manta, V. (1990). Specificity of restriction endonucleases and DNA modification methyl transferases—A review (edition 3). *Gene* 92: 1–248. The *Sau*3A1 and *Mbo*I enzymes are isoschizomers, but only *Mbo*I is blocked by *dam* methylation.

strand is methylated (a recently replicated DNA duplex), the methylase activity is stimulated. If the DNA is unmethylated (foreign DNA), the enzyme remains bound to the recognition sequence and uses energy from the ATP to pull in DNA from both sides, forming loops of DNA. If a second enzyme complex also has bound the DNA, there is eventually a collision between the two complexes. At that time (loop size between 1000 and 6000 bp) the enzymes produce a gap of about 75 nucleotides in the duplex, releasing the excised bases as single nucleotides. The result of all of this activity is a double-strand break in the DNA at a distance of 1000 or more nucleotides from the original recognition sequence. Because the double-strand break appears randomly located at any suitably distant point, the process is considered nonspecific.

In contrast to the mode of action just described, type II restriction endonucleases always make cuts at a definite point relative to the recognition sequence and have nonassociated methylase and restriction proteins. Because of their specificity of action, it has been easier to obtain information about their mode of action, some of which are summarized in Table 2.2. (As there are more than 225 commercially available restriction endonucleases, only a selected few are presented here.)

As might be expected from the fact that the recognition sequences are generally four to seven bases in length, enzymes recognizing the same sequences have been isolated from different organisms. These duplicate enzymes are called **isoschizomers**. The first three-letter abbreviation assigned is considered to have priority in preference to any other. However, no evolutionary relationship between isoschizomers is implied because their respective amino acid sequences are quite different, and they do not necessarily cleave the recognition sequence at the same site (in which case they are neoschizomers). Examples of isoschizomers are enzymes *Asu*II and *Bst*BI, which cleave the sequence TT/CGAA, where the slash indicates the point of cleavage.

The specific restriction endonuclease activities have been divided into three subgroups, based on the nature of the base sequence that each enzyme recognizes. For *Eco*RI (see Table 2.2), the sequence is considered symmetric because it has a twofold rotational symmetry. If the base sequence of both strands is written out, the symmetry can be more readily seen:

5′ GAATTC 3′
3′ CTTAAG 5′

Note that the sequence is a **palindrome**, one that reads the same regardless of the direction in which it is read. If the recognition sequence contains an odd number of bases, the symmetry is maintained if the center base is nonspecific (represented by the letter N). Symmetric enzymes are the most prevalent class, but asymmetric enzymes are also known. A few enzymes such as *Eco*RII actually require two copies of the same recognition sequence for full activity. The enzyme must bind to a noncleaved site up to 1 kb away from the cleaved site, or the cleavage is inefficient. DNA molecules that contain only a few, widely separated *Eco*RII recognition sites are refractory to cleavage.

In the case of type III enzymes, the modification protein functions alone or is complexed, but the endonuclease must be complexed to be functional. Therefore, the modification protein provides sequence recognition to the endonuclease. Type III enzymes make cuts at a definite position some distance from the recognition site. Only a few type III enzymes are known, and the best-studied is *Eco*P1, the restriction enzyme produced by phage P1. In this case, the recognition site is 5 bp, but the base sequence on the two strands of DNA is not the same when read from the two respective 5'-ends. The enzyme makes its cuts some 24–26 bases from the 3'-sides of the recognition site.

The various endonucleases can produce three types of fragments when they cut DNA. If they cut in the middle of the recognition site (e.g., *Alu*I), all fragments have blunt ends. If they cut a symmetric site off center, each fragment has a single-strand tail. If the cut is to the left of the center (e.g., *Bam*HI) the tails have 5'-ends, whereas if the cut is to the right of the center (e.g., *Pst*I) the tails have 3'-ends. Note that if a symmetric enzyme produces single-strand tails, all of the DNA fragments have identical tails. These single-strand tails can act as "sticky ends" to join various fragments by hydrogen bonding. The same is not true for the asymmetric enzymes, which seem to cut at some distance from the recognition site. The mechanism of this asymmetric cutting is not known, but its effect is to generate unique ends for each fragment (e.g., all *Hga*I fragments have five-base single-strand tails of nonspecific sequence).

Although methylation of bases is protective in the typical restriction system discussed earlier, there are exceptions, which constitute a fourth group of restriction enzymes. For example, *Hpa*II and *Msp*I are isoschizomers in the sense that they recognize the same base sequence. However, *Hpa*II cuts when the sequence is not methylated, while *Msp*I cuts when the sequence is methylated. In *E. coli* K-12, the *mcrBC* function restricts DNA containing 5-methylcytosine, hydroxymethylcytosine, and N^4-methylcytosine.

Bacteriophages often have defenses against restriction enzymes. There is obviously a strong selection against having restriction sites in phage DNA. In addition, many phages produce proteins that inhibit restriction enzymes. T7

Ocr protein blocks type I enzymes. Some *Bacillus* phages make their own methylation enzymes. The *E. coli* phage λ codes for a Ral protein that greatly stimulates the modification activity of type I enzymes toward totally unmethylated DNA.

Use of Restriction Fragments to Make New DNA Molecules

The terms **DNA joining**, **gene splicing**, and **gene linking** refer to the construction of new types of DNA molecules from DNA fragments such as those produced by the symmetric restriction endonucleases. The process is commonly described in the literature as the production of **recombinant DNA**. However, this type of DNA rearrangement is not the same as the recombination of classic genetics, and it seems better for a textbook representation to use the term recombination for the naturally occurring in vivo process and to use the terms conjoining or linking for the in vitro process. The term DNA splicing should not be confused with RNA splicing, a naturally occurring process.

A complete discussion of DNA linking requires information that is yet to be presented in this book. At the same time, the general outline of the procedure is readily understood. Therefore, a brief discussion is presented here, and a more thoroughgoing treatment can be found in Chapter 16.

The purpose of linking fragments of DNA is to produce a DNA molecule coding for a combination of traits that would be difficult or impossible to obtain by more conventional genetic techniques. If the new molecule is to be useful, it should be able to replicate itself (i.e., it must be a plasmid). Therefore, the first step in any DNA splicing procedure is selection of a suitable plasmid **vector** (carrier) into which one or more DNA fragments can be inserted. Typical vectors are phage M13 (see Chapter 7), lambda phage (see Chapter 8), ColE1 (see Chapter 12), and various derivatives of R plasmids (see Chapter 12).

In the simplest form of experiment, the vector DNA and the DNA to be linked are both cut with the same restriction endonuclease, one that generates single-strand tails or "overhangs." Upon mixing the two sets of DNA fragments together, the single-strand regions can pair, re-forming the original molecules or creating new combinations (Fig. 2.6). However, the process also allows the vector to close back on itself (form hydrogen bonds) without any extra DNA inserted.

There are two possible ways to avoid this problem. One more advantageous arrangement is to cut the vector with two enzymes so that a small piece of DNA is released. The larger vector piece is combined with an appropriately

Joining DNA after a single enzyme has cut it

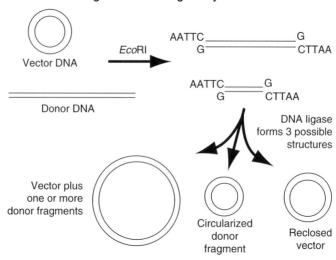

Joining DNA after two enzymes have cut it

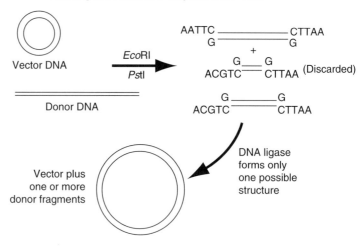

Figure 2.6. Simple DNA linking. (*Top*) Donor DNA (from any source) and vector DNA (a plasmid) are cut into pieces with the same restriction endonuclease. The ends of each fragment have the same sequence of single-strand DNA and are therefore complementary. Double-strand circles of DNA can be formed by joining the complementary ends of one or more fragments and using DNA ligase to restore the phosphodiester bond. Only if the vector DNA is included will the new circular molecule be able to replicate. (*Bottom*) If the same DNA is cut with two different restriction endonucleases, the single-strand ends of each fragment are different, and no fragment can self-close. If donor DNA is mixed with the large fragment of the vector, all plasmid genetic functions are preserved, but only if the vector fragment is linked to one or more donor fragments will a circular DNA molecule form.

cut piece of extra DNA. Under these conditions, the vector can only close if DNA is inserted, and the newly base-paired DNA has a definite orientation relative to the vector. The new molecules can be stably joined by the enzyme DNA ligase in the presence of ATP. This step completes the actual linking process. A second approach is to use a phosphatase enzyme to remove the terminal phosphates from the cut vector. DNA ligase can no longer seal the vector because it cannot provide the missing phosphate groups. However, the DNA fragments intended for insertion still have their phosphates and can allow ligase to form the DNA construct.

When the linked DNA is introduced into a cell in which the vector can replicate, it reproduces and transmits itself to all of the daughter cells. As a group of cells descended from one original cell is considered a clone, the process of obtaining a set of cells that carries newly joined DNA of a particular sequence is called **cloning** of the DNA.

Analytical Techniques for DNA Molecules

One of the major changes in the field of genetics over the past decade has been the implementation of methods that permit a geneticist to work directly with the DNA molecules under study. These techniques have rapidly become so fundamental that a basic knowledge of them is essential for understanding most of the contemporary work.

Electrophoresis

Electrophoresis is the separation of molecules using an electric field. The extent to which a particular molecule moves depends on the net charge on the molecule and its size and shape. Large DNA molecules, either double or single-stranded, are separable on the basis of size and shape because DNA has a uniform charge per unit length. The DNA undergoing electrophoresis is embedded in an agarose (a high molecular weight polysaccharide) gel with large pores through which DNA molecules can pass. The DNA molecules to be examined are placed in wells at one end of the gel slab, and electric current is applied. The charged DNA molecules migrate through the agarose gel but are subjected to retarding forces from the agarose latticework. The result is that smaller molecules and molecules that are coiled tend to travel faster than longer or linear molecules. Typically, the bands of DNA in a gel are visualized by staining the gel with a fluorescent dye such as ethidium bromide or Syber Green and viewing the gel under ultraviolet illumination.

A successful linking reaction of the type described in Fig. 2.6 produces a new DNA molecule that is larger than either of the original DNA molecules. This procedure can be demonstrated by means of standard agarose gel electrophoresis as just described above if the DNA molecules are not too large. Otherwise, a more elaborate system such as pulsed field gel electrophoresis must be used. In that procedure, the direction of the electric field is varied several times per hour. Larger molecules have difficulty reorienting in the electric field, and they do not move much. Smaller molecules progress more rapidly. Conventional electrophoresis has difficulty separating DNA molecules that are larger than 20–30 kb (Fig. 2.7a), but pulsed field electrophoresis can separate molecules up to 1000 kb. A sample of such a gel is shown in Fig. 2.8.

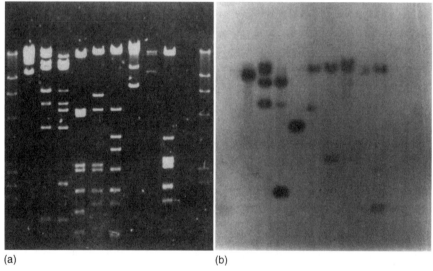

(a)   (b)

Figure 2.7. Southern blotting. (a) Agarose gel stained with ethidium bromide and viewed under ultraviolet radiation. The two outermost lanes are pure phage λ DNA digested with *Hind*III and used as size standards. The other lanes are a human DNA clone in a phage λ vector that has been digested with various restriction enzymes singly and in pairs. (b) An autoradiogram of a blot of the gel from (a) probed with radioactive human DNA. The bands from the original gel that show up darkly on the blot are the human DNA, and the bands that do not appear on the blot are the vector DNA. (Courtesy of L.G. Pearce.)

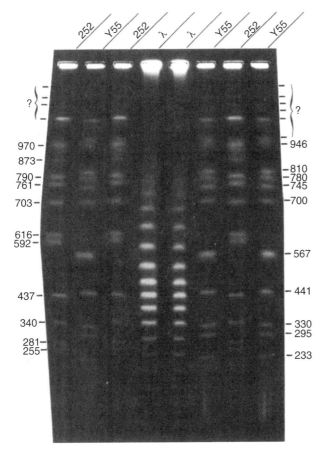

Figure 2.8. Pulsed field agarose gel. Conventional agarose gels such as that shown in Fig. 2.7a have difficulty separating DNA fragments larger than about 20 kb. If the gel is run using a pulsed field in which the direction of the voltage gradient is switched by rotating the gel through a 90° angle, larger fragments have difficulty reorienting themselves in the agarose and move more slowly than smaller fragments. The gel shown here presents individual chromosomes from two strains of yeast, 252 and Y55, separated by pulsed field gel electrophoresis. The sizes of the chromosomes in kb are indicated where they are known. Note the size variation between strains. The lanes labeled λ contain DNA molecules extracted from phage λ and joined end-to-end, to make a series of size standards of predictable lengths. (Reprinted by permission of the publisher from Gemmill, R.M., Coyle-Morris, J.F., McPeek, F.D., Jr., Ware-Uribe, L.F., Hecht, F. [1987]. Construction of long-range restriction maps in human DNA using pulsed field gel electrophoresis. *Gene Analysis Techniques* 4: 119–132. Copyright ©1987 by Elsevier Science Publishing Co. Inc.)

Atomic Force Microscopy

Examination of DNA and/or RNA molecules by conventional transmission electron microscopy is laborious in terms of sample preparation and extremely finicky in terms of equipment operation. A new technique, atomic force microscopy, offers substantial improvement in quality of image and ease of sample preparation. The sample DNA is placed on a mica surface and examined either under water or in the air. The examination consists of scanning a very fine probe over the surface of the mica (Fig. 2.9). The probe is subject to atomic forces from the specimen that result in deflection of a cantilever arm and the light beam reflecting off it. A computer then converts the deflection into an image such as that seen in Fig. 1.9.

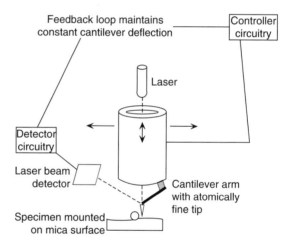

Figure 2.9. Atomic force microscopy. The microscope depends on a very fine tip that interacts with the specimen to generate attractive or repulsive forces at the atomic level. A laser beam reflects off the cantilever arm holding the tip and provides information on changes in the position of the tip. A feedback circuit moves the tip (in this diagram) to scan across the surface of the specimen and to compensate for changes in specimen height. A computer takes the information on the x-, y-, and z-axis position of the tip and uses it to create an image of the specimen. A sample micrograph is presented in Fig. 1.9.

Detection of Specific DNA Sequences

There are many instances in genetics when the identification of DNA sequences that are the same or similar to a particular reference sequence is useful. In the past, it was not feasible to completely sequence all candidate DNA molecules; thus a rapid screening procedure was required. Methods for such screening took advantage of the natural complementarity of nucleic acids to form **heteroduplexes**, which are DNA duplexes comprising strands of DNA from different sources.

Natural DNA molecules contain two perfectly complementary DNA strands (each adenine is paired to thymine, and each guanine is paired to cytosine). If the temperature of a solution of DNA is raised to near 100°C or the pH is raised to about 12.3, hydrogen bonds that link the two complementary strands are broken, and individual strands **denature**, or separate, but covalent bonds remain intact. If the temperature or pH is reduced rapidly, hydrogen bonds do not have time to re-form properly, and single strands of DNA remain in solution but become coiled on themselves. This state is not stable, and the DNA strands will gradually attempt to **rehybridize** (make as many hydrogen bonds as possible with neighboring molecules). However, at room temperature it is a slow process.

Maximum hydrogen bonding is obtained when the DNA is held at approximately 60°C or kept in a buffer with a mild denaturing agent (e.g., formamide). Either of these conditions prevents formation of mismatches in the double helix but is not sufficient to break correctly formed hydrogen bonds. The exact incubation procedure to be used depends on the DNA base composition, as G•C base pairs have three hydrogen bonds and are thus stronger than A•T pairs, which have only two. Thus, an A:T-rich region denatures more easily than a G:C-rich one.

Molecules to be tested are first separated by agarose gel electrophoresis (Fig. 2.7). Testing for particular DNA sequences is generally done by means of **blotting**, during which DNA from a gel is transferred (blotted) to a solid support such as a nylon membrane and denatured in place. The blot is then soaked in a solution containing a single-stranded nucleic acid probe under conditions that allow hydrogen bonds to form. If the corresponding sequences are present, the probe hybridizes to the blotted DNA. The probe is usually made radioactive or linked to biotin. Tests for the presence of the probe are easily performed. These procedures include autoradiography or addition of avidin, a chemical that binds very strongly to biotin. Usually, avidin is conjugated to an enzyme molecule that can be localized by addition of a substrate that yields a colored product. A typical blot prepared using a radioactive probe is shown in Fig. 2.9. (This technique was invented by Earl Southern using a DNA gel and is therefore usually referred to as a Southern blot. If an RNA gel is used, the procedure is known as Northern blotting, although no one by that name was involved in developing the procedure.)

Molecules of any size can be made to form heteroduplexes. Whenever large molecules such as entire plasmids or viral genomes are used, the resulting heteroduplexes can be visualized with a transmission electron microscope. A basic protein such as cytochrome C is used to neutralize the charge on the DNA molecule and allow it to spread out on the surface of the support film. Formamide is used to stabilize any single-stranded regions that may occur. An electron-dense material such as uranyl acetate is applied to stain the DNA so that it can be easily seen. For a set of electron micrographs showing some examples of heteroduplex molecules and giving their interpretation, see Fig. 7.1.

Polymerase Chain Reaction

A problem with Southern blotting is that the experimenter must be able to isolate DNA in sufficient quantity so that a radioactively labeled fragment is visible on an autoradiogram. If the desired sequence occurs only rarely in a population of molecules, there may not be enough probe bound to expose the film. The patented technique of polymerase chain reaction (PCR) helps to amplify rare sequences.

The basic concept of PCR is very simple (Fig. 2.10). A pair of short primers is introduced into a mixture of denatured DNA and is allowed to renature. If the primers match any sequences, they bind and each one forms a short duplex. The pairs of duplexes define a region of DNA whose outer boundaries are the 5'-ends of the primers. Further, DNA polymerase can use these primers to synthesize complementary strands. The reaction is stopped after a short time, the DNA denatured, and the process repeated. In theory, the number of desired sequences should double with each cycle. Therefore, 20–30 cycles should amplify the desired sequence by more than a millionfold, yielding a product that is readily visible on a normal agarose gel.

A major drawback to the experimental plan is that the DNA denaturation step also tends to denature DNA polymerase, requiring addition of a new enzyme after each cycle. The great conceptual leap by Kary Mullis was to realize that DNA polymerase from thermophilic (growing at high temperature) organisms is often stable at temperatures that denature DNA. The DNA polymerase from *Thermus aquaticus* (*Taq* polymerase) is particularly suitable and commonly used,

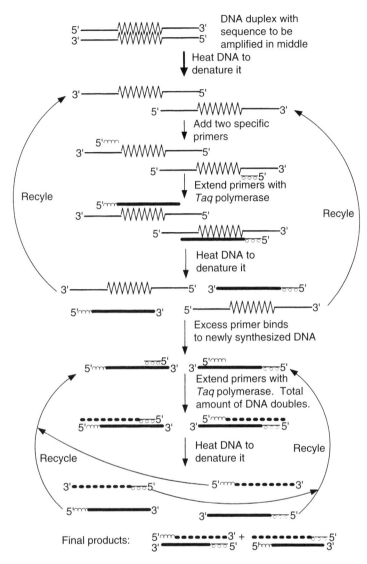

Figure 2.10. Polymerase chain reaction. DNA is denatured, and a primer of specific sequence is added. When the primer is bound, it can be extended by DNA polymerase in the usual semiconservative manner. If all original DNA molecules bound a primer, the number of sequences of interest doubles. The next denaturation step releases the product DNA, and each of these strands can bind an appropriate primer. The next extension reaction doubles the amount of new DNA. This process is then repeated many times, with the number of sequences approximately doubling each time. After extensive PCR, the original DNA molecules and the first products are lost in the myriad of short duplexes with defined end points.

although several other polymerases are also commercially available. The experimenter adds an excess of polymerase and primer at the beginning of the experiment and places the reaction mixture tube in a thermal cycler that first raises the temperature to denature DNA (approximately 90 °C), then lowers it to allow polymerase function (approximately 60–70 °C), repeating as many times as needed. Higher polymerization temperatures make for greater stringency in primer binding.

PCR is a powerful methodology but prone to experimental error. The primer will amplify any sequence to which it corresponds, including DNA molecules that are introduced accidentally into the DNA preparation by careless laboratory technique. Even a single contaminating molecule can be amplified to yield enough DNA to be visible on a gel.

DNA Sequence Analysis

There are two primary methods to determine the nucleotide sequence of a DNA molecule. One of the methods has been developed by Fred Sanger and the other by Alan Maxam and Walter Gilbert (Fig. 2.11). Sanger and Gilbert shared the Nobel Prize for their efforts. Both methods are based on the principle of generating from a DNA molecule a set of labeled oligonucleotides (short chains of DNA) of random length but with a specific base at the 3′-end of the fragment. The individual fragments are then separated by electrophoresis on a polyacrylamide gel (PAGE) or on a specially formulated agarose gel.

With this process, a very thin gel of polyacrylamide is prepared, a mixture of oligonucleotides from either procedure is layered on the top, and a potential gradient of several thousand volts is applied across the gel. The oligonucleotides migrate toward the anode at a rate inversely proportional to their size. By using a gel of appropriate composition, it is possible to separate oligonucleotides that differ in length by only one base.

Detection of the DNA bands within a gel originally used one of three methods. If the bands are labeled with either ^{32}P or ^{35}S, autoradiography is possible. Alternatively, DNA bases can be biotinylated (have a biotin moiety attached). Avidin and related molecules bind strongly to biotin, and their binding ability remains even in the presence of an attached enzyme such as a phosphatase. Addition of an appropriate substrate gives either a colored or a chemiluminescent band. In the latter case, a comparatively brief exposure to x-ray film provides an image of the gel. The most recently developed method is one using PCR tech-

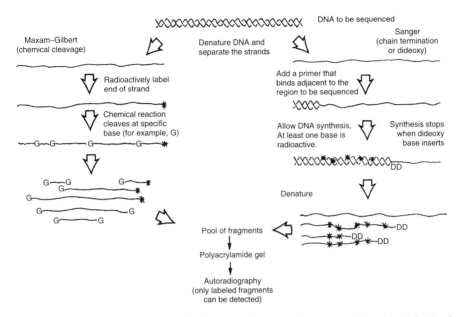

Figure 2.11. DNA sequencing. Both original protocols require that the individual DNA strands be separated and that a label (usually radioactive) be introduced. The label (*asterisk*) must be at the end of the fragment for the Maxam–Gilbert procedure to guarantee that all fragments observed on the autoradiogram begin at the same base. However, in the dideoxy procedure any base can be labeled because the specificity of the fragment end is ensured by the primer used. Typical autoradiograms produced by these procedures are shown in Fig. 2.13.

nology to prepare large quantities of dideoxy (see later) fragments. After separation, the fragments are colored with a silver-based stain.

Obviously, for a DNA sequencing technique to be interpretable, only a single-strand DNA can be used as a source of DNA fragments; otherwise, it would not be possible to tell which fragments came from which strand. The Maxam–Gilbert and Sanger methods differ from each other in their approach used to generate the single-stranded oligonucleotide fragments. Interpretation of the polyacrylamide gel patterns requires knowledge of how the fragments were generated.

In the Sanger (or dideoxy) technique, the fragments are the result of interruption of DNA synthesis by insertion of an inappropriate base. Figure 1.4 shows that for a base to be accommodated within the DNA structure it must have hydroxyl groups on both its 3′- and 5′-carbons. A dideoxy base (Fig. 2.12)

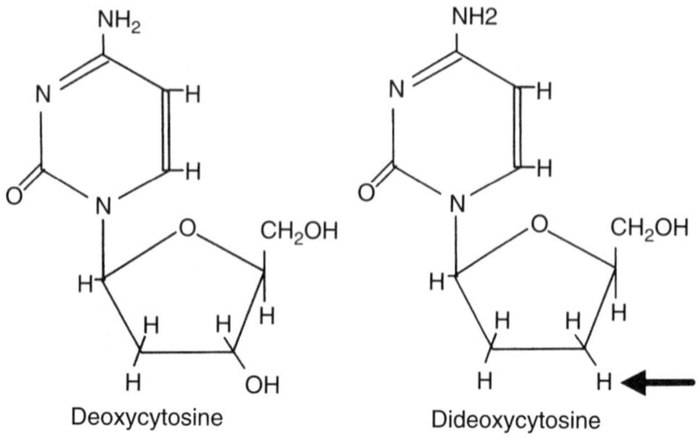

Figure 2.12. A dideoxy base compared with the normal deoxy base. The *arrow* indicates the position of the missing oxygen atom.

has a 5'-hydroxyl group but not one on the 3'-carbon. Therefore, whenever such a base is incorporated into a DNA molecule, it becomes the last base in the fragment. In practice, a shortened protein chain from DNA polymerase I known as the Klenow fragment is used to replicate purified single strands of DNA. The Klenow fragment has lost all error-correcting ability ($3' \rightarrow 5'$ exonuclease activity) and thus inserts an incorrect base but cannot remove it. Recently, DNA polymerases from extreme thermophiles have also been used. The advantage of these enzymes is that higher temperatures reduce secondary structure within the single-strand DNA being copied. In the case of repetitive sequences, DNA looping can occur, and the polymerase may read across the bottom of the loop instead of opening it and copying the entire strand.

The point at which replication begins is controlled by the base sequence of the primer used. Modern high-speed technology uses trace amounts of four different dideoxy bases, each labeled with a different fluorescent dye. Whenever a dideoxy base is inserted into a growing DNA chain, all further synthesis must stop. For example, when the concentration of dideoxy adenosine is adjusted properly, a series of fragments of all possible lengths ending in adenine is generated. The fragments can be made radioactive by adding one or more appropriately labeled nucleoside triphosphates to the reaction mixture so that they are incorporated into the oligonucleotides.

When the four reaction mixtures are subjected to electrophoresis, a ladder of individual bands is produced (Fig. 2.13). Because each lane on the gel corresponds to fragments ending in a particular base, the DNA sequence is

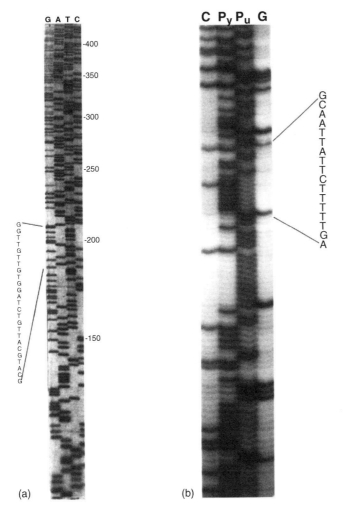

Figure 2.13. Sample autoradiograms of DNA sequencing reactions. Base abbreviations are same as those given in Fig. 2.1. (a) The pattern produced by the Sanger method (courtesy of Karen Denzler, School of Life Science, Arizona State University). (b) The pattern produced by the Maxam–Gilbert method (courtesy of E. Goldstein, School of Life Science, Arizona State University).

determined by beginning with the smallest band and reading up the gel, noting which lane has the next smallest fragment.

The Sanger dideoxy method is readily adaptable to PCR technology, thereby eliminating the necessity for strand separation. In this modification, the sequencing reactions are run as PCR reactions but using a single primer instead of a pair of primers. The product of the reaction is complementary to only one of the two strands in the original helix. The fidelity of the DNA polymerase used is important for the accuracy of the sequence determination, and the enzyme chosen must be heat-stable and have proofreading capability.

In the Maxam–Gilbert method, double-strand DNA is first made radioactive at its 5'- or 3'-end using an appropriate enzyme. The molecule is then denatured to give single strands that are separated, usually by electrophoresis. The physical basis for the separation is unclear, but the technique yields two strands of identical length but complementary sequence. Choosing one of the two separated strands, chemical reactions are used to randomly cleave specific bonds within the DNA molecules. Four standard reactions cleave the DNA either at one of the guanine residues, any purine residue, one of the cytosine residues, or any pyrimidine residue.

Conditions of the reaction are controlled so that complete cleavage does not occur, and once again a series of fragments of different lengths is obtained. Since the fragments are visualized by autoradiography, only those fragments that include the labeled end are detectable. Separation of the fragments is achieved via the same electrophoretic technique used in the Sanger method.

Reading the sequence is a trifle more complex than in the case of the dideoxy gels (Fig. 2.13) because some of the reactions cut at two bases. Thus, a band present in both the (C) and (C + T) lanes indicates that a cytosine residue is present, but a band present only in the (C + T) lane indicates that a thymine residue is present. One advantage of the Maxam–Gilbert procedure is that the complementary strand can also be sequenced and used to confirm the sequence deduced for the original strand. The dideoxy method would require that the sequence be determined and a primer prepared that is the same as the 3'-end of the sequenced DNA. This primer could then be used to initiate DNA replication of the complementary strand.

The practical limitation on either manual DNA sequencing technique is the number of bands resolvable on the electrophoretic gel. For routine work, 200–400 bases of DNA per gel is a reasonable expectation. Most laboratories now use an automated DNA sequencer based on the Sanger method. In this procedure, a specific primer is added to the DNA to be sequenced so that strand purification is no longer necessary. The strand whose sequence is determined is controlled by the choice of primer. The bases used to terminate the sequencing reactions are bonded covalently to a fluorescent dye, with a different color for each base. Therefore, all the reactions can be run in the same tube and the resulting DNA fragments run through a capillary column. As each droplet elutes from the column, a laser beam strikes it, and any fluorescent dye present emits its characteristic color. The information obtained is presented as a series of four overlapping graphs of fluorescence as a function of time of elution (Fig. 2.14). The computer program that scans the gel automatically tries to read the base sequence and prints its results at the top of the graph. This automated method yields sequence information for more than 600 bases at a time. The National Center for Biotechnology Information web site at http://www.ncbi.nlm.nih.gov/entrez/query.fcgi?db=Genome provides access to complete genome sequence information for over 1000 viruses and over 100 bacteria.

Custom Synthesis of DNA

Microprocessor-controlled instruments that can prepare short pieces of single-strand DNA of predetermined sequence are now available. The chemistry of the system is somewhat complicated, but the basic outline is simple enough (Fig. 2.15). A small bead of material has attached to it the first base of the desired sequence. A solution containing a highly reactive form of the next base is then added and given time to react. The bead serves as an anchor to allow complete washing before the next base is added. One complete step can be carried out in about 15 min under optimal conditions.

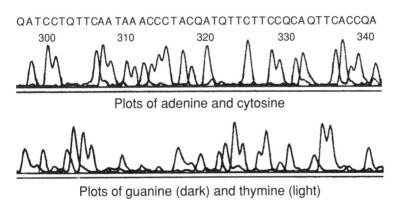

QATCCTQTTCAA TAA ACCCT ACQATQTTCTT CCQCA QTTCACCQA

Plots of adenine and cytosine

Plots of guanine (dark) and thymine (light)

Figure 2.14. Results from an automatic DNA sequencer. The output is normally in color with all four graphs superimposed. Each graph represents the absorbance of a different dye attached to a dideoxy base. For this black-and-white rendering, only two of the graphs are shown in each panel. The letters at the *top* of the figure indicate the deduced base sequence of the combined graphs using the abbreviations of Fig. 1.4. A color version of this figure is available on the web site for this book, http://lifesciences.asu.edu/bactgen/.

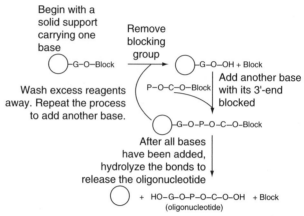

Figure 2.15. Custom DNA synthesis. The reactions are carried out in a small tube with the product held in place via a chemical linkage to an inert material. Reagents are passed sequentially through the tube. To prevent unwanted side reactions, each new base has its 3′-hydroxyl group blocked chemically. This block must be removed before the next base can be added. The final hydrolysis step removes the last blocking group and cleaves the chemical bond to the solid support so that the single strand of DNA can be isolated in pure form. In the reaction diagrammed, a dinucleotide (G–C) has been synthesized.

A limitation to this process is the completeness of addition of each base. If the chemical reaction is not 100% complete, some DNA molecules have an incorrect base sequence. Obviously, the proportion of incorrect DNA molecules increases with the length of the DNA synthesized. Thus, for practical reasons most custom DNA molecules have a length of about 100 bases or less. Technological developments have brought down the cost of custom-synthesized DNA to less than US $1 per base.

Footprinting

The technology used for DNA sequencing also facilitates precise determination of a particular molecule's binding site on DNA (Fig. 2.16). In general, if a protein is bound to a specific DNA base sequence, it effectively covers that region and protects the DNA against degradative enzymes. If pure end-labeled DNA is digested randomly with DNase I, an endonuclease that makes double-strand cuts in the molecule, a family of DNA fragments of all possible lengths is produced.

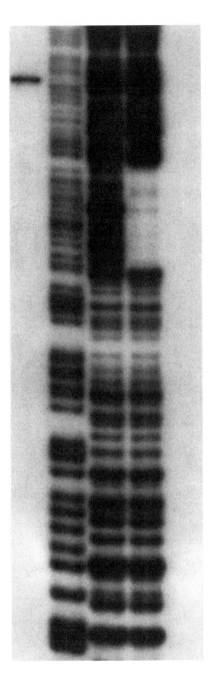

Figure 2.16. Footprinting. The gel shows the results of an experiment designed to demonstrate the binding of a protein to the regulatory region for *recA* (see Chapter 5). DNA fragments were made radioactive at their 5′-end prior to the experiment. Lane 1: Fragment after partial degradation by the restriction enzyme *Alu*I. Lane 2: Maxam–Gilbert reaction, which yields all possible fragments ending in a purine. Lane 3: Partial digestion with DNase I in the absence of added protein. Lane 4: A similar digestion in the presence of the *lexA* protein. The *lexA* protein was bound to the region of the *recA* DNA just upstream from the translation start site. (From Little, J.W., Mount, D.W., Yanisch-Perron, C.R. [1981]. Purified *lexA* protein is a repressor of the *recA* and *lexA* genes. *Proceedings of the National Academy of Sciences of the USA* 78: 4199–4203.)

However, if the digestion is carried out on DNA to which protein is bound, certain fragment lengths are not produced because the bound protein prevents the endonuclease from reaching the DNA. When the fragments are later separated by electrophoresis on a polyacrylamide gel, certain bands are missing completely. If sequencing samples are run on the same gel, the protected base sequence can be read directly.

Gel Mobility Shift Analysis

The alternative way to identify DNA regions that bind specific proteins is the gel mobility shift. The movement of DNA through an agarose gel depends on

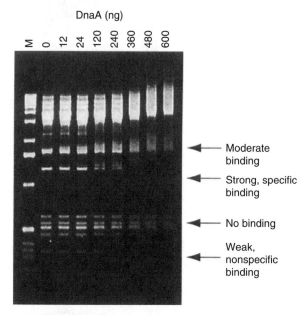

Figure 2.17. Gel mobility shift assay. Phage λ DNA was digested with a combination of *Eco*RI and *Pvu*II restriction enzymes. Varying amounts of DnaA protein were added to individual samples, and the DNA was separated on an agarose gel. Lane M is a series of molecular weight standards. Four types of mobility shifts are indicated. (Reproduced with permission from Szalewska-Palasz, A., Weigel, C., Speck, C., Srutkowska, S., Konopa, G., Lurz, R., Marszalek, J., Taylor, K., Messer, W., Wegrzyn, G. [1998]. Interaction of the *Escherichia coli* DnaA protein bacteriophage λ protein. *Molecular and General Genetics* 259: 679–688.)

the shape of the DNA and the voltage applied. If the voltage is constant, then changes in the DNA shape are reflected by changes in position. Specifically, if a DNA fragment binds to a protein, its shape is different and its mobility is changed. Conceptually, gel mobility shift analysis can have four possible outcomes (Fig. 2.17): (1) The DNA might not bind any of the added protein, retaining the same mobility. (2) It might bind the protein to a single very specific site. This will result in a sharp band migrating at a different position at a low level of added protein. (3) It might bind the protein weakly, giving a diffuse band with altered migration. (4) It might bind the protein weakly and at many sites, leading to the disappearance of the original DNA band. No new band appears because the binding is so variable that only an extremely diffuse smear results.

Yeast Two-Hybrid Systems

Researchers may want to determine what proteins interact to give a particular effect, as in the case of DNA replication proteins. Cloning technology provides a fairly simple method to look for such interactions. The assay is based on the fact that in eukaryotes, yeast in this case, transcription of a reporter gene occurs as a result of the binding of a protein to two distinct sites, a protein binding site and a transcription activation site. As long as these two sites are bound, transcription activates regardless of whether the binding is due to one or a pair of molecules. The experimenter clones each separate binding domain to a different protein of interest. If the proteins interact, they trigger the transcription process to produce a detectable gene product (Fig. 2.18).

Summary

Replication of bacterial DNA molecules occurs in the usual semiconservative manner, beginning at a single point of origin. In most cases, there are two oppositely directed replication forks to reduce the time required for replication. The primer for DNA synthesis is a short RNA molecule. DNA polymerase III holoenzyme can work only in the $5' \rightarrow 3'$ direction and thus can lengthen one primer indefinitely to form the leading strand of the newly replicated DNA. However, the opposite, or lagging strand can be synthesized only as relatively short Okazaki fragments. DNA polymerase I replaces RNA primers with DNA, and then DNA ligase joins the resulting pure DNA fragments.

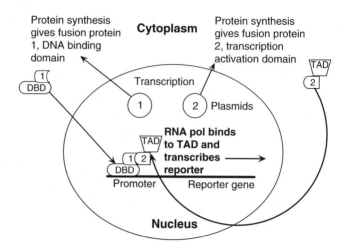

Figure 2.18 Yeast two-hybrid system. Two proteins that may bind to each other are cloned as fusion proteins, one to a specific DNA binding domain (1) and the other to a transcription activator (2). Both plasmids are introduced into the same yeast cell. After transcription and translation of the plasmid mRNA, the fusion proteins return to the nucleus. If they bind to one another, RNA polymerase is able to attach itself and transcribe the reporter gene. An example of a reporter gene is Gal-6.

Modern molecular biology has contributed techniques that have become fundamental for bacterial geneticists. Among them are the abilities to determine the sequence of a DNA molecule and to use the technique of heteroduplex formation to identify specific sequences within a DNA molecule. The most far-reaching discovery is that of restriction and modification enzymes within the bacteria. These enzymes normally protect against virus infection within a group of closely related organisms. However, certain restriction enzymes make specific and reproducible cuts within DNA molecules, and it is possible to create families of DNA fragments, all of which have identical single-strand ends. These fragments can then be reassorted in any desired order. If one or more of the fragments derives from a plasmid, the resulting DNA construct is capable of self-replication and can be transmitted along a cell line.

Questions for Review and Discussion

1. Why does DNA replication gradually stop when protein and/or RNA synthesis stop? Inhibition of which enzyme functions would cause a rapid stop?

2. How would you detect whether a particular microorganism produced a restriction enzyme? Given that such an enzyme exists, how would you determine the base sequence that it recognizes?

3. Both PCR and Southern blotting serve to identify specific sequences in DNA. What are the advantages and disadvantages of both the techniques?

4. How would you determine whether a particular protein interacted with DNA or with another protein?

References

General

Baker, T.A, Bell, S.P. (1998). Polymerases and the replisome: Machines within machines. *Cell* 92: 295–305.

Bickle, T.A. (2004). Restricting restriction. *Molecular Microbiology* 51: 3–5.

Coates, P.J., Hall, P.A. (2003). The yeast two-hybrid system for identifying protein–protein interactions. *Journal of Pathology* 199: 4–7.

Delagoutte, E., von Hippel, P.H. (2003). Helicase mechanisms and the coupling of helicases within macromolecular machines. Part II: Integration of helicases into cellular processes. *Quarterly Reviews of Biophysics* 36: 1–69.

Draper, G.C., Gober, J.W. (2002). Bacterial chromosome segregation. *Annual Review of Microbiology* 56: 567–597.

Giraldo, R. (2003). Common domains in the initiators of DNA replication in Bacteria, Archaea and Eukarya: Combined structural, functional and phylogenetic perspectives. *FEMS Microbiology Reviews* 26: 533–554.

Kornberg, A. (2003). *Enzymatic Synthesis of DNA*. Temecula, CA: Textbook Publishers.

Messer, W. (2002). The bacterial replication initiator DnaA, DnaA and oriC, the bacterial mode to initiate DNA replication. *FEMS Microbiology Reviews* 26: 355–374.

Margolin, W. (2003). Bacterial division: The fellowship of the ring. *Current Biology* 13: R16–R18.

Murray, N.E. (2002). Immigration control of DNA in bacteria: Self versus nonself. *Microbiology* 148: 3–20.

Pogliano, K., Pogliano, J., Becker, E. (2003). Chromosome segregation in eubacteria. *Current Opinion in Microbiology* 6: 586–593.

Specialized

Barionovi, D., Ghelardini, P., Di Lallo, G., Paolozzi, L. (2003). Mutations arise independently of transcription in non-dividing bacteria. *Molecular and General Genomics* 269: 517–525.

Choudhary, M., Mackenzie, C., Nereng, K., Sodergren, E., Weinstock, G.M., Kalpan, S. (1997). Low-resolution sequencing of *Rhodobacter sphaeroides* $2.4.1^T$: Chromosome II is a true chromosome. *Microbiology-UK* 143: 3085–3099.

Loewe, L., Textor, V., Scherer, S. (2003). High deleterious genomic mutation rate in stationary phase of *Escherichia coli. Science* 302: 1558–1560.

Mulugu, S., Potnis, A., Shamsuzzaman, Taylor, J., Alexander, K., Bastia, D. (2001). Mechanism of termination of DNA replication of *Escherichia coli* involves helicase–contrahelicase interaction. *Proceedings of the National Academy of Sciences of the USA* 98: 9569–9574.

Noirot-Gros, M.-F., Dervyn, E., Wu, L.J., Mervelet, P., Errington, J., Ehrlich, S.D., Noirot, P. (2002). An expanded view of bacterial DNA replication. *Proceedings of the National Academy of Sciences of the USA* 99: 8342–8347.

Robinson, N.P., Dionne, I., Lundgren, M., Marsh, V.L., Bernander, R., Bell, S.D. (2004). Identification of two origins of replication in the single chromosome of the archaeon *Sulfolobus solfataricus. Cell* 116: 25–38.

Sunako, Y., Onogi., T., Hiraga, S. (2001). Sister chromosome cohesion of *Escherichia coli. Molecular Microbiology* 42: 1233–1241.

Woldringh, C.L. (2002). The role of cotranscriptional translation and protein translocation (transertion) in bacterial chromosome segregation. *Molecular Microbiology* 45: 17–30.

Yamazoe, M., Adachi, S., Kanaya, S., Ohsumi, K., Hiraga, S. (2005). Sequential binding of SeqA protein to nascent DNA segments at replication forks in synchronized cultures of *Escherichia coli. Molecular Microbiology* 55: 289–298.

Yates, J., Aroyo, M., Sherratt, D.J., Barre, F.-X. (2003). Species specificity in the activation of Xer recombination at *dif* by FtsK. *Molecular Microbiology* 49: 241–249.

3

Mutations and Mutagenesis

The first problem facing early bacterial geneticists was to prove that bacteria did have inherited traits. The earliest presumption was that bacteria and other microorganisms were too small to have any phenotypic traits that could be studied. That concept was disabused by the work of George Beadle and Edward Tatum, who demonstrated that biochemical reactions could be used as phenotypic traits and then developed the famous "one gene–one enzyme" hypothesis. There was, however, one remaining area of uncertainty regarding bacterial genetics. Many workers thought that the hypothesis of Lamarck regarding inheritance of acquired traits was true for bacteria even though it had already been disproved for animals and plants. The first task of the fledgling science of bacterial genetics was to prove that the same processes of mutation that had already been shown to occur in eukaryotes also occurred in prokaryotes. Recently, this controversy has arisen again in a new and particularly interesting form.

Major topics include:

- Techniques available to study the appearance of mutations in populations of bacteria
- Mutations, their occurrence, and potential effects
- Nature of the controversy over adaptive evolution
- Factors that can affect the expression of mutations and the importance of conditional mutations for geneticists

Bacterial Variation

The original experiments dealing with acquisition of new traits focused on bacterial resistance to some sort of selective agent, usually a bacteriophage or an antibiotic such as streptomycin. Researchers agreed that treating bacteria growing on the surface of an agar plate with a selective agent killed most of the cells. However, a few resistant colonies would grow. The descendants of these resistant colonies were also resistant, meaning that the change was stable and inherited (i.e., genetic). The theoretical interpretation of these results was in considerable dispute. The Lamarckian theorists maintained that any cell in the culture had a small but finite probability of surviving the selective treatment. If a cell did survive, the acquired trait of resistance was passed to all subsequent members of the clone. In other words, the nature of the mutation was directed by the applied selection.

The opposing Darwinian, theory was one of preexisting mutations. Certain bacteria in a culture were assumed to be naturally resistant to the selective agent because of a preexisting change in their genetic material (a mutation) that had occurred in the absence of the selective agent. Treatment with the selective agent did not cause the change, but rather demonstrated its existence by eliminating sensitive cells. Most of the subsequent experiments were designed to show that mutant cells were present before the selection was applied.

Plate Spreading

H.B. Newcombe developed a directly visible method of demonstrating that mutations preexisted selection. The basic experiment consisted of taking several agar plates and spreading a uniform lawn of bacteria on each one of them. After various periods of incubation, Newcombe sprayed the plates with a solution of virulent bacteriophage (virus) and incubated the plates until bacteriophage-resistant colonies appeared. However, before spraying, some plates had the bacteria on one-half of their surfaces spread around with a glass rod. If phage resistance were due to acquired immunity, the spreading should have had no effect because the same number of cells were treated in each case. On the other hand, if there were preexisting mutant cells, their number would double at each cell division, but all cells in a clone would remain in the same physical location unless spreading redistributed clonal members. Under a Darwinian hypothesis, the spread side of a plate would be expected to have more resistant colonies than the unspread side. The actual results were fully in accord with this expectation. One set of six plates, for example, had a total of 28 colonies on the unspread sides but 353 on the spread sides.

Replica Plating

An entirely different experiment from Joshua and Esther Lederberg contributed evidence in favor of preexisting mutations and a valuable new technique for bacterial genetics. The technique is called **replica plating** and is a type of printing process (Fig. 3.1). A piece of sterile velvet or velveteen is placed across the top of a cylinder with the pile surface of the fabric facing up; a band holds the velvet tightly to the cylinder. The experimenter then takes a master plate, marks its orientation, and presses it gently against the surface of the velvet. As the fabric fibers poke into the agar, bacteria stick to their sides. The experimenter then carefully removes the master plate and presses in turn as many as 10–12 uninoculated plates that are also marked for orientation. After incubation each plate carries a faithful replica of the master plate because some bacteria that adhered to the fabric transferred to the agar at each replication.

The Lederbergs' experiment consisted of replicating master plates of sensitive cells to two or more plates containing either streptomycin or T1 virus. When the replica colonies had grown, the Lederbergs compared their positions and marked any colonies that appeared in the same relative position on all replica plates. They cut out the corresponding area of the master plate and resuspended the bacteria on it in a liquid medium. If the Darwinian hypothesis were correct, the culture derived from those cells would be enriched for resistant mutants by virtue of the fact that only a small piece of agar (and therefore, a limited number of cells) was removed from the master plate. The Lederbergs then repeated the entire process using the newly prepared culture.

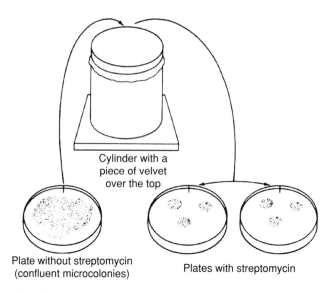

Cylinder with a
piece of velvet
over the top

Plate without streptomycin
(confluent microcolonies)

Plates with streptomycin

Figure 3.1. Use of replica plating to demonstrate undirected, spontaneous appearance of streptomycin-resistant mutants. The Lederbergs spread about 10^5 sensitive cells on a plate of drug-free solid medium and allowed them to reach full growth (10^{10} to 10^{11} cells). Sterile velvet covering the end of the cylindrical block was pressed lightly on this continuous heavy lawn ("master plate") and was then pressed successively on two plates of medium containing streptomycin at a concentration that killed sensitive cells. A few colonies of resistant cells appeared on each plate, usually in coincident positions. (From Davis, B.D., Dulbecco, R., Eisen, H.N., Ginsberg, H.S. [1980]. *Microbiology*, 3rd ed., p. 31. Philadelphia, PA: Lippincott.)

After many repetitions, the final result was a master plate that contained nothing but resistant bacteria, even though the cells and their progenitors had never been directly subjected to selection. This process of indirect selection is entirely in accord with the Darwinian idea of preexisting mutations, but not with the idea of acquired immunity.

Fluctuation Test

Max Delbrück and Salvador Luria performed what was considered to be the definitive experiment in the analysis of bacterial mutations. They developed an experimental system known as the **fluctuation test** that not only greatly contributed to answering the mutation question, but also finds use today. The basic protocol is simple. The experimenter takes a small number of cells from the same liquid culture and inoculates 10–100 tubes of sterile medium. After incubating for sufficient time so that the cultures reach stationary growth phase, the experimenter spreads either aliquots or the entire contents of a tube on selective medium (in Luria and Delbrück's case, agar saturated with T1 phage). The number of resistant colonies arising on each plate is then determined.

Interpretation of the data is reasonably clear. If selective treatment induces a change in the cells (acquired immunity), there should be no difference between applying the selective treatment to ten samples of the same culture or one sample from each of the ten cultures. In both cases, the number of treated cells is approximately the same, and therefore, approximately the same number of resistant colonies is expected. On the other hand, if the resistant colonies reflect the number of preexisting mutations in the cultures, the ten samples from the same culture should be similar, whereas samples from different cultures should be less similar because some cultures have mutations that occurred early and hence gave rise to a larger number of resistant progeny compared to others. The way to quantify the difference is to measure the variance of the population, which is the square of the standard deviation.

Table 3.1 shows the type of data obtained by Luria and Delbrück. It is obvious that there is a dramatic difference in the variances of the two sample populations. Also note the existence of certain **jackpot tubes**, those whose samples contained substantially more resistant cells than the average. For decades these experiments were taken as definitive proof of the correctness of the Darwinian hypothesis. Recent experimental results, however, have shown that the results of Luria and Delbrück are more limited in scope than originally thought and have compelled a reevaluation of the idea of directed mutation.

Beginning in 1988, workers in the laboratories of John Cairns, Barry Hall, Ulrike Wintersberger, Patricia Foster, and others made observations that called into question some of the assumptions underlying the analysis of Luria and Delbrück. They specifically examined the occurrence of mutations during starvation in *Escherichia coli* and other microorganisms.

The general outline of the experiments was that exponentially growing cells were washed and shifted to a medium lacking an essential nutrient or containing a sugar that could not be metabolized (i.e., a selective medium). Standard Darwinian analysis as propounded by Luria and Delbrück assumes that any population of cells contains certain preexisting mutants. When these cells encounter selective conditions, only the mutants can grow. However, the number of mutant cells in a population is supposed to be independent of the selection applied.

Table 3.1. Results of a typical fluctuation test.

	Number of Resistant Bacteria[a]	
	Same Tube	Different Tubes
	46	30
	56	10
	52	40
	48	45
	65	183
	44	12
	49	173
	51	23
	56	57
	47	51
Mean	51.4	62
Variance	27	3498

[a]The values in each column indicate the number of T1-resistant bacteria observed per 0.05 ml sample. The total volume of the individual cultures was 10 ml. (Data are from experiment 11 of Luria, S.E., Delbrück, M. [1943]. Mutations of bacteria from virus sensitivity to virus resistance. *Genetics* 28: 491–511.)

The original observations of Cairns were that under starvation conditions the frequency of mutations that would by themselves allow the starved cells to grow was greatly increased, while other mutations that would not permit growth occurred at the same rates as during exponential growth. This process has been called **adaptive evolution**.

Adaptive evolution as an experimental result is now generally accepted. The phenomenon seems limited, however, in that it occurs only under very specific conditions. For example, a repetition of Cairns' original experiments requires that the experimenter use the exact same mutation and that the mutation be located on a plasmid and not on the chromosome.

Similar experiments have demonstrated genetic loci that undergo adaptive evolution in both prokaryotes and eukaryotes. Examination of cells from the portion of the colony not included in the papillae has shown that nonadaptive mutations have occurred in these cells. Therefore, one current model is that when stressed, cells enter a transient hypermutable state (Bjedov et al. 2003; Cairns and Foster 2003). If an adaptive mutation occurs, a papilla develops (Fig. 3.2). The effect seems to require the presence of oxygen and the absence of a carbon source.

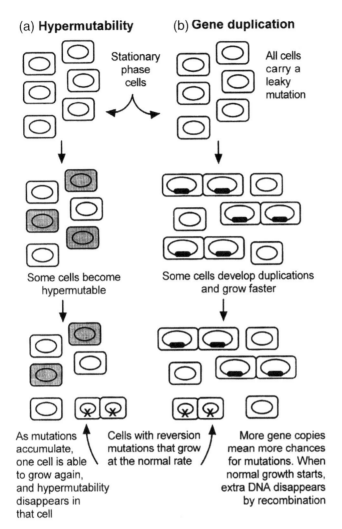

(a) **Hypermutability**   (b) **Gene duplication**

Stationary phase cells

All cells carry a leaky mutation

Some cells become hypermutable

Some cells develop duplications and grow faster

As mutations accumulate, one cell is able to grow again, and hypermutability disappears in that cell

Cells with reversion mutations that grow at the normal rate

More gene copies mean more chances for mutations. When normal growth starts, extra DNA disappears by recombination

Figure 3.2. Contrasting theories for the origin of adaptive mutations. (a) The hypermutability theory that postulates normal amounts of DNA in a cell but much higher than normal mutation at all sites (Hersh et al. 2004). (b) The gene duplication theory of Roth and Andersson (2004). This theory postulates that adaptive mutation requires leaky (partially functional) mutations that allow the cells to grow slowly. Such a condition favors gene duplication because more partially functional gene copies will allow faster growth. Eventually, an adaptive mutation occurs (at the normal frequency) in a duplicated region, thereby allowing rapid cell growth. Under rapid growth conditions, gene duplications are quickly lost by recombination, and only the functional gene copy remains.

However, the daughter cells do not become permanently hypermutable. Indeed, some of them carry antimutator mutations. Roth and his coworkers have developed a countermodel in which gene duplication results in sufficient production of partially functional gene product to give microcolonies. When one of the duplicated genes reverts, that clone of cells takes over rapidly. The duplicated genes are unstable and thus are lost rapidly due to a lack of selective pressure (Fig. 3.2). For a recent summaries of the controversy, see Brisson (2003) and Hersh et al. (2004). The latter group proposes that both pathways described in Fig. 3.2 occur during adaptive evolution. Roth and Andersson (2004) discount this possibility.

Bayliss and Moxon (2002) summarize evidence indicating that the *mutS* gene, a member of the mismatch repair system, has significant DNA sequence heterogeneity in different strains of *E. coli*. The suggestion is that different bouts of hypermutability and its eventual restoration of normal function have led to this rapid sequence divergence of the gene. If the presence of hypermutability rapidly leads to a beneficial mutation, both itself and the *mutS* mutation will be maintained in the population for a significant period.

Measurement of Mutation Rate

Measurement of mutation rates in microorganisms is more complicated than that for larger organisms owing to their rapid rate of division and high population sizes. For bacteria, the unit by which the mutation rate for a particular trait is expressed is mutations per bacterium per cell division, and the observable quantities are total number of bacteria at the beginning and at the end of the experiment as well as the total number of mutant cells.

To calculate the number of cell divisions that have occurred in a culture since an arbitrary zero time, it is possible to make use of the following relation:

Cells (N)	1	2	4	8	16	32	64,	etc.
Cell divisions (n)		1	3	7	15	31	63,	etc.

Note that the total number of binary cell divisions required to obtain a particular number of cells is always one less than the number of cells. Hence, because the number of bacterial cells is generally large, the total number of cell divisions that has occurred since time zero is closely approximated by the increase in total number of cells during the experiment. After many generations, this approximation can use just the final number of cells, as the initial number of cells is negligible.

Over many culture doublings, the average number of cells present is merely the final number of cells divided by two because, as noted, the initial number of cells is negligible. However, for shorter intervals such as one doubling, the situation is more complex because the cells are dividing in an asynchronous manner described by a smooth exponential curve rather than a step function. The rate of change in bacterial number in a culture per unit time is

$$\frac{dN}{dt} = \mu N \tag{3.1}$$

where N is the number of cells in a culture at the beginning of the time interval and μ is a proportionality constant designated as the growth rate constant. Integration of Eq. (3.1) yields

$$N_t = N_0 e^{\mu(t - t_0)} \tag{3.2}$$

where N_t is the number of cells in the culture at time t and N_0 is the number of cells in the culture at time zero. The time required for the number of cells in the culture to double (i.e., $N_t = 2N_0$) can be represented by g, the average doubling time, or generation time. Appropriate substitutions in Eq. (3.2) yield the relation:

$$g = \frac{\ln 2}{\mu} \tag{3.3}$$

The above equations can be converted into a form suitable for the calculation of $\bar{N}$, the average number of cells in a culture during a particular time interval. The basic equation is

$$\bar{N} = \frac{\int_{t_1}^{t_2} N_t \, dt}{\int_{t_1}^{t_2} dt} \tag{3.4}$$

After integration, for one generation (one doubling), Eq. (3.4) becomes

$$\bar{N} = \frac{2N_g - N_g}{\ln 2 \,(1)} = \frac{N_g}{\ln 2} \tag{3.5}$$

The point is that the average number of bacteria in one generation can be obtained by dividing the number of bacteria at the beginning of the time period by the natural logarithm of 2, provided the cells are dividing by binary fission and randomly with respect to time. Using this information, it is possible to demonstrate how to calculate a mutation rate.

Luria and Delbrück developed two methods for calculating mutation rates based on the fluctuation test. The first of these methods uses the Poisson distribution. A fluctuation test can be prepared so that, unlike the experiment shown in Table 3.1, some of the tubes have no resistant mutants. This test can be done using small inocula and sampling a smaller number of cells than usual. Assuming that mutant cells are distributed randomly throughout the culture, this system is ideal for applying the Poisson distribution (see Appendix 1) because the sample size is large, the probability of success (a mutation) is small, and thus the average number of mutant cells per sample is moderate.

In the specific case presented by Luria and Delbrück (their experiment 23), the samples from the fluctuation test contained 2.4×10^8 bacteria per tube, and 29 of 87 tubes had no resistant cells. Solving the Poisson distribution for the zero case gives

$$\frac{29}{87} = 0.33 = \frac{e^{-m} m^0}{0!} \quad \text{and } m = 1.10 \text{ mutants per tube.}$$

This number must be divided by the number of bacteria per cell division, which can be calculated by using the above approximations as the average number of bacteria during the last cell division (which actually occurs on the selective plate) and is equal to $(2.4 \times 10^8)/\ln 2 = 3.4 \times 10^8$. Then, the actual mutation rate is $1.10/(3.4 \times 10^8) = 3.2 \times 10^{-9}$ mutations per bacterium per cell division.

The principal disadvantage of this calculation is that it wastes much available information from the fluctuation test because it does not take into account the frequency distribution of tubes that do contain resistant cells. By making some additional assumptions, it was possible for Luria and Delbrück to develop a graphic method to measure the mutation rate (now computerized by Arthur Koch). To use this method, it is necessary to assume that there is a certain density of cells in a culture such that the probability of at least one mutation occurring somewhere in the culture is high. If this assumption is correct, then whenever this density is surpassed, the sum of the mutational events at each cell division results in the same number of resistant cells in the final population (Fig. 3.3). Therefore, if one designates the time at which the proper density is reached as time zero and ignores all mutations that occur earlier, it is possible to describe mathematically the behavior of the population of mutant cells.

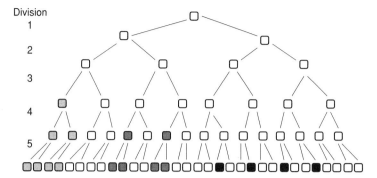

Figure 3.3. Constancy of the number of mutant progeny in a large population. When the number of cells in a culture reaches a certain value, one or more mutations are expected to occur during each doubling of the cell number. Nonmutant cells are represented by *open shapes*, and mutant cells by *filled shapes*. No mutations arose during division 1 or 2, but one mutation occurred during division 3 (*light shading*). In division 4, there were two mutations (*darker shading*), and in division 5 there were four mutations (*darkest shading*). Each round of mutation produced four mutant cells in the final population. Note that the overall proportion of mutant cells in the population is increasing, as the possibility of back-mutation has been neglected.

The final method of calculating mutation rate discussed here is based on the plate spreading experiment. Using Newcombe's experimental protocol, several plates are prepared. It is possible to spray them at different times and to take some control plates and wash off all the bacteria for counting and testing for the presence of mutant cells. These values yield the change in the total cell number and the change in the number of mutant colonies. The mutation rate is then

$$d = \frac{\text{Change in the number of resistant colonies}/\text{change in the total number of cells}}{\ln 2}$$

In this case, the mutation rate is approximately 5.8×10^{-8} mutations to phage resistance per bacterium per cell division.

For most mutable sites on the normal *E. coli* genome, the mutation rate is found to vary between 10^{-6} and 10^{-10}. Mutation rates have been determined for specific genes in a number of organisms (Table 3.2). Drake et al. (1998) have taken experimental data that meet specific experimental and statistical criteria and used them to calculate genomic mutation rates. Note that for the bacteria and viruses for which data are available, the total mutation rate per genome per

Table 3.2. A constant rate of spontaneous mutation in DNA-based microbes.

Genome Organism	Size (bp)	Target	Mutation Rate per Replication per bp (μ_b)	per Genome (μ_g)
Bacteriophage M13	6.41×10^3	lacZα	7.2×10^{-7}	0.0046
Bacteriophage λ	4.85×10^4	cI	7.7×10^{-8}	0.0038
Bacteriophage T2 and T4	1.60×10^5	rII	2.4×10^{-8}	0.0040
E. coli	4.639×10^6	lacI	4.1×10^{-10}	0.0019
		lacI	6.9×10^{-10}	0.0033
		hisGDCBHAFE	5.1×10^{-10}	0.0024
		Average of all values	5.4×10^{-10}	0.0025
Sulfolobus acidocaldarius (pH 3.5, 75°C)	2.243×10^6	pyrE	7.8×10^{-10}	0.0018
S. cerevisiae	1.2068×10^7	URA3	2.8×10^{-10}	0.0038
		CAN1	1.7×10^{-10}	0.0024
		Average of all values	2.2×10^{-10}	0.0027
Neuospora crassa	4.19×10^7	ad-3AB	4.5×10^{-11}	0.0019
			1.0×10^{-10}	0.0042
		Median		0.0033
		Arithmetic mean		0.0033
		Geometric mean		0.0031

Note: The details of the calculations are given in the original paper. Adapted from Drake, J.W. (1991). A constant rate of spontaneous mutation in DNA-based microbes. *Proceedings of the National Academy of Sciences of the USA* 88: 7160–7164; Drake, J.W., Charlesworth, B., Charlesworth, D., Crow, J.F. (1998). Rates of spontaneous mutation. *Genetics* 148: 1667–1686 (whose averages include data not shown in this table); and Grogan et al. (2001). The median and means do not include the *Sulfolobus* data.

replication (μ_g) is relatively constant at 0.003, which means that the relative mutation rate per base must vary inversely with genome size. By way of comparison, Drake et al. report the μ_g value for the RNA bacteriophage Qβ as 6.5.

It is also important to remember that this discussion has dealt only with **forward mutations**. The term "forward" is used to imply a change from some arbitrarily designated original genetic state to a new mutated state. The reverse process (i.e., a change from the mutant to the original state) is called **backward mutation** or **reversion**. In practice, cultures grown for many generations tend to drift toward **genetic equilibrium**. The proportion of mutant cells becomes constant as the number of forward mutational events equals the number of backward mutational events (allowing for possible differences in growth rates). In all of the foregoing discussion about mutation rates, it has been assumed tacitly that the rate of reversion was negligible compared to the (forward) mutation rate. If it were not, all calculated mutation rates would be too low. Furthermore, Bharatant et al. (2004) have shown that mutations at certain loci may occur only during a particular growth phase.

Expression and Selection of Mutant Cell Phenotypes

Expression

In the previous section, mutation is discussed as an all-or-none phenomenon: the cell is either mutant or it is not. However, strictly speaking, there is a period of transition during which the new phenotype is expressed (i.e., appropriate macromolecules are synthesized). The nature of the transition period depends in part on the nature of the final gene product. All cells have a gradual turnover of proteins and RNA molecules as existing molecules are degraded or diluted during cell growth and new molecules are produced according to current needs of the cell. The rates of turnover of macromolecules vary widely, with mRNA being relatively unstable and proteins and ribosomal RNA molecules being relatively stable (except during starvation conditions when considerable protein degradation occurs). Turnover rates may affect the timing, but not the nature, of events described later. The types of events that occur following changes in regulatory regions of the DNA (i.e., DNA that does not have a macromolecular product) are discussed in Chapters. 4 and 14.

One reason for the existence of a demonstrable transition period following a mutational event is the multiple genome copies found in an actively growing cell. As a general rule, a mutation occurs in only one copy of a particular genetic locus, leaving the cell with several DNA copies coding for unmutated product and only one coding for mutated product. When the normal process of transcription (and of translation, in the case of a protein) has occurred, the cytoplasm contains two kinds of macromolecules, mutant and nonmutant. Such a cell is said to be a transient merodiploid, and the question of dominance of the mutation arises.

Dominance of a bacterial mutation is due to exactly the same biochemical processes as those occurring in eukaryotes. If the mutation confers the ability to carry out some biochemical process, it normally has a dominant effect and the phenotype of the cell changes as soon as enough new product is present to permit the reaction to take place at a significant rate. However, Fig. 3.4 indicates that even though the phenotype of the cell has changed, the number of cells with the mutant phenotype cannot increase for several generations because of the large number of genome equivalents and the fact that most of the DNA currently being replicated actually does not segregate until several cell divisions later (compare with Fig. 1.3). This type of delay is called segregation lag, and its duration depends on the number of genome equivalents in the cell. An example of segregation lag is shown in Fig. 3.5. The first recombinant cells appeared at about 10 min, but they did not begin to divide at the same rate as the rest of the cells until roughly 120 min had elapsed. Therefore, in this case the lag amounted to approximately 110 min, or 2.5 doublings, suggesting the presence of slightly more than four genome equivalents per original cell.

Recessive mutations experience a different sort of lag called phenotypic lag. In this case the cytoplasm still contains nonmutant dominant-type product until after the segregation process shown in Fig. 3.4 is complete and the genome is homogeneous. At that time, all new product being produced is mutant, but there is still some nonmutant product remaining in the cytoplasm. As this nonmutant product decays away or is diluted, the phenotype of the cell changes to mutant. Note that whereas segregation lag consists of only one process, phenotypic lag consists of two processes: segregation and replacement of macromolecules.

Lag phenomena can have important consequences in terms of applying selective treatment to a bacterial culture. After a genetic transfer or a mutational event, sufficient time must be allowed for expression of the new phenotype. If selection is applied too early, potentially recombinant or mutant cells may have their metabolism shut off before the DNA rearranging process has reached the point at which an altered gene product can be produced (see Chapter 5). In such a case, the cell is not scored as a recombinant or mutant because it fails to produce a colony under selective conditions.

Evelyn Witkin demonstrated clearly the necessity for metabolic activity in the expression of a mutation. Working with a *trp* strain of *Salmonella typhimurium*, she set up parallel cultures. She treated one culture with only ultraviolet radiation

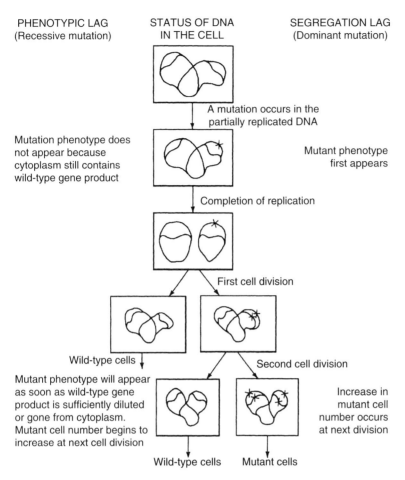

PHENOTYPIC LAG
(Recessive mutation)

STATUS OF DNA
IN THE CELL

SEGREGATION LAG
(Dominant mutation)

A mutation occurs in the
partially replicated DNA

Mutation phenotype does
not appear because
cytoplasm still contains
wild-type gene product

Mutant phenotype
first appears

Completion of replication

First cell division

Wild-type cells

Second cell division

Mutant phenotype will appear
as soon as wild-type gene
product is sufficiently diluted
or gone from cytoplasm.
Mutant cell number begins to
increase at next cell division

Increase in
mutant cell
number occurs
at next division

Wild-type cells   Mutant cells

Figure 3.4. Phenotypic and segregation lag. The replicating DNA molecule is shown within the cell in much the same fashion as in Fig. 1.3. A mutation is assumed to occur within one duplicated region at the point marked X. If it is a dominant mutation, the phenotypic effect of the mutation is observed immediately, but only one of the two daughter cells produced after the first division is mutant (segregation lag). If the mutation is recessive, the appearance of the phenotype is delayed until the second division when the entire DNA in the daughter cell is homogeneous (phenotypic lag). It is important to remember that the duration of either type of lag is a function of the number of genome equivalents of DNA in the cell.

3 Mutations and Mutagenesis

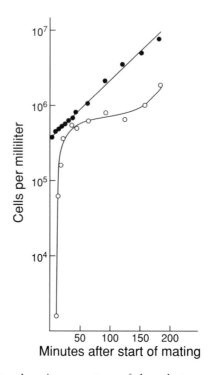

Figure 3.5. Segregation lag. A comparison of the relative growth of an *E. coli* culture (*filled circles*) and of Lac$^+$ cells newly arisen within the culture as a result of conjugation (*open circles*). The initial increase in the number of Lac$^+$ cells seen during the first 30 min of the experiment was due to the transfer of new DNA into the cells. After 30 min no further transfer was possible because nalidixic acid was added to the culture (see Chapter 11). The increase in the number of Lac$^+$ cells after 125 min marks the end of the segregation lag. Note that the growth rates of the Lac$^+$ and Lac$^-$ cells are essentially identical.

(UV) (see "Mutagens," later in this chapter) and the other with both UV and generalized transducing phage grown on a UV-irradiated wild-type donor strain. The first culture would be expected to yield *trp*$^+$ mutants induced by UV, and the second to yield both UV-induced *trp*$^+$ mutants and phage-induced *trp*$^+$ recombinants. After appropriate treatments, cells were plated on several agar plates containing sufficient tryptophan to allow one to six generations of residual growth, the plates were incubated, and the number of Trp$^+$ colonies was determined.

The maximum number of recombinant Trp$^+$ colonies was produced if sufficient tryptophan was present to allow one generation of residual growth, but the maximum number of mutants did not appear unless sufficient tryptophan to allow six generations of residual growth was present, even though the growth rates of mutant and recombinant Trp$^+$ cells were shown to be identical. Further experiments involving the shifting of cells from one type of medium to another indicated that expression of both mutant and recombinant phenotypes was complete after one generation, even though the mutated cells needed enough tryptophan to allow six doublings. Apparently, mutagenic processes require a higher level of metabolic activity, especially protein synthesis, than do recombinational processes.

USING WHAT YOU HAVE LEARNED

Why is it reasonable in terms of biochemistry that mutation rates would be higher when cells are growing rapidly than when cells are growing slowly?

Other workers have obtained similar results that taken together indicate that most but not all mutants or recombinants must be allowed time for expression of their new phenotype prior to selection. However, it is possible to have too much of a good thing, and excessive amounts of nutrients produce microscopic colonies from even nonmutant cells.

Selection

Until now most of the discussion has dealt only with the type of mutants that are dominant and easily selected. There are, however, many useful and important mutations that do not fall into this category, and a method is needed to find them. In desperate cases, it is possible to find a particular mutant by the brute-force method of checking individual colonies until the appropriate phenotype is found. Paula De Lucia and John Cairns used this method when they were searching for the original *polA* mutant in a culture that had been treated with a chemical to enhance the rate of mutation. In that case, it was necessary to test approximately 5000 individual colonies to find one mutant. It is a tedious method of searching, and a more efficient technique is generally used unless there is no alternative.

One alternative method was developed by Luigi Gorini and Harriet Kaufman and is based on the well-known ability of penicillin or related antibiotics to preferentially attack growing cells. The main effect of penicillins is the inhibition of formation of new peptide cross-bridges that tie the peptidoglycan structure of the cell wall into a cohesive unit. Without cross-bridges, the peptidoglycan sacculus cannot be maintained and the cell, having lost its wall, becomes osmotically fragile.

Penicillin selection (indirect selection) is of general applicability, requiring only that it be possible to stop growth of the desired mutant by starvation, extreme temperature, or similar treatment. It is usable with either liquid cultures or agar plates and, despite its name, with any drug that blocks cell wall biosynthesis. For isolation of an auxotrophic mutant, a culture of cells in exponential growth phase is placed in a defined medium under starvation conditions in which prototrophs can grow but auxotrophs cannot. After sufficient time has elapsed to allow the cells to exhaust any intracellular pools of nutrients the experimenter adds penicillin, ampicillin, D-cycloserine, or various combinations of them to the culture and continues incubation until cell lysis occurs. Removal of penicillin from the liquid culture is accomplished either by pelleting the cells in a centrifuge or by filtration, followed by resuspension in fresh complete medium. Penicillin can be removed from agar plates by adding the enzyme penicillinase. The addition of fresh nutrients allows mutant cells to grow again, giving rise to a new culture that is greatly enriched for auxotrophic mutants.

This method has several limitations. Rossi and Berg (1971) showed that different types of auxotrophs are recovered with highly variable efficiency after penicillin selection. They theorized that the result was due to incomplete shutdown of metabolism during starvation. Moreover, inherent limitations of penicillin selection are that nonauxotrophs that are growing slowly for any reason tend to survive the treatment, and if growing cells lyse during penicillin treatment, they may release enough nutrients to allow auxotrophic cells to grow and to be affected by penicillin. Gorini and Kaufman controlled the last problem by raising the osmotic strength of their medium with sucrose until after the penicillin had been removed.

Another potential problem is that of **syntrophy** (cross-feeding), which can occur in either liquid or solid medium but is most easily visualized on agar plates. Auxotrophic cells do not grow on a minimal medium lacking a required nutrient when plated as a pure culture. However, when plated as a mixed culture, it is possible that the second type of cell may release a substance into the medium that permits the auxotroph to grow. The phenomenon may occur even if the second cell type is auxotrophic for the same nutrient, provided that the block in the metabolic pathway of the second cell type is at a step that occurs later than the blockage of the first cell type. For example, in the case of a simple biochemical pathway involving two enzymes:

$$\text{Substrate} \xrightarrow{\text{Enzyme 1}} \text{Product 1} \xrightarrow{\text{Enzyme 2}} \text{Product 2} \qquad (3.2)$$

A cell with a mutation in enzyme 2 tends to accumulate product 1. If product 1 can be released into the medium, it can act as a nutrient for a cell defective only in enzyme 1, permitting it to grow. In some cases, as product 2 is produced, some of the molecules diffuse through the medium to the other cell, also permitting it to grow. This type of analysis can be used to order the steps of a biochemical pathway by observing which cell types are cross-feeders and which are cross-fed. To the extent that cross-feeding occurs during penicillin selection, auxotrophic cells are also affected by penicillin and hence are selected against.

Genetic Code

The universal form of the genetic code is presented in Table 3.3 in both its RNA and DNA forms. In the absence of any evidence to the contrary, geneticists assume that all organisms conform to this code. There are, however, some notable exceptions, particularly among the eukaryotic organisms. In human mitochondria, the RNA codon UAA codes for tryptophan instead of termination, and in both human and yeast mitochondria AUA is often used as an initiator codon instead of AUG.

By contrast, the mitochondria of *Chlamydomonas reinhardii* seem to follow the universal code, so not all mitochondria are identical. Among the prokaryotes, *Mycoplasma capricolum* has been shown to use the codon UGA for tryptophan instead of UGG. Thomas Jukes theorized that the coding change is a consequence of the high (A + T) content (75%) for the DNA of this organism. The substitution of DNA codon ACT for the normal tryptophan DNA codon ACC serves to help raise the (A + T) content slightly.

Francis Crick has codified the general tendency of the base in the third position of the codon to have little coding importance in his **wobble hypothesis**, in which various tautomeric forms of the bases are assumed to "wobble" back and forth to generate hydrogen bonds between codon and anticodon. There is some indication that the wobble base normally counteracts the G + C content of the first two bases in the codon. In a related vein, it is clear that where more than one codon can code for a particular amino acid each organism exhibits a definite preference for certain codons, and codon preference tables of the characteristic of each organism can be prepared. Jayaram (1997) has proposed a model that says if the first two bases in a codon form six hydrogen bonds, the third base is irrelevant. If there are only four bonds, the third base must be specified. With five bonds and the second base a pyrimidine, the third base is also irrelevant.

Simplified versions of the genetic code often present AUG as the only codon used to initiate protein synthesis. This statement is not strictly correct, as AUG is merely the most common and most efficient initiator codon. It is also possible to use GUG (the next most common), UUG, or AUU to start a protein (still with a methionine). Not all finished proteins begin with methionine, however, because there is an enzyme that specifically removes amino-terminal methionine residues.

An obvious problem with the use of methionine as the initiator amino acid is that there is only one possible codon for that amino acid. Therefore, an AUG codon at the beginning of the mRNA must mean something different from an AUG codon located in the interior of the mRNA. The answer apparently lies in so-called **context effects**, alterations in coding properties of a codon by neighboring bases. Base sequence analysis has shown that the beginning of an mRNA is unlikely to self-pair to form a stem-and-loop structure, whereas the internal sequences have a strong tendency to do so. Therefore, the initiator AUG codon is more likely to be exposed to a ribosome in a single-stranded region than is an internal codon.

Kinds of Mutation

The term **mutation** has been used throughout this chapter in the classic genetic sense of an abrupt, inherited change in an observable trait of an organism. It is now time to consider what the term mutation means at the molecular level. However, it is first necessary to redefine mutation. A mutation is henceforth considered to be any alteration in the base sequence of a nucleic acid comprising the genome of an organism, regardless of whether there is any phenotypic effect from the alteration. This definition is deliberately broad to encompass the tremendous variety of mutational types. Except where specifically indicated, the following discussion concerns only mutations in DNA sequences coding for polypeptides.

Base Substitutions

The easiest type of mutation to visualize is that of **base substitution**, in which a single nucleotide base is replaced by another. If a purine is replaced with a purine (e.g., adenine for guanine) or a pyrimidine with a pyrimidine (e.g., thymine for cytosine), the change is termed as a **transition** (for an RNA virus the transition would be uracil for cytosine). If a purine is replaced by a pyrimidine or vice versa, the change is considered a **transversion**.

During base substitution, a transient DNA heteroduplex forms, and thus these mutations can occur only if the natural repair systems in the cell (see Chapter 5) do not correct the error. Some errors are apparently more easily detected than others. For example, x-ray diffraction studies on G•A base pairs indicate that they have adenine in the *syn* configuration and guanine in the *trans* configuration. A•G•A base pair thus turns out to be less symmetric than normal pairing but much more symmetric than, for example, a G•T pair. The added symmetry of a G·A base pair may make it less susceptible to standard repair mechanisms than some other combinations of bases, thereby producing transversion mutations.

Although a single base change is the simplest kind of mutation to visualize, it may also be one of the more difficult to detect. This result follows from the redundancy in the genetic code. For example, see the codons for leucine listed in Table 3.3. A DNA codon that was originally GAA could change to GAG with no change in the amino acid sequence of the coded polypeptide. This type of change is called a silent mutation.

Another possible effect of a base substitution is what is known as a **missense mutation**. Instead of the original amino acid in the polypeptide chain, a different one is substituted. The phenotypic effect of the substitution can range from nonexistent to devastating. Certain types of amino acids (e.g., threonine and alanine) can often be substituted for one another with little effect on the secondary and tertiary structure of the protein. However, substitution of a proline into what is normally a helical region of protein destroys the remainder of the helix and possibly the activity of the polypeptide.

A unique class of substitution mutations is the **nonsense mutations** (terminators), which lack any corresponding tRNA. Table 3.3 shows that there are three such RNA codons: UAG, UAA, and UGA. These codons normally act as punctuation marks within the genetic code, signaling the end of a polypeptide chain. When any one of them appears within the coding sequence for a polypep-

Table 3.3. Genetic code.

DNA	A	G	T	C
A	AAA, AAG — Phe; AAT, AAC — Leu	AGA, AGG, AGT, AGC — Ser	ATA, ATG — Tyr; ATT, ATC — Term	ACA, ACG — Cys; ACT — Term; ACC — Trp
G	GAA, GAG, GAT, GAC — Leu	GGA, GGG, GGT, GGC — Pro	GTA, GTG — His; GTT, GTC — Gln	GCA, GCG, GCT, GCC — Arg
T	TAA, TAG, TAT — Ile; TAC — Met	TGA, TGG, TGT, TGC — Thr	TTA, TTG — Asn; TTT, TTC — Lys	TCA, TCG — Ser; TCT, TCC — Arg
C	CAA, CAG, CAT, CAC — Val	CGA, CGG, CGT, CGC — Ala	CTA, CTG — Asp; CTT, CTC — Glu	CCA, CCG, CCT, CCC — Gly

tide, it results in premature termination of the growing peptide chain and formation of a truncated polypeptide consisting of the amino-terminus and a specific number of amino acids determined by the physical site of the mutation. The opposite class of mutations, those that result in a new AUG start signal, is also known but is much rarer because start signals require a ribosome binding site as well as a single triplet codon.

Nonsense mutations have been given specific names in what began as a pun on the name of the person who identified the first mutant of this type. The codon UAG is an amber codon (a translation of the German name Bernstein). By analogy, UAA is the ochre codon, and UGA has been referred to as opal or umber.

Table 3.3. (*Continued*)

	U		C		A		G	
					RNA			
U	UUU UUC	Phe	UCU UCC	Ser	UAU UAC	Tyr	UGU UGC	Cys
	UUA UUG		UCA UCG		UAA UAG	Term	UGA UGG	Term Trp
C	CUU CUC	Leu	CCU CCC	Pro	CAU CAC	His	CGU CGC	Arg
	CUA CUG		CCA CCG		CAA CAG	Gln	CGA CGG	
A	AUU AUC	Ile	ACU ACC	Thr	AAU AAC	Asn	AGU AGC	Ser
	AUA		ACA ACG		AAA AAG	Lys	AGA AGG	Arg
	AUG	Met						
G	GUU GUC	Val	GCU GCC	Ala	GAU GAC	Asp	GGU GGC	Gly
	GUA GUG		GCA GCG		GAA GAG	Glu	GGA GGG	

Note: Each base is represented by a single letter: A, adenine; C, cytosine; G, guanine; T, thymine; U, uracil. Amino acid abbreviations are listed in Table 1.1. Term means translation termination. The leftmost base of the codon is the 3′-end of the DNA, but the 5′-end of the RNA.

Because peptide chain termination is often associated with release of the ribosome from an mRNA molecule, nonsense mutations frequently exert **polar effects** (i.e., they prevent translation of regions located downstream (3′) on the same mRNA molecule). The degree of polarity is apparently a function of the probability that untranslated RNA will form a terminator loop and thereby prevent further transcription. Consequently, terminator mutations near the amino end of a polypeptide are much more polar than those close to the carboxy-terminus because of the greater length of untranslated mRNA in the former case.

Finally, there are base substitutions affecting DNA sequences that do not code for proteins. Some of this DNA codes for structural RNA molecules (tRNA, rRNA). Changes in structural RNA can result in altered ribosome function or in tRNA molecules that have altered amino acid or anticodon specificity. Changes in nontranscribed DNA may result in altered regulatory functions such as those discussed in Chapters 4 and 14. Some nontranscribed DNA acts merely as a spacer between transcribed regions, and changes here are expected to have little effect.

Insertion and Deletion Mutations

A **deletion mutation** is the removal of one or more base pairs from DNA, whereas an **insertion mutation** is the addition of one or more base pairs. In practice, deletion and insertion mutations usually involve considerably more than one base pair. Insertion or deletion of base pairs in multiples of three results in the addition or elimination of amino acids in the polypeptide chain (Fig. 3.6). All other insertions or deletions result in frameshift mutations. The formation of at least some insertion and deletion mutations appears to be intimately con-

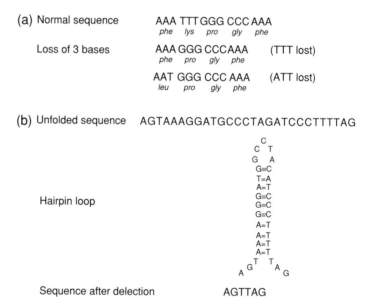

Figure 3.6. Production of deletions. (a) Removal of bases in multiples of three results in the loss of one or more amino acids (and the possible change of one of the amino acids flanking the deletion). (b) If a hairpin loop should form in the DNA duplex, it is possible for DNA polymerase to replicate across the hairpin and create a deletion. Shown here is a single strand of DNA. A similar loop would form in the complementary strand.

nected to transposons, which are discussed later in this chapter as well as in Chapter 15. In addition, certain types of DNA synthetic mutations such as *polA* appear to increase the probability of spontaneous deletion mutations. Many deletions arise from hairpin looping followed by DNA replication across the stem (Fig. 3.6) or by "stuttering" in a region where the same base is repeated many times. Inversions can, on occasion, be large. Genomic studies of *E. coli* K-12 have shown that some strains carry a large inversion amounting to some 10% of the genome.

Frameshift Mutations

When a ribosome translates an mRNA molecule, it must accurately determine the **reading frame**, as all possible triplet codons are meaningful. Proper framing is accomplished when the ribosome binds to the Shine Dalgarno box adjacent to the initial AUG codon. It then moves along the molecule in three base jumps. If an insertion or deletion of base pairs in other than multiples of three has occurred, the reading frame shifts and gibberish is produced instead of the predicted amino acid sequence. For example:

Normal RNA sequence	AUG	AGU	UUU	AAA	GAC	etc.
Normal amino acids	met	ser	phe	lys	asp	etc.
Deleted RNA	AUG	A*UU	UUA	AAG	ACU	etc.
Gibberish sequence	met	ile	phe	lys	thr	etc.

All amino acids downstream from a frameshift are incorrect, although the amino-terminus is normal. Commonly, frameshift mutations also result in downstream chain-terminating codons, leading to production of both gibberish and a truncated protein. It is also possible, of course, to have a frameshift such that normal terminators are bypassed and the abnormal protein molecule is much longer than usual. Frameshift mutations are categorized according to the extent that the reading frame differs from the normal multiple of three: +1, +2, −1, or −2.

Barry Glickman and his collaborators studied spontaneous mutations in the *lacI* gene. Of the 144 mutations they characterized, roughly two-thirds occurred in a frameshift hotspot with the sequence TGGCTGGCTGGC in which the usual mutation was the addition of four base pairs. Other repeated sequences did not give frameshift mutations, so whatever causes spontaneous frameshifts requires more than simple repeats. When they examined the remainder of the mutations, 37% were deletions away from the hotspot, mainly associated with other repeated sequences. Another 34% were base substitutions with no obvious bias toward any particular base pair. The remaining mutations were mainly single base frameshifts, a few tandem duplications, and some 12% insertion mutations due to transposition (see "Mutagens," later in this chapter). Thus, most of the mutations examined were either frameshifts or deletions.

Whether the same would be true for all coding regions is not yet known. However, a similar study of mutations in the phage T4 rII region (see Chapter 6) found that about half of the frameshifts were consistent with a model in which misalignment of DNA with repeated sequences had occurred. In that particular instance, 25% of all frameshifts were found to be the addition or loss of a single base.

Suppressors of Mutations

A **suppressor** is something that eliminates the phenotypic effect of another mutation but not its genotypic effect; that is, a cell that carries one mutation has a mutant phenotype, but a cell that carries the original mutation plus a suppressor mutation has a normal phenotype but a doubly mutant genotype. There are a variety of ways in which suppression can occur, but only a few common types are discussed here.

Suppression of a mutation is often based on an alteration of the codon:anticodon interaction that occurs during translation. It may take the form of an error in the translation mechanism itself. In one case certain types of ribosomal mutation make it susceptible to perturbation by low concentrations of streptomycin. When the antibiotic is present, there is a strong tendency for misreading to occur in the wobble base. Given an appropriate initial mutation in a protein, it may be possible to make a streptomycin-induced error that restores functionality to the protein. Under other conditions ribosomal misreading may not occur at all (stringent translation), or even frameshift mutations may be suppressed.

THINKING AHEAD

What are some possible molecular mechanisms that would allow suppression of a frameshift mutation?

The use of a second-site mutation for suppression is called **intergenic suppression**. The most common type of intergenic suppression involves changes in a tRNA molecule. If the original type of mutation were a nonsense mutation, any change in an anticodon that would allow pairing with the terminator codon would tend to alleviate the polar effects of the terminator mutation and possibly also restore function to the protein as well. For example, $_{gly}$tRNA^{GAA} can become $_{gly}$tRNA^{UAA} and thereby insert a glycine residue wherever the ochre codon occurs. Other nonmutated tRNA molecules would be used by the cell to supply glycine for the normal codons.

There are apparently strong context effects on the ability of a tRNA anticodon to correctly translate a codon. Hanna Engelberg-Kulka has shown that it is possible to have a normal $_{trp}$tRNA^{UUG} translate a UGA codon, provided the codon is followed by another codon beginning with A. Glenn Bjørk and coworkers have shown that the ability to ignore context effects in some measure depends on the correct modification of the base just before the anticodon (i.e., the 3′-base). When it is unmodified, suppressor tRNA molecules function much less efficiently. Bertrand et al. (2002) have extended these observations to show that six bases are actually important for the ability of a ribosome to slip at a repeated sequence to give a −1 reading frameshift.

Missense mutations are also susceptible to suppression by the same sort of mechanisms as those just described. In addition, it is known that in the case of tRNA$_{lys}$ misacylation (attaching an amino acid other than lysine) can occur. If the incorrect amino acid can compensate for the original mutation, protein function can be regained. Misacylation is, however, not common.

The efficiency of suppression naturally cannot be high. If it were, many normal proteins would be damaged by the same mechanisms that result in mistranslation of the mutated protein. The extent of suppression is rarely more than 10% and frequently less than 1%. Emanuel Murgola estimated that in the case of a particular *trpA* mutation, a suppression efficiency of 2% allows a mutated cell to grow. In a study of alterations in tRNA sequence created by genetic engineering, Michael Yarus found that the anticodon loop could be modified to maximize suppressor function, but changes in the stem region usually minimized suppression. Moreover, changes in an anticodon that cause it to match another codon in the same row of Table 3.3 give a better efficiency of suppression than do changes that result in a match with another codon in the same column.

Another type of intergenic suppression is the development of an entirely new biochemical pathway to replace the one that has been blocked by mutation. Examples in *E. coli* are the *sbc* mutations that reverse the effect of *recBCD* mutations (see Chapter 5) and the *ebg* mutation isolated by Barry Hall and coworkers that codes for a new enzyme to replace the defective β-galactosidase in *lacZ* strains.

Examples of **intragenic** suppression are also known where the second mutation occurs in the same gene as the first. One of the simplest is the occurrence of a second frameshift mutation opposite in sign to that of the first. The combination of frameshift mutations results in a polypeptide with normal amino acid sequences at the amino- and carboxy-termini, but with a region of gibberish somewhere in the interior of the molecule. Experiments involving this type of suppression first verified the triplet nature of the genetic code.

Intragenic suppression may also occur for missense mutations. In this case, a compensating amino acid change elsewhere in the same polypeptide restores normal secondary and tertiary structures of the molecule and hence its enzymatic activity. Charles Yanofsky and coworkers provided a good example with their work on mutations affecting the enzyme tryptophan synthetase.

Suppressible mutations are important for working with bacteriophage because they are **conditional** (i.e., the mutant phenotype is expressed only under certain conditions). Mutant phage can be grown normally on a suppressor-carrying host strain and then be transferred to a suppressor-free strain for genetic crosses. The phenotype accordingly shifts from normal to mutant, and the experimenter is able to prepare large quantities of phages for analysis even though the mutations they carry should be lethal. In phage work, suppressible mutations are frequently referred to as "sus" mutations.

In the case of bacteriophages, L. Paolozzi and P. Ghelardini have shown that it is possible to devise a selection that targets a defect in a specific function. It can be done by using a cell with a plasmid as a host that carries a duplication of the gene to be mutated. Any phages that carry a suppressible mutation located in the region of cloned DNA are able to grow using the normal gene product provided by the cloned DNA.

When talking of conditional mutations, it is also important to remember that temperature, either high or low, can modify some mutations. In effect, environmental factors are the suppressors, and the phenotypes can be changed in the middle of an experiment. If a phenotype is mutant at high temperature, the polypeptide is thermolabile or temperature-sensitive (ts). If a phenotype is mutant at low temperature, the polypeptide is cold-sensitive (cs).

THINKING AHEAD

What kinds of mutations in DNA polymerase III would affect the spontaneous mutation rate for a cell?

Mutagens

A **mutagen** is anything that increases the mutation rate of an organism. Mutagens are used frequently to increase the probability of finding a mutation by some selective process. In this section, a variety of mutagens is discussed and some indication of their modes of action given (summarized in Table 3.4). The

Table 3.4. Some common mutagens and their properties.

Mutagen	Structure	Mode of Action
X-rays	5 nm wavelength	Single- and double-strand breaks
UV	254 nm wavelength	Pyrimidine dimers
Nitrous acid	HNO_2	Deamination of A, C, and G residues
Hydroxylamine	NH_2OH	Hydroxylation of cytosine
N-Methyl-N'-nitro-N-nitrosoguanidine	$O{=}N{-}N{-}\overset{\overset{H}{\mid}\;\overset{N}{\mid}}{C}{-}N{-}NO_2$, with CH_3 and H	Production of 6-methyl guanine at replication fork; gives mainly transitions
Ethyl methane sulfonate	$CH_3SO_3CH_2CH_3$	Alkylation of purines; gives mainly transitions
Methyl methane sulfonate	$CH_3SO_3CH_3$	Alkylation of purines
2-Aminopurine		May replace adenine; may hydrogen-bond to cytosine
5-Bromouracil		May replace thymine; may hydrogen-bond to guanine
Acridine orange		Production of frameshifts
ICR 191 (a nitrogen mustard)	$NH(CH_2)_3NH(CH_2)_2Cl$, OCH_3, Cl	Production of frameshifts

list of mutagens is intended to be illustrative but by no means comprehensive. As might be expected, all mutagens inflict various types of damage on cellular DNA, damage that either cannot be repaired properly or is so extensive that it would overwhelm repair mechanisms. A more detailed discussion of repair processes can be found in Chapter 5, along with some discussion about mechanisms of mutagenesis.

Radiation

Two types of radiation are commonly used: UV and x-rays. They differ greatly in terms of the energy involved and therefore, in their effects. X-rays are extremely energetic, and when they interact with DNA the result is usually a break in the phosphodiester backbone of the DNA. UV, on the other hand, catalyzes a reaction in which adjacent pyrimidine bases (on the same strand) form dimers. The presence of a dimer prevents the various polymerases from using that region as a template. Mutations may occur during the repair process. Chapter 5 provides further information.

Chemical Modifiers

One of the earliest mutagens used on bacteria was nitrous acid, whose primary effect is to oxidize amino groups to keto groups. The significant conversions are adenine to hypoxanthine, cytosine to uracil, and methyl cytosine to thymine. Each of the converted bases hydrogen-bonds in such a way as to cause transitions, and the G•C to A•T type has been reported most frequently. However, even transversions are observed occasionally. Nitrous acid can also make nitroso compounds that are strong alkylating agents. Hydroxylamine is a moderately specific mutagen that reacts primarily with the amino group on cytosine but may also attack uracil or adenine. A typical effect of the chemical is to replace a cytosine with a thymine residue (i.e., a G•C to A•T transition).

Numerous **alkylating agents** have mutagenic activity. Their most common effect is to make 7-methyl guanine, which is not mutagenic by itself but destabilizes a base so that it may separate from the deoxyribose moiety, leaving a "hole" in the DNA strand. They can also attach ethyl or methyl groups at the 2- or 6-position on a guanine ring or the 3-position on an adenine ring, which results in mispairing of the base. Some bifunctional alkylating agents can actually cross-link two strands of a DNA helix. Examples of alkylating agents are ethyl methane sulfonate (EMS), methyl methane sulfonate (MMS), and N-methyl-N'-nitro-N-nitrosoguanidine (MNNG). The last is an extremely potent mutagen that produces 6-methylguanine and has its greatest effect at a replication fork. In a culture treated with this mutagen, as many as 15% of the cells may be mutated for a specific trait such as maltose utilization. In fact, the greatest problem with MNNG mutagenesis is its tendency to produce multiple mutations. In a study parallel to that mentioned earlier, Barry Glickman and coworkers examined 167 MNNG-induced mutations of *lacI*. All but three were G•C→A•T transitions.

Base Analogs

A **base analog** is a chemical that has a ring structure similar to one of the normal bases found in a nucleic acid but does not have the same chemical properties. Some base analogs such as 5-bromouracil (5-BU) or 2-aminopurine (2-AP) are also structural analogs and are incorporated directly into DNA in place of normal bases (thymine and adenine, respectively). They tend to be more variable in their hydrogen-bonding properties (Fig. 3.7) and thus may induce errors during replication, either by inserting themselves in the wrong position or by causing an incorrect pairing when acting as a template. In addition, base analogs may increase sensitivity of the molecule to other mutagenic treatments (e.g., 5-BU makes DNA more sensitive to UV).

Other types of base analogs act as **intercalating agents**. To intercalate means to slip between two things, and these chemicals have a flattened ring structure similar to a base but no deoxyribose phosphate with which to be linked into DNA. An intercalating agent can transiently insert between two bases, unwinding the helix somewhat and thereby interfering with proper base stocking during replication. Depending on whether the agent inserts in the template strand or in a new strand, it may cause a gap in the newly synthesized strand or may result in the newly replicated strand having an extra base at a position corresponding to the point of intercalation. If the gap left by the departing mutagen is repaired, a base substitution may result. Examples of intercalating agents are acridine orange, proflavin, and nitrogen mustards. Regions of DNA containing runs of a single base, especially guanine, are particularly susceptible to frameshifting.

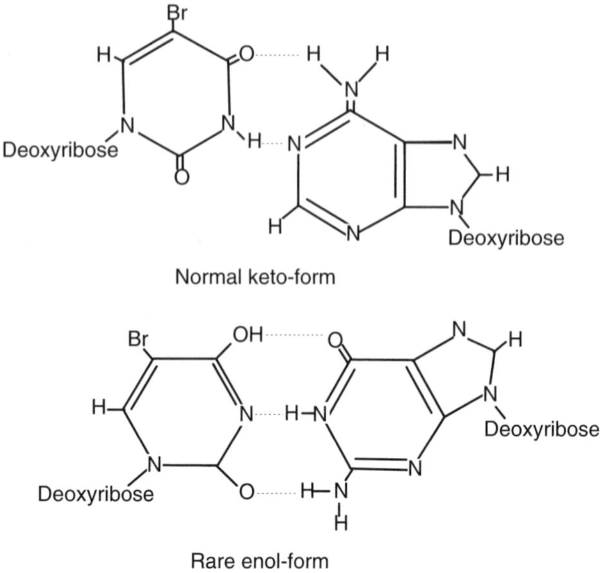

Figure 3.7. Possible hydrogen bonding relations of 5-bromouracil. In each case, the base on the left is 5-bromouracil. In its keto-form it pairs with adenine (*top*), but in its enol-form it can pair with guanine (*bottom*). The *dotted lines* indicate hydrogen bonds, and the *solid lines* indicate covalent bonds.

Cross-Linking Agents

Certain chemicals produce interstrand cross-links in DNA that obviously prevent DNA replication until they are repaired. Examples of cross-linkers are mitomycin C and dimethyl psoralen. The latter compound has been widely used because it must be activated by exposure to 360 nm light. It gives the experimenter good control over the timing of cross-linking events.

Transposons

Transposons are units of DNA that move from one DNA molecule to another, inserting themselves nearly at random. They are also capable of catalyzing DNA rearrangements such as deletions or inversions. An excellent example is bacteriophage Mu (see Chapter 8), which acts as a mutagen owing to its propensity for inserting itself randomly into the middle of a structural region of DNA during lysogenization, causing loss of the genetic function encoded by that stretch of DNA. The mutations produced are stable, as normal Mu unlike many other temperate phages, is not inducible and therefore, rarely leaves the DNA again. This mode of insertion is in marked contrast to that of a phage such as lambda, which has a specific site of integration for its DNA.

Mutator Mutations

Certain types of mutations that affect the DNA replication machinery have mutagenic effects. These mutations affect the fidelity of the replication process but do not appear to significantly impair the polymerization reactions. Mutations have been isolated in *E. coli* that tend to produce transitions, transversions, deletions, or frameshifts. For example, mutations in the *mutS, H,* and *U* functions affect mismatch repair and lead primarily to transition mutations. Mutations in *dnaQ* affect the epsilon subunit of DNA polymerase III, which is the $3' \rightarrow 5'$ exonuclease activity. Some of these mutations have been designated *mutD*. One study of *dnaQ*-induced mutations found that 95% were of the type G•C→T•A or A•T→T•A. An alternative class of mutations affecting the same components of the replication system has also been identified. These are antimutators that alter the replication machinery so as to reduce the error rate below the normal 10^{-10} per base incorporated.

THINKING AHEAD

Why would both mutator and antimutator mutations put their host organisms at a selective disadvantage? (Hint: see Chapter. 17)

Site-Directed Mutagenesis

The availability of cloned DNA has made possible **site-directed mutagenesis** (in vitro mutagenesis of a specific base or sequence of bases). The general approach is to separate a single strand of DNA that includes the region of interest. An artificial primer DNA sequence is then prepared whose base sequence differs from the original at one or more points (Fig. 3.8). These differences represent the specific mutations to be introduced. They can be base substitutions,

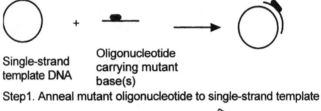

Single-strand template DNA

Oligonucleotide carrying mutant base(s)

Step1. Anneal mutant oligonucleotide to single-strand template

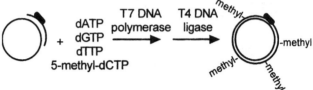

dATP
dGTP
dTTP
5-methyl-dCTP

T7 DNA polymerase

T4 DNA ligase

Step 2. Extend the primer with phage T7 DNA polymerase in the presence of 5-methyl dCTP and ligate to produce double-strand heteroduplex

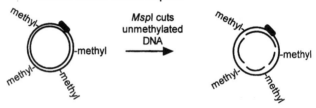

MspI cuts unmethylated DNA

Step 3. Nick the unmethylated (parental) strand with restriction enzyme *MspI*

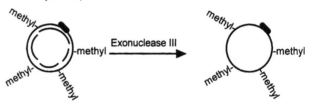

Exonuclease III

Step 4. Remove the parental strand with an exonuclease leaving a circular DNA suitable for electroporation to clone the mutant gene

Figure 3.8. One method for site-directed mutagenesis developed by a commercial laboratory. Step 1 requires a DNA replication primer with one or more mismatches (*thick bar*). During DNA synthesis (step 2) methylated cytosine is used instead of normal cytosine, so the new DNA strand is fully methylated, while the old strand is not. Step 3 requires a restriction endonuclease that cuts only at unmethylated sites, meaning that the original template strand is cut many times. Addition of an exonuclease removes all the original DNA, leaving a single strand circle carrying a mutation.

insertions, or deletions. The primer is then extended using the Klenow fragment of DNA polymerase I as in the case of DNA sequencing (lack of repair activity is essential in preserving the mutation in the primer). The resulting double-stranded DNA molecule is transformed into an appropriate host, and the culture is screened to identify mutants of the appropriate type.

A frequent difficulty with site-directed mutagenesis is that the mis-matched bases are repaired by the host bacterium after transformation but before replication, thereby negating the effect of the primer mismatch. One commercially available system uses an interesting method to circumvent this problem. The new strand is synthesized using 5-methyl cytosine while the template strand is unmethylated. Restriction enzyme *MspI* will cut only the unmethylated strand. *HhaI* enzyme degrades incompletely synthesized molecules, and exonuclease III degrades the unmethylated fragments. The remaining DNA is an intact circle that contains the desired mutation and that can be used as a template in additional reactions.

As an example of an application of this technique, Zhou and Lutkenhaus (2004) used site-directed mutagenesis and the yeast two-hybrid system to explore the interactions of two proteins involved in cell division. They were able to show that some mutations in the MinD protein affected only its ability to bind to the MinC protein, while other mutations did not affect binding of MinC but did affect the other properties of MinD.

In the case where the nature of an amino acid substitution is known, the single letter code for amino acids (Fig. 17.6) is used to indicate the change. For example, the phrase K87E indicates that the amino acid at position 87 counted from the amino terminus has changed from lysine (K) to glutamic acid (E).

Summary

Genetic variation in bacteria occurs by mutation in the same manner as it does in eukaryotic organisms. Fluctuation tests, indirect selection, or plate spreading experiments have been used to demonstrate this fact. The rates at which mutations occur in a bacterial culture can be calculated from either the direct plate counts or the fluctuation test. In the latter case, assumptions must be made as to the nature of the distribution of the mutations within the population. Mutations in stationary phase *E. coli* or *Saccharomyces cerevisiae* cells do not seem to follow the usual Darwinian rules, but there is no consensus as to whether the molecular mechanism involves gene amplification or hypermutability.

Mutant cells can be found by manual screening, but usually some enrichment procedure is used. Direct phenotypic selection is possible in some cases, but in others an indirect selection using penicillin to attack growing cells is the only feasible method. In any case, it is always necessary to allow sufficient time for segregation lag or phenotypic lag. These lag periods are due to multiple-genome copies present in growing cells and involve dominance relations of mutant and nonmutant products.

The genetic code is fundamentally a universal one, but exceptions are known within the mitochondria of certain organisms and within at least one bacterial species. Many types of mutation are known, including base substitutions, insertions, deletions, and frameshifts. An important subclass of base substitutions is the terminator class in which the mutation results in production of a shortened polypeptide. Modifications in tRNA molecules can lead to suppression of base substitutions or frameshifts, but large insertions and deletions can be suppressed only by the development of new metabolic pathways or additional insertions and deletions.

A plethora of mutagens has been identified. All produce some kind of damage to the DNA that cannot be repaired properly, thereby giving rise to a mutation. Radiation treatments, base analogs, cross-linking agents, chemical modifiers, mutations within the DNA synthetic apparatus, and transposable elements such as phage Mu have all been shown to have mutagenic effects. The efficiency and specificity of the mutagens vary widely. A modification of the procedure used for DNA sequencing allows the site-directed mutagenesis of cloned DNA.

Questions for Review and Discussion

1. Give the experimental evidence in favor of a simple Darwinian model for bacterial mutation. What is the experimental evidence that argues for a more complex model?
2. What are the possible types of mutations? How can each type be identified experimentally?
3. How can an experimenter select specific phenotypes? What factors influence how soon after a genetic manipulation an experimenter can apply a selection to a culture?
4. Give some examples of mutagens. How does each one of them cause a mutation?

References

General

Bayliss, C.D., Moxon, E.R. (2002). Hypermutation and bacterial adaptation. *ASM News* 68: 549–555.

Brisson, D. (2003). The directed mutation controversy in an evolutionary context. *Critical Reviews in Microbiology* 29: 25–35.

Hersh, M.N., Ponder, R.G., Hastings, P.J., Rosenberg, S.M. (2004). Adaptive mutation and amplification in *Escherichia coli*: Two pathways of genome adaption under stress. *Research in Microbiology* 155: 352–359.

Maki, H. (2002). Origins of spontaneous mutations: Specificity and directionality of base-substitution, frameshift, and sequence-substitution mutageneses. *Annual Review of Genetics* 36: 279–303.

Specialized

Bertrand, C., Prere, M.F., Gesteland, R.F., Atkins, J.F., Fayet, O. (2002). Influence of the stacking potential of the base 3' of tandem shift codons on –1 ribosomal frameshifting used for gene expression. *RNA* 8: 16–28.

Bharatant, S.M., Reddy, M.N., Gowrishankar, J. (2004). Distinct signatures for mutator sensitivity of *lacZ* reversions and for the spectrum of *lacI/lacO* forward mutations on the chromosome of nondividing *Escherichia coli*. *Genetics* 166: 681–692.

Bjedov, I., Tenaillon, O., Gérard, B., Souza, V., Denamur, E., Radman, M., Taddei, F., Matic, I. (2003). Stress-induced mutagenesis in bacteria. *Science* 300: 1404–1409.

Cairns, J., Foster, P.L. (2003). The risk of lethals for hypermutating bacteria in stationary phase. *Genetics* 265: 2317–2318.

Grogan, D.W., Carver, G.T., Drake, J.W. (2001). Genetic fidelity under harsh conditions: Analysis of spontaneous mutation in the thermoacidophilic archaeon *Sulfolobus acidocaldarius*. *Proceedings of the National Academy of Sciences of the USA* 98: 7928–7933.

Jayaram, B. (1997). Beyond the wobble: The rule of conjugates. *Journal of Molecular Evolution* 45: 704–705.

Lombardo, M.J., Aponyi, I., Rosenberg, S.M. (2004). General stress response regulator RpoS in adaptive mutation and amplification in *Escherichia coli*. *Genetics* 166: 669–680.

Rossi, J.J., Berg, C.M. (1971). Differential recovery of auxotrophs after penicillin enrichment in *E. coli*. *Journal of Bacteriology* 106: 297–300.

Roth, J.R., Andersson, D.I. (2004). Amplification – mutagenesis — How growth under selection contributes to the origin of genetic diversity and explains the phenomenon of adaptive mutation. *Research in Microbiology* 155: 342–351.

Slechta, E.S., Bunny, K.L., Kugelberg, E., Kofoid, E., Andersson, D.L., Roth, J.R. (2003). Adaptive mutation: General mutagenesis is not a programmed response to stress but results from rare coamplification of dinB with *lac*. *Proceedings of the National Academy of Sciences of the USA* 100: 12847–12852.

Zhou, H., Lutkenhaus, J. (2004). The switch I and II regions of MinD are required for binding and activating MinC. *Journal of Bacteriology* 186: 1546–1555.

4

Transcription and Translation: Processes and Basic Regulation

This chapter summarizes processes involved in using genetic information and synthesizing RNA and proteins. These are highly energy-intensive syntheses, and their proper regulation is very important to a cell's competitiveness. The final section of this chapter presents basic transcription regulatory mechanisms and the concepts of the operon and regulon. Advanced regulatory topics are found in Chapter 14.

Major topics include:

- Operon and its regulation
- Control of transcription
- Control of translation
- Basic mechanism of protein biosynthesis

RNA Structure

Cellular RNA molecules are predominantly single-stranded, although some examples of double-stranded RNA molecules are present among animal and bacterial viruses. Single-strand nucleic acids are always available to form hydrogen bonds with other nucleic acids, and thus it is easy to produce folded-back double-strand RNA molecules or RNA•DNA hybrid molecules.

Because formation of hydrogen bonds gives a lower energy configuration to a molecular system, single-strand RNA has a strong tendency to form internal hydrogen bonds that hold it into a defined three-dimensional configuration. One of the most common of these structures is the stem-loop (Fig. 4.1). Complex RNA structures can be held together according to the usual hydrogen bonding rules plus the possibility of a single G•U hydrogen bond. For long RNA molecules, the number of slightly different folded configurations that can be drawn is large. Presumably, there is a favored structure that has a maximum number of hydrogen bonds and a minimum energy configuration. Unfortunately, rules that might be used to calculate the minimum energy configuration for a

...AUGAAAAACGGACAUCACUCCAUUGAAACGGAGUGAUGUCCGUUUUAC...

...AUGAA:UACUA...

Figure 4.1. Formation of a stem-loop structure in an RNA molecule. A•T pairs can form two hydrogen bonds. G•C pairs can form three hydrogen bonds, and G•U pairs can form one hydrogen bond. The more hydrogen bonds that form, the lower is the overall energy state of the molecule and the more stable is that particular configuration.

large RNA molecule are not perfect, and scientists may disagree as to the exact structure that might be obtained in a given situation.

Transcription

Transcription is synthesis of single-strand RNA from double-strand DNA using a DNA-dependent RNA polymerase. Eukaryotic cells have three RNA polymerase activities designated I, II, and III that function in specific ways. RNA polymerase I resides in the nucleoli and synthesizes ribosomal RNA. RNA polymerases II and III are situated in the nucleoplasm. RNA polymerase II synthesizes mRNA and some small RNA molecules. RNA polymerase III synthesizes tRNA, 5S RNA, and some other small RNA molecules. All polymerases consist of multiple subunits (as many as 16 in certain organisms), some of whose genes have substantial base sequence similarity among the three enzymes.

In Bacteria there is only one RNA polymerase core or apoenzyme consisting of four protein chains: two α, one β, and one β', all of which have some sequence similarity to the equivalent eukaryotic proteins. A **holoenzyme** or complete enzyme is assembled by adding appropriate protein(s) to regulate core function. The Archaea also have only one RNA polymerase, but it most closely resembles eukaryotic RNA polymerase II (Reeve 2003), particularly in the number of subunit proteins (≥10). Many of the archaeal RNA polymerase subunits have no corresponding subunit in the Bacterial RNA polymerase. However, the three-dimensional structures of the enzymes are quite similar (Severinov 2000).

RNA polymerases bind to specific **promoter** sites on double-strand DNA molecules. A promoter has an intrinsic directionality to it and is located upstream ($5'$) of the first nucleotide of the transcript. Yeast promoters for RNA polymerase II, whose function parallels most of the Archaeal functions to be discussed, have three elements: an upstream promoter element (UPE), an AT-rich region called the TATA box whose consensus sequence is $5'$TATA(A or T)A(A or T)$3'$, and the initiator site for transcription. Binding of yeast RNA polymerases to promoters requires a specific pair of transcription factors for each polymerase. In a similar vein, Archaeal promoters also have TATA boxes and an initiator site and require two distinct transcription factors (Reeve 2003).

A conventional Bacterial promoter, the type used for transcription during exponential phase growth, has two regions of contact between the DNA molecule and the polymerase (Fig. 4.2). In Bacteria one contact region is centered at base -10 (the so-called Pribnow–Schaller box, which is roughly equivalent to

```
        -35                              -10         +1
      GGATTGACACCCTCCAATTGTATGTTATGTTGTGTGGATG
```

Figure 4.2. Promoter site on a DNA molecule. The bases comprising the promoter site are given negative numbers to indicate that they are not a part of the final transcript. Base +1 is the start of the transcript, and thus there is no base numbered zero. The *wavy lines* indicate the points of contact between the promoter and an RNA polymerase holoenzyme.

the yeast TATA box), and its *Escherichia coli* consensus DNA sequence is 5'TATAAT3'. The other is centered at −35, and its *E. coli* consensus DNA sequence is 5'TTGACA3'.

Other types of promoters are known to function in the Bacteria under special circumstances and will be discussed in subsequent chapters. In the Archaean *Methanococcus thermolithotrophicus*, the promoter element at −25 has the sequence TTATATA. Archaeal promoters do not have the intrinsic directionality of the two-site Bacterial promoters. Their directionality is determined by TBB (TATA-box-binding-protein), which bridges between the promoter and the RNA polymerase.

Bacterial promoters not matching their respective consensus sequences can be functional but at a reduced efficiency. Studies measuring the amount of holoenzyme binding to variant promoter sequences have shown that spacing between the Bacterial sites is important (mostly 17 ± 1 base pairs, range 15–20), but the actual bases in the interval from −17 to −26 are not critical. For optimal promoter activity, it is also important that template DNA be negatively supercoiled, which has the effect of slightly loosening the DNA helix. Transcription primarily begins at base +1, as is the case in yeast, but the spacing between the point of contact in the promoter and the first base in the mRNA molecule in prokaryotes (5–9 bp) is much less than in yeast (60–120 bp). There is generally some heterogeneity in the specific start site, with minor forms of mRNA beginning at +2 or −1 (there is no base 0).

Proteins that are base sequence-specific control recognition of promoter sites. The Archaea have their TFB protein, but their RNA polymerase does not have intrinsic DNA binding activity. By contrast, RNA polymerase in Bacteria has such an intrinsic activity that needs to be modified to suit the occasion (Reeve 2003). Proteins called σ (sigma) factors first bind to the core polymerase to give a holoenzyme. They show notable structural similarities among Bacteria but have no corresponding protein in the Archaea. The holoenzyme then searches for a promoter whose −35 sequence is appropriate for the σ factor. During exponential growth, the primary sigma factor has a molecular weight of 70 kDa, and therefore is known as σ^{70}. The amino terminus of σ^{70} is important for DNA binding and complex formation.

Initial binding to DNA results in formation of a **closed complex**, a normal DNA duplex with an RNA polymerase attached. The closed complex exists in two distinct forms (Coulombe and Burton 1999). In closed complex 1, DNA has begun to wrap around the holoenzyme, giving a footprint from base −70 to −10 on the promoter. After a protein conformation change, the DNA passes through a channel in the holoenzyme, and the holoenzyme clamps down on it to assure processivity. The footprint now extends from base −70 to +20 relative to the promoter. Hydrogen bonds between the 12 bp from base −9 to +3 are then broken, and the RNA polymerase settles into the DNA major groove to form an **open complex**. Broken hydrogen bonds in the open complex allow nucleoside triphosphates to pair with the template strand of the DNA helix to start the transcript (Fig. 4.3). Sequence analysis indicates that the first base is a purine in over 80% of sequences examined. As in DNA synthesis, a new strand of RNA is synthesized in a 5′→3′ direction. The act of initiation "triggers" RNA

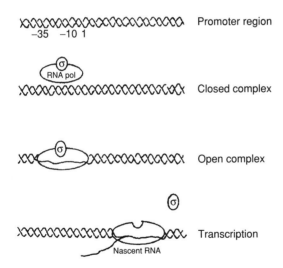

Figure 4.3. RNA transcription. RNA polymerase first binds to the promoter to form a closed complex. Melting of hydrogen bonds yields an open complex, and transcription begins at base +1. The sigma factor is released shortly after the transcription begins.

polymerization and results in release of σ factor after ten nucleotides have been added. As synthesis begins, the RNA polymerase forms a **ternary complex** with the DNA helix that is more resistant to denaturation by high salt or attack by proteolytic enzymes. Chapter 14 presents some ways in which this process is modified in Archaea.

As a polymerase travels along DNA, hydrogen bonds are broken in front and re-formed behind so that the size of the ternary complex remains essentially the same. This movement of the ternary complex causes the DNA helix ahead (3′) of the ternary complex to become overwound (supercoiled). These excess supercoils must be released by a topoisomerase. Studies using enzyme-specific inhibitors have shown that DNA gyrase catalyzes the necessary unwinding. Bonita Brewer examined many *E. coli* transcripts and found that the majority (433/613) are oriented in such a manner that transcription proceeds in the same direction as the replication forks. For 97 highly transcribed regions (like ribosomal RNA), the bias is even more pronounced (92/97).

In bacteria, movement of RNA polymerase is jerky, with numerous pauses at regions of dyad symmetry. Pause sites cause RNA polymerase to halt or to slow down and are a necessary prelude to transcription termination in bacteria. Bacterial systems use two general types of terminators—factor-dependent and factor independent. Proteins bound to the core enzyme can control the type of terminator recognized in any particular case. An example of such a protein is NusA, which is discussed in connection with phage lambda (see Chapter 8). Proper termination is very important because if a transcript happens to extend into the *ori* region of a replicon, it can trigger replication of that DNA molecule. RNA transcripts begin to fold into an internally hydrogen-bonded structure as soon as they attain a reasonable length. The three-dimensional configuration of recently synthesized RNA has an important impact during a pause. If a stem-loop structure of appropriate type has formed, transcription terminates.

There are basically two kinds of stem-loop structures that can cause termination. One is an intrinsic or unconditional terminator, and the other depends, for its action, on the presence of a protein factor called rho (ρ). In order for rho to function, the RNA polymerase complex must pause long enough for several events to occur. Rho must bind to an untranslated (~70 nucleotide) region (*rut* site), relatively rich in cytosine residues and free of secondary structure that is upstream from the terminator site, and then move toward the paused polymerase. Rho is an RNA–DNA helicase composed of a hexameric ring of identical proteins that can unwind RNA–DNA duplexes in a 5′→3′ direction using energy from ATP, an effect that eventually eliminates one set of hydrogen bonds stabilizing the transcription complex, thereby releasing the transcript.

An intrinsic (ρ-independent terminator) in the Bacteria has an RNA sequence that will form a stem-loop structure with 7–10 G•C pairs followed by a run of four to eight uridine residues. The secondary structure both reduces the affinity of the RNA transcript for the template DNA strand and causes the RNA polymerase complex to pause. The weak hydrogen bonds formed by A•U pairs are readily broken to release the transcript. Adding G•C pairs to the stem strengthens termination efficiency, and decreasing the number of uridine residues weakens it. von Hippel (1998) has described a thermodynamic model for the process. Yarnell and Roberts (1999) have shown that annealing of an oligonucleotide that mimics the upstream hairpin structure is sufficient to cause release of the transcript. Among the halophilic archaebacteria, ρ-independent termination can occur at a run of uracil residues in the RNA transcript that does not appear capable of forming a stem-loop structure. Both types of terminators can function in the absence of supercoiling but can be blocked by closely coupled translation of the RNA. The presence of a ribosome is assumed to prevent formation of the stem-loop or binding of ρ.

RNA Processing

In eukaryotic cells most RNA processing must occur before the transcripts are ready to be exported from the nucleus into the cytoplasm. In prokaryotic cells, most of this activity is not needed because there is no physical boundary between the site of RNA synthesis and the cytoplasm. The type of RNA processing that occurs in both systems depends on the type of the transcript.

Transfer RNA (tRNA) (Fig. 4.4) is used to carry amino acids for protein synthesis (see later). In *Saccharomyces cerevisiae* there are about 360 genes coding for 28 types of tRNA molecules. These genes are distributed throughout all the chromosomes and are not adjacent to one another. They do contain **intervening sequences (introns)**, segments of RNA located on the 3′-side of the anticodon that are not found in the mature product. In yeast the 5′-end is first trimmed, then the 3′-end is trimmed and the base sequence CCA is added to it. Finally, the intron (if present) is removed by an RNA splicing reaction, and certain bases are chemically modified (e.g., a uracil may be converted to a pseudouracil).

Bacteria do not usually carry more than one copy of a gene, but in *E. coli* this is specifically not true for tRNAs. Bacteria in general differ from yeast in that

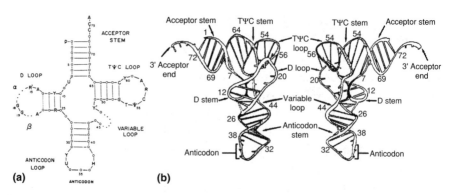

Figure 4.4. (a) Secondary and tertiary structures of all tRNA sequences except initiator tRNAs. The position of invariant and semi-invariant bases is shown. The number system is that of yeast tRNA^{Phe}. Y, pyrimidine; R, purine; H, hypermodified purine; R^+_{15} and Y^+_{48} are usually complementary. Positions 9 and 26 are usually purines, and position 10 is usually G or a modified G. The dotted regions α and β in the D loop and the variable loop contain different numbers of nucleotides in various tRNA sequences. (b) Two side views of yeast tRNA^{Phe}. The ribose–phosphate backbone is depicted as a coiled tube, and the numbers refer to nucleotide residues in the sequence. Shading is different in different parts of the molecule, with residues 8 and 9 in black. Hydrogen-bonding interactions between bases are shown as cross-rungs. Tertiary interactions between bases are shown as *solid black rungs*, which indicate either one, two, or three hydrogen bonds between them. The bases that are not involved in hydrogen bonding to other bases are shown as shortened rods attached to the coiled backbone. (From Rich, A., RajBhandry, U.L. [1976]. Transfer RNA: Molecular structure, sequence, and properties. *Annual Review of Biochemistry* 45: 805–860. Copyright © by Annual Reviews Inc.

there are rarely intervening sequences. Introns are located in tRNA molecules both from Bacteria (e.g., *Agrobacterium*) and from Archaea, in which introns are more common. Deletion of the intron results in failure to properly modify the base located next to the deletion site. Additional discussion about introns is presented in Chapter 6. The tRNA genes in the Bacteria tend to be clustered. An extreme case is that of *Bacillus subtilis* in which 21 tRNA genes are contiguous in one instance and 16 genes in another. In *E. coli*, the tRNA genes are often found as spacers within ribosomal RNA (rRNA) transcripts (see later). There is one report of an *E. coli* transcript that contains tRNA, rRNA, and messenger RNA (mRNA). In *B. subtilis* the two large sets of tRNA genes occur at the ends of rRNA transcripts. Genes for tRNA are not transcribed at a uniform rate, and some tRNAs (especially those associated with rRNA transcripts) are much more common.

In any of these cases in which tRNA is part of a larger molecule, it is necessary to cut out the tRNA precursor. This step is accomplished at the 5′-end by the endonuclease RNase P. This is an unusual ribonucleoprotein enzyme that in the case of *E. coli* consists of two identical protein chains and a 377-base RNA molecule that has endonuclease activity (a **ribozyme**). The protein component enhances the enzymatic activity and provides substrate specificity. The *B. subtilis* RNA molecule will function in *E. coli* to replace a defective RNA, even though there is only 45% sequence similarity. The endonucleolytic cut at the 3′-end of the processed molecule can be made by one of the several endonucleases. Often the released tRNA is still too long and must be trimmed to size from the 3′-end. In *E. coli*, all tRNA genes include the CCA 3′-end and do not require any additional synthesis. They do, however, have the same requirement for base modification as is seen in yeast.

Processing of rRNA is less complicated than that of tRNA. Each rRNA transcript contains the information for three rRNA molecules and may carry one or more tRNAs that act as spacers. In *E. coli*, endonuclease processing releases one to three tRNA molecules and the various rRNAs (5S, 16S, and 23S). In *E. coli*, there are six sets of rRNA genes, three of which appear to be identical in terms of the single tRNA spacer used. *B. subtilis* has ten sets of rRNA genes, but only two contain spacer tRNA genes. *Halobacterium* sp., on the other hand, generally has only one or two copies of the rRNA genes. It seems that there is no correlation between number of copies of rRNA genes and the maximum growth rate. The tRNA molecules included in the rRNA transcripts are, of course, processed as described earlier. After processing, the rRNA is assembled into the appropriate ribosome subunits: a single 16S rRNA molecule associates with 21 proteins to give a 30S ribosomal subunit, and a 5S and a 23S rRNA molecule associate with 33 different proteins to give a 50S ribosomal subunit.

As is often the case, the Archaea have an intermediate position between the Bacteria and the Eukaryota. In this instance the difference is seen in their ribosomal RNA, which is methylated in the same manner as is found in the eukaryotes. The process is directed by a series of small guide RNA molecules (reviewed by Omer et al. 2003). In yeast, all cytoplasmic mRNA molecules are polyadenylated (carry a chain of adenyl residues on their 3′-end). The 5′-end of the mRNA is capped with a 7-methylguanosine triphosphate linked via its three phosphate groups in a 5′→5′ bond with the first base of the mRNA coding sequence. These modifications are involved in preparing mRNA for export to the cytoplasm and in facilitating its translation. They are not found in prokaryotes. The original mRNA transcript in eukaryotic cells is longer than the final translated version and must be processed to remove the intervening sequences, as in the case of the tRNA molecules. Once again the RNA splicing reaction occurs at specific sites characteristic of each type of mRNA. As a general rule, RNA splicing reactions are not found in prokaryotic cells, although some of their viruses use them (e.g., phage T4, discussed in Chapter 6).

Translation of the Genetic Message
Ribosome Function

The lack of a physical boundary between a bacterial nucleoid and its cytoplasm means that translation can proceed simultaneously with transcription. The net effect is that mRNA is rarely found as a separate entity in a bacterial cell. Instead, it is promptly located by ribosomes that attach and follow closely behind the synthesizing RNA polymerase. As each ribosome moves along the growing mRNA molecule, a new ribosome binds to the ribosome-loading site. The complex of mRNA and translating ribosomes is designated as **polysome**. Specific tRNA molecules bring individual amino acids to the polysome.

A typical two-dimensional structure for tRNA is shown in Fig. 4.4a. The anticodon pairs with complementary bases on an mRNA molecule during translation, and the translational efficiency of the tRNA molecule is determined primarily by the sequence of the anticodon arm. The characteristic amino acid is attached to the CCA or 3′-end of the molecule via its carboxyl group. The specificity for this attachment lies in the structure of the tRNA molecule, especially the anticodon and/or the amino acid-accepting stem. The actual determinant varies with the tRNA molecule studied.

The mRNA binding site for a ribosome is not at the 5′-end of an mRNA molecule but is located internal to it. In Bacteria the binding site is a region of some 9 bp known as the Shine–Dalgarno box and is complementary to the 3′-end of the 16S ribosomal RNA. The Shine–Dalgarno box is located within a larger region of nonrandom sequence that extends from base −20 to base +13 and is not the major nonrandom element within that region.

The spacing between the Shine–Dalgarno box and the initiator (first) codon is important. Formation of the 30S initiation complex requires the following: a 30S ribosome; protein initiation factors IF1, IF2, and IF3; mRNA; and a special initiator tRNA molecule, usually carrying the amino acid methionine. A formyl group blocks the amino group of the initiating amino acid so that protein synthesis can proceed in only one direction. Thus, the first tRNA that is bound usually carries the amino acid N-formylmethionine (fMet). Then, the 50S ribosomal subunit is added to form an active 70S ribosome. By then, the IF proteins are released. Although the first amino acid is methionine, the **codon** AUG (triplet of bases coding for the amino acid methionine; see Chapter 3) is not always first. The initiator tRNA is unusual in the sense that wobble can occur at the first base in the codon, and therefore, GUG and UUG codons sometimes (< 5%) begin a coding sequence.

The Archaea display a mixture of properties in their translation process. The protein initiation factors resemble those that are found in the eukaryotes, including the equivalent of eIF2, eIF4, and eIF5, depending on the organism. However, there are two physical processes for recognizing the start codon. One is fundamentally the same as in the Bacteria. There is a Shine–Dalgarno analog located 3–10 nucleotides upstream of the first codon that can pair with the 3′-end of the 16S rRNA. Moreover, the possible start codons are similar. However, like eukaryotes the Archaea use methionine instead of formylmethionine as the first amino acid. The second mechanism for translation involves mRNA molecules where the start codon is right at the end of the molecule. In this case the 30s chromosomal subunit must be preloaded with the appropriate tRNA (Londei 2005)

tRNA^{fMet} is located at the so-called peptidyl or **P site**, on the ribosome (Fig. 4.5). Adjacent to it is a tRNA carrying the amino acid corresponding to the next mRNA codon. This binding site is the **A site**, or aminoacyl site. The fidelity of codon–anticodon binding is mediated not only by the usual base-pairing rules but also by the ribosomes themselves (Hosaka et al. 2004). The complex that enters the A site is composed of a temperature-unstable elongation factor (EF-Tu), GTP, and the aminoacyl-tRNA. Noncognate (incorrect) tRNA molecules quickly dissociate from the ribosome before EF-Tu can hydrolyze the GTP, whereas cognate tRNA molecules remain bound. When the GTP has hydrolyzed, the complex is ready for peptide bond formation. However, even at this stage noncognate tRNAs are unlikely to remain bound to the ribosome for the duration required for this step to occur. The result of these differences in stability is that the intrinsic error rate of tRNA binding to ribosomes is dramatically improved to about one error per 3000 codons translated.

In the **translocation** step, the ribosome together with elongation factor G forms a peptide bond between the two amino acids, perforce releasing the fMet from its tRNA molecule and moving the ribosome one codon down the message. The two tRNA molecules remain bound to the ribosome. The uncharged tRNA is now located at the **E site**, or exit site, originally postulated by Knud Nierhaus and coworkers; the other tRNA, which is still attached to the growing peptide chain, is located in the P site. Wilson and Nierhaus (2003) have summarized the three-site

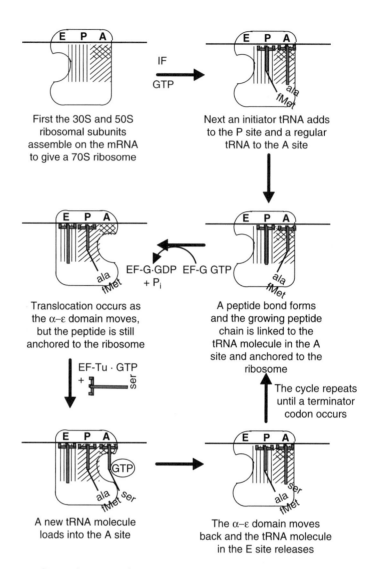

First the 30S and 50S ribosomal subunits assemble on the mRNA to give a 70S ribosome

IF
GTP

Next an initiator tRNA adds to the P site and a regular tRNA to the A site

Translocation occurs as the α–ε domain moves, but the peptide is still anchored to the ribosome

EF-G·GDP   EF-G GTP
+ P$_i$

A peptide bond forms and the growing peptide chain is linked to the tRNA molecule in the A site and anchored to the ribosome

EF-Tu · GTP
+ ser

The cycle repeats until a terminator codon occurs

A new tRNA molecule loads into the A site

The α–ε domain moves back and the tRNA molecule in the E site releases

Figure 4.5. General pattern for protein synthesis using the α–ε model of Wilson and Nierhaus (2003). There are three binding sites on a 70S ribosome: A, aminoacyl; P, peptidyl; E, exit. In the presence of initiation factors (IF) and GTP, an initiator tRNA carrying formyl methionine binds to the P site, and a normal tRNA carrying alanine binds to the A site. The δ or decoding portion of the A site is shown by crosshatching. A peptide bond forms as the ribosome transfers the formyl methionine from its tRNA to the alanine. Elongation factor G (EF-G) and GTP catalyze the sliding motion of a portion of the ribosome (the α–ε domain). Note that the peptide chain remains anchored in its position so that the attached tRNA must bend in a new direction. The tRNA that is ready to be discarded is now in the E site. Elongation factor Tu (EF-Tu) and GTP catalyze the addition of a new tRNA plus its associated amino acid into the A site. GTP hydrolysis provides the energy to slide the α–ε domain back to its original position. The cycle can go on until a terminator codon (no corresponding tRNA) reaches the A site.

model for translation. In this model the A, P, and E sites are fixed, and the tRNA molecules actually slide from one side of the ribosome to the other while the nascent peptide remains anchored to the ribosome (Fig. 4.5).

APPLICATIONS BOX

Certain ribosomal mutations that confer resistance to the antibiotic streptomycin also affect the fidelity of the codon–anticodon binding. What are the potential phenotypic effects of such mutations?

The next round of amino acid addition commences when a newly charged tRNA molecule binds to the A site and in the presence of EF-Tu causes release of the tRNA bound to the E site. Wilson and Nierhaus (2003) proposed that tRNA occupation of the E sites greatly reduces the binding activity of the A site. The effect of this reduction is to prevent noncognate and near cognate tRNA molecules from binding. When a correct tRNA binds to the A site, the tRNA in the E site is evicted.

The explicit assumption in the original model for protein synthesis was that the ribosome moves down the mRNA molecule three bases at a time so as to maintain the correct reading frame. However, subsequent chapters will present examples of reading frameshifts that occur not as a result of mutation (as discussed in Chapter 3), but rather as a result of deliberately programmed translational frameshifting. There seem to be two basic mechanisms that can cause translational frameshifts. One mechanism depends on special tRNA molecules that contain unusual anticodons that pair with two bases instead of the usual three, using the strength of the G•C base pairs to maintain binding. The frameshifting can be triggered by overproduction of certain tRNA molecules or starvation for particular amino acids, which improves the competitive advantage of the frameshifting tRNA molecules.

The second frameshifting mechanism requires so-called slippery DNA. A primary example in bacteria is the gene $dnaX$, which codes for both the τ and γ subunits of the DNA polymerase III holoenzyme complex. In this case, the normal reading frame codes for the τ protein. However, roughly 80% of all ribosomes shift their reading frame at a specific site to produce the shorter γ protein in a −1 frameshift (the new reading frame has a UGA codon two codons downstream from the point of shift). The slippery heptamer in this case is A-AAA-AAG, where the hyphens indicate the codon divisions in the normal reading frame. The −1 frameshift generates two lysine codons where there was only one before. Chapter 7 presents a more extensive discussion about how multiple proteins can arise from a single coding region.

There are three specific codons for termination of protein synthesis (see Chapter 3). When one of these codons arrives at the A site, there is no corresponding tRNA to act as a translator, and the protein chain is released from the ribosome in a process that requires protein release factors. RF1 interacts with codons UAA and UAG, while RF2 interacts with codons UAA and UGA. RF3 releases RF1 or RF2 from the A site and triggers disassembly of the ribosome. Nonsense suppressor mutations alter a tRNA so that it translates the appropriate terminator codon. After termination, the N-formyl group, or in a minority of cases the entire N-formylmethionine, is cleaved from the end of the protein, making it ready to function.

There is a potential problem with translation termination. There are times when an mRNA molecule loses its 3′-terminus along with the corresponding stop codon. In Bacteria, a ribosome cannot terminate without the stop codon but can't stay bound to the mRNA molecule forever either. Moreover, if the ribosome does manage to terminate, the truncated protein that is released might have toxic or inhibitory properties within the cytoplasm. The solution to the problem is a small RNA molecule that occurs in all Bacteria (but is not necessarily essential). This RNA molecule is variously known as 10Sa RNA, SsrA, or tmRNA. The last term is the most descriptive because the RNA has properties of both tRNA and mRNA and is involved in a process known as trans translation.

Trans translation is a process that separates the nascent protein from a stalled ribosome and in the process tags the protein for destruction (Fig. 4.6). The tmRNA molecule carries an alanine whose addition is catalyzed by SmpB protein. Initially it binds to the A site and adds the alanine residue, just like a tRNA molecule. After translocation, however, the tmRNA switches positions, displacing the original mRNA and inserting its small open reading frame (ten codons). The ribosome translates these codons and terminates in the normal fashion. The newly synthesized protein is tagged for destruction (usually by ClpX protease) by its 11 amino acids at the carboxy end.

The fate of a ribosome after termination of translation depends on the nature of the mRNA molecule being translated. Unlike eukaryotic cells, many prokaryotic mRNA molecules are polyinformational (i.e., contain information for making more than one protein). In such a case, the 70S ribosome may remain attached to the mRNA and move in either direction until it arrives at another initiator codon. This action is subject to a distance phenomenon. The greater the separation between the terminator signal for one protein and the initiator codon

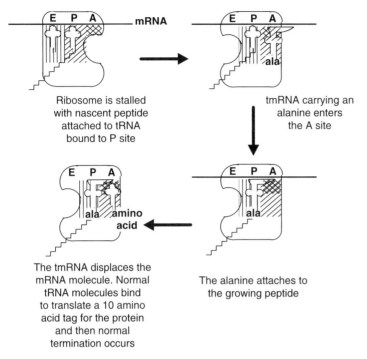

Ribosome is stalled
with nascent peptide
attached to tRNA
bound to P site

tmRNA carrying an
alanine enters
the A site

The tmRNA displaces the
mRNA molecule. Normal
tRNA molecules bind
to translate a 10 amino
acid tag for the protein
and then normal
termination occurs

The alanine attaches to
the growing peptide

Figure 4.6. Effect of tmRNA on stalled translation. A stalled ribosome results when a properly charged tRNA is not provided to the A site. The tmRNA has features resembling both a tRNA and an mRNA molecule. It enters the A site regardless of the codon present and behaves like a tRNA carrying an alanine residue. Insertion of the alanine restarts the translation process. However, instead of exiting the ribosome, the tmRNA ejects the mRNA molecule and takes its place. The open reading frame in the tmRNA codes for ten amino acids and a translation terminator. The released protein is tagged for immediate degradation.

for another, the greater the probability that the ribosome will release from the mRNA. When the ribosome does release, it again dissociates into subunits before reinitiating.

Protein Folding

The simple production of a polypeptide molecule is not necessarily sufficient to provide the cell with an appropriate function. Proteins must assume the correct tertiary, and in some cases quaternary, structure. The correct structures represent minimum energy states, but not always the only minimal energy states. The tunnel that traverses the large ribosomal subunit may serve as the initial area of protein folding. Molecules, particularly mutated molecules, often fold more or less incorrectly. All types of cells produce proteins called molecular chaperones whose function is to facilitate proteins in folding correctly. In *E. coli*, the two most common chaparonins are GroEL and GroES. Similar proteins occur in other Bacteria as well as some Archaea, although they cannot necessarily substitute for each other.

Both proteins are members of a group of proteins produced in response to heat shock (see Chapter 14). The S and L designations stand for small protein (10 kDa) and large protein (57 kDa), respectively. The GroEL/S complex consists of a toroid of 14 GroEL monomers capped with a ring of seven GroES monomers to create a chamber. They have an ATPase activity to provide the energy required to unfold misaligned amino acids. The Archaea can also use a different mechanism designated as a thermosome. This structure is homologous to the eukaryotic TRiC or CCT proteins. Unlike GroEL/S, they function as monomers.

All RNA turnover is important for regulatory processes. Implicit in the discussion that follows is the fact that protein and RNA molecules do not last indefinitely in the cell. Proteases and nucleases are produced that slowly degrade those molecules. It is this slow turnover of macromolecules that makes the gradual alteration of phenotypes possible. For example, mRNA molecules do not last indefinitely. A typical *E. coli* mRNA molecule may remain functional for only 0.5–2.0 min. Degradation of mRNA seems to occur by various means, as it may proceed from either the 3'- or the 5'-end of the molecule. Degradation competes with translation, and untranslated regions of mRNA are more susceptible to nuclease activity than the translated ones. Recent work in several laboratories has identified a **degradosome**, a multienzyme complex whose function is to attack mRNA and rRNA molecules. The complex contains ribonuclease E (endonuclease), polynucleotide phosphorylase (exonuclease), RhlB (an RNA helicase), enolase, and various RNA fragments (Bernstein et al. 2004). Depending on the mRNA, individual components or the entire degradosome may be required for normal decay.

Degradation of mRNA seems to involve two steps. First, the structure of the mRNA is altered so that it can no longer be translated, even though blotting experiments can show that the coding sequences are still present. Evidence from studies with *lac* operon mRNA (see later) indicate that cuts may occur in regions that are not being translated effectively (unprotected by ribosomes). Later, the large fragments of RNA are digested and the bases reused. About 25% of all *E. coli* RNA transcripts contain a highly conserved sequence designated as REP (repetitive extragenic palindromic) at their 3'-end. The REP moiety is capable of forming a stem-and-loop structure that prevents single-strand-specific exonucleases from attaching to the RNA and thereby helps to stabilize the upstream RNA.

Protein degradation normally occurs at a low rate, but is considerably more rapid at slower growth rates (when nutrients are in short supply and recycling is important). An exception to this is the case of abnormal proteins. Proteins that are truncated by a nonsense mutation, misfolded as the result of a missense mutation, or the result of tmRNA termination are often unstable in a cell. Apparently, some proteases are capable of specifically recognizing aberrant structures and attacking them. For example, missense mutations that unfold phage λ Cro protein have been shown to make the protein susceptible to proteolytic attack. The *E. coli lon* gene codes for such a protease. These gene products can impact the success of DNA cloning experiments involving foreign genes.

Regulation

In a sense, regulatory mechanisms can be considered energy-conserving processes because synthesis of every kind of macromolecule requires input of energy from ATP or an equivalent molecule. To the extent that a cell synthesizes unneeded protein, DNA, RNA, or other molecules, it handicaps itself in its competition with other cells for nutrients and space. A cell has several mechanisms or levels of regulation available to it that vary in sensitivity. The most basic, but least sensitive, control is at the level of transcription. A less dramatic effect is seen with regulation at the level of translation, and the most sensitive control is posttranslational—the determination of whether the gene product is allowed to act. The latter types of controls are mediated by a mixture of biochemical and genetic processes, and are considered in subsequent chapters.

It can be difficult to study expression of a particular gene. Often, the protein is not enzymatically active, is unstable, or is inconvenient to assay. Most of the current experimentation in regulatory genetics involves the use of **gene fusions**, where a known indicator gene, coding for a large stable protein, is joined to the amino terminus of another. A gene fusion may result from DNA cloning procedures, or it may involve insertional mutagenesis by phage Mu derivatives to deliver indicator DNA more or less randomly to a fusion site.

Examples of Regulation in Simple Functional Units

Operon: The Basic Regulatory Unit

The original concept of the operon came from François Jacob and Jacques Monod in Paris, for which they won a Nobel Prize. It grew out of some observations Monod had made while still a graduate student. When he grew a culture of *E. coli* in a medium containing both glucose and lactose as potential carbon and energy sources, he observed a biphasic growth pattern. During the first burst of growth, the cells used all of the glucose and none of the lactose. After a variable lag period that began when the supply of glucose in the medium was exhausted, the cells began to grow again, this time using lactose. Monod called this phenomenon **diauxie** and reasoned that the lag period represented a time of change in the regulatory state of the cell. Enzymatic analysis of cell extracts indicated that cells had only low levels of enzymes necessary for lactose utilization while glucose was present, but enzyme levels increased in a coordinated fashion upon exhaustion of glucose. In a similar vein, later experiments showed that enzymes for tryptophan biosynthesis were produced in the absence of tryptophan but not in its presence. Phrased in contemporary terminology, the conclusion to be drawn from the lactose and tryptophan experiments was that genes affecting the same process (e.g., lactose utilization) are coordinately regulated.

Groups of coordinately regulated *E. coli* genes have definite orientations, or polarity, that can be demonstrated by introducing polar mutations. These nonsense mutations have the effect of causing loss of function in genes downstream from the site at which they are located. For example, if a group of genes has the map order *ABCDE*, a typical observation might be that a polar mutation in *A* may reduce or eliminate functions in *B, C, D,* and *E,* whereas a polar mutation in *C* would have no effect on *A* or *B.* The implication of these results is that the mutation is interrupting some process that begins at gene *A* and proceeds in a linear fashion through *B, C,* and *D* toward *E,* thereby establishing a functional gradient.

Taking all these observations into account, Jacob and Monod in 1961 proposed a new genetic unit, the operon. In molecular terms, the operon consists of a group of genes, usually coding for related functions that are transcribed as a unit (beginning at *A* in the example above) to produce a polygenic mRNA molecule. Coordinate regulation of the enzymes encoded within an operon is ensured because translation of the mRNA yields all of the enzymes in a sequential manner, beginning with A and ending with E.

If translation or transcription stops prematurely owing to the presence of an abnormal terminator signal (such as a nonsense mutation), production of enzymes whose genetic information is located distal to (transcribed after) the point of mutation depends on the ability of translation/transcription to restart. Failure to restart leads to the observed polar effects. However, note that $E.\ coli$ is a comparatively simple case in which coordinately regulated genes are contiguous and usually transcribed as a unit. There are bacteria such as $Pseudomonas$ that do not have clusters of metabolically related genes. Moreover, even simple genetic contiguity can be misleading. Many related genes in $Halobacterium$ are also contiguous but transcribed into separate pieces of RNA.

Conceptually, the operon contains certain definite genetic elements. It must have one or more structural genes that are transcribed into RNA (which may be rRNA, mRNA, or tRNA). Transcription must begin and end at a definite site or sites. There must be an opportunity for some sort of regulator molecule(s) to interact with the operon and affect transcription. There are several ways to fulfill these criteria, and the rest of the discussion in this section deals with specific examples.

Lactose Operon

For many years the lac operon was the one studied most intensively. It is a fairly simple operon consisting of three structural genes designated Z, Y, and A. The $lacZ$ gene codes for the enzyme β-galactosidase that catalyzes hydrolysis of the disaccharide lactose to glucose and galactose. The $lacY$ gene codes for a galactoside permease that provides transport functions for a variety of sugars, including lactose, melibiose, and raffinose. The $lacA$ gene codes for thiogalactoside transacetylase, an enzyme of uncertain function that may play a role in detoxifying certain thiogalactosides. All three proteins are normally present in trace amounts in a cell, but when that cell is growing and using lactose, enzyme levels increase up to 1000-fold. The process of stimulating the increase is called induction, and enzymes of the lactose operon are therefore considered inducible. Any compound such as lactose whose presence in the medium results in induction is said to be an inducer. After the supply of inducer in a medium is exhausted owing to the action of β-galactosidase, synthesis of lactose enzymes is once again repressed, and the cell returns to its original state.

The opposite of an inducible-repressible enzyme is a constitutive enzyme, an enzyme that is produced at a constant rate under all conditions. Constitutive production of an enzyme implies a lack of control mechanisms, and certain types of mutations can render production of lactose enzymes constitutive instead of inducible. Among them is a class of mutations that map in a gene called *lacI*, which is located adjacent to the lactose *Z*, *Y*, and *A* genes. Nonsense mutations have been observed to occur within the *lacI* gene, and because standard tRNA suppressors can suppress these mutations, the RNA transcribed from the *lacI* gene must be translated into protein. Although a *lacI* cell is constitutive for *lacZ*, *Y*, and *A* expression, a merodiploid cell that is F' *lacI*$^+$/F$^-$ *lacI* is inducible (i.e., the gene is transdominant). These observations confirm that the *lacI* gene codes for a protein repressor that exerts negative control over the lactose operon (i.e., prevents transcription), even if the operon is not located on the same piece of DNA as the *lacI*$^+$ gene. By extension, a general principle is established that a protein produced by transcription and translation on one DNA molecule can diffuse through the cytoplasm to act on a different DNA molecule.

The repressor must interact with the operon in some fashion to prevent transcription. The site at which this interaction occurs is called the operator and is defined genetically by another class of constitutive mutations. These mutations (called O^c) that map between the *lacI* and *lacZ* genes are *cis*-dominant; that is, the phenotype of a *lacO*^{c} cell cannot be affected by the presence of a functional *lacI* gene in the cell. The interpretation given to this observation is that an operator mutation prevents repressor binding to the operator, and hence transcription continues unabated. Note that the operator is only a binding site and produces no diffusible product.

If the presence of repressor protein on an operator prevents transcription and the absence of a repressor permits it, induction must consist of removing the repressor from an operator, presumably owing to a change in properties of the repressor resulting from interaction with inducer. Various molecules having a β-galactoside linkage, such as that found in lactose, function as inducers in vivo and in vitro. Some of these molecules, unlike lactose, are not degraded by β-galactosidase and, therefore, are termed gratuitous inducers. Among these compounds are thiomethyl-β-D-galactopyranoside (TMG) and isopropyl-β-D-thiogalactopyranoside (IPTG), both of which have frequently been used for studies on induction. Because they are not degraded, their concentration does not change even if the cells grow for many generations. One class of *lacI* mutations, I^S or super-repressor, increases the relative ability of the repressor to bind to an operator rather than an inducer such that induction of the *lac* operon is no longer possible, although basal enzyme levels are maintained. Other *lacI* mutations include I^Q and I^{SQ} (quantity and super quantity) that overproduce the repressor protein. Such mutants are difficult to induce because of the problems involved in raising the cytoplasmic inducer concentration to sufficient levels.

The *lac* repressor was the first example of a class of proteins that Jacques Monod and Jean-Pierre Changeaux called allosteric proteins. Such proteins have at least two stable minimum energy configurations, and each configuration has a characteristic activity (or lack of activity) associated with it. In the case of the *lac* repressor, one allosteric configuration of the protein binds to the operator as a tetrameric complex and the other does not. The shift in configuration of an allosteric protein is triggered by an allosteric effector molecule, a small molecule that binds to a special site on an allosteric protein. For the *lac* repressor, the effector is any of the inducer molecules listed above. The inducer does not actually compete with the operator for binding of the protein, but can shift the allosteric equilibrium regardless of the amount of operator DNA available, and thus the inducer needs to be present only in small amounts. A functional repressor has the classic helix-turn-helix protein-folding motif typical of DNA binding proteins. It is normally present in a tetrameric form.

The low inducer concentration required for an allosteric shift is important because the actual chemical inducer of the lactose operon is not the sugar lactose, but a derivative of it called allolactose. β-Galactosidase produces trace amounts of allolactose from lactose by shifting the glycosidic bond joining the glucose and galactose moieties from carbon 4 to carbon 6 on the galactopyranoside ring. In other words, the enzyme does not cleave the glycosidic bond; it merely alters the position of the bond.

One implication of this observation is that a minimal amount of β-galactosidase must be present at all times, or induction of the lactose operon cannot occur. Another is that a sugar does not have to be broken down by β-galactosidase to be an inducer. In addition to the gratuitous inducers mentioned earlier, melibiose is an inducer that requires only galactoside permease function, as it is broken down via a different enzymatic pathway. Raffinose, on the other hand, is not an inducer, but is transported by galactoside permease. Only cells constitutively expressing the *lac* operon are able to grow on raffinose.

The original operon model assumed that RNA polymerase bound to the operator to begin transcription, and that repression was a simple competition between polymerase and repressor for the same DNA site. However, Agnes Ullman and Jacques Monod isolated *lac* mutations that affected the level to which lactose enzymes could be induced (the maximum amount of enzyme that could be produced) but not the actual inducibility of the operon. These properties are expected if the mutation affects the ability of the polymerase to bind to DNA in order to initiate transcription.

A convention developed that "up" promoter mutations are those that bind polymerase more efficiently (more transcript produced), and "down" promoter mutations are those that bind polymerase less efficiently and thus ultimately produce fewer transcripts. The promoter mutations map between the *lacI* gene and *lacZ* operator sites and are designated *lacZp*. The Z designation indicates that the promoter is "upstream" from the Z gene.

The last major regulatory element to be discovered was also the one that finally explained the first observations. This phenomenon was shown by Boris Magasanik to be part of a larger group of regulatory events called catabolite (or glucose) repression. The basic concept was that some component released during breakdown (catabolism) of glucose caused inhibition of ancillary enzyme systems such as the lactose operon. However, the actual mediator of catabolite repression had not been discovered until Earl Sutherland and coworkers identified a small molecule, 3′,5′cyclic adenosine monophosphate (cAMP), as a regulatory element in animal cells and bacteria. Workers later showed that addition of this compound to growing *E. coli* cultures relieved catabolite repression and allowed induction of a variety of operons including lactose, although the cells grew poorly under these conditions. The concentration of cAMP does not vary significantly in *E. coli* during exponential growth on either glucose or lactose, and *B. subtilis* does not produce it at all. Therefore, cAMP concentration cannot be the mechanism responsible for catabolite repression.

Experiments on glucose transport (Hogema et al. 1998) have provided an explanation for the phenomenon. Glucose enters a cell via the phosphotransferase system (PTS). One of the enzymes in that pathway is enzyme IIA(Glc), a protein that is alternately phosphorylated and dephosphorylated during glucose transport because it supplies the phosphate group to make glucose 6-phosphate. In its dephosphorylated state, enzyme IIA causes inducer exclusion of lactose by binding to the rare lactose permease present in the cell membrane, meaning that the *lac* operon cannot turn on while glucose is being transported.

Once again, mutations served to define the role of the new regulatory element in the *lac* operon. Mutations lying in the promoter region could relieve the requirement for cAMP (i.e., the cells became insensitive to catabolite repression). Another type of mutation that mapped well away from the lactose operon resulted in the inability of cAMP to activate the lactose operon (as well as some others). This new genetic locus (*crp*) was shown to code for a protein variously called the catabolite activation protein (CAP), cAMP receptor protein (CRP), or catabolite gene activation protein (CGA). The requirement for this protein could also be alleviated by mutations mapping in the promoter region. The CRP and cAMP moieties act as positive regulatory control elements because in their absence transcription cannot be increased above basal levels.

The regulatory region of the lactose operon has been completely sequenced, as has been the *lacZ* gene. Comparisons with other known transcription regions have been made (Table 4.1), and a comprehensive model (Fig. 4.7) has been developed. Normal promoters have a consensus sequence at −10 bases from the RNA start and another consensus sequence at −35 bases from the start. These regions must be separated by 17 bp for normal function. A careful examination of the region between the end of the *lacI* gene and the beginning of *lacZ* shows that there are in fact several candidates for promoters in this region. Promoter 1 has a good match to the consensus sequence in the −10 region (see Table 4.1), but a relatively poor match in the −35 region. Promoter 2 has mismatches in both regions and would be expected to operate at low efficiency.

Table 4.1. Some regulator sequences found in *E. coli* operons.

Promoters	−10 Sequence	−35 Sequence
Consensus	TATAAT	TTGACA
lac Operon		
Promoter 1	TAT$\overset{-10}{G}$TTG	GCT$\overset{-35}{T}$TACACT
Promoter 2	TTACA$\overset{-30}{C}$T	TCA$\overset{-55}{C}$TCATT
gal Operon		
Promoter 1	TAT$\overset{-10}{G}$GTT	TGTCA$\overset{-35}{C}$ACTTT
Promoter 2	TAT$\overset{-15}{G}$CTA	ATG$\overset{-40}{T}$CACACTT

CRP binding		
Consensus	AANTGTGANNTNNNTCANATW	
lac	C$\overset{-75}{A}$ATTAATGTGAGTTAGCTCA$\overset{-55}{C}$T	
gal	A$\overset{-50}{A}$TTTATTCC$\overset{-40}{A}$TGTCACACTTTTCG	

Note: The usual base abbreviations are used, plus N for any base and W for adenine or thymine. Only the antisense strands are shown. The numbers above the sequence are the number of bases prior to the start of the mRNA when promoter 1 is used.

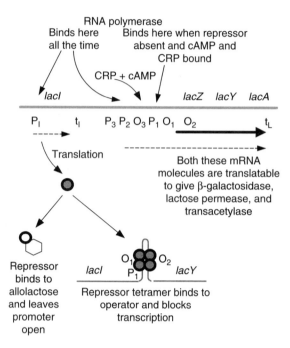

Figure 4.7. Regulation of the lactose operon. The *gray lines* represent duplex DNA in the region of the *lac* operon. Horizontal *arrows* represent mRNA transcripts, with amount of transcript indicated by line thickness. Promoter sites are indicated by P, transcription terminator sites are indicated by t; and repressor binding sites are indicated by O. The *lacI* transcript is synthesized at all times, but the resulting repressor protein is only active in the absence of allolactose (product of lactose metabolism). An active repressor tetramer binds to O_1 and either O_2 or O_3 to form a loop that blocks access to promoter 1. Even if promoter 1 is accessible, RNA polymerase will only bind when cAMP and the CRP proteins are present. Promoters 2 and 3 are active at low levels at all times and provide background synthesis of mRNA.

In vivo, its functionality is detectable if promoter 1 is inactivated. It overlaps promoter 1 in the sense that the −10 site of promoter 2 is located in the −35 site of promoter 1 (see Fig. 4.7). A third promoter sequence operating at even lower efficiency is also detectable by similar experiments.

Protein binding experiments have demonstrated that *lac* repressor binds to three sites within the *lac* operon region. O_1 is the original operator site; and there are two auxiliary operators. O_2 is a site within the *lacZ* gene itself (401 bp downstream), and O_3 is a site 92 bp upstream near the cAMP binding site. The most stable configuration is when a tetramer of repressor protein can bind to O_1 and either O_2 or O_3 simultaneously. This binding results in formation of a U-shaped structure by the DNA helix and makes promoter 1 unavailable to RNA polymerase. At realistically low levels of repressor protein, a mutant repressor that forms only dimers and binds only to O_1 is 60-fold less effective than a tetramer (Oehler et al. 1994), and the auxiliary operators seem to serve to increase the local concentration of the repressor protein. O_3 binds repressor more weakly than O_2, but O_3 is closer to O_1. Therefore, the net effect of either O_2 or O_3 on repressor binding to O_1 is about the same.

The operator sequence is located so that it overlaps the −10 site for promoter 1. When the repressor is bound to the operator, RNA polymerase cannot attach. Even if the repressor is not bound, RNA polymerase still may not attach because of the defect in the −35 site of promoter 1. Just upstream (toward the 5'-end) of the promoter 1 sequence is a binding site for CRP that overlaps the −35 site of promoter 2. Because the binding site for CRP is upstream, this is a class I CRP-dependent promoter (Lawson et al. 2004). CRP attaches to this site in the presence of cAMP. Bound CRP blocks promoter 2 because it induces an 80° bend in the DNA and binds to the α subunit of RNA polymerase, enabling it to initiate effective contacts with the DNA and form a closed transcription complex using promoter 1. Transcription continues until the supply of inducer is exhausted as a result of β-galactosidase activity. Site-directed mutagenesis of promoter 2 has shown that it has little or no role to play in the activation of promoter 1 by CRP.

In the absence of CRP and cAMP, promoter 1 is minimally functional, and transcription may also initiate at promoter 2. This situation provides essential low levels of β-galactosidase to process inducer molecules and of galactoside permease to bring substrate molecules into the cell. Successful transcription from promoter 2 in the presence of *lac* repressor indicates that, once bound, RNA polymerase is capable of displacing repressor bound to the *lac* operator. When lactose is present in the medium, a small amount is transported into the cytoplasm, and some of it is converted to allolactose. The allolactose causes a shift in structure of the *lac* repressor so it can no longer bind to the operator.

Note that although both positive and negative regulatory elements are present, the *lac* operon (defined as the promoter, operator, Z, Y, A genes, and terminator sequence) is considered to be under negative control because a protein repressor is produced. The *lacI* gene is not part of the *lac* operon, as it has its own promoter and terminator sequences. Normally it is constitutively expressed at a low level. As noted earlier, several up-promoter mutations have been isolated that cause a considerable increase in the amount of repressor produced. (Additional sequence details of both the operons can be found in the book entitled *The lac Operon* by Müller-Hill [1996].)

LOOKING AHEAD

Suppose that the repressor protein does not bind operator DNA strongly enough to force it into a looped configuration. What strategy might a cell use to magnify the effect of the bound repressor on DNA configuration?

The *lacZ* gene plays a major role in molecular biology. It is used frequently as an indicator of operon activity (a reporter gene) by cloning it into a site adjacent to the promoter and operator under study because it is very stable and quick to assay with colorimetric reactions. Sometimes phage Mu derivatives are used to create gene or operon fusions. Mu ends are provided to insert the *lacZ* gene coding for β-galactosidase near an existing promoter. β-Galactosidase is a large, stable protein and can have extra amino acids added to its amino-terminus without major loss of enzymatic activity.

Galactose Operon

The monosaccharide galactose is important to *E. coli* not only because it is a source of carbon atoms and metabolic energy but also because it is an important component of the polysaccharide chains attached to the surface of the outer membrane. Therefore, cells must have some of the sugar always present for biosynthetic purposes. Use of galactose as a sole carbon and energy source involves three proteins coded within the *gal* operon. First, a kinase enzyme encoded by *galK* converts galactose to galactose-1-phosphate; next, a transferase enzyme encoded by *galT* attaches the phosphorylated galactose to uridine diphosphoglucose (UDPG) to give uridine diphosphogalactose (UDPgal) plus glucose-1-phosphate; finally, the *galE*-encoded epimerase enzyme converts UDPgal to UDPG (uridine diphosphoglucose) and the cycle repeats. The *galU* gene, which is not located in the *gal* operon, codes for a pyrophosphorylase enzyme that forms UDPG from UTP and glucose-1-phosphate to initiate the cycle. When the cells are growing on glucose, the epimerase acts in reverse to convert UDPG to UDPgal that serves as a metabolite in outer membrane biosynthesis.

Genetic analysis of the *gal* operon reveals that, like the *lac* operon, it is a negative control system possessing most of the same properties as the *lac* operon, but with several different features. The *galR* gene encodes a repressor protein and maps at a considerable distance from the *gal* operon itself (near *lysA*, see Fig. 11.3). One prediction from this observation is that binding within the *gal* repressor–operator interaction is probably tighter than that of the *lac* repressor. Any protein is synthesized near its coding site, as translation follows quickly on transcription in prokaryotes. If the binding site for the protein is nearby, it is not difficult for the protein to find it. However, if the binding site is located at a substantial distance from the coding site, the protein must be able to readily identify the site from among a large mass of heterogeneous sequences. Therefore, more efficient binding to the site is required when it is finally located. The *gal* repressor is a dimeric protein, unlike the tetrameric *lac* repressor.

Footprinting experiments have shown that there are two gal repressor binding sites, one external operator located 5′ to promoter 2 at base −60 (O_E) and the other internal operator located within the *galE* gene at base +55 (O_I). Both the operator sites are fully functional, and appropriate mutations in those sites can prevent repressor binding, thereby raising the background level of GalE product eight- or tenfold. The *gal* operon remains somewhat inducible even after the introduction of a single operator mutation because of the effect of the second operator site. McClure and coworkers have shown that *gal* repressor, CRP/cAMP, and RNA polymerase form a complex that binds to the promoter region. The mechanism of repressor function is DNA loop formation that blocks both promoters. GalR protein and DNA binding protein HU (see Chapter 2) bind cooperatively to the operator sequences to create the transcription-inhibitory loop that probably consists of antiparallel strands, provided that the DNA is negatively supercoiled (Virnik et al. 2003). In the absence of supercoiling and HU, the repressor alone can block promoter 1.

Once again there are two promoter sequences observed, one providing for a background level of synthesis (Fig. 4.8). The two promoters overlap by about 5 bp, with promoter 2 preceding promoter 1. Presence of the *gal* repressor in the RNA polymerase holoenzyme favors open complex formation at promoter 2. Promoter 1 has all the properties of *lac* promoter 1, except that the presence of the repressor has only a 10- or 15-fold effect instead of the 1000-fold effect seen for *lac*. In part, this difference is due to the fourfold more efficient translation of mRNA transcribed from promoter 2, and it demonstrates the influence of context on the translation process. The relatively large amount of UDPgal

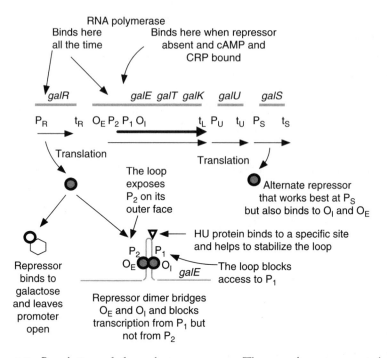

Figure 4.8. Regulation of the galactose operon. The *gray lines* represent duplex DNA in four different transcription regions. Horizontal *arrows* represent mRNA transcripts, with amount of transcript indicated by line thickness. Promoter sites are indicated by P, transcription terminator sites are indicated by t, and repressor binding sites are indicated by O. The *galR* and *galS* transcripts are synthesized at all times, but the resulting repressor proteins are only active in the absence of galactose. Binding of a GalR repressor dimer to each of the operators forms a stem-loop structure that hides promoter 1 but exposes promoter 2 on its outer surface. While protein HU is not required for that binding, it does significantly stabilize the "repressosome." The transcription from promoter 2 maintains baseline levels of galactose metabolic enzymes. The GalS repressor autoregulates itself but can also bind with lower affinity to O_E and O_I. Similarly, GalR protein binds weakly to the other operators in the regulon.

epimerase production is necessary to permit conversion of UDPG to the UDP-gal required for outer membrane biosynthesis.

The base sequence for the binding site for CRP (see Table 4.1) lies upstream of promoter 1 and within promoter 2, making this a Class II CRP-dependent promoter. When the CRP binds to the DNA, it physically blocks promoter 2, but it activates promoter 1. A second CRP molecule then binds upstream of the first, presumably to further stabilize the RNA polymerase complex and aid in formation of the open complex.

Results from Adhya (1996) indicate that an additional level of control is present in the *gal* operon. The model presented above predicts that deletion of *galR* should result in a constitutive phenotype, and indeed *galR* mutants are derepressed. However, addition of galactose results in an ultrainduction phenomenon in which presence of galactose increases the enzyme levels an additional two- or threefold. A completely constitutive phenotype results only when the strains are *galR* and *galS* double mutants. Purified GalS protein functions as an isorepressor, a second repressor that binds to both the *gal* operators, but not so strongly as the *galR* protein. The *galS* protein is a negative regulator for both the *gal* and *mgl* (methyl galactoside) operons that are separated by about 30% of the chromosome. The *galS* and *galR* repressors are complementary in the sense that their maximal effects are mirror images of one another. The GalR protein has its greatest effect on the operators of the *gal* operon and a lesser effect on the *mgl* (major galactose transport system) while the GalS protein is the reverse. The two proteins are homologous as 55% of their amino acids are identical and over 80% are similar. Systems in which two or more discrete operons are under the control of the same genetic element are often referred to as **regulons** (see Chapter 14).

When *E. coli* is growing on glucose, the GalK protein is not needed, although the epimerase is still necessary. This situation leads to a disparate expression of the genes in the operon. Møller et al. (2002) have shown that a small RNA molecule encoded by the *spf* gene (Spot 42) can bind to the Shine–Dalgarno region of *galK* and prevent its translation. This is an example of regulation by **antisense RNA**, RNA that is complementary to part of the mRNA.

The genetic organization of the *gal* operon in *E. coli* is mirrored in other bacteria. For example, Ajdic and Ferretti (1998) have shown that *Streptococcus mutans* has a similar arrangement of genes and functions, except that *galR* is located adjacent to the *gal* operon and transcribed in the opposite direction.

Tryptophan Operon

Biosynthesis of the amino acid tryptophan is a complex process that begins with the compound chorismic acid (a product of enzymes encoded by the *aro* genes) and is catalyzed by three enzymes whose subunits are encoded within five genes (*A–E*). The products of the *trpA* and *trpB* genes form the enzyme tryptophan synthetase, and the products of the *trpE* and *trpD* genes form the enzyme anthranilate synthetase. The *trpC* gene codes for the enzyme indole–glycerophosphate synthetase.

THINKING AHEAD

How can a cell use the amount of tRNA that has its cognate amino acid attached as an indicator of whether transcription of the genes coding for enzymes to synthesize that amino acid is needed?

The *E. coli* genetic map of the *trp* operon seems conventional, although the structural genes are arranged in reverse alphabetical order (Fig. 4.9). As in the case of the *gal* operon, the *trpR* gene, which codes for a protein repressor, is located at some distance from the operon it regulates. Once again mutations have defined discrete promoter and operator sites, but in this case the operator lies within the promoter. Promoter strength is affected by two A•T-rich blocks of DNA located upstream from the promoter (at −50 and −90). Deletion of these regions greatly reduces promoter efficiency, making these blocks similar in function to the enhancer sequences of eukaryotes. Examination of strains carrying polar mutations in gene *E* or the first part of *D* shows that a second low-

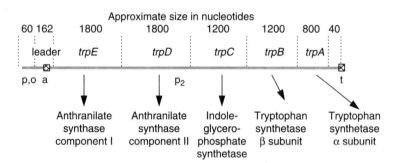

Figure 4.9. Regulation of the tryptophan operon. The operon is drawn approximately to scale with the nucleotide size indicated above each gene. Abbreviations: p, promoter; o, operator; a, attenuator; t, terminator. Note the existence of a secondary promoter located near the right-hand end of the *D* gene (Redrawn from Platt, T. [1978]. Regulation of gene expression in the tryptophan operon of *E. coli*, pp. 263–302. In: Miller, J.H., Reznikoff, W.S. (eds.) *The Operon*. Cold Spring Harbor, NY: Cold Spring Harbor Laboratory.)

efficiency promoter is present that can provide constitutive transcription of the *trpCBA* genes.

Despite its similarity to the *lac* and *gal* operons, the *trp* operon has several novel genetic features. Unlike sugar operons, there is no involvement of cAMP or CRP in regulation of the *trp* operon. Instead, there is regulation of production of biosynthetic enzymes according to the amount of tryptophan available to a cell. This regulation requires a means of detecting the presence of tryptophan in the cytoplasm and stopping mRNA transcription whenever the concentration of tryptophan is sufficiently high.

One detection system is by means of the repressor molecule. Binding studies show that the repressor protein, as coded by the *trpR* gene, is an aporepressor and does not bind to the *trp* operator unless it is complexed with tryptophan itself or a structural analog of it. Thus, formation of the inhibitory complex is dependent on the presence of the end product of the biochemical pathway, tryptophan. The repressor complex competes directly with RNA polymerase for promoter binding and also acts on *aroH* (one of several genes coding for isoenzymes that catalyze the first step in chorismic acid biosynthesis) and on its own gene, *trpR* (autoregulation). In cells supplied with excess tryptophan, radioimmunoassay reveals the presence of 120 repressor dimers, but in the absence of tryptophan there are 375 repressor dimers, most of which are not functional owing to a lack of tryptophan. It should be noted in passing that tryptophan also acts as an allosteric effector to inhibit directly the enzyme anthranilate synthetase, thereby providing two levels of regulation. When tryptophan is present in excess, the first enzymatic step in the biochemical pathway is blocked, and the newly formed repressor complex prevents further synthesis of the *trp* operon mRNA. Repression of the operon reduces the amount of enzyme present by about 70-fold.

It would seem that the *trp* operon has sufficient regulatory mechanisms for its needs, but Charles Yanofsky and his coworkers, who were engaged in sequence analysis of operator and promoter DNA, discovered yet another mechanism. In almost all mRNA molecules there is a region called the **leader sequence** that comes before the code for the first major protein product of the operon, and the *E. coli trp* leader consists of 162 bp lying between the end of the promoter–operator region (site at which RNA polymerase binds) and the start codon for the *trpE* gene (the site where ribosomes bind). Appropriate footprinting and sequencing experiments have defined specific genetic elements within this region (see Fig. 4.9).

Yanofsky's group showed that if purified *trp* operon DNA obtained from an appropriate plasmid was transcribed by an in vitro system, there were two

RNA products. One product was the expected long RNA molecule that carried information from the *trp* genes. The second, unexpected product consisted only of the first 140 bases of leader sequence. In other words, premature termination of the RNA transcript had occurred. A comparison of the relative amounts of the two transcripts indicated that 85–90% of all transcripts initiated in the presence of excess tryptophan terminated at the early site. They called this site an **attenuator** after a similar site identified in the histidine operon by T. Kasai, and the termination process was denoted attenuation. By analogy to operators and promoters, the attenuator is designated *trpEa*. When tryptophan is absent, attenuation is rare, and 75–90% of all transcripts are completed. The combination of repression and attenuation allows for regulation of tryptophan enzymes over a 600-fold range.

Several lines of evidence indicated that attenuation was not just an experimental artifact. Deletion mutations that removed the region of the DNA near the attenuation site increased basal enzyme levels eight- to tenfold without affecting the inducibility of the operon. Small RNA molecules that seemed to correspond to attenuated transcripts could be isolated from normal cells. Furthermore, some mutations in the *rho* gene that affected normal RNA transcription termination also seemed to prevent attenuation. Moreover, cells carrying nonsense-suppressing tRNA molecules produce more tryptophan enzymes than do cells that lack suppressors. Both of the latter observations suggested that attenuation occurred via the standard mRNA termination systems.

Sequence analysis of leader RNA has shown that it can fold in several ways immediately after transcription to yield various stem-loop structures (Fig. 4.10). At the beginning of the first potential loop (bases 27–68) there is an open reading frame coding for a small polypeptide of 14 amino acids. It is not well translated in vitro owing to its secondary structure, which blocks access by the ribosome, but the polypeptide has been detected in vivo. Among the amino acids contained within this polypeptide are two adjacent tryptophan residues near the carboxyl terminus. Tryptophan is a relatively little used amino acid, and two tryptophan residues in succession are rare. Yanofsky and his colleagues suggested that this polypeptide is the key to regulation by attenuation.

Their basic model is presented in Fig. 4.10. After transcription is initiated, RNA polymerase pauses at base 92 of the leader sequence. During this pause the RNA already synthesized may form one of several loops. The loops are created from the sequences numbered 1–4 in the figure. The 1–2 loop (the protector loop) is assumed to be the most stable and to form if possible. It can occur if protein synthesis is totally blocked, and there is no ribosome translating the leader

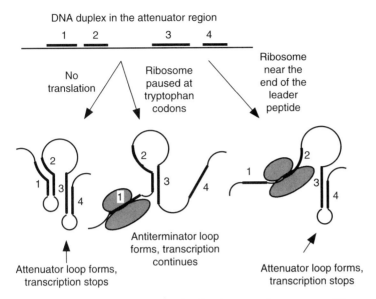

DNA duplex in the attenuator region

Figure 4.10. Model for attenuation in the *E. coli* tryptophan operon. The top linear structure is the initial portion (leader sequence) of the tryptophan mRNA that lies between the promoter site and the translation start site. Potential regions of hydrogen bonding are numbered and shown by thicker lines. The shaded structure represents a ribosome. The leader sequence has multiple ways in which it can fold on itself. The simplest is shown on the *left* where regions 1 + 2 and regions 3 + 4 form stem-loop structures. The 3–4 loop is a terminator loop, and its formation causes attenuation. This condition would prevail when protein synthesis is inhibited, and no ribosomes are bound to the mRNA. When translation does occur, a ribosome binds to region 1 and begins to translate the leader peptide sequence. When the tryptophan concentration is limited, the situation shown in the middle prevails. The ribosome has paused at the pair of tryptophan codons, so region 1 is unavailable for hydrogen bonding. Region 2 then forms a stem-loop with region 3. This is an antiterminator loop because when it is present, the terminator loop cannot form. When there is excess tryptophan, the ribosome passes quickly across the leader peptide sequence, blocking regions 1 and 2 from pairing. Under these conditions the terminator loop forms and attenuation occurs. (Adapted from Oxender, D.L., Zurawski, G., Yanofsky, C. [1979]. Attenuation in the *E. coli* tryptophan operon. Role of RNA secondary structure involving the tryptophan codon region. *Proceedings of the National Academy of Science of the USA* 76: 5524–5528.)

polypeptide. Regions 3 and 4 then form a terminator loop of the usual type, and the paused RNA polymerase terminates transcription. If tryptophan is present in excess and translation is allowed, a ribosome has no difficulty translating two successive tryptophan codons and physically covers both regions 1 and 2. Under these conditions the terminator loop forms as before. However, if tryptophan is in short supply, much of the tRNA^{trp} will not have an associated tryptophan. Consequently, when a ribosome translating the leader polypeptide reaches the two consecutive tryptophan codons, it will have to wait for a charged tRNA to arrive and will lag behind the polymerase. In that event, regions 2 and 3 can pair to form an antiterminator loop, and the paused RNA polymerase continues on to make a full-length mRNA molecule. In accord with this model, mutagenesis of the 1–2 loop to reduce its stability by reducing its hydrogen-bonding capability results in shorter RNA polymerase pauses and thus less termination.

Yanofsky and his collaborators have demonstrated that appropriately sized transcripts are produced in vivo. They suggested that pausing may serve the general function of preventing a ribosome from lagging too far behind the RNA polymerase complex. If a large gap develops, there is a chance that random folding of the RNA might stimulate ρ to terminate the transcript. Normal translation of the mRNA by a series of ribosomes would forestall access of ρ to the RNA. Increasing the length of the leader peptide by 55 codons showed that there was no effect on regulation. Nonsense mutations that prevent synthesis of leader peptide increase attenuation and block induction of the operon during tryptophan starvation.

Polar mutations in the tryptophan operon demonstrate some effects related to the concept of keeping the ribosome-RNA polymerase gap small. Specifically, polar mutations in *trpE* have a tenfold greater effect on *trpD* than on *trpCBA*. The difference results from a separate and efficient ribosome binding site located just before *trpC*, whereas *trpD* has a relatively inefficient one that has a 1-base overlap with the *trpE* termination codon. Apparently the *trpD* ribosome binding site is difficult to locate unless ribosomes release their nascent protein chain right beside it (i.e., translate to the end of the *trpE* gene). Oppenheim and Yanofsky proposed that this phenomenon of one gene needing correct translation of the preceding gene for its own translation be called translational coupling. A similar effect is observed for the *trpA* and *trpB* genes.

The tryptophan operon in *B. subtilis* has a similar pattern of regulation by attenuation, but control of the attenuation process is totally different (reviewed by Babitzke 2004). Once again several loops are possible in the leader sequence

(Fig. 4.11). In the region of base 1–30, a stem-loop structure always forms. Alternative loops are possible with bases 60–111 (antiterminator) or 108–133 (terminator). In this case, the regulatory protein again is an apoprotein requiring tryptophan for functionality. It is designated as a *trp* RNA-binding attenuation protein (TRAP) and is encoded by the *mtrB* gene. The TRAP protein is an 11-mer that, when bound to L-tryptophan, binds to bases 34–91 of the leader, preventing

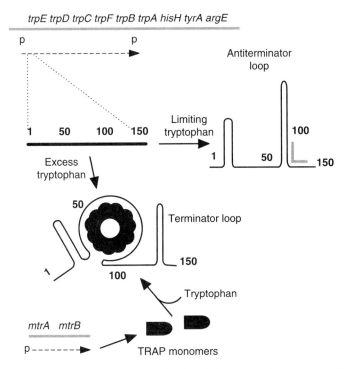

Figure 4.11. Model for attenuation in the *B. subtilis* tryptophan operon. The *shaded line* represents the tryptophan operon. Two promoters (p) are shown. Only the first promoter is subject to attenuation control. The magnified view shows an approximately 150 base leader sequence. In the presence of limiting amounts of tryptophan, the leader mRNA folds as shown on the *right* to give an antiterminator loop. The shaded region is the area that could form a terminator loop. Meanwhile, transcription and translation of the *mtrB* gene produces subunits of TRAP (*trp* RNA-binding attenuation protein). This protein, in the presence of tryptophan, forms a 11-meric β-wheel structure. The β-wheel binds specifically to leader region between bases 40 and 100, thereby preventing formation of the antiterminator loop. The terminator loop then forms and attenuation occurs. (Based on the model of Babitzke [2004].)

formation of the antiterminator loop and allowing the terminator loop to form. Binding of the regulator can be graphically demonstrated by cloning the leader sequence in a high-copy-number plasmid. Cells carrying the clone are de-repressed for tryptophan synthesis because nearly all of the available repressor is bound to the cloned DNA, leaving the normal tryptophan operon fully expressed.

The TRAP 11-mer binds to a leader region that has 11 trinucleotide repeats (either GAG or UAG), each separated from its neighbor by two nonspecific bases. The crystal structure for TRAP indicates that it is a cir-cle of protein (a β-wheel), and physical studies suggest that the leader RNA binds to the outside edge of the TRAP circle. Eventually, the TRAP recycles as the exoribonuclease polynucleotide phosphorylase degrades the bound RNA.

The first gene transcribed in the *trp* mRNA is *trpE*, and its ribosome bind-ing site is also sequestered by TRAP. Therefore, TRAP acts as a regulator of translation, having a 13-fold effect on levels of TrpE synthesis in addition to its 90-fold effect on transcription. The *trpG* gene is located in a different operon, but its biosynthesis responds to variations in tryptophan levels as well. In this case the TRAP protein binds to a shorter region containing repeated trinu-cleotides and again blocks the ribosome binding site. This process allows con-trol of *trpG* expression even though the gene is located physically in an operon otherwise dedicated to folic acid biosynthetic genes.

In addition to its effects on the *trp* operon, the TRAP protein regulates the translation of several other genes. By binding to the Shine–Dalgarno sequence, it can prevent the translation of *pabA*, *trpP*, and *ycbK*.

In *Lactococcus lactis* there is yet another mechanism of attenuation in the *trp* operon (van de Guchte et al. 1998). In this case the leader region of the mRNA also has two possible conformations: termination and antitermination. There are four large loops that form in the leader sequence during transcription (Fig. 4.12). The first loop includes a region that binds to the anticodon of tRNA^{trp}. Down-stream the fourth loop is a terminator loop. The intervening region has the potential to bind to the portion of the tRNA^{trp} molecule that is the amino acid accepting end of the molecule, which is not available for binding if the tRNA is charged (carries a tryptophan molecule). An uncharged tRNA^{trp} binds to both sites on the leader region of the mRNA and stabilizes a new loop whose forma-tion prevents terminator loop formation (no attenuation). Presence of a charged tRNA^{trp} indicates that further tryptophan biosynthesis is not needed, and its fail-ure to bind to the second site on the leader sequence allows the terminator loop

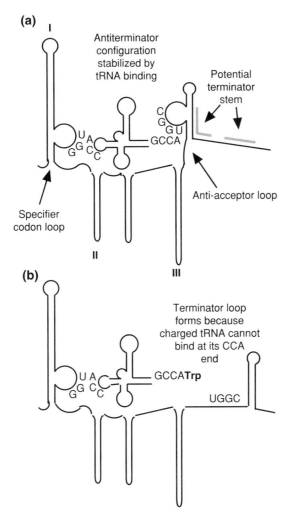

Figure 4.12. Model for attenuation in the *L. lactis* tryptophan operon. Only the leader region is shown. Loops I, II, and III are conserved in a variety of organisms that use this mode of attenuation. (**a**) Antitermination configuration. An uncharged tryptophan tRNA has formed a bridge between the two portions of the leader sequence. The anticodon is bound to a specifier codon loop, and the acceptor stem of the tRNA is bound to an anti-acceptor loop. The anti-acceptor loop prevents formation of the terminator loop (*shaded area*). (**b**) Termination configuration. While the codon of the tRNA molecule is still bound to the leader RNA, the acceptor stem now has an amino acid attached and is not available for binding. Hence, the terminator can form and attenuation occurs. (Based on the model of van de Guchte et al. [1998].)

to form. If the first loop of the leader sequence is deleted, attenuation is not possible because the tRNA^{trp} has nowhere to anchor itself. Genes whose expression is induced by interaction with uncharged tRNA molecules are considered to be members of the T box family.

Elf et al. (2001) have prepared mathematical models of repressor- and attenuator-regulated operons. They suggest that both systems behave in vivo like Boolean systems, either totally on or totally off. However, they also conclude that repressor systems offer better performance.

Summary

Transcription is an energy-expensive process for a cell and thus is highly regulated. The most obvious control is at the level of RNA polymerase binding to a promoter. Some promoter sequences are intrinsically better able to bind RNA polymerase than others. Protein repressors can bind to operator sequences and physically block access to the promoter and/or bend that region of DNA so that the promoter is not accessible. Some promoters require binding of a protein activator before significant binding of RNA polymerase can occur. Even after RNA polymerase is bound to a promoter and transcription has begun, completion of the process may not be certain. Some amino acid biosynthetic operons include within the leader regions of their mRNA molecules, attenuators that can form transcription terminator loops under appropriate conditions. Different organisms use different strategies to affect terminator loop formation, including translation of a peptide, binding of a protein complex to the leader region, or binding of the uncharged tRNA.

After an appropriate portion of mRNA has been produced, it is translatable by ribosomes even before synthesis is completed. Ribosomes bind sequentially to a specific site or sites on an mRNA molecule and start synthesis at a specific initiator codon. The binding of proteins or RNA molecules can prevent ribosome attachment. The first tRNA to bind to a Bacterial ribosome–mRNA complex carries N-formylmethionine and enters the P site on the ribosome. The second RNA must have an anticodon matching the next codon on the mRNA and binds to the A site on the ribosome. As each peptide bond is synthesized, the ribosome translocates along the mRNA and the tRNA molecules move into the P and E sites on the ribosome. Arrival of a new tRNA at the A site displaces the uncharged tRNA in the E site.

Questions for Review and Discussion

1. Predict the phenotypes of *E. coli* cells mutated in the following functions:

   galS

   Adenylate cyclase

   Tryptophanyl tRNA synthetase

2. Describe the molecular means by which repressors prevent transcription.

3. Enumerate some of the advantages of having more than one promoter in an operon.

References

General

Adhya, S. (1996). Negative Control of Transcription, pp. 1503–1512. In: Neidhardt, F.C., Ingraham, J.L, Low, K.B., Magasanik, B., Schaechter, M., Umbarger, H.E. (eds.), *Escherichia coli and Salmonella typhimurium. Cellular and Molecular Biology*. 2 vols. Washington, DC: American Society for Microbiology.

Babitzke, P. (2004). Regulation of transcription attenuation and translation initiation by allosteric control of an RNA-binding protein: The *Bacillus subtilis* TRAP protein. *Current Opinion in Microbiology* 7: 132–139.

Coulombe, B., Burton, Z.F. (1999). DNA bending and wrapping around RNA polymerase: A "revolutionary" model describing transcriptional mechanisms. *Microbiology and Molecular Biology Reviews* 63: 457–478.

Lawson, C.L., Swigon, D., Murakami, K.S., Darst, S.A., Berman, H.M., Ebright, R.H. (2004). Catabolite activator protein: DNA binding and transcription activation. *Current Opinion in Structural Biology* 14:10–20.

Londei, P. (2005). Evolution of translational initiation: New insights from the archaea. *FEMS Microbiology Reviews* 29: 185–200.

Mooney, R.A., Artsimovitch, I., Landick, R. (1998). Information processing by RNA polymerase: Recognition of regulatory signals during RNA chain elongation. *Journal of Bacteriology* 180: 3265–3275. (This review compares RNA polymerase to a Turing machine and presents a model for function.)

Müller-Hill, B. (1996). *The lac Operon: A Short History of a Genetic Paradigm*. Berlin: Walter de Gruyter. (A very readable summary of the history of the *lac* operon and our present state of knowledge.)

Omer, A.D., Ziesche, S., Decatur, W.A., Fournier, M.J., Dennis, P.P. (2003). RNA-modifying machines in Archaea. *Molecular Microbiology* 48: 617–629.

Reeve, J.N. (2003). Archaeal chromatin and transcription. *Molecular Microbiology* 48: 587–598.

Severinov, K. (2000). RNA polymerase structure–function: Insights into points of transcriptional regulation. *Current Opinion in Microbiology* 3: 118–125.

von Hippel, P.H. (1998). An integrated model of the transcription complex in elongation, termination, and editing. *Science* 281: 660–665. (A summary of findings from *E. coli*.)

Wilson, D.N., Nierhaus, K.H. (2003). The ribosome through the looking glass. *Angewandte Chemie International Edition* 42:3464–3486.

Withey, J.H., Friedman, D.I. (2003). A salvage pathway for protein synthesis: tmRNA and trans-translation. *Annual Review of Microbiology* 57: 101–123.

Yanofsky, C., Konan, K.V., Sarsero, J.P. (1996). Some novel transcription attenuation mechanisms used by bacteria. *Biochimie* 78: 1017–1024.

Specialized

Ajdic, D., Ferretti, J.J. (1998). Transcriptional regulation of the *Streptococcus mutans gal* operon by the galR repressor. *Journal of Bacteriology* 180: 5727–5732.

Bernstein, J.A., Lin, P.-H., Cohen, S.N., Lin-Chao, S. (2004). Global analysis of *Escherichia coli* RNA degradosome function using DNA microarrays. *Proceedings of the National Academy of Sciences of the USA* 101: 2758–2763.

Deikus, G., Babitzke, P., Bechhofer, D.H. (2004). Recycling of a regulatory protein by degradation of the RNA to which it binds. *Proceedings of the National Academy of Sciences of the USA* 101: 2747–2751.

Elf, J., Berg, O.G., Ehrenberg, M. (2001). Comparison of repressor and transcriptional attenuator systems for control of amino acid biosynthetic operons. *Journal of Molecular Biology* 313: 941–954.

Geanacopoulos, M., Adhya, S. (1997). Functional characterization of roles of GalR and GalS as regulators of the *gal* regulon. *Journal of Bacteriology* 179: 228–234.

van de Guchte, M., Ehrlich, S.D., Chopin, A. (1998). tRNA^{Trp} as a key element on antitermination in the *Lactococcus lactis trp* operon. *Molecular Microbiology* 29: 61–74.

Hogema, B.M., Arents, J.C., Bader, R., Eijkemans, K., Inada, T., Aiba, H., Postma, P.W. (1998). Inducer exclusion by glucose 6-phosphate in *Escherichia coli*. *Molecular Microbiology* 28: 755–765.

Hosaka, T., Tamehiro, N., Chumpolkulwong, N., Hori-Takemoto, C., Shirouzu, M., Yokoyama, S., Ochi, K. (2004). The novel mutation K87E in ribosomal protein S12 enhances protein synthesis activity during the late growth phase in *Escherichia coli. Molecular and General Genomics* 271: 317–324.

Lewis, D.E.A., Geanacopoulos, M., Adhya, S. (1999). Role of HU and DNA supercoiling in transcription repression: Specialized nucleoprotein repression complex at *gal* promoters in *Escherichia coli. Molecular Microbiology* 31: 451–461.

Márquez, V., Wilson, D.N., Tate, W.P., Triana-Alonso, F., Nierhaus, K.H. (2004). Maintaining the ribosomal reading frame: The influence of the E site during translational regulation of release factor *Cell* 118: 45–55.

Møller, T., Franch, T., Udesen, C., Gerdes, K., Valentin-Hansen, P. (2002). Spot 42 RNA mediates discoordinate expression of the *E. coli* galactose operon. *Genes & Development* 16: 1696–1706.

Panina, E.M., Vitreschak, A.G., Mironov, A.A., Gelfand, M.S. (2003). Regulation of biosynthesis and transport of aromatic amino acids in low-GC Gram-positive bacteria. *FEMS Microbiology Letters* 222: 211–220. (An example of how to use computer technology to identify regulatory elements.)

Spitzfaden, C., Nicholson, N., Jones, J.J., Guth, S., Lehr, R., Prescott, C.D., Hegg, L.A., Eggleston, D.S. (2000). The structure of ribonuclease P protein from *Staphylococcus aureus* reveals a unique binding site for single-stranded RNA. *Journal of Molecular Biology* 295: 105–115.

Virnik, K., Lyubchenko, Y., Karymov, M.A., Dahlgren, P., Tolstorukov, M.Y., Semsey, S., Zhurkin, V.B., Adhya, S. (2003). "Antiparallel" DNA loop in Gal repressosome visualized by atomic force microscopy. *Journal of Molecular Biology* 334: 53–63.

Yarnell, W.S., Roberts, J.W. (1999). Mechanism of intrinsic transcription termination and antitermination. *Science* 284: 611–615.

5

DNA Repair and Simple Recombination

All organisms and some viruses have their own mechanisms for maintaining the integrity of their nucleic acid (i.e., for repairing any damage). Nevertheless, most organisms can undergo some sort of genetic transfer or exchange process(es). The two mechanisms may seem antithetical because recombination, the movement of genetic information from one molecule of nucleic acid to another, implies that a nucleic acid molecule loses its integrity and undergoes some kind of structural alteration. However, as is discussed in this chapter, one way to view the recombination process is that it has appropriated the essential DNA repair processes for a function in which the potential for damage to the genetic information contained in a nucleic acid molecule is outweighed by the potential benefit to be derived from new genetic information. Hence, the genetic transfer processes that are discussed in subsequent chapters are in fact afterthoughts that trigger preexisting repair pathways to accomplish DNA recombination.

Practically speaking, the genetic analysis of RNA molecules has been difficult since they are comparatively unstable or imprecisely synthesized. On the other hand, repair and recombination of DNA molecules have been studied extensively. The processes that lead to complete repair of damaged DNA can be divided into two groups: those that correct (reverse) the actual chemical alteration and those that first remove a damaged DNA segment and then replace or resynthesize it in a corrected manner.

Both recombination and repair are discussed in this chapter. In order to simplify their explanation, *Escherichia coli* and its bacteriophages have been used since they provide almost all the examples that need to be discussed. Note, however, that genes with similar names in other organisms generally carry out the same processes, as considerable effort has been made to maintain a uniformity of genetic nomenclature among bacteria.

Major topics include:

- Types of possible DNA damages
- SOS global regulatory network
- Major recombination deficiency phenotypes in *E. coli*
- Elements of the general model for recombination
- Major mechanisms of DNA repair and their relationship to recombination

---

THINKING AHEAD

When does the process of DNA repair occur during the cell cycle?

DNA Repair

DNA repair refers to the correction of simple mismatched bases, or more serious problems like chemically altered bases, ribonucleic acid bases instead of deoxyribonucleic acid bases, or cases where the deoxyribose moiety is present but the actual base is missing. The time during the reproductive process when DNA damage is likely to occur could be either **extrareplicational** or **intrareplicational**.

Intrareplicational damage refers to the errors committed by various polymerase molecules, and it results in the insertion of incorrect bases, or the omission or addition of a base. The fundamental structure of the DNA duplex is unaltered, but a heteroduplex condition exists in the damaged area. If the damage is not repaired, the next replication fork that passes through the area yields two different homoduplexes, but each homoduplex is perfectly normal in structure.

Extrareplicational damage, on the other hand, refers to physical damage caused by external agents, such as the mutagens described in Chapter 3. The result of a mutagenic treatment may be a chemically modified normal base (e.g., alkylated or deaminated) or a damage to the actual structure of the DNA itself. Modified bases lead to intrareplicational mistakes due to mispairing. As a general rule, they do not prevent further replication. Physical damage may take different forms, but each type prevents the occurrence of normal replication. For example, x-rays often cause double-strand breaks in a DNA molecule. The problem of correctly reassociating the broken ends is not a trivial one, and recombinational repair using a second normal molecule may be necessary. The damage caused by ultraviolet (UV) radiation is of two types: a pyrimidine dimer in which two adjacent pyrimidines are joined to form a cyclobutane ring, or a pyrimidine-$(6{\rightarrow}4)$-pyrimidone (Fig. 5.1). Both dimers are examples of intrastrand cross-links.

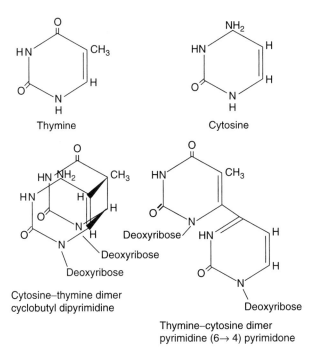

Figure 5.1. Effects of UV radiation on adjacent pyrimidine bases. Two types of structure are formed: cyclobutane dimers (*left*) and pyrimidine-$(6{\rightarrow}4)$-pyrimidones (*right*).

Another type of extrareplicational damage is induced by agents such as mito-mycin C that form interstrand cross-links. It is not possible for DNA polymerase III to replicate through areas of physically damaged DNA, although it can restart after the damaged region has been bypassed (Fig. 5.2). Paul Howard-Flanders and coworkers have shown that replication of cross-linked DNA yields daughter molecules containing gaps, approximately the size of one or more Okazaki fragments. These gapped molecules cannot be replicated further until repair has been done.

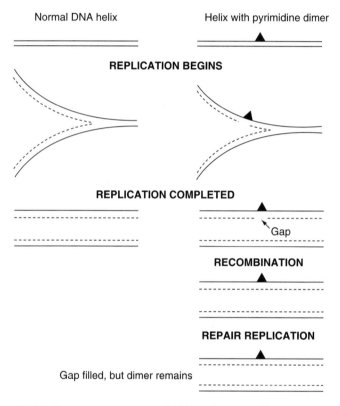

Normal DNA helix   Helix with pyrimidine dimer

REPLICATION BEGINS

REPLICATION COMPLETED

Gap

RECOMBINATION

REPAIR REPLICATION

Gap filled, but dimer remains

Figure 5.2. DNA damage can prevent DNA replication. The presence of pyrimidine dimers causes the DNA polymerase to stop before the dimer and to resume its synthetic activities at some distance. The result is a gap in the newly replicated DNA strand. If the companion DNA molecule has normal DNA in the corresponding region, recombination can be used to fill in the gap, although the dimer still exists and must be removed later. However, if the gapped regions overlap, no repair is possible, and the cell does not have intact DNA molecules.

Mismatch Repair: An Example of Intrareplicational Repair

The measured error rate of DNA polymerase III activities in various bacteria is substantially higher than the observed spontaneous mutation rate by a factor of 100–1000. This finding suggests that there is a **mismatch repair system** that corrects polymerase mistakes before they are converted to permanent changes in the DNA by another round of replication. The accuracy of mismatch repair is different for the leading and lagging strands, with the lagging strand apparently being more accurately repaired than the leading strand. Also, the enzymes cannot repair C•C mismatches.

Mutations affecting the mismatch repair system have a mutator phenotype and are localized in the *mutS*, *mutH*, and *mutL* genes. All of them code for proteins that work as a concerted unit. MutS protein binds to the mismatched bases. MutL, in a process requiring ATP hydrolysis, binds to the MutS–heteroduplex DNA complex. Finally, MutH, an endonuclease with sequence homology to the restriction enzyme *Sau*3A1, binds to the nearest unmethylated GATC sequence (which may be thousands of base pairs away), and cuts 5′ to the G. A second nick is generated either at the heteroduplex site itself or at another GATC site downstream. The unmethylated (most recently replicated and therefore incorrect) strand is removed and replaced. Removal requires DNA helicase II encoded by *uvrD* (also known as *mutU*). MutS and MutL proteins are essential for all editing modes in replication and recombination. DNA polymerase III and SSB then replace the missing DNA bases. The MutS protein from *Thermus aquaticus* has a crystal structure very similar to that from *E. coli*.

Mutations in the *dam* gene (DNA adenine methylase) result in under-methylation of DNA. Mutant cells have a higher than normal spontaneous mutation rate and are more sensitive to various base analogs. The role of *dam* is to differentiate between the old and the new strand. Thus, normally methylated duplex DNA, soon after replication, has one strand of the duplex correctly methylated, but not the other. Correction normally favors preservation of the methylated strand. Artificially prepared heteroduplexes that are introduced into a cell by genetic transformation generally confirm this hypothesis. For example, a heteroduplex that carries no GATC sites undergoes no significant repair. In such cases, the MutHLS mismatch repair system removes one strand randomly. A phage such as λ that has its DNA incompletely methylated (only about 75% under normal circumstances) shows some mismatch repair but a higher mutation rate than normal E. coli because the enzymes frequently cannot distinguish an old strand from a new one. If λ DNA is fully methylated in vitro, normal mismatch repair is observed. The mismatch repair process does not work with single-strand loops and has difficulty with transversion mutations, both of which disrupt the normal helical structure.

Other mismatch repair processes are also known. These include a system to convert A•G mismatches to C•G base pairs and another to convert T•G to C•G pairs. The latter requires only MutL and MutS. The *hex* system discussed in Chapter 10 is an example of another mismatch repair process. Similar mismatch repair systems are known in yeast. The *PMS1* (postmeiotic segregation) locus encodes a protein, with 32% amino acid sequence similarity to MutL from *Salmonella typhimurium*.

Ultraviolet-Irradiated DNA as a Model Repair System

E. coli cultures exposed to UV radiation are used as standards for discussions on structurally damaged DNA repair. The treatment is easy to quantitate, the damage it does is well characterized, and potential hazards to the experimenter can be controlled easily. Some pyrimidine dimers can be completely reversed (>99%) by a process called **photoreactivation**. An enzyme designated as a photolyase consists of a single protein chain and two chromophores and is isolatable as a blue complex. It absorbs energy from long wave (more than 300 nm, maximally at 384 nm) UV radiation, uses it to cleave cyclobutyl dimers into monomers, and restores the DNA to its original state (an example of error-free repair) (Fig. 5.3). Although, both long and short wave UV can catalyze dimer formation and photolyase activation, short wave UV leads predominantly to dimer formation, and long wave yields photolyase activation. Again, a similar enzyme with striking sequence similarity (36%) occurs in yeast, and plant blue-light photoreceptors also share homology with photolyases.

An alternative error-free repair process that can occur in the dark is termed **short-patch repair** or **nucleotide excision repair**. This process involves

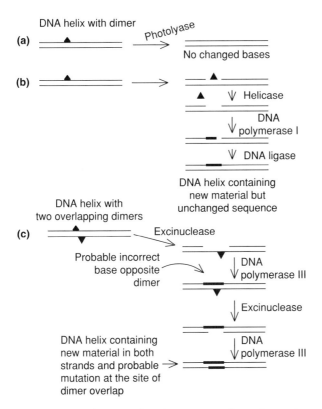

Figure 5.3. DNA repair mechanisms for damage away from the replication fork. (**a**) Photoreactivation. The chemical reaction is reversed using the energy derived from long wave UV or blue light. (**b**) Short-patch repair. A specific nicking enzyme acts next to a pyrimidine dimer, and the damaged region is replaced by DNA polymerase I using the intact strand as a template. (**c**) Long-patch or SOS repair. This is the same system as that for short-patch repair except that under SOS induction the patch is longer. In this example, the longer patch encounters a dimer on the other strand, resulting in translesion DNA synthesis.

the removal of a part of one DNA strand, including the dimer, resulting in a gap of about 20 bases that is then filled in. Because no radiant energy is required, this type of repair is sometimes called dark repair. It is particularly effective against the pyrimidine-(6→4)-pyrimidone structure.

Short-patch repair begins with recognition of a region of damage on a DNA molecule. This damage may have been due to modified bases or may have resulted from the removal of a damaged base by a glycosylase enzyme in a separate step. A glycosylase generates an apurinic or a apyridinic acid, a site where the deoxyribose sugar is present but without a base. Such a molecule is known as AP DNA (apurinic or apyridinic DNA), and the process is called **base excision repair**. Enzymes that remove the damaged region can be considered as AP endonucleases.

An **excinuclease** (excision endonuclease) is responsible for the crucial enzymatic activity for short-patch repair. In *E. coli* this enzyme consists of multiple subunits encoded by the *uvrA, B,* and *C* genes. Two molecules of UvrA protein and ATP bind to the damaged DNA, either to structures like pyrimidine dimers or to AP DNA. UvrB and UvrC then add to the complex to introduce two nicks into the DNA, one 8 bases upstream and the other four or five bases downstream from the lesion. The Uvr complex is displaced by UvrD helicase II that unwinds the damaged region of the DNA. After a gap has been created in a process of nucleotide excision (as opposed to base excision), polymerase I can use its $5' \rightarrow 3'$ exonuclease activity to degrade any additional damaged DNA and resynthesize an undamaged replacement using the existing 3'-end as a primer. Complete integrity of the DNA is restored when DNA ligase seals the nick. Note that this repair process removes entire nucleotides from the damaged area.

Excinuclease enzymes occur in eukaryotes as well. In yeast, the *Rad3* locus has strong functional correlations to the UvrA and UvrD loci in *E. coli*. However, it has structural similarities with the corresponding human protein, not *E. coli*.

Photoreactivation, short-patch repair, base excision repair, and mismatch repair enzymes are produced constitutively. They represent the primary cellular defense against DNA damage. However, they could be overwhelmed if the amount of DNA damage is excessive. In the case of short-patch repair, for example, it is necessary that dimers appearing in opposite DNA strands be located sufficiently far apart that no two regions needing to be patched overlap. However, if overlaps occur, polymerase I lacks a proper template and is unable to complete the excision repair process. In such a case, unrepaired DNA tends to accumulate in the cell, and some other mechanism must be used to eliminate it.

APPLICATIONS BOX

There have been reports that the extremely halophilic bacteria do not have a dark repair system, that is, they use only photoreactivation. How would you test whether these organisms are capable of repair in the dark?

The SOS Global Regulatory Network

The process by which an *E. coli* cell handles large amounts of DNA damage is called inducible or **SOS repair** because it occurs only when a cell cannot complete repairs via the short-patch pathway. The term SOS repair was coined by Miroslav Radman to signify that accumulation of damaged DNA in a cell leads to the equivalent of a distress signal that triggers induction of new enzymes. SOS repair is not a single discrete function, and it includes diverse responses such as the ability to repair pyrimidine dimers, induce various prophages, delay septum formation during cell division, shut off respiration, and increase protein degradation (Table 5.1). Because all these dissimilar properties are coordinately regulated, they constitute a **global regulatory network** (see Chapter 14).

THINKING AHEAD

How would you identify unknown genes that were members of a global regulatory network if you had already identified some members of the network?

The easiest way to demonstrate the existence of an inducible repair system is via the **W-reactivation** phenomenon first observed by Jean Weigle. If UV-irradiated λ bacteriophages are used to infect normal *E. coli*, the yield of cells that produce more virus and lyse is low because the host cells are not

Table 5.1. Selected functions of the SOS global regulatory network for which the LexA protein is a repressor.

Gene	Function
lexA	Repressor of SOS genes (autoregulated)
recA	DNA synaptase, enhancer of proteolysis
sulA	Inhibitor of septation, normally degraded by Lon protease but overproduced after SOS induction
dnaG	DNA primase, may trigger new rounds of replication
uvrABC	Excinuclease, long-patch repair
umuCD	UV mutability

able to repair all the damage in time to permit normal phage functions before the infection aborts. However, if the *E. coli* cells are first given a dose of UV radiation and then infected, damage to the irradiated λ DNA is repaired rapidly and the yield of infectious virus particles is greatly increased. It can be demonstrated that it is not UV radiation per se but the DNA damage it causes that leads to repair induction and hence to W-reactivation. For example, if a donor cell is UV-irradiated and then conjugated to an unirradiated recipient cell, the newly transferred, damaged DNA causes repair system induction in the recipient cell and leads to W-reactivation. A period of about 30 min is required for maximal enzyme levels to be achieved.

A search of various types of *E. coli* mutant cells identified two loci, *lexA* and *recA*, that seemed to be regulators of the SOS response. The LexA protein functions as a repressor that binds to a variety of operators and is the major controlling element in the global regulatory network. Temperature-sensitive, constitutive, and inactivating mutations are known. Unlike the case with the operons discussed in Chapter 4, the LexA protein is not allosterically changed during induction. John Little showed that LexA is physically cleaved at a specific alanine–glycine bond near the middle of the molecule. This bond is apparently naturally unstable, as cleavage can occur in vitro in the absence of other factors. However in a cell, cleavage seems to occur as a result of the action of another protein, RecA.

The *recA* locus is the major recombination proficiency locus for the *E. coli* cell, and its protein is normally produced at low levels. Low-level synthesis is regulated by LexA repressor acting at the *recA* operator. The role of RecA in recombination is discussed later. Its role in the SOS response is to display a coprotease activity that rapidly promotes cleavage of LexA protein at its unstable bond. In this regard, it behaves as an allosteric effector. This activity is apparently triggered by an accumulation of large amounts of single-strand DNA as might occur following abortive attempts at short-patch repair. At high temperature, temperature-sensitive mutants in *recA* do not require an inducer to express SOS responses, so their coprotease activity is more easily stimulated. Following cleavage of the LexA protein, a wide variety of genes is derepressed and the SOS response occurs. As repair progresses, the stimulus of single-strand DNA is lost, the RecA coprotease activity is lost, and LexA protein regains control of the network.

Members of the SOS network can be identified by one of two ways. They may be found by making fusions of *lacZ* DNA to random promoters using a phage Mu derivative to create insertion mutations (see Chapter 3). Fusions to genes in the SOS network would express β-galactosidase activity after exposure to UV radiation. They may also be found by running computer searches on

DNA sequence databases for LexA binding sites (consensus sequence TACTG-TATATA–A–ACAGTA). Some of the sites identified can be found in Table 5.1. Basically, they are associated with the inhibition of septation (which prevents segregation of unrepaired DNA and possible death of daughter cells), enhanced recombination (suitable for certain types of repair; see later), and enhanced excision repair (long-patch repair). Long-patch repair seems to involve the same enzymes as short-patch repair, but the length of the patch is now about 100 bases. The inhibitor of septation, *sul*, is unusual in that it is normally synthesized at a reasonable rate, but rapidly degraded by an ATP-dependent protease called Lon. The SOS response greatly increases the amount of Sul protein and leads to the inhibition of cell division. As LexA regains control, Lon protein cleaves the inhibitory Sul protein and allows cell division to resume.

Whereas non-SOS repair is basically error-free, SOS repair is characteristically error prone. It is this trait that makes UV irradiation mutagenic. The ability to perform mutagenic repair is controlled by two loci, *umuC* and *umuD*, that code for proteins whose function is **translesion DNA synthesis** (replication past a missing or damaged base in a DNA strand). An in vitro system composed of purified DNA polymerase III, RecA, UmuC, SSB, and two copies of UmuD′ (RecA coprotease clipped UmuD) can replicate a substrate containing an apurinic or apyridinic site, usually inserting an adenine at the site corresponding to the missing base. Members of the Y family of translesion DNA polymerases from humans (Polymerase η) or *Sulfolobus* (Polymerase Dpo4) also show this effect. This result preserves the length of the DNA strand. However, Maor-Shoshani et al. (2003) have shown that in the case of a DNA molecule containing a hydrocarbon insert, DNA PolV (another member of the Y family) can either hop across the insert (creating a deletion) or add several bases from the hydrocarbon in the template strand (creating an insertion).

The presence of the Umu proteins eventually blocks the ability of the RecA nucleoprotein filament to form heteroduplexes, and cleave LexA protein. In the absence of the Umu proteins, successful replication results in a −1 deletion at the site of the lesion. Thus, if either of the *umu* loci is mutated, normal mutagens that require activation of the SOS response do not work.

Similar but more mutagenic functions, identified as *muc*, have been identified on certain plasmids such as ColI or R46. Some bacteria such as *Hemophilus influenzae* or *Streptococcus pneumoniae* seem to be natural *umu* mutants and are not mutable by UV radiation. Those mutagens that are independent of SOS (nitrosoguanidine, ethylmethanesulfonate, ICR191) are still operative because they induce lesions that result in direct mispairing.

A Conceptual Model for Generalized Recombination

Lack of meiosis and formal mitosis means that obligatory pairing of homologous DNA molecules does not occur during cell division in bacteria. Nonetheless, genetic exchange does occur between DNA molecules in the same cell, and thus, some mechanism of properly aligning homologous sequences must be present. Presumably, the DNA structures at the actual moment of physical exchange are similar in prokaryotes and eukaryotes, even though the methods by which the structures are initiated may not be the same.

The model for recombination presented here stems from a long line of recombination models presented by many workers and owes much to the work of Charles Radding. He and his collaborators first demonstrated the special properties of RecA protein. A general outline of their model is presented in Fig. 5.4. It begins with the known properties of RecA protein and requires a piece of single-strand DNA as an initiator. One way in which such a piece of DNA might be obtained is via strand displacement during DNA synthesis, as shown at the upper right of the figure. For this reason, nicks in DNA are recombinogenic, and nuclease invasion of double-strand breaks to leave a single-strand tail is also recombinogenic. Gap formation (upper left) is a natural consequence of excision repair and can also lead to strand invasion. After suitable degradative steps, both pathways result in the production of a single-strand exchange. The difference lies in whether DNA synthesis occurs before or after strand exchange. The joint molecule common to both the pathways is two essentially complete DNA helices connected by a single DNA strand that originates in one molecule and terminates in the other. Its formation requires homology at the 3'-end of the invading strand.

The isomerization step indicated at the lower right is a crucial one. After isomerization, the positions of all the members of each DNA duplex are reversed as indicated. At first glance, it seems that it is an unlikely event, but in fact the isomerization is merely a rotation of the DNA strands about the longitudinal axis of the paired helices, as shown in Fig. 5.5. This rotation can be accomplished without displacing any bases from their normal helical configuration. Cox (2003) has reviewed the evidence that RecA protein is in fact a motor that causes strand rotation. The final structure is important because it now shows an exchange of two DNA strands in the manner corresponding to a model first proposed by Robin Holliday and sometimes called a **Holliday structure**. Note that the structure may still have nicks and in the region of the original strand,

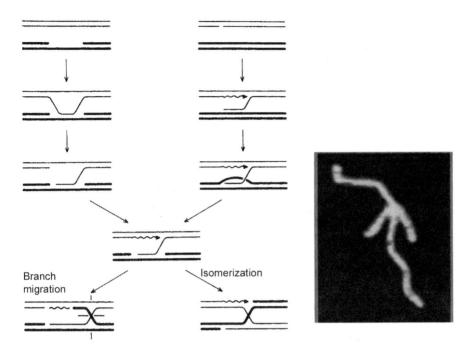

Figure 5.4. Meselson–Radding model (lower three diagrams) and two hypothetical pathways of initiation of strand transfer. Gaps or nicks are assumed to result from repair or other processes. The *wavy line* in the pathway on the upper right indicates new synthesis displacing an existing strand from a nick, and the *arrowheads* indicate growing 3′-ends. The horizontal and vertical marks at the lower left show endonucleolytic cuts that would lead, respectively, to the parental or recombinant configurations of the flanking arms of DNA. (From Radding, C.M. [1982]. Homologous pairing and strand exchange in genetic recombination. *Annual Review of Genetics* 16: 405–437.) The *inset* shows a Holliday structure visualized by atomic force microscopy. (From Lushnikov, A.Y., Bogdanov, A., Lyubchenko, Y.L. [2003]. DNA recombination: Holliday junctions dynamics and branch migration. *Journal of Biological Chemistry* 278: 43130–43134.)

exchange events have been nonreciprocal. A similar Holliday structure (lower left) can be obtained by displacing one strand from the recipient molecule into the donor molecule so as to continue the strand exchange already in progress.

The two bottom structures are susceptible to the process known as **branch migration** in which the site of a physical crossover moves along the DNA duplexes in a sliding motion, thereby forcing an additional exchange of strands. Such movement, of course, requires that hydrogen bonds be broken to allow the exchange; but for each base that loses its hydrogen bonds because it is being

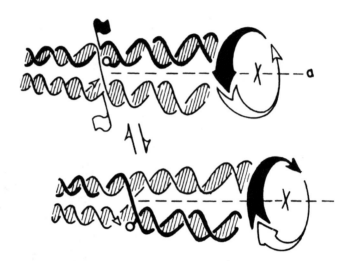

Figure 5.5. Hypothetical isomerization of a one-strand crossover to produce a two-strand crossover. The *top* diagram represents DNA molecules linked by the crossover of one strand. The structure is identical to the central intermediate in Fig. 5.4. The *arrow-head* represents a 3′-end, and the *open circle* represents a 5′-end. Isomerization, which can be demonstrated with molecular models, proceeds by rotation of the two arms to the right of the flags about an axis, *a*, that is between the arms and parallel to them. The *heavily shaded arm* would rotate above the plane of the paper, and the *lightly shaded arm* would rotate behind the plane of the paper. A phosphodiester bond at the point signaled by the dark flag becomes the cross-connection closer to the viewer, and a bond signaled by the light flag becomes the other cross-connection. The isomerization is reversible unless branch migration supervenes, as in Fig. 5.4, or the nick is sealed where the 3′-end (*arrowhead*) abuts on the 5′-end (*open circle*). (From Radding, C.M. [1978]. Genetic recombination: Strand transfer and mismatch repair. *Annual Review of Biochemistry* 47: 847–880.)

exchanged from one helix to another, a second base is restored to a condition in which it can again hydrogen-bond (albeit with a new partner). Although there is no net energy change in the state of the molecule, the two duplex molecules require some energy input in order to rotate during the strand exchange.

Three facts are of importance regarding the process of branch migration: (1) The exchanges it produces are all reciprocal. (2) If one base in ten is mismatched, branch migration stops. (3) If the molecule is linear, the branch can migrate right off the end of the molecule and the structure can resolve itself into two intact but recombinant DNA molecules. Migration off the end cannot occur in the case of *E. coli* as the DNA is circular, and it is thus necessary to have nucleases that resolve the structure into two separate molecules by introducing appropriate nicks as indicated in Fig. 5.4. The bottom structures in this figure are consistent with these observations. The overall result of the recombination process is two molecules showing a large region of reciprocal exchange, with nonreciprocal exchange at one or both ends.

In *E. coli*, the relevant nuclease activity comes from a complex of the RuvAB proteins (helicase) and the RuvC protein (resolvase). Ayora et al. (2004) have shown that *Bacillus subtilis* has a *recU* gene with functions similar to *E. coli* *ruvC* although both of them are genetically unrelated. There are no known analogs of RuvAB in the archaea, although several laboratories are working on the problem (e.g., Bolt and Guy [2003]).

Electron micrographic evidence is basically in accord with the model. It is possible to simulate a Holliday junction by the formation of a cruciform structure (Fig. 5.6) or actually examine short Holliday structures (*inset* to Fig. 5.4). These structures allow an experimenter to test the specific binding properties of proteins thought to be involved in resolution of the Holliday structure.

It is important to remember that never has a model for generalized recombination accommodated all known experimental data. For example, Leonard Norkin crossed *E. coli* *lacZ* mutants in conjugation experiments to test whether there was a direct relationship between recombination frequency and distance between genetic markers as assumed in the process of constructing genetic maps. The general tendency was that the amount of recombination between individual mutations increased as the physical distance between the mutations increased, but there were clearly exceptions (Fig. 5.7). Therefore, the present recombination model is considered broadly, if not entirely, accurate.

Genetic and Functional Analysis of Recombination

RecA and Similar Proteins

Mutations affecting the *recA* gene were first identified by A.J. Clark, Paul Howard-Flanders, and coworkers, and it was immediately apparent that the preeminent feature of these mutations was their pleiotropic effect. The *recA* gene is considered primary because mutations in it reduce generalized recombination to low levels, usually less than 10^{-6}. Mutants in *recA* are sensitive to UV radiation because they cannot induce SOS repair; they prevent lysogenic phages such as λ from being stimulated to lyse their host cells; and they lower viability of a

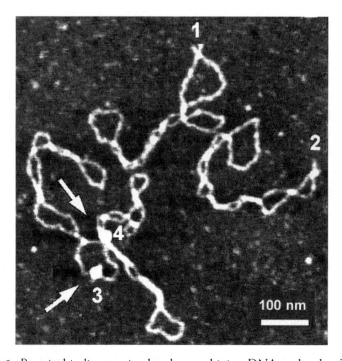

Figure 5.6. Protein binding to simulated recombining DNA molecules. The plasmid DNA molecules shown contain a 106 bp inverted repeat that can adopt a cruciform (cross-shaped) structure with 53 bp long hairpin arms, depending on the extent of DNA supercoiling. The cruciform structure is essentially the same as the Holliday junction of two recombining DNA molecules. This picture, taken with an atomic force microscope, shows several supercoiled plasmid molecules. Each numbered structure is a cruciform. RuvA protein, which participates in the resolution of Holliday structures, specifically binds to the cruciform structures (numbers 3 and 4). (Micrograph courtesy of Yuri L. Lyubchenko, Department of Pharmaceutical Sciences, University of Nebraska Medical Center.)

mutant cell to about 50% of normal. Genes with essentially similar functions are found in many other bacteria, including *Erwinia carolovora*, *Pseudomonas aeruginosa*, *Rhizobium melliloti*, *Neisseria gonorrhoeae*, and *B. subtilis* (*recE*), as well as in the bacteriophage T4 (*uvsX*), the fungus *Ustilago maydis* (*rec-1*), and yeast (RAD51 and DMC1). Cox (2003) has reviewed the similarities.

RecA protein is known to have a molecular weight of about 38,000 Da. The coprotease activity of RecA discussed previously is independent of its recombinase activity. Large quantities of RecA protein do not necessarily guarantee SOS induction. Conversely, SOS induction can occur in the absence of

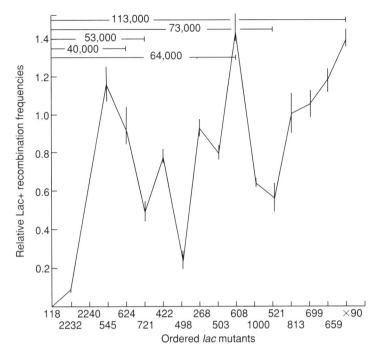

Figure 5.7. Intragenic recombination frequencies are not necessarily proportional to physical distances between the markers. In this experiment, an Hfr strain carrying the *lacZ118* mutation (which maps close to the *lac* operator region) was crossed to various F⁻ strains carrying other *lacZ* mutations. The precise order of the *lacZ* mutations, but not their relative spacings, was known from deletion mapping experiments. The relative *lac⁺* recombination frequency for each mating was determined by normalizing the number of colonies observed on plates selective for Lac⁺ to the number of colonies observed on plates selective for Leu⁺ (a marker transferred some minutes after *lacZ*). Note that although the recombination frequency increases with increasing map distance, there are several dramatic exceptions to it. Similar results were obtained if the Hfr strain carried mutation X₉₀ instead of *118*. The molecular weights of the polypeptide fragments produced by some of the terminator mutations used in this experiment are presented at the top. These numbers provide some indication of the physical distance between the various markers. (From Norkin, L.C. [1970]. Marker specific effects in recombination. *Journal of Molecular Biology* 51: 633–655.)

overproduction of RecA. RecA protein has important direct functions in repair and recombination of DNA molecules. Its activities are driven by its enzymatic property of being a DNA-dependent ATPase. In fact, during SOS repair a noticeable drop in cellular ATP is seen in all strains except *recA* or *lexA* mutants. Charles Radding noted that while interacting with DNA, the RecA protein performs two distinct functions that are important for recombination. The first is its renaturation activity, taking two complementary single-strand DNA molecules and promoting their renaturation into a duplex. The second is its **recombinase** activity of promoting **strand invasion**.

During strand invasion, RecA protein coats a single strand of DNA in an asymmetric fashion to form a nucleoprotein filament. The RecO and RecR proteins act to promote filament formation and stabilize it against disassembly. This filament displays the coprotease activity of the RecA protein, and LexA protein binds to its deep groove.

Recombination proceeds when the filament enters into an association with a normal DNA duplex to form a transient triple-strand structure in regions where the DNA molecules possess similar sequence. The strand invasion occurs at the 3'-end of the single strand in the filament. The net result is displacement of one original DNA strand by the invading single strand. Even low levels of sequence divergence reduce efficiency of strand invasion.

Strand invasion is subdivided into three phases. During the first phase called **presynapsis**, RecA protein uses energy from ATP to comparatively slowly polymerize itself along a single-strand DNA molecule (nearly one RecA protein molecule for every three nucleotides). This step is facilitated by helix-destabilizing proteins such as Ssb that break up random pairing within the molecule. Ssb is needed only in the proportion of one molecule for every eight nucleotides, and a strand fully coated with Ssb is not available for RecA binding. The nucleoprotein filament that results from presynapsis is polyvalent with multiple weak binding sites for double-strand DNA and is resistant to the action of DNase I.

It then enters into a **synaptic phase** during which the nucleoprotein filament binds to a DNA duplex, first nonspecifically and then specifically with a homologous region. This phase is surprisingly rapid, probably because the multiple filament-binding sites ensure that the two molecules remain closely associated at all times, even though binding at any single site is relatively weak and transient.

Finally, there is a slow **postsynaptic** or strand exchange phase during which the single-strand DNA forms a displacement loop, or **D-loop** (Fig. 5.8). Paradoxically, Gupta et al. (1998) have shown that strand exchange proceeds in

Figure 5.8. D-loop formation. The figure shows a portion of a DNA duplex (*thick lines*) and one end of a single strand of DNA (*thin line*). Various genetic markers are indicated by letters. The end of the single strand of DNA may insert into the homologous portion of the DNA duplex and loop out one of the existing strands. Because the resulting loop was caused by insertion of an extra DNA strand, the loop is designated a D-loop. If the loop had been produced by insertion of an RNA strand, it would have been designated an R-loop.

the 5′–3′ direction although it is the 3′-end of the filament that inserts into the D-loop. The helix is unwound, and the single-strand DNA physically replaces a portion of the duplex to create a heteroduplex region called a joint molecule. The displaced strand is of the same polarity as the original single strand and becomes nuclease-resistant owing to protein binding. The complementary strand need not be a perfect match to the invading strand and base mismatches up to 10% are tolerable.

The fidelity of the strand displacement reaction after the initial invasion is not high. If one of the participating DNA molecules is supercoiled, the average density of the mismatched base pairs can be as high as 30%, but it could still give a heteroduplex joint molecule. This capability may allow RecA protein to drive a strand exchange reaction beyond a damaged region, allowing for recombinational repair. Recombination is an important element for repair because it provides a new template strand that can be used to synthesize DNA past a lesion(s). It is, however, error prone because of the allowable mismatches during strand displacement.

Comparable proteins are found in many organisms. The *Saccharomyces cerevisiae* Rad51 and the archaeal RadA proteins carry out similar functions with respect to strand exchange, nucleoprotein filament formation, and DNA pairing (Cox 2003).

Gowrishankar and Harinarayanan (2004) have taken up the problem of such polar mutations that lead to the nontranscription of genes downstream from a nonsense mutation. In other words, they have tried to answer why it is necessary that transcription and translation be coupled? They suggest that free mRNA has the ability to form R-loops (single strand of RNA invades a DNA duplex, see Fig. 5.8). The R-loops may have the ability to stop further transcription, which in turn may block the DNA polymerase from replicating the region. Note that this would just be an aberration of the normal mechanism for triggering recombination.

Other Genes Affecting Recombination

The original investigations into recombination deficiency produced not only *recA* mutants but also *recB* and *recC* mutants. The latter two have similar phenotypes (moderate UV sensitivity, only 1% of normal recombination proficiency, and cell viability of 30%) and map at essentially the same spot on the *E. coli* genome. The affected enzyme is exonuclease V, an ATP-dependent enzyme with multiple activities. It consists of three nonidentical subunits, α, β, and γ, encoded by *recB*, *recC*, and *recD*, respectively, and possesses both helicase and endonuclease activities. A nonessential but stimulatory RNA molecule copurifies with the enzyme. The B and D subunits are both helicases, B acting slowly on the 3'-end and D acting rapidly on the 5'-end (Taylor and Smith 2003).

When presented with a linear double-strand DNA molecule, exonuclease V invades the helix at the rate of 300 bp/s. Owing to the differences in helicase rates, the 3'-strand coils in front of the B subunit while the 5'-strand is completely displaced. When exonuclease V encounters a specific base sequence known as a Chi site (see Chapter 8), its nuclease activity cuts 4–6 bases on the 3'-side of the Chi site. Further movement by exonuclease V generates a single-strand tail of DNA (Fig. 5.9). Mutations in *recD* yield cells that lack the nuclease activity but are nonetheless recombination-proficient. These cells had previously been designated *recB*‡.

Exonuclease V participates in DNA excision–repair systems as well as in recombination, although it is not induced as part of SOS repair. In UV-irradiated cells, exonuclease V degrades the damaged DNA, but its action is held in check by one of the SOS functions. In the absence of RecA coprotease activity, extensive DNA degradation is observed ("recless degradation," as one experimenter has put it).

Researchers studying recombination have discovered new facts by looking at ways in which the *recBC* mutations can be suppressed. Conceptually, two types of extragenic suppressor mutations are possible. The first type is to supply

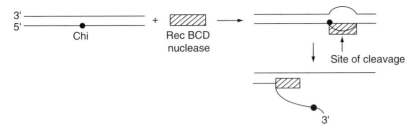

Figure 5.9. The action of the RecBCD nuclease as a traveling helicase is changed to a nuclease when a Chi site is encountered. (Reproduced from Adams, R.L.P., Knowler, J.T., Leader, D.P. [1992]. *The Biochemistry of the Nucleic Acids*, 11th ed. London: Chapman & Hall.)

an equivalent enzyme from a different source. The second type is to turn on a biochemical pathway that substitutes for the one that is blocked by the loss of exonuclease V activity. Both types of suppressors have been identified, and they are called *sbcA* and *sbcB*.

The *sbcA* mutations are located within a 27 kb remnant of a viral genome known as the *rac* (recombination activation) region that is found in many *E. coli* chromosomes. The *rac* virus can no longer produce functional progeny owing to the large amount of genetic material that has been deleted, but it does have a functional origin for DNA replication and keeps its genes turned off by means of a protein repressor. The *sbcA* mutations activate the *recE* and *recT* genes that code for exonuclease VIII and RecT protein, respectively (Noirot et al. 2003). Exonucleases V and VIII are functionally interchangeable for purposes of recombination, and RecT protein is a member of the recA protein family. Therefore, *recBC⁻* recipient cells that also carry an *sbcA* mutation are functionally recombination-proficient. During conjugation, the *rac* locus transfers along with other bacterial genes, but the protein repressor does not. Hence, the genes comprising the *rac* locus activate in recipient cells, even in the absence of an *sbcA* mutation. The effect of this activation is to make *recBC⁻* transconjugants phenotypically Rec⁺ until the protein repressor of *rac* can be synthesized in them.

The *sbcB* mutations, instead of providing a replacement exonuclease, actually result in the loss of yet another enzyme. In this case, the new deficiency is in exonuclease I, but the result is the same as in the case of *sbcA* mutations: recombination proficiency. This apparent contradiction arises because loss of exonuclease I activity prevents degradation of a recombination intermediate that feeds into a new recombination pathway: the *recF* pathway. The *recF* pathway is not of critical importance in wild-type cells except that it acts as a mediator of plasmid recombination, as $recF^-$ $recBCD^+$ strains are recombination-proficient, although cleavage of LexA protein is delayed after UV irradiation. However, $recF^-$ $recBCD^-$ $sbcB^-$ strains are again recombination-deficient.

Multiple steps are required for completion of recombination via the $recF$ pathway, and many of these steps have been defined by mutations. The role of the $recF$ gene seems similar to that of $lexA$, as the $recF^-$ phenotype (in $recBC^-$ $sbcB^-$ cells) includes UV sensitivity, deficient gap filling, deficient double-strand break repair, and failure of $recA$ protein induction, as well as blockage of recombination. The combination of $recF$, $recO$, and $recR$ genes is necessary to stabilize replication forks that have stopped due to UV radiation damage (Chow and Courcelle 2004).

A Model for General Recombination

The basic recombination model (Fig. 5.10) takes a look at the events that occur at one end of a linear DNA molecule as it recombines with the resident DNA. The process requires three elements: a linear, double-strand DNA molecule; Chi sites (Chapter 8), which are linear octamers of the sequence 5'-GCTGGTGG-3'; and the $recBCD$ exonuclease V. Exonuclease V attaches to each end of the DNA and penetrates into the helix, initially creating a simple loop in one strand, followed by adjacent loops in both the strands (Fig. 5.10a–c).

The single-strand binding protein (SSB) stabilizes these loops and prevents excessive degradation by exonuclease V. The enzyme (and associated loops) moves into the DNA helix, allowing the DNA at the end to recoil, until the enzyme encounters a Chi site in the appropriate orientation. It then cuts the 3'-strand a few bases before the Chi site (Fig. 5.10d). The nicked strand unwinds, generating a recombinogenic single strand of DNA that initiates formation of a D-loop (Fig. 5.10g). Isomerization is followed by resolution of the Holliday structure.

Note that in this model, recombination events always occur in pairs. Once exonuclease V has reacted with a Chi site, it becomes unreactive for other such sites; therefore, each molecule generates only one integrative event. A computer search of the E. coli genome shows 1006 Chi sites in either orientation, which amounts to roughly one site per 5 kb. Therefore, the assumption that there is a Chi site near the end of any transferred DNA fragment seems reasonable. Combinations of Chi sites on the two DNA molecules are synergistic.

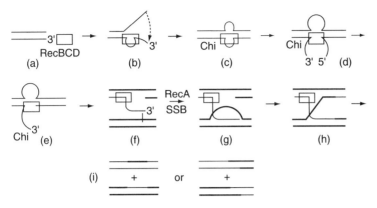

Figure 5.10. Model for homologous recombination by the RecBCD pathway. In standard bacterial genetic transfer processes, a fragment of DNA recombines with a circular chromosome, necessitating an even number of strand exchanges to regenerate the circle. This model was developed for phage λ, in which one recombinational exchange yields an intact chromosome because the parental chromosomes (pairs of *thin and thick lines*) are complete. RecBCD enzyme (exonuclease V, *open box*) binds to a ds DNA end (**a**) and unwinds the DNA by forming a loop-tail (**b**) and then a twin-loop (**c**) structure, which moves along the DNA as the ssDNA loops enlarge. Upon reaching a properly oriented Chi recombinational hotspot (marked chi), the enzyme cuts one DNA strand (the one containing 5′-GCTGGTGG-3′) to generate a 3′-ended single-strand DNA tail (**d**), which is elongated on further unwinding (**e** and **f**). Aided by RecA and SSB proteins, this 3′ tail invades a homologous dsDNA molecule to form a D-loop (**g**). Cleavage of the D-loop, possibly by exonuclease V, and strand pairing, aided by RecA and SSB proteins, generate a Holliday junction (**h**). Resolution of the Holliday junction, probably by RecG (and other proteins) or by RuvABC, produces a pair of recombinant molecules with hybrid DNA flanked by DNA with parental (**i**, *left*) or recombinant (**i**, *right*) configuration. (Modified from Smith [1991]. Conjugational recombination in *E. coli*: Myths and mechanisms. *Cell* 74: 19–27.)

All recombination pathways require some sort of termination event to resolve the X-shaped Holliday structure shown in Figs. 5.4 and 5.10. The RuvAB proteins promote ATP-dependent branch migration of Holliday structures, and the RecG protein has a similar effect. RuvC is an endonuclease that specifically resolves (cleaves) the Holliday junction in combination with RuvAB, thereby restoring two normal DNA duplexes (see Fig. 5.6). The RusA protein has a similar effect with RecG. Protein HU is essential for both these processes.

Interrelation of Repair and Recombination Pathways

In addition to the repair mechanisms discussed earlier, there is yet another variety of repair–recombinational repair. It takes advantage of gaps in newly synthesized DNA corresponding to damaged regions. These gaps might be filled by a gap-filling repair using a repair polymerase like Pol I, resulting in long-patch repair. McGlynn and Lloyd (2002) have suggested that the RecG protein may function in cases where leading strand synthesis cannot continue but lagging strand synthesis can. They proposed that the lagging strand is degraded back to the point of stoppage on the leading strand. Then RecG catalyzes denaturation of the newly synthesized strands from their separate duplexes and renaturation of the two strands to form another duplex. This structure is now a Holliday structure that can be resolved in a manner leading to synthesis of a normal DNA strand across from the lesion. Figure 5.11 shows three possible models for the

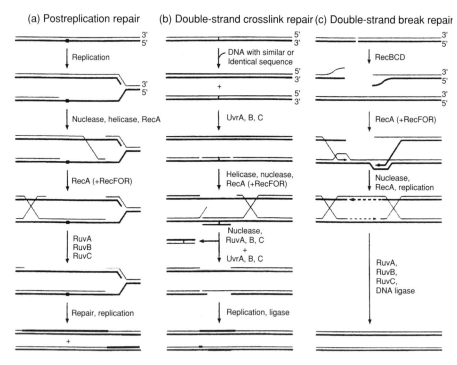

Figure 5.11. Models of possible pathways for recombinational repair of DNA lesions. Each column shows a model for dealing with a particular type of DNA repair problem. The *thick lines* at the bottom of the first two columns indicate newly replicated DNA. (Adapted from Roca and Cox [1997] and used by permission.)

recombination repair of different types of lesions. The terminology used to describe this repair relies heavily on the terminology developed for recombination mutants. However, in an evolutionary sense it is likely that repair antedates recombination. In other words, the accidental triggering of repair pathways by the free ends of the transferred DNA might result in recombination.

As shown in Fig. 5.11a, the postreplication repair pathway takes up the task of recombination first and that of repair second because the thymine dimer remains until after recombination is complete. All that recombination accomplishes is to fill in the gap left by the replication fork so that the damaged base(s) can be excised and replaced by a mismatch repair system. On the other hand, in Fig. 5.11b and c, both strands of the DNA duplex are damaged at the same site, and therefore neither intramolecular repair nor replication is possible. Instead, a basically identical, undamaged DNA duplex provides normal sequence for recombination to replace one strand of the damaged region. Replication of the recombined region completes the duplex later. Since these repair pathways depend on *recA* function, they are error-prone because the RecA-coated DNA filament does not require perfect sequence alignment in its pairing. Note that replication is essential in each of these repair pathways.

The experiments of Paul Howard-Flanders and coworkers provide strong evidence in favor of the existence of repair pathways using recombination mechanisms. They have shown that if a cell is infected with two phage particles, one carrying nonradioactive damaged DNA and the other carrying undamaged radioactive DNA, the latter is cut during the repair process, even though it is undamaged. Moreover, heterologous DNA, such as that from an unrelated phage, present in the same cell is unaffected. The process has been called **cutting in** *trans* and has been shown to require the presence of the RecA protein in vitro.

Based on the repair pathways shown in Fig. 5.11, it is clear that recombination is essential for cell survival as certain types of DNA damage cannot be repaired without it. The recombination of transferred DNA that has been discussed in subsequent chapters is in fact incidental to this repair process.

Summary

DNA is a molecule that is stable both chemically and genetically. Chemical stability is maintained by various repair systems that correct structural errors. Newly replicated DNA is initially hemimethylated, allowing repair systems to distinguish between the new and old strands, thus facilitating appropriate repairs. The most intensively studied repair system is UV-irradiated *E. coli*.

UV radiation causes two types of pyrimidine dimers to form. In the presence of long wave UV radiation, some dimer formation can be reversed. In its absence, excision repair occurs. Excision repair removes a length of DNA including the pyrimidine dimer. DNA polymerase then synthesizes new DNA. In the presence of excessive amounts of single-strand DNA, an inducible repair system, SOS repair, activates. This global regulatory network has multiple effects that act to inhibit cell division until the repairs are completed. Genetic stability results from error-free DNA repair, whereas mutation results from error-prone (SOS) repair.

DNA molecules can also undergo recombination with one another if there is sufficient base sequence homology. Various mutations have been isolated that affect the recombination process. The most common mutants in *E. coli* are *recA*, *recBCD*, and *recF*. RecA protein coats single-strand DNA and encourages its pairing with homologous DNA. The RecBCD complex constitutes exonuclease V that cleaves DNA molecules in such a way as to create single-strand (recombinogenic) ends. The RecF pathway is an alternative recombination pathway that can be activated by mutation.

Many models for recombination are possible. A general model begins with the invasion of a DNA duplex by a single-strand region of DNA to create a D-loop. The region of the D-loop becomes the point of exchange of strands between the two DNA molecules. Exchanges may be reciprocal or nonreciprocal. Sometimes an isomerization reaction is necessary to bring about a double-strand exchange. Such a structure is a Holliday structure and is subject to branch migration. During branch migration, the actual site of crossover moves to yield reciprocally recombinant DNA molecules. There are several possible mechanisms for resolving Holliday structures.

Questions for Review and Discussion

1. What are the various types of DNA damage? How can they be repaired? Are any of them irreparable?
2. What is a global regulatory network? How does it function? Can you think of other cellular responses that might be global regulatory networks? How would you test your hypothesis?
3. What are some of the possible phenotypes of *rec* mutants?
4. What is a Holliday structure? How is it generated? Give at least two examples of ways in which a Holliday structure might be converted into two separate DNA duplexes.

5. Why do evolutionary biologists think that the DNA repair pathways preceded those of recombination?

References

General

Cox, M.M. (2003). The bacterial RecA protein as a motor protein. *Annual Review of Microbiology* 57: 551–577.

Friedberg, E.C., Walker, G.C., Siede, W. (1995). *DNA Repair and Mutagenesis.* Washington, DC: ASM Press.

Gowrishankar, J., Harinarayanan, R. (2004). Why is transcription coupled to translation in bacteria? *Molecular Microbiology* 54: 1365–2958.

McGlynn, P., Lloyd, R.G. (2002). Genome stability and the processing of damaged replication forks by RecG. *Trends in Genetics* 18: 413–419.

Schofield, M.J., Hsieh, P. (2003). DNA mismatch repair: Molecular mechanisms and biological function. *Annual Review of Microbiology* 57: 579–608.

Specialized

Amundsen, S.K., Taylor, A.F., Smith, G.R. (1998). A stimulatory RNA associated with RecBCD enzyme. *Nucleic Acids Research* 26: 2125–2131.

Arnold, D.A., Handa, N., Kobayashi, I., Kowalczykowski, S.C. (2000). A novel, 11 nucleotide variant of χ, χ^*: One of a class of sequences defining the *Escherichia coli* recombination hotspot χ. *Journal of Molecular Biology* 300: 469–479.

Ayora, S., Carrasco, B., Doncel, E., Lurz, R., Alonso, J.C. (2004). *Bacillus subtilis* RecU protein cleaves Holliday junctions and anneals single-stranded DNA. *Proceedings of the National Academy of Sciences of the USA* 101: 452–457.

Bolt, E.L., Guy, C.P. (2003). Homologous recombination in Archaea: New Holliday junction helicases. *Biochemical Society Transactions* 31: 703–705.

Chow, K.-H., Courcelle, J. (2004). RecO Acts with RecF and RecR to protect and maintain replication forks blocked by UV-induced DNA damage in *Escherichia coli. Journal of Biological Chemistry* 279: 3492–3496.

Gupta, R.C., Golub, E.I., Wold, M.S., Radding, C.M. (1998). Polarity of DNA strand exchange promoted by recombination proteins of the

RecA family. *Proceedings of the National Academy of Sciences of the USA* 95: 9843–9848.

Maor-Shoshani, A., Ben-Ari, V., Livneh, Z. (2003). Lesion bypass DNA polymerases replicate across non-DNA segments. *Proceedings of the National Academy of Sciences of the USA* 100: 14760–14765.

Noirot, P., Gupta, R.C., Radding, C.M., Kolodner, R.D. (2003). Hallmarks of homology recognition by RecA-like recombinases are exhibited by the unrelated *Escherichia coli* RecT protein. *The EMBO Journal* 22: 324–334.

Quillardet, Q., Rouffaud, M.-A., Bouige, P. (2003). DNA array analysis of gene expression in response to UV irradiation in *Escherichia coli*. *Research in Microbiology* 154: 559–572.

Stambuk, S., Radman, M. (1998). Mechanism and control of interspecies recombination in *Escherichia coli*. I. Mismatch repair, methylation, recombination, and replication functions. *Genetics* 150: 533–542.

Taylor, A.F., Smith, G.R. (2003). RecBCD enzyme is a DNA helicase with fast and slow motors of opposite polarity. *Nature* 423: 889–893. (The article following this one presents more information by different authors.)

van Loock, M.S., Yu, X., Yang, S., Galkin, V.E., Huang, H., Rajan, S.S., Anderson, W.F., Stohl, E.A., Seifert, H.S., Egelman, E.H. (2003). Complexes of RecA with LexA and RecX differentiate between active and inactive RecA nucleoprotein filaments. *Journal of Molecular Biology* 333: 345–354.

6

T4 Bacteriophage as a Model Genetic System

This chapter, in combination with Chapters 7 and 8, presents an overview of some extensively studied bacteriophages. The intent of the three chapters is to illustrate the basic nature of genetic processes using relatively simple genetic systems as examples. In Chapter 2, the distinction is made between phages that are always virulent and those that can temper their lytic response to form lysogens. Chapters 6 and 7 discuss the virulent (i.e., intemperate) phages, and Chapter 8 discusses problems associated with lysogeny.

The early work on bacteriophages, like that of many other facets of bacterial genetics, received its impetus and direction from Max Delbrück. It was at his insistence that during the formative years of phage genetics most phage workers concentrated on just one or two viruses. Most of the early studies were devoted to bacteriophage T4, a member of the T series of phages that are numbered 1–7. The remainder of this chapter is devoted entirely to this phage, whose metabolism demonstrates an amazing variety of genetic regulation. The rest of the T-series phages, including the similar phages T2 and T6, are discussed in Chapter 7.

Major topics include:

- Techniques used to study the genetics and molecular biology of a bacteriophage
- Mechanisms by which a virus controls the metabolism of a host cell
- Mechanism by which a virus ensures that there is sufficient time for DNA replication before assembly of new viruses starts to deplete the DNA pool
- Requirement for host proteins for successful phage infection

Morphology and Composition

T4 is a complex phage, familiar to most students via textbook pictures. It possesses an oblong head, 78 × 111 nm in size, with a contractile tail measuring 113 × 20 nm (Fig. 6.1a). The various specific anatomic structures of which the **virion** (phage particle) is composed are indicated in Fig. 6.1b. Most of the mature virion consists of associations of protein subunits, but there are several important nonprotein constituents that have been identified. Included among them are a large, linear, double-stranded DNA molecule with a molecular weight of 1.21×10^6 Da (1.69×10^5 nucleotide pairs) contained within the head; certain polyamines (putrescine, spermidine, and cadaverine) associated

Figure 6.1. (a) Electron micrograph of a T4 virion negatively stained with potassium phosphotungstate. The *bar* indicates a length of 25 nm. (b) Structure of bacteriophage T4, based on electron microscopic structure analysis to a resolution of about 2–3 nm. Near the head and tail are the locations of the known major and minor proteins. The icosahedral vertices are made of cleaved gene 24 protein (gp24). The gene 20 protein is located at the connector vertex, bound to the upper collar of the neck structure. The six whiskers and the collar structure appear to be made of a single protein species, gp*wac*. The gp18 sheath subunits fit into holes in the baseplate, and the gp12 short tail fibers are shown in a stored position. The baseplate is assembled from a central plug and six wedges; and although the locations of several proteins are unknown, they are included here with the plug components. (Diagram from Eiserling, F.A. [1983]. Structure of the T4 virion, pp. 11–24. In: Mathews, C.K., Kutter, E.M., Mosig, G., Berget, P.B. [eds.], *Bacteriophage T4*. Washington, DC: American Society for Microbiology.)

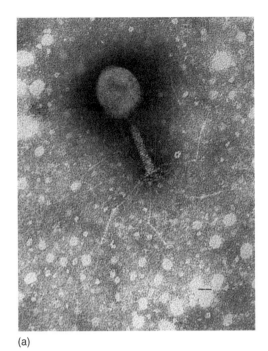

(a)

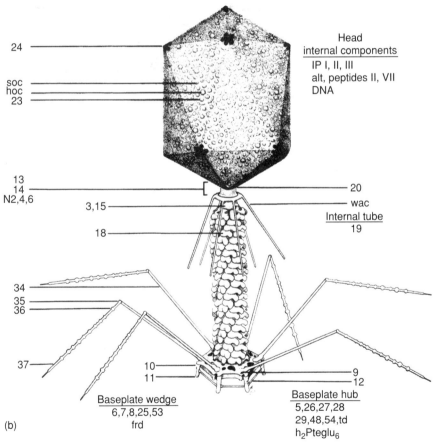

24

soc
hoc
23

Head
internal components
IP I, II, III
alt, peptides II, VII
DNA

13
14
N2,4,6

3,15

18

20

wac

Internal tube
19

34

35
36

37

10
11

9
12

Baseplate wedge
6,7,8,25,53
frd

Baseplate hub
5,26,27,28
29,48,54,td
$h_2Pteglu_6$

(b)

with the DNA; ATP and calcium ions found associated with the tail sheath; and dihydropteroylhexaglutamate associated with the baseplate.

THINKING AHEAD

What are the advantages to a virus that can distinguish viral DNA from host DNA?

A unique feature of T4 phage is its DNA composition. The DNA has had all of its cytosine residues replaced by hydroxymethylcytosine (Fig. 6.2), a base that normally does not occur in *Escherichia coli*. This base has the same hydrogen bonding characteristics as normal cytosine. Thus, its presence does not affect the genetic code, but it does offer a reactive site to which one molecule of glucose is attached (70% of the time in an α-linkage and 30% of the time in a β-linkage). These two differences permit phage DNA to be unambiguously differentiated from host cell DNA in various experiments. They also have important physiologic functions. For example, the glucose moieties prevent restriction of phage DNA by host cell restriction endonucleases and also add some hydrogen bonding (and stability) to the phage DNA. The added stability is useful because the base composition of T4 DNA is 65.5% (A + T) compared to host: 50% (A + T).

The T4 genome has ~300 **open reading frames** (regions that appear to be genes because they have promoters and terminators) and can be expected to pro-

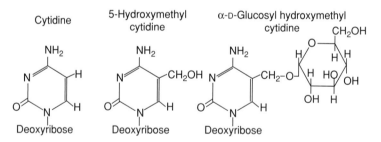

Figure 6.2. Forms of cytidine found in T4-infected cells. The structural formula for the normal base is shown at the *left*, and the progressive modifications made by T4 are shown at the *right*. Although the glucosyl moiety is shown in an α-linkage, a β-linkage also occurs. Phages T2 and T6 may add a second glucosyl moiety attached in an α-linkage to the first one.

duce a large number of proteins, many of which are enzymatically active. Most of them are, of course, associated with intracellular activity of the phage. However, certain enzymatic activities have been detected within free virions as well as in the cytoplasm of infected cells. Some of these enzymes are listed in Table 6.1, along with their presumptive functions in the infectious process. In many cases, a functional enzyme is not required for normal infectivity, suggesting that there are other means for the same function. Modifications in protein structure, however, do produce physiologic changes. For example, strains T4B (Benzer) and T4D (Doermann) exhibit slight differences in their dihydrofolate reductase activity that appear to be reflected in a requirement by T4B, but not T4D, for the presence of tryptophan in order to unfold tail fibers from around the sheath.

Experimental Methods Used to Study Phage Infection

It is possible to study phage-infected cells in a variety of ways. The presence of virions in a sample can be demonstrated by adding it to a culture of phage-sensitive bacteria and then plating the mixture in soft agar (0.6%) overlaid on

Table 6.1. Virus-specific enzymatic activities associated with T4 virions.

Dihydrofolate reductase (*frd*)	Located in baseplate wedge; may have role in unfolding tail fibers
Thymidylate synthetase(*td*)	Found in baseplate; necessary for infectivity
Lysozyme-like activity (two proteins, gp5 and gp25)	Found in baseplate; penetration through cell wall
Gamma glutamyl hydrolase (gp28)	Found on distal surface of baseplate central hub
Phospholipase	Possible role in host cell lysis
ATPase	Associated with the tail sheath; presumed to be involved with the contractile process
Endonuclease V (*denV*)	Excision repair (see Chapter 5) of phage or host DNA
alt Protein	Associated with viral DNA; reversible ADP ribosylation of host RNA polymerase subunit α

normal agar. Because T4 is a virulent phage, each infected bacterial cell eventually lyses, releasing virions that can infect other sensitive cells. As the bacteria increase in number, so do the virions by constantly infecting new bacterial cells. The multiple rounds of infection give rise to a hole, or **plaque**, in the lawn of bacterial cells; hence, this technique is called a plaque assay. Each plaque is assumed to represent one original, phage-infected cell, and thus each plaque is considered analogous to a bacterial colony for counting purposes. Experimenters have developed four specialized techniques for obtaining more precise information about events that occur during phage infection: the **one-step growth experiment**; the single burst experiment; the premature lysis experiment; and the electron microscopic observation experiment.

The one-step growth experiment was developed by Ellis and Delbrück and depends on production of a synchronous phage infection (all phage-infected cells at the same stage in the infectious process at the same time). The synchronization is accomplished by either (1) restricting attachment of phage particles to bacterial cells to a short period of time or (2) first treating the bacterial culture with a reversible metabolic poison such as potassium cyanide and then adding the virions to the culture. In the latter case, the phages go through the early stages in their life cycle, but the cyanide prevents any macromolecular synthesis from taking place. In either case, after a suitable period of time, any phages that have not attached to host cells are eliminated either by diluting the culture to the point where collisions between phages and bacteria become improbable or by neutralizing the mixture with phage-specific antiserum. If cyanide is used, it is washed out of the culture and metabolism is allowed to start. The time of washing or dilution becomes the zero time, after which the experimenter removes samples from the culture at intervals, mixes them with indicator (phage-sensitive) bacteria, and assays for **infectious centers** (plaque-forming units or PFU).

A typical curve from such an experiment is shown in Fig. 6.3. The curve is triphasic, with an initial **latent phase** during which the number of infectious centers is constant, followed by a **rise phase**, and then a **plateau phase**. The interpretation of these results is that during the latent phase an infectious center is a phage-infected cell that lyses some time after being mixed with indicator bacteria and immobilized in agar. The soft agar ensures that released virions do not diffuse more than a few micrometers, and only a single plaque is formed for each originally infected cell. During the rise phase, infected bacteria begin to lyse before the sample is removed from the culture so that an infectious center may be either an infected cell or a free virion. During the plateau phase an infec-

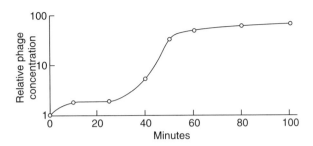

Figure 6.3. One-step growth curve. Bacteriophage T4 was added to a bacterial culture at time zero. After 10 min of adsorption, the entire culture was diluted 10,000-fold so that there was little probability of further collisions between virus particles and bacterial cells. When release of new phage particles began, the culture was further diluted tenfold to prevent a new step in the growth cycle. The period prior to 25 min is the latent phase, between 25 and 45 min is the rise phase, and later than 45 min is the plateau phase. The entire experiment was carried out at 37°C (Redrawn from Fig. 3 of Ellis, E.L., Delbrück, M. [1939]. The growth of bacteriophage. *Journal of General Physiology* 22: 365–384.)

tious center is only a free virion. The ratio between number of infectious centers during plateau phase and number of infectious centers during latent phase represents the average **burst size** (the average number of phage particles released per infected cell).

The **single burst experiment**, also devised by Ellis and Delbrück, can be applied to the study of individual infected cells rather than using averages obtained in a one-step growth experiment. The experimental goal is for each culture tube to have only one infected cell. After lysis is complete, all phage particles found in a tube result from a single burst cell. To achieve this goal, a culture of bacteria is infected with a small number of phages such that the **multiplicity of infection** (MOI, the average number of phages per bacterium) is less than 1.0. Then, the Poisson distribution can be used to describe the distribution of infected cells among small samples taken from a culture (see Appendix 1). If the sample size is appropriately chosen, the probability of obtaining two or more phage-infected cells per sample is low. Samples are diluted to reduce the cell density and then incubated for several hours. Any infected cells lyse during incubation, but released phage particles are unable to infect new cells because of the low culture density (for cultures with fewer than 10^6 bacteria/ml, the probability of collision between any two objects is almost negligible).

The data obtained by Ellis and Delbrück are presented in Table 6.2. Assuming one infected cell per tube, the range of individual burst sizes is enormous,

Table 6.2. Single burst experiment of Ellis and Delbrück.

Culture No.	No. of Plaques	Culture No.	No. of Plaques
1	0	21	0
2	130	22	0
3	0	23	0
4	0	24	53
5	58	25	0
6	26	26	0
7	0	27	48
8	0	28	1
9	0	29	0
10	0	30	72
11	123	31	45
12	83	32	0
13	0	33	0
14	9	34	0
15	0	35	0
16	31	36	0
17	0	37	190
18	0	38	0
19	5	39	9
20	0	40	0

Note: A culture of *E. coli* was infected with a dilute phage solution, allowed to stand for 10 min, and then diluted more than 100-fold. Aliquots of 0.05 ml were placed in separate tubes and incubated for 200 min. The entire contents of each tube were then used in a plaque assay.

varying from 1 to 190. The average burst size, as determined from the 15 infected tubes, is 883 total phage particles per 15 infected tubes, or about 59 phage particles per burst.

A more accurate determination of the average burst size is possible if one takes into account the probability that there was more than one infected cell per tube. This can be done using the zero case of the Poisson distribution to determine m, the average number of infected cells per tube (which works out to be 0.47). Because 40 tubes were used in the experiment, the number of infected cells expected in the entire experiment is $40 \times 0.47 = 18.8$ infected cells, which would have been expected to be distributed among the 15 tubes that produced phage particles. The excess of the calculated number of infected cells over the observed 15 infected tubes suggests the existence of samples with two or more infected cells (which are expected to occur 8% of the time). The new average burst size then becomes 883 total progeny phage particles per 18.8 virions that infected the cells, or approximately 47 phages per infecting virus. Both values for the average burst size are in accord with values determined in one-step growth experiments. Phage infection is apparently a complex process and is subject to numerous outside influences that affect its efficiency because even if the highest three values of Table 6.2 are discarded on grounds that they represent multiple infections, the burst sizes still range from 1 to 83.

The **premature lysis experiment** was designed by A.H. Doermann, who took a standard one-step growth experiment and altered it. In this experiment the samples removed at various times are treated to lyse all cells in the culture. The lysates can then be assayed for the presence of free phage particles or their components. Cells can be lysed by shaking with chloroform or by superinfecting the culture with a high multiplicity (~100) of T6 phages whose simultaneous attempts to infect cells causes those cells to lyse before any T6 infection can be initiated (**lysis from without**, in contrast to lysis from within, caused by a completed phage infection).

A graph similar to that obtained by Doermann is shown in Fig. 6.4. Under experimental conditions, no infectious T4 phage particles could be detected until about 12 min postinfection. This particular phase of the growth cycle, a period when there are no detectable infectious phages in the culture, is referred to as the **eclipse phase**.

An experiment of Hershey and Chase, using phage particles carrying different radioactive labels in the protein (^{35}S) and DNA (^{32}P), indicated that only phage DNA enters a host cell. A premature lysis experiment demonstrates that pure phage DNA is not infectious and the length of the eclipse phase, therefore, must represent the minimum time required to form a new protein coat for the virus. When infectious phage particles do appear, their rate of production is not exponential, as with the release of phages in the one-step growth experiment but arithmetical. The arithmetical growth rate during the remainder of the latent period indicates that infectious phage particles appear singly, as though they were produced by an assembly-line type of process, rather than by some sort of fission as would be observed in a bacterial culture in which two new cells arise from an old one.

Electron microscopy has proved to be a powerful tool for understanding interactions between phages and bacterial cells. Two general types of procedure can be used. Free phages, or portions thereof, can be examined by negative staining techniques such as that used to prepare the specimen shown in Fig. 6.1. Partially purified fractions of cells or phages, such as pieces of cell wall or **ghosts** (phage particles devoid of DNA), can be interacted with the intact attachment

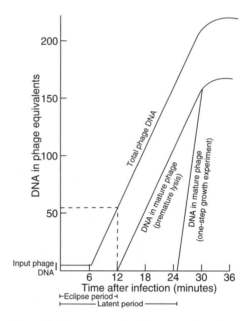

Figure 6.4. Idealized graphs illustrating the kinetics of synthesis of phage DNA in bacteria infected with one of the T-even phages. The amount of DNA observed in various experiments is expressed as phage equivalents, which in the case of the DNA found in mature phage correspond to the actual number of infectious virions present. "Total phage DNA" refers to both the DNA within the virions and the DNA in the bacterial cytoplasm. "DNA in mature phage" refers to those molecules found within infectious virions. (From Hayes, W. [1968]. *The Genetics of Bacteria and Their Viruses.* Oxford: Blackwell Scientific.)

site to study early events in the adsorption of virions to the bacteria. Phage-infected cells can be thin-sectioned to demonstrate internal interactions of virus components with the cell wall and membrane.

Rabinovitch et al. (1999) have tried to develop a series of mathematical expressions to describe T4 development under various growth conditions. Their model can account for >95% of the observed experimental variations.

Genetic Organization of T4

Types of Mutations Observed in the T4 Genome

Most biochemical reactions of the eclipse phase have been defined genetically by appropriate mutations. In nearly all cases these mutations have turned out to be conditional, usually chain terminators, as the normal enzymatic activities are required for successful infection. These mutations, although indubitably useful to a geneticist, produce phenotypes that are frequently difficult to assay. Consequently, the earliest genetic studies relied on mutations that affected general physiologic traits and gave rise to readily observed phenotypes. These mutations are usually referred to as **plaque morphology mutations**.

A standard T4 plaque has a clear center with a halo or turbid area surrounding it. The halo is an area where lysis was not complete. Some cells lysed, but many others did not. This failure is due to a phenomenon called **lysis inhibition**, which results when an *E. coli* strain B cell that has been infected by a T4 phage is **superinfected** (reinfected) by another T4 phage prior to cell lysis. The new phage delays (inhibits) lysis of the host cell so that intracellular production of phage may continue for 60 min or more. The burst size, of course, is enormous and may be as large as 1000 phages per cell. Hershey isolated mutant T4 phages called *r* (for rapid lysis) that were not susceptible to lysis inhibition and were therefore producing plaques with sharp edges (Fig. 6.5). Slavcev and Hayes (2003) have shown that lysis inhibition depends on normal tmRNA function, suggesting that there are conditions under which the translation process stalls.

Progeny from crosses between *r⁺* and *r* phages are easy to categorize phenotypically, as all that is required is the inspection of the resulting plaques. The experiments by Benzer, which are discussed later, dealt exclusively with rapid lysis mutants.

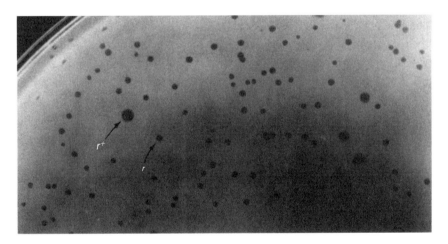

Figure 6.5. The *r* phenotype of phage T4B. Note the several large plaques with fuzzy edges that were produced by wild-type phages. The small, sharp-edged plaques were produced by *rIIA* phages.

THINKING AHEAD

How does a geneticist decide whether two genes are close together or far apart on a DNA molecule?

Genetic Crosses

The basic strategy for phage crosses is to infect individual cells simultaneously with two or more genetically distinct phages, allow the viruses to complete their life cycle and lyse the host cells, and then examine the genotypes of progeny phages. The recombination frequency is calculated as the quotient of the number of recombinants divided by the number of minority parental phages recovered, assuming no selective pressure is applied. The tacit assumption is that the greater the genetic distance between any two mutations, the more frequently genetic exchanges occur between them. Therefore, closely linked markers show little recombination (are inherited simultaneously), whereas the unlinked markers exhibit the 50% frequency characteristic of independently inherited traits.

Table 6.3 presents data from a typical set of T4 crosses. The three mutations, r47, r51, and tu41 (a temperature-sensitive r mutation), appear to be located near one another and tend to comprise one **linkage group** (i.e., a group of markers that tend to be coinherited). The gene order for the linked mutations is r47-r51-tu41, with greater genetic distance between tu41 and r51 than between r47 and r51. Each member of this linkage group recombines at equal frequency with tu44, which is therefore considered to lie in a second linkage group. If it did not, either r47 or tu41 would be expected to show a lower recombination frequency than the rest of the members of that linkage group. Using the data presented, it is impossible to tell whether tu44 lies to the right or to the left of the other mutations.

Two important problems may arise when certain types of mutants are used in phage genetic crosses. One is that many mutant phenotypes are controlled by a group of nonallelic genes, implying that a mutation at any one of several independent sites produces the same phenotype. A good example is the case of the r mutations just discussed. Although r47 and ts44 are unlinked (and thus not identical), they confer the same phenotype. The second problem is that using mutations affecting the adsorption characteristics of a phage may cause phenotypic mixing. **Phenotypic mixing** results in a virion whose DNA does not

Table 6.3. Results of some T4 crosses.

Cross	Percentage Recombination	Cross	Percentage Recombination
r47 × *r51*	5.4	*tu44* × *r47*	36.1
r51 × *tu41*	23.7	*tu44* × *tu41*	37.0
r47 × *tu41*	25.0	*tu44* × *r51*	38.8

Deduced map order

Note: Two *r* mutants and two turbid halo (*tu*) mutants were subjected to pairwise crosses and the percentage of recombinant progeny was determined by examining the plaque morphology shown by progeny phages. It is not possible to use these data to determine the map position for *tu44* (see text). (Data from Doermann, A.H., Hill, M.B. [1953]. Genetic structure of bacteriophage T4 as described by recombination studies of factors influencing plaque morphology. *Genetics* 38: 79–90.]

code for all of the virion proteins encasing it. The existence of an eclipse phase indicates that virions are assembled from pools of constituents rather than by division of preexisting virus particles. Therefore, in a cell that has a mixed infection (infection from two genetically distinct phages), it is possible that the DNA that codes for one phenotype is accidentally packaged into a phage particle with a second phenotype. If the resultant particle is taken through more than one cycle of infection, as in a standard plaque assay, the discrepancy between genotype and phenotype makes no difference. After the next infection cycle, the DNA will be correctly packaged and have the appropriate genotype for the virion proteins. However, if the experimental protocol calls for a phenotypic selection during the first growth cycle, it is possible that phage particles of the correct genotype will, nonetheless, not be recovered because they have the wrong phenotype.

Using calculated recombination frequencies as shown in Table 6.3, it would be possible to construct a genetic map for T4, although pairwise matings between all mutations in a given region would be required for the determination of the map order. Before such extreme measures became necessary, Seymour Benzer developed a mapping system that could be used to reduce the number of pairwise matings to a few or none and yet allow the unambiguous ordering of each mutation on the genetic map.

Benzer began by adapting the standard **cis–trans test** of eukaryotic genetics for use with bacteriophages. He infected bacterial cells with different mutant phages having the same phenotype (one that would normally prevent their growth) and investigated whether phage infection was successful (Fig. 6.6). This step was the *trans* portion of the test because the mutations were located on different DNA molecules. The *cis* portion of the test was a control experiment that consisted of infecting a cell with one phage that carried both the mutations and another that carried none. In the *cis* case, wild-type phages were expected to grow and lyse their host cells unless mutant proteins interfered with normal function. In the *trans* case, if the mutations affected different genes, each phage contributed a different functional gene product. Here, **complementation** occurs because the combination of all viral proteins in the cell includes a functional member of every type (the success of the *cis* test has already shown that cells having a mixture of normal and mutant phage proteins can produce virus

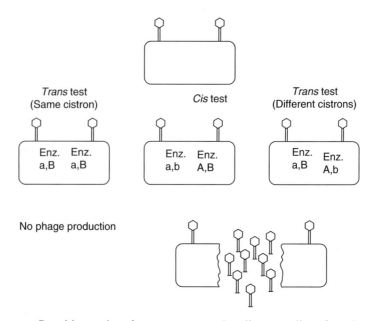

Figure 6.6. Possible results of a *cis–trans* test. A cell is initially infected with two genetically distinct phage particles. In the second row are shown three arrangements of mutations among the infecting DNA molecules, and in the third row are the two possible results of the simultaneous infection. The site of a mutation on the phage DNA is indicated by an X. Mutant proteins are indicated by lowercase letters, and functional proteins are indicated by uppercase letters. Necessary assumptions are that protein products manufactured from the genetic information are freely diffusible within the cell and that functional A protein and functional B protein are both required for production of new virions and cell lysis.

particles). However, if both the mutations affect the same gene, the combination of all viral proteins does not include a functional version of one particular protein, and no phage is able to grow. If complementation occurs during the *trans* portion of the test, the *cis–trans* test is considered positive; if it does not, the test is considered negative.

The fundamental *cis–trans* test assumes that no recombination occurs. If recombination does occur between the two mutations used in the *trans* test, the results may be somewhat muddled but can be sorted out by an examination of the phenotypes of progeny phages. Complementation gives rise only to parental types of phages, whereas recombination produces some wild-type phages.

The results of a large number of *cis–trans* tests were sufficiently unambiguous for Benzer to propose a new term, **cistron**, to describe the region of a genome within which all mutations give negative *cis–trans* tests. The term cistron is much more precise than the term gene, as the latter has no operational definition. However, the situation is not as clear-cut as Benzer had hoped, and examples of discrete polypeptide-encoding regions that do not complement are known. The term gene has not disappeared and probably will not. In most cases when the term gene is used, it is possible to substitute it with the term cistron, and it is in this way that the two terms have been used in the remainder of this book. Thus, the "one gene–one enzyme" hypothesis of Beadle and Tatum could also be the "one cistron–one polypeptide" hypothesis, and so on.

After the assignment of mutations to specific cistrons, the next hurdle was to order mutations within each cistron. During the demonstration of this procedure, Benzer dealt specifically with the *r*II region of T4, which he showed to consist of two cistrons, but the technique has general applicability. All that is required is a large number of deletion mutations in the region of interest. Deletion mutations can be readily identified because they can never revert to wild type. That is, they suffer from a lack of genetic information rather than an alteration. If two deletion mutations overlap (i.e., have lost the same portion of a genetic coding region) it is impossible to recombine the two deleted DNA molecules to produce a wild-type DNA molecule. Therefore, in a cross between two overlapping r deletion mutants, no r^+ progeny would be expected; whereas if there were no overlap, r^+ progeny would be expected. By a similar argument, when a point mutation (missense, nonsense, frameshift) is crossed to a deletion mutation, no r^+ recombinants are possible if the point mutation lies within the deleted region. If it lies outside the deletion, r^+ recombinants are produced at a frequency that depends on the distance between the point mutation and the nearest end of the deletion. The greater the distance, the more recombinants are expected.

Benzer's deletion-mapping protocol requires a series of overlapping deletions whose map order has been determined by crossing them to known point mutations and to each other, in a manner as described in Table 6.3 (realizing that it is impossible to distinguish left from right because there is no landmark such as a centromere on the DNA molecule). The particular set of deletions used by Benzer is displayed in Fig. 6.7. The upper group of deletions divides the two cistrons of the rII region in seven discrete segments. The lower group of deletions can be used to subdivide any required segment. Crosses of point mutations and deletions could be carried out quickly by means of a spot test. All rII mutants grow on *E. coli* strain B, but not on *E. coli* strain K-12 (λ^+) (a phage lambda lysogen). All r^+ phages grow on either strain. Various T4 phage lysates prepared on strain B can be tested in a pairwise manner by spotting phages carrying a point

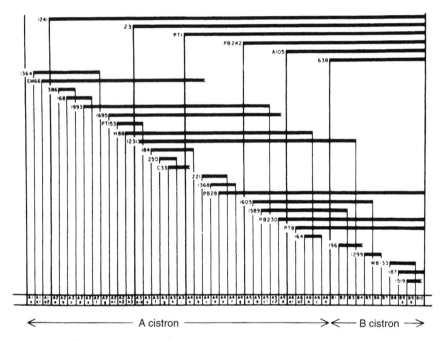

Figure 6.7. Deletions used to divide the rII region of the T4 genome into 47 small segments. Each horizontal line represents the material that has been deleted in a particular mutant. If the end of the deletion is shown as fluted, its position is not precisely known, and it has not been used to define a segment. The *A* and *B* cistrons, which are defined by the *cis–trans* test, coincide with the indicated portions of the recombination map. (From Benzer, S. [1961]. On the topography of the genetic fine structure. *Proceedings of the National Academy of Sciences of the USA* 47: 403–415.)

mutation and a deletion mutation together on a plate with a lawn of *E. coli* K-12 (λ^+). If recombination yields r^+ progeny, lysis occurs and a clear area develops in the lawn, which can be scored visually (Fig. 6.8).

Using his deletion-mapping procedure, Benzer was able to assign a given *r*II mutation to one of the 47 segments on the T4 DNA molecule in just two experimental steps by first crossing it to the seven large deletions and then to some of the smaller deletions. Final map order was obtained by two- and three-point (marker) crosses as given above. By the time the work was completed,

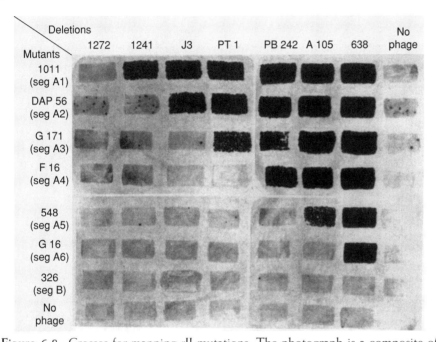

Figure 6.8. Crosses for mapping *r*II mutations. The photograph is a composite of four plates. Each row shows a given mutant tested against the reference deletions shown at the top of Fig. 6.7. The test consisted of allowing the two mutants to infect *E. coli* B and then using a paper strip to spread some of the culture onto a lawn of *E. coli* K cells that had been spread on the plate. Under test conditions, neither parental phage was able to grow on *E. coli* K. However, if recombination occurred during one cycle of growth in *E. coli* B, the progeny phage would be able to grow on *E. coli* K, and would manifest itself as a clear (dark) area on the plate. Isolated plaques appearing in the blank spaces represent spontaneous revertants in the phage stock. As an example of the interpretation of the data, mutant 1011 formed no *r*II+ recombinants with deletion 1272 but did with all other deletions tested. Therefore, the mutant site in strain 1011 is included within the region uniquely deleted in strain 1272 (i.e., the region at the far left in Fig. 6.7). (From Benzer [1961].)

Benzer had mapped more than 2400 independently derived mutations (i.e., from more than 2400 separate cultures), representing 304 sites on the T4 genome. A map presenting the altered sites and frequency distribution of 1612 spontaneous mutations is shown in Fig. 6.9. This was genetic analysis in the purest sense because Benzer completed the entire study without knowing anything about the biochemistry of the gene products he was studying. In fact, more than 10 years elapsed before two proteins appearing in the cell wall of infected bacteria were identified as the products of the *r*II cistrons.

Benzer's work still stands as the ultimate in genetic fine structure analysis, and it is never likely to be equaled. Today a study of this type would not be necessary because it is actually easier to sequence the DNA molecule and use computer programs to identify open reading frames (i.e., regions that should code for

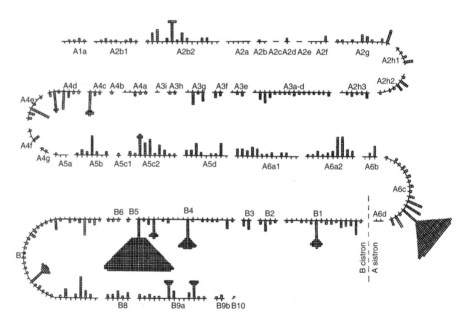

Figure 6.9. Topographic map of the *r*II region for spontaneous mutations. Each mutant arose independently in a plaque of either standard-type T4B or various *r*II mutants. Each square represents one occurrence observed at the indicated site. Sites with no occurrences indicated are known to exist from induced mutations and from a few other selected spontaneous mutations. Each segment of the map is defined by combinations of the deletions shown in Fig. 6.7. The arrangement of sites within each segment is arbitrary but could be absolutely established by crosses such as those shown in Table 6.3. (From Benzer [1961].)

proteins). However, in addition to providing a thorough genetic map of the *r*II region, Benzer developed several other important concepts from an analysis of his data. Although it is commonly accepted today that genetic maps are linear structures, the question of linearity was by no means resolved when Benzer began his experiments. In fact, until electron microscopy of DNA molecules came into vogue, maps such as the one shown in Fig. 6.9 were the only real evidence for linearity.

By comparing the frequency distribution of mutations at different sites in the *r*II region in a manner similar to that of Fig. 6.9, Benzer showed that spontaneous mutations tend to occur at sites other than those of mutagen-induced mutations. Moreover, it is possible to test if the mutations are randomly distributed over the cistrons by comparing predicted Poisson distributions with the observed distribution of mutant sites. Highly mutable sites occur too frequently, indicating that mutations are not entirely random, but sites represented only once or twice in the distribution fit the Poisson distribution well. Because the sizes of the $r = 1$ and $r = 2$ classes were known, Benzer was able to use the Poisson distribution to estimate the probability of the $r = 0$ case, the number of sites within the cistrons at which no mutations had been observed. The calculated value was 28%, giving a map saturation value of 72% (i.e., 72% of all possible mutable sites had been identified). These data, taken in conjunction with some estimates of the physical size of the *r*II region, provide an experimental basis for the tacit assumption that a **muton** (the smallest mutable unit of a genome) is a single base. Moreover, they indicate that a **recon** (the smallest unit of recombination) is also a single base.

Using deletion mapping, pairwise crosses, and some physical mapping techniques (see Chapters 7 and 8), it is possible to construct a genetic map for T4 (Fig. 6.10). The zero point is taken as the junction between the *r*IIA and *r*IIB cistrons, as they have been precisely mapped. One map unit represents 1% recombination in a standard phage cross, with the proviso that map distances greater than 1.0 unit (1% recombination) are generally not additive. For example, two markers separated by 9.5 map units actually have a recombination frequency of 4% (see the review article by Stahl [1989]). The fact that the genetic map is circular but the DNA inside a virion is linear can be explained by the physical structure of viral DNA during infection, which is discussed in the next section. Individual proteins are often designated according to the gene (cistron) that encodes them. Thus, the protein product of gene 32 is known as gene product 32 or gp32.

The *r*II data forced a reassessment of the traditional genetic term 'allele'. Two or more markers that affect the same phenotype and map at the same locus

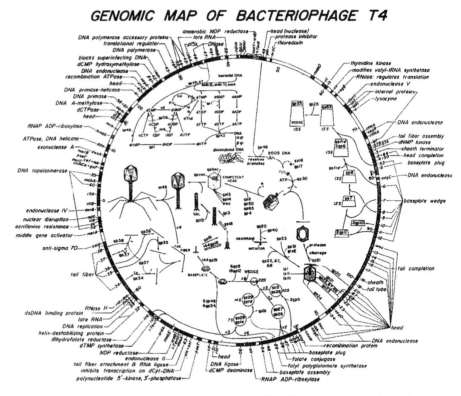

Figure 6.10. Genomic map of bacteriophage T4. This map is based on the complete DNA sequence. The diagrams in the center of the map show the key pathways for assembly of the virus. (Figure courtesy of Burton Guttman and Elizabeth Kutter, The Evergreen State College, Olympia, WA 98505.)

are considered allelic in both a functional and a structural sense. However, the data of Benzer clearly demonstrated that functional alleles (those occurring in the same cistron and thus affecting the same phenotype) were not necessarily structural alleles because they could be separated by recombination.

Visconti–Delbrück Analysis

Visconti and Delbrück endeavored to provide a mathematical model of the recombination process in T4 as it occurs in crosses such as those described in Table 6.3. They began with four basic observations:

1. If cells with mixed infections are separated from one another in a single burst experiment, it can be shown that both recombinant and non-recombinant (parental-type) phages are released in the same burst. This result implies that recombination can occur after DNA replication has begun because if the two parental DNA molecules recombined in a reciprocal fashion before replication, only recombinant progeny could result.

2. In a triparental cross, recombinant phages can carry markers from each parent, which could result only if the DNA molecules recombined more than once.

3. If a mixed infection is set up with one parent definitely being in the minority, it is possible to find more recombinant phages than there were minority parents. This result again suggests that genetic exchange occurs repeatedly and also after the DNA replication has begun. Note that this result invalidates the assumption made in Chapter 1.

4. If the number of recombinant phages observed after premature lysis is plotted as a function of the sampling time, the resulting curve shows a drift toward genetic equilibrium rather than a continuous increase. Therefore, recombination occurs at all times during the infectious cycle and may include exchanges between recombinant DNA molecules that regenerate parental DNA molecules.

The model developed to explain these observations assumed that replicating (vegetative) DNA generated a large pool of T4 DNA molecules within the cell. For any given DNA molecule within the pool, three processes might occur: It might replicate again; it might be packaged up into a phage particle (**maturation**); or it might undergo recombination before replicating or maturing. Because premature lysis experiments indicated that the pool of DNA molecules reaches a certain size and then remains constant, it was assumed that replication and maturation occurred at the same rate. It was further assumed that any recombination that occurred was random with respect to time and partner.

The previous section discussed the T4 genetic map and pointed out that 1% recombination was equal to 1 map unit. Another way to devise a map would be to measure distance in terms of the average number of recombination events (genetic exchanges) that occur between two markers, regardless of whether the recombination resulted in a phenotypic change. Then a "true linkage" would be expressed by the parameter l, which is equal to the average number of genetic exchanges that occur between a pair of markers. If genetic exchanges are distributed randomly, the Poisson distribution can describe their distribution. However, as J.B.S. Haldane pointed out, only an odd number of genetic exchanges between two markers produces a phenotypically recombinant DNA molecule, as an even number of exchanges has no effect on the linkage of flanking markers. Therefore, the probability of a phenotypically expressed recombinant, given two markers x and y a distance l apart, is just the sum of the odd terms of the Poisson distribution. Thus, when l is large, the recombination frequency K approaches 50%. However, when l is small (i.e., the expected number of exchanges is less than 1.0), most samples contain either 1 or 0 exchanges and K is approximately equal to l. Of course, when l is zero, K is also zero because if two mutations have occurred at an identical site, they cannot recombine.

An equation can be set up to describe the number of recombinant phages inside a cell. Each time two nonidentical DNA molecules undergo a genetic exchange (a mating), either new recombinants are formed or old recombinants are lost. If there are M rounds of mating, then, in the small interval dM, the increase in recombinants is $(1/2)(1 - R_{xy})^2 K_{xy} dM$; where R_{xy} is the proportion of recombinants for markers x and y in the population, and K_{xy} is the probability of an exchange occurring between the markers if a mating does take place. The factor $1/2$ is necessary because in an equal-input cross, only half of the matings are between nonidentical DNA molecules. By a similar argument, the decrease in recombinants is $(1/2)R^2_{xy} K_{xy} dM$. Then, the change in the proportion of recombinants is

$$dR_{xy} = (1/2)(1 - R_{xy})^2 K_{xy} \, dM - (1/2)R^2_{xy} K_{xy} \, dM \qquad (6.1)$$

Integration of Eq. (6.1) gives

$$R_{xy} = 1/2\left(1 - e^{-K_{xy} M}\right) \qquad (6.2)$$

Unfortunately, the only observable quantity in the equation is R_{xy}. If the markers x and y are chosen so that they are far apart, K approaches 0.5 and it is possible to solve for M after measuring R_{xy}. When this step is accomplished, the values of M range from 2 to 5. Because five to eight rounds of DNA replication are necessary to account for the observed amount of T4 DNA in a cell, the number of rounds of mating (the average number of times a particular DNA molecule can be expected to undergo recombination) approximates the number of rounds of replication.

THINKING AHEAD

How would you interpret a T4 phage plaque whose margin is partly very sharp and partly fuzzy?

Phage Heterozygotes

An important clue to the structural nature of T4 DNA was discovered accidentally during an examination of the progeny of r^+ and r matings. When the plaque morphologies were scored, about 2% of them were mottled (parts of the margin looking sharp like an r mutant and the remainder with the turbid halo characteristic of r^+). When the phages comprising a mottled plaque were retested, they were observed to be a mixture of r^+ and r, each of which subsequently bred true. The original phages that gave mottled plaques were thus behaving either as **hets** (heterozygotes), in which segregation was occurring after infection, or as cases of multiple infection of an indicator bacterium by genetically distinct phages.

The possibility of dual infection was easily ruled out by repeating the cross and subjecting the progeny phages to various doses of ultraviolet radiation (UV) before plating. Phages exhibit characteristic inactivation kinetics under these conditions. A particular dose of UV destroys the infectivity of a certain percentage of virions. However, if two viable phage particles are required to form a heterozygous plaque, the number of such plaques should decrease as the square of the probability of inactivating a single virus because the inactivation of either phage particle would prevent production of a mottled plaque. In fact, the phage hets were inactivated at exactly the same rate as the rest of the population and thus were not due to multiple infections.

A careful examination of phage hets showed that they could occur for any pair of markers and that the length of the heterozygous region was on the order of a few genes. For a cross in which the heterozygous locus on a genome was flanked by other markers (e.g., $a\ r\ b$), two types of hets were observed: those that were recombinant for the flanking markers and those that were not. Some of the recombinant hets were apparently true recombination intermediates that were accidentally packaged into phage particles (strand-mismatch hets; Fig. 6.11, bottom left). This type of intermediate could be trapped inside a phage particle if a molecule where strand invasion was complete but replication had not yet occurred was packaged. However, the remaining hets were due to an

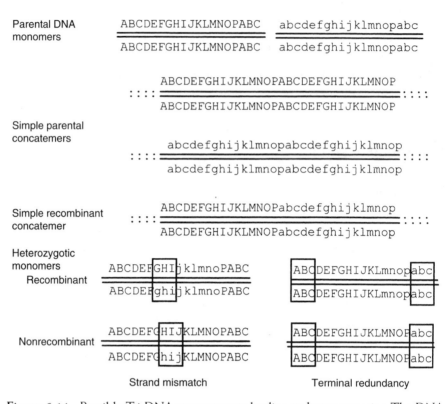

Figure 6.11. Possible T4 DNA arrangements leading to heterozygosity. The DNA monomers are shown as they would be isolated from a virion with their terminal redundancies intact. The DNA sequence of each monomer strand is indicated by letters, with lowercase and uppercase being used to identify genetically different markers. Simple concatemers are assumed to be assembled by recombination in the terminally redundant region. Several types of recombinant concatemers are possible, although only one is shown. The heterozygotic monomers are assumed to have been cut from concatemers such as those shown above or others generated by recombination between simple concatemers. If the recombination process had not completely finished prior to the cutting of the concatemer, strand-mismatch heterozygotes would be generated. If the recombination event that generated a strand mismatch consisted only of formation of a D-loop (see Chapter 5), a nonrecombinant heterozygote would be generated. It is particularly important to note that in this case the term nonrecombinant refers not to the region of heterozygosity, but rather to the genetic markers flanking the heterozygotic region.

unusual feature of T4 DNA: **terminal redundancy** (e.g., if there were nine numbered genes comprising the T4 genome, their sequence might be 1234567891'2'). A terminally redundant DNA molecule is essentially an incomplete concatemer (Fig. 6.11). Phage hets could arise in terminally redundant DNA molecules if the redundancy came from a different parent than the bulk of the molecule. Again, there could be obvious evidence of recombination or merely a simple end-to-end joining.

Today, terminal redundancy would be confirmed by sequencing the DNA molecule in question. At the time under discussion, that technique was not an option. Nevertheless, using electron microscopy of nuclease-treated DNA it was possible to show that about 2% of the genetic information present at the left-hand end of a linear T4 DNA molecule is repeated at the right-hand end. Further information about the physical structure of T4 DNA can be obtained by making random heteroduplexes. This procedure involves denaturing (breaking the hydrogen bonds of) a mixture of T4 DNA molecules and then allowing the separated strands to reanneal (form hydrogen bonds with one another). Regions of DNA that are identical form a perfect double helix. Regions of nonidentity form various types of loops or single-stranded tails. If T4 DNA molecules were all identical, the experiment would yield perfect DNA duplexes. However, the experiment actually produced structures such as the ones diagrammed in Fig. 6.12, indicating that T4 DNA molecules are not identical. The apparently random length of the observed single-strand tails indicates that T4 DNA molecules are not only terminally redundant but also circularly permuted.

Conceptually, circular permutations in DNA could be generated by having a sequence of genes or DNA bases in the form of a closed circle and then cutting the circle at different points. Each new cut generates a linear molecule representing a different circular permutation. Note, however, that such a model will not generate a terminally redundant molecule. The existence of circularly permuted, terminally redundant DNA molecules offers another explanation for how a linear DNA molecule can give rise to a circular genetic map. Individual molecules may circularize, or different permutations may recombine to give longer-than-unit-length molecules. The important point is that because the DNA has no fixed end points, neither can the genetic map. Although it is not difficult to conceive of mechanisms by which circular permutations might be generated in vivo, it is less obvious how terminal redundancy can be produced. The answer to this question lies in the way in which T4 DNA is replicated and packaged.

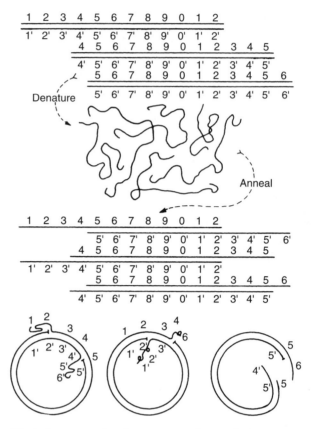

Figure 6.12. Circle formation by denaturing and annealing a permuted collection of duplexes. Each horizontal line represents a single strand of DNA. The numbers are used to indicate the arrangement of different pieces of genetic information. Note that each permutation is also terminally redundant. These redundant ends cannot find complementary partners during circle formation (bottom) and are left out of the resulting duplex molecule. The separation of the single strands within any circular duplex depends on the relative permutations of the partner chains. (From MacHattie, L.A., Ritchie, D.A., Thomas, C.A., Jr. [1967]. Terminal repetition in permuted T2 bacteriophage DNA molecules. *Journal of Molecular Biology* 23: 355–363.)

DNA Replication

It is obvious that normal replication, as envisioned by Watson and Crick, does not suffice to produce DNA molecules with the properties attributed to T4. From a conceptual standpoint, the easiest way to produce terminal redundancy is to have a DNA molecule that consists of two or more genomes linked end to end (a concatemer; Fig. 6.11) and then to cut the appropriately sized piece of DNA to fit the phage head from the concatemeric structure. This mechanism is apparently the correct one, as the size of the packaged DNA molecule is 2% larger than the sequenced genome.

The problem that arises is how to produce a concatemeric structure. One method known to be used by viruses is the rolling circle DNA replication mechanism (see Chapter 7). It is used by phages φX174 and λ. However, because T4 DNA molecules do not need to circularize in order to replicate, this mechanism is not appropriate.

An alternative model proposed by Mosig and coworkers to describe how linear concatemers can develop is shown in Fig. 6.13. Basically, the model assumes that there is a single replication origin that starts the usual bidirectional

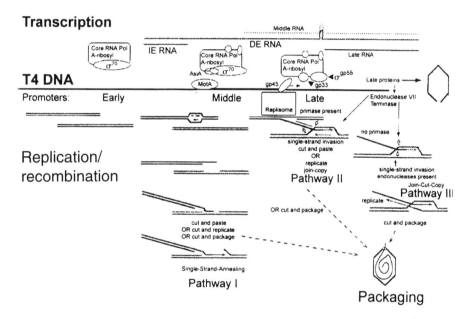

Figure 6.13. Cartoon representation of different T4 recombination mechanisms and programmed gene expression, DNA replication, and packaging. The upper panel shows transcripts initiated from early, middle, and late promoters by sequentially modified forms of host RNA polymerase that recognize these promoters. Hairpins in several early and middle transcripts inhibit early translation of several late genes that are cotranscribed from early or middle promoters. (From Mosig, G. [1998]. Recombination and recombination-dependent DNA replication in bacteriophage T4. *Annual Review of Genetics* 32: 379–413.)

replication immediately after infection. Unlike the case with a circular molecule, when a replication fork reaches the end of a DNA molecule, it is generally not able to complete replication of the lagging strand because there is no site located right at the end of the leading strand at which a new primer could be synthesized. This single-stranded region of the leading strand is recombinogenic (see Chapter 5) and stimulates exchange between two DNA molecules via D-loop formation. The various breaks found in recombination intermediates (e.g., see Fig. 5.7) act to prime additional replication forks, and a large, multiply branched concatemer begins to develop.

Mosig (1998) distinguished three replication/recombination pathways (Fig. 6.13). Pathway I is a single-strand annealing pathway, where the single strands result from exonucleolytic "erosion" of DNA ends even in the absence of replication. Pathway II is a join and copy pathway in which DNA replication occurs from the normal origin and single strands result as noted earlier. Pathway III is a join–cut–copy pathway that functions even in the absence of primase. In this case, after single-strand invasion to give a D-loop, an endonuclease cuts the nonlooped DNA strand to provide a new free end. Any one of these pathways can initiate replication or recombination under the appropriate conditions.

T4 DNA replication begins about 6 min after infection at 30°C, with the first concatemeric structures appearing several minutes later. The initial priming event is catalyzed by host RNA polymerase, but the modifications to the polymerase noted below require that the mechanism of Mosig be invoked for later replication initiations. RNA primers for Okazaki fragments are synthesized by a combination of T4 RNA primase (gp61) and T4 DNA helicase (gp41) using the template sequences GTT or GCT. The primers are four to five bases long and always start at the middle base of the recognition sequence. The number of phage genome equivalents in the cell increases to 40–80 by 12 min, when the maturation begins. Equilibrium is quickly achieved, so the rate of maturation is approximately equal to that of replication, causing the size of the DNA pool to remain constant even though the number of infectious phage particles is increasing.

Molecular Biology of T4 Phage Infection

Initial Stages of Infection

Using information derived from experiments such as those described earlier in the chapter, it is possible to present a reasonably clear picture of the course of

T4 infection. These events are summarized in Fig. 6.14. The activity begins when the tips of the tail fibers (gp37) attach reversibly to the core polysaccharide of lipopolysaccharide chains located on the outer membrane of a Gram-negative cell. One common model proposes that the six tail fibers bind and release randomly so as to allow the virion to "walk" across the surface of a cell until an appropriate site is reached and binding of the virion becomes irreversible. The tail fibers bend to allow the baseplate to touch the cell surface. Electron microscopy shows a **zone of adhesion** (physical connection) between the inner and outer cell membranes at the site of phage attachment, although it is not known whether the zone is caused by the presence of the phage or is preexisting.

When a baseplate touches such a zone, the tail sheath contracts and the plug located in the center of the baseplate (see Fig. 6.1) is removed, allowing the DNA to be ejected from the head. The ejection process is triggered when the tail core touches the cell membrane and requires gp5 and gp27.

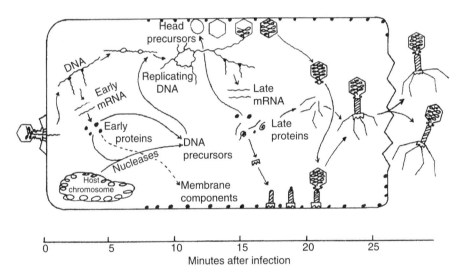

Figure 6.14. Temporal sequence of events during a T4 infection. The infecting virion is assumed to attach to the left-hand end of the cell at time zero. The position of each event or structure along the horizontal axis of the cell is reflective of the time at which it usually occurs. Vertical displacements are used to indicate the existence of separate pools of assembled head and tail substructures. (From Mathews, C.K. [1977]. Reproduction of large virulent bacteriophages, pp. 179–294. In: Fraenkel-Conrat, H., Wagner, R.R. [eds.], *Comprehensive Virology*, vol. 7. New York: Plenum Press.)

Kanamaru et al. (2002) have described their structure and prepared a movie showing rotation of the tail sheath and extrusion of the tail core during infection. The shape of the head is unchanged throughout. For the viral DNA to actually enter a cell, the normal, inwardly directed proton gradient across the cell membrane must be intact. Indeed, phage ghosts are capable of significantly depolarizing a cell membrane. It thus seems that cells cooperate in their own infection. However, phages with contracted tails can infect cells without cell walls and with no proton gradients, so the proton gradient is not directly used for transport across cell membranes.

THINKING AHEAD

What are the means that a virus might use to change the pattern of transcription as it prepares to assemble new virus particles?

Regulatory Issues

The major goal of gene regulation in a virulent phage is to ensure that its DNA has sufficient time to replicate before virus assembly begins. Clearly, if maturation begins too soon, burst size will be adversely affected. Despite the small size of the virus, the gene density is twice as large as that for E. coli, about 95% of the DNA sequence coding for a gene product. Miller et al. (2003) have reported functions for only half of the identified open reading frames.

A clue to the regulatory strategy comes from observations made on phage-specific mRNA molecules. As phage DNA enters a cell, RNA transcription begins. Phage-specific RNA molecules are assigned to four temporal classes (Fig. 6.13) based on the time at which abundant synthesis first occurs at 30°C. Luke et al. (2002) have used microarray technology to characterize the traditional mRNA classes: **immediate early** (30 s), **delayed early** (2 min), **middle** (3 min), and **late** (12 min). In each case, the synthesis of the preceding group of mRNA molecules shuts off as the next group takes over. As might be expected, early transcripts are involved with establishing the infection and late transcripts with the assembly of progeny virions. The two groups of early RNA molecules include eight species of tRNA molecules not normally produced in large quantities by E. coli as well as sundry mRNA molecules. Nearly all early transcripts are complementary to only one strand of the phage DNA helix, the l strand or minus strand, which is defined as the strand most readily binding to an RNA copolymer composed of guanine and uridine residues (poly UG). By contrast, late mRNA is predominantly complementary to the r or plus strand. Middle mRNA is a mixture of unique transcription units and some early transcription units that can be expressed from two distinct promoters.

The difference between immediate early and delayed early RNA is one of genetic position. The genetic information for delayed early RNA is physically located downstream (3′) to the immediate early information, and between the two groups lies one or more ρ-dependent termination signals. There are two possible ways to permit transcription of delayed early genes, and both occur in T4-infected cells. One possibility is the location of a second promoter so as to allow direct transcription of delayed early genes but not immediate early genes. The mRNA molecules produced by this process would be shorter than those including immediate early genes and would be considered middle RNA. The other option is to prevent the terminator signals from functioning. This process is called **antitermination**. Both these processes are protein-based because if the antibiotic chloramphenicol is added to prevent translation of early mRNA, only short transcripts are produced. For more discussion of antitermination, see Chapter 8.

One of the early phage-specific proteins is Ndd, which causes disruption of the host nucleoid. The disruption is in the nucleoid structure, not a result of direct damage to the host DNA, because there is no induction of DNA repair systems. When cloned, *ndd* is lethal to the host cell unless it carries a temperature-sensitive mutation.

The host RNA polymerase undergoes major changes during the phage growth cycle, and these changes correlate with changes in the viral transcription pattern. The *alt* protein ADP-ribosylates an α subunit of RNA polymerase. The modified enzyme binds very strongly to early promoters, particularly those having an A-rich upstream element at position −42, called the UP element. Later during the infection, ModA protein (an early protein) modifies the other α subunit of RNA polymerase. Meanwhile the Alc protein causes termination of rapid transcription on cytosine-containing (host) DNA (Nechaev and Severinov 2003).

Other RNA polymerase changes also occur. Phage protein RpbA binds tightly to the polymerase core. An additional two proteins bind less strongly to the core, and a fourth protein binds either to the core or to normal *E. coli* σ factor (σ^{70}). T4 protein gp55 acts as a substitute for host σ factor. The cumulative effect of these modifications to the host transcription system is to prevent host promoters from being recognized while making phage middle and late promoters recognizable. One immediate effect of this change is that the virus can no longer initiate DNA replication in the normal way because the σ^{70} promoter used for primer synthesis is no longer available. Figure 6.15 shows the differences between the various types of T4 promoters. During the transitional stages, RNA polymerase may actually recognize more promoters than usual. Loss of the ability to recognize host promoters means that infected cells cannot express any new functions not encoded by T4 and thus cannot respond to changes in their environment in the manner discussed in Chapter 4.

True middle promoters do not have the typical −35 sequence but instead contain a special sequence centered at base −30 that is known as the MotA box (Fig. 6.15). These promoters are activated by the binding of MotA and its coactivator AsiA to the MotA box, somewhat analogous to the binding of CRP to the

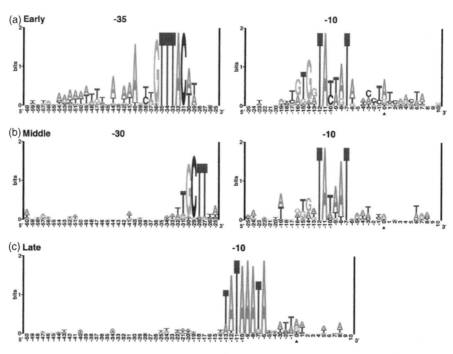

Figure 6.15. Logo of T4 promoters. Results of sequencing various T4 promoters are presented visually with the relative size of the letter indicating the proportion of promoters having that base in that position. The triangle marks the +1 transcription start site. (a) 39 early promoters; (b) 30 middle promoters; (c) 50 late promoters. (Reproduced with permission from Miller et al. [2003].)

lac promoter. In this case MotA is not a sigma factor but rather binds directly to σ^{70} and has been called an anti-σ factor. Orsini et al. (2004) report that neither AsiA nor the complete ribosylation of the α-subunit of RNA polymerase is responsible for the shutoff of early mRNA synthesis.

Synthesis of true late mRNA requires not only promoter recognition but also the presence of specific proteins needed for DNA replication as well as binding of another protein to an enhancer site. In the absence of these two pre-conditions, late mRNA is produced at low levels or is not produced at all. The late promoter sequence has only an extended −10 region as a conserved area (Fig. 6.15). The accessory replication proteins involved in late transcription include gp55 (σ^{55}), gp33, and gp45, which form a complex that facilitates formation of a particularly stable open promoter complex. Proteins gp44 and gp62 form a complex that acts to load a gp45 trimer onto the DNA molecule at an enhancer site, where it forms a sliding clamp to hold the replisome in place (Fig. 6.16). An enhancer site in phage T4 is not the same as an enhancer sequence in a eukaryotic cell. Instead of being a specific sequence of bases, it is rather a nick or gap in the DNA that can be located either upstream or downstream from the promoter. The accessory replication proteins bind to the enhancer, and RNA polymerase with attached gp33 and σ^{55} then binds to the accessory proteins. Kamali-Moghaddam and Geiduschek (2003) have shown that the open complex formed by the holoenzyme complex is remarkably stable.

Because the accessory replication proteins can also be found at replication forks, a replication fork can be considered as a mobile enhancer. The sliding clamp is not stable and must be renewed for each transcript. The enhancer can either be 1000 bp upstream or downstream of the promoter and still have its effect. The model for activation of late RNA transcription is shown in Fig. 6.16.

T4-infected cells demonstrate a phenomenon called RNA splicing, which was originally thought to be unique to eukaryotic cells. T4 codes for at least four distinct proteins that contain an intron: *td* (thymidylate synthase), *nrdB* (ribonucleotide reductase B subunit), *nrdD* (*sunY*), and gene *60* (DNA topoisomerase subunit).

The gene *60* intron is not actually removed. Instead there is an example of context effects on translation called **translational bypassing** in which an mRNA molecule coding for a single peptide is not fully translated in its middle region. Codons 46 and 47 are the same (GGA) but have an optimum 50 bases separating them. Codon 46 is followed by a terminator codon (UAG) and a short stem-loop sequence (Fig. 6.17). Herr et al. (2004) have examined some of

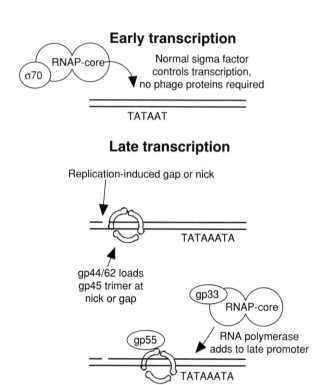

Figure 6.16. Enhancement of late transcription. Early transcription occurs from promoters that are fundamentally similar to host promoters and are recognized by host σ^{70}. During replication and recombination, numerous nicks or gaps form, and any of them can serve as an enhancer of transcription. Each site can serve as a loading site for a trimer of gp45 (the sliding clamp) as mediated by the gp44/62 clamp loader complex. The sliding clamp maintains processivity of the phage DNA polymerase. As the clamp moves along the DNA, it can acquire gp55, the T4 late sigma factor. The combination of core RNA polymerase and gp33 then adds to the gp55/ sliding clamp complex, and late transcription begins.

the factors that encourage bypassing, such as base pairing and stringency of ribosome P site binding. Nearly all ribosomes stop translation at codon 46. About 50% of them resume translation at codon 47. The remainder either terminate translation at the UAG codon or fail to resume translation at codon 47.

In the case of *nrdD*, base sequence analysis of the coding region for the enzyme has shown that there is a 1033-bp intron located within it that

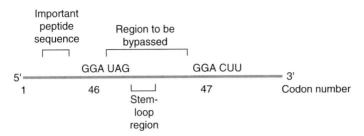

Figure 6.17. Translational bypass. T4 gene 60 codes for an untranslated region of 50 bases. The figure shows an mRNA molecule (*gray line*), with codon numbers below the line. Notice that codons 46 and 47 are the same. In the presence of the appropriate upstream peptide sequence and an appropriate stem-loop in the region to be bypassed, most ribosomes skip from codon 46 to 47 without inserting another amino acid. Ribosomes that do not skip terminate at the next codon.

contains an open reading frame located at the 3'-terminus that codes for a homing endonuclease (see Chapter 15). Like all group IA2 introns, *nrdD* contains a ribozyme core (a region of catalytically active RNA) that is responsible for removal of the intron by mRNA splicing. Proper excision requires pairing of two peripheral regions of the intron. This type of reaction is characteristic of class I splicing systems such as those found in *Saccharomyces*, *Neurospora*, or *Tetrahymena* mitochondria, and there is significant homology between all group I introns. The self-splicing reaction is not always factor-independent. Some introns require specific protein(s) for in vivo functionality because addition of chloramphenicol to an infected culture also blocks mRNA splicing. In addition, nonspecific RNA-binding proteins such as ribosomal protein S12 may function as chaperones to promote correct RNA folding. A guanosine nucleoside is required for splicing.

    After mRNA molecules have been produced, their code must still be translated into protein before they could have any effect on metabolism. It can easily be shown that various mRNA molecules may be present in the cytoplasm of a T4-infected cell and not be translated even though they are readily translated in an uninfected cell. Lack of translation is certainly seen for most *E. coli* transcripts because even phage ghosts are able to inhibit translation of cellular mRNA. A similar phenomenon has been demonstrated for at least some other phages when the cell they infected becomes superinfected by T4. There is substantial disagreement as to the molecular mechanisms underlying the phenomenon. Some workers suggest a direct interaction between T4 proteins and the ribosomes or their protein cofactors, but others say that they are unable to verify the association.

Gene 32 and gene 43 proteins regulate their own translation. Gene 32 protein is normally a single-strand-DNA-binding protein. When present in excess, however, it binds to its own mRNA and covers over the ribosome-binding site, thereby preventing excessive accumulation of gene 32 protein. Gp43 has a similar effect.

One T4 protein that does seem to have a more general effect on translation is the *regA* protein. The observation is that *regA* mutant strains generally overproduce a wide variety of T4 proteins, even though their transcription rates are essentially unaltered. This finding suggests that *regA* acts as a **translation repressor**, a protein that prevents mRNA molecules from attaching to ribosomes. Fitting in with this hypothesis are observations that most, if not all, mRNA transcripts subject to *regA* control have the same or similar base sequence in the region of the ribosome-binding site (Shine–Dalgarno sequence), whereas those mRNA transcripts that are not under *regA* control do not carry the sequence. Some host proteins are also regulated (positively and negatively) by *regA*. The common element in regulated mRNA molecules seems to be a uridine-rich region (5′AAAAUUGUUAUGUAA3′).

In the case of the gene coding for T4 lysozyme, the basis for translational control is structural. Appropriate mRNA molecules for lysozyme are present in large quantity during both early and late transcription, but under normal circumstances only late mRNA is translated.

Comparison of the sequences of early and late mRNA transcripts shows that early mRNA is part of a larger transcript that is predicted to have the ability to form an internal hairpin loop that embeds both the Shine–Dalgarno sequence and the first codon for translation of the lysozyme within the duplex region (Fig. 6.18). A fundamental principle of translation is that ribosomes are capable of disrupting mRNA secondary structure only after they have already bound to appropriate initiator sites and are actively translating the mRNA. Therefore, the mRNA transcribed early is predicted to be untranslatable unless some outside influence disrupts hydrogen bonds within the stem. On the other hand, the late mRNA transcribed from eP1 or eP2 should be incapable of forming the stem-and-loop structure and consequently be readily translated.

Many of the T4 proteins make fundamental changes in the metabolism of the host cell. Enzymes are produced to degrade cytosine and uridine triphosphates and diphosphates to their monophosphoric derivatives (nucleoside diphosphatase and triphosphatase); produce hydroxymethylcytosine (deoxy-

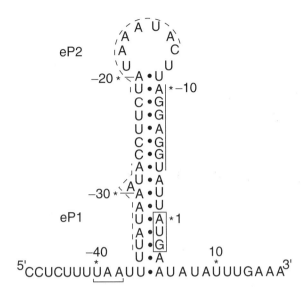

Figure 6.18. Proposed secondary structure of early lysozyme mRNA. The stem-and-loop structure that is predicted to form in early lysozyme transcript(s) is shown. *Dots* indicate Watson–Crick base pairing. The initiator AUG is boxed, and the Shine–Dalgarno sequence is indicated by a *line* next to the bases. The two sequences homologous to the conserved T4 late promoter sequence are indicated by *dashed lines* and labeled eP1 and eP2. The termination codon of the upstream open reading frame is bracketed. (From McPheeters et al. [1987]. *Nucleic Acids Research* 14: 5813–5826.)

cytidylate hydroxymethylase); produce thymine, guanine, and hydroxymethyl-cytosine triphosphates (deoxynucleoside monophosphate kinase); glucosylate the hydroxymethyl cytosine residues (glucosyltransferases); and methylate certain adenine residues (DNA adenine methylase). Also produced are eight endonucleases, many of which are involved in DNA repair, and a new DNA polymerase molecule. The DNA polymerase is a multidomain protein with structural similarity to *E. coli* DNA polymerase I.

Jan Drake and his collaborators have shown that the takeover of the host cell by T4 is so complete that mutator mutations such as *mutHLST* or *uvrD* have no impact on the mutation rate of T4. An interesting assortment of enzymes whose roles in the infective process are uncertain is also produced: DNA ligase (necessary but normally supplied by the host cell), RNA ligase (a dual-function protein that attaches tail fibers to the virion and can also join single-stranded polynucleotides), polynucleotide kinase, several DNA phosphatases, and a type II topoisomerase.

All the events described above occur prior to the end of the eclipse phase. They result in accumulation of the building blocks necessary for construction of new virions. The actual assembly of infectious phages proceeds in a linear manner and signifies the end of the eclipse phase.

Morphogenesis and Maturation

Three separate biochemical pathways generate heads, tails, and tail fibers, which are then joined to make a virion (see Fig. 6.14). The process of assembly of each component is similar to that of crystallization. A cluster of proteins forms the starting structure (a nucleus), and other subunits spontaneously arrange themselves around the nucleus in an orderly manner. At least two host cell chaperonins, *groEL* and *groES*, are necessary for this process. If one viral gene product is missing or is nonfunctional, the assembly process stops at the point at which that product is required, and all proteins that would have been added subsequently remain in solution. If a mutation alters but does not remove certain proteins, long structures called polyheads or polytails may develop. Chain terminator mutations in the same proteins involved in polyhead formation may lead to the formation of normally shaped "giant" or "petite" heads, which can produce infectious phages carrying correspondingly greater or lesser amounts of DNA.

DNA maturation requires the presence of a mature prohead, which consists of the outer proteins visible in electron micrographs as well as a complex of internal "scaffold" proteins (products of genes *22*, *III*, *II*, and *I*). The mature prohead is some 16% smaller than the head of a viable phage particle. The increase in size is associated with the cleavage of gp23 by a T4-specific protease and may occur in the absence of any DNA. The process is shown in Fig. 6.19.

DNA packaging occurs by a "headful" method and requires ATP and DNA ligase. Restriction digests of T4 DNA isolated from virions indicate that there is no preference to the end sequences as is seen in the case of phage P22 (see Chapter 8), and thus, the initial cut must be made randomly by an endonuclease. This random cutting gives rise to the observed circular permutations of the genome. The terminase enzyme that catalyzes the cutting and translocation reactions is composed of the products of the overlapping genes *16* and *17*, where the 3'-terminus of gene *16* overlaps the beginning of gene *17*.

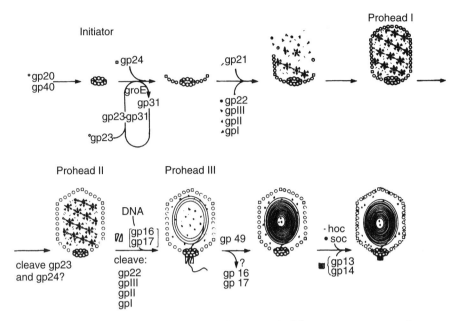

Figure 6.19. Morphogenetic pathway of T4 capsids. The various gene products are represented by geometric figures (*circles, squares, triangles*, etc.) shown beside the gene product designation. A dash between two components indicates association between them. The DNA is shown as a *thin line* that begins to fill prohead III. In phage T4, tails are synthesized in an independent pathway and then attached to complete heads. (From Murialdo, H., Becker, A. [1978]. Head morphogenesis of complex double-stranded deoxyribonucleic acid bacteriophages. *Microbiological Reviews* 42: 529–576.)

Electron microscopic experiments, in which a beam of ions is used to erode away the capsid, show that the DNA is arranged in a spiral-fold model such as that shown in Fig. 6.20. In this model the DNA is packed from inside toward outside, leaving both ends free so that either end can be the first one to be ejected. DNA is added to the prohead until it is full. Because the head can hold more than a genome equivalent of DNA, terminal redundancy is ensured. Presumably, after the initial packaging, the endonucleolytic cut that serves to release the first virion also serves as the initiator cut for the next prohead.

Tails are attached to filled heads, and tail fibers are then added. All that now remains is for the phage to lyse the host cell. This step is accomplished by a phage-specific lysozyme (product of gene *e*) in conjunction with gp5 (located in the baseplate of the virion). The product of gene *t* is a **holin**, a membrane protein

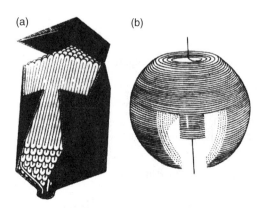

Figure 6.20. Models for the arrangement of DNA inside the head of phage T4. (a) Spiral-fold model described in the text. (b) An older, concentric shell model in which the DNA is wrapped like a ball of string beginning at the center. (From Black, L.W., Newcomb, W.W., Boring, J.W., Brown, J.C. [1985]. Ion etching of bacteriophage T4: Support for a spiral-fold model of packaged DNA. *Proceedings of the National Academy of Sciences of the USA* 82: 7960–7964.)

that provides a channel across the cell membrane. When the primary particles have been released, the infectious cycle can restart.

Summary

T4 phage is a large and structurally complex virus. During initiation of infection, tail fibers form the attachment to a cell wall, first reversibly then irreversibly. DNA is ejected from the virion through the tail core as the tail sheath contracts and is transported into the cytoplasm of the cell concomitantly with depletion of the proton gradient across the cell membrane. Once inside the cell, the virus is in the eclipse phase, as no infectious virus is present when cells are prematurely lysed. Phage-specific protein synthesis begins immediately, producing enzymes to initiate replication, to expand transcription of viral DNA, and to supply proteins needed to modify certain cellular components.

 Both transcription and translation are highly regulated: the former by modifications to RNA polymerase holoenzyme complex and the latter by mRNA structure as well as competitors for ribosomal binding. Most proteins that are synthesized late in infection are structural components of the virion. DNA synthesis begins at about 6 min at 30°C and generates a large concatemeric structure by multiple rounds of recombination. At 12 min, condensation of heads, tails, and tail fibers begins. When mature proheads are ready, an endonuclease makes a random cut in the DNA, and the phage head is stuffed with DNA until full. The "headful" mechanism leads to production of a viral DNA molecule that is linear, circularly permuted, and terminally redundant.

Genetic crosses with T4 phages are carried out by infecting *E. coli* cells with two or more genetically distinct phages (a mixed infection). The easiest phenotype for study is the one involving plaque morphology, but more than 130 genetic loci affecting all types of functions have been mapped. The most carefully mapped portion of the genome is the *r*II region, which was studied intensively by Benzer. From his work comes the conclusion that many mutations are random in their distribution and that the unit of mutation and recombination is a single base. Roughly 2% of all progeny phages formed in a cross are heterozygous for one or more traits. Some of these hets are due to terminal redundancy, whereas others seem to be normal recombination intermediates. The way in which recombination occurs during T4 infection has been the subject of a mathematical model prepared by Visconti and Delbrück.

Phage T4 exhibits a number of unusual genetic features. These include the presence of introns in some genes and gene regulation at the levels of transcription and translation. The phage extensively modifies host RNA polymerase and also changes ribosome specificity. The net result is a complete subversion of host metabolism to viral function.

Questions for Review and Discussion

1. Predict the effects of a T4 head protein mutation that yields a phage particle containing exactly one genome length of DNA.
2. What evolutionary advantage accrues to a virus that can regulate its mRNA translation?
3. What new types of regulation (beyond those discussed in Chapter 4) are seen in phage T4?
4. Explain why Benzer was not able to achieve 100% map saturation in his *r*II mutagenesis experiments.
5. Construct a genetic map for this T4 virus using the given data. If there are any ambiguities in the data, be sure to point them out.

There is a standard set of deletion mutants. The extent of each deletion is shown by the lines below.

Δ1 ──────────

Δ2 ─────────────────

Δ3   ──────────

Mutation A does not recombine with Δ2, but does recombine with all other deletions.

Mutation B does not recombine with any deletion mutation.

Mutation A does not recombine with B, but gives 45% recombination with C.

Mutation D recombines 15% with C and 35% with A.

In a *cis–trans* test, mutation E complements mutation A and C but not D.

References

General

Ackermann, H.W., Krisch, H.M. (1997). A catalogue of T4-type bacteriophages. *Archives of Virology* 142: 2329–2345.

Karam, J.D. (ed.) (1994). *Molecular Biology of Bacteriophage T4*. Washington, DC: American Society for Microbiology.

Miller, E.S., Kutter, E., Mosig, G., Arisaka, F., Kunisawa, T., Rüger, W. (2003). Bacteriophage T4 genome. *Microbiology and Molecular Biology Reviews* 67: 86–156. (A particularly useful article with over 1,000 references.)

Nechaev, S., Severinov, K. (2003). Bacteriophage-induced modifications of host RNA polymerase. *Annual Review of Microbiology* 57: 301–322.

Stahl, F. (1989). The linkage map of phage T4. *Genetics* 123: 245–248. (A retrospective analysis of problems associated with determining the first complete genetic map for T4. Historical memoir by one of the participants.)

Specialized

Herr, A.J., Wills, N.M., Nelson, C.C., Gesteland, R.F., Atkins, J.F. (2004). Factors that influence selection of coding resumption sites in translational bypassing: Minimal conventional peptidyl-tRNA:mRNA pairing can suffice. *The Journal of Biological Chemistry* 279: 11081–11087.

Kamali-Moghaddam, M., Geiduschek, E.P. (2003). Thermoirreversible and thermoreversible promoter opening by two *Escherichia coli* RNA polymerase holoenzymes. *The Journal of Biological Chemistry* 278: 29701–29709.

Kanamaru, S., Leiman, P.G., Kostyuchenko, V.A., Chipman, P.R., Mesyanzhinov, V.V., Arisaka, F., Rossmann, M.G. (2002). Structure of the cell-puncturing device of bacteriophage T4. *Nature* 415: 553–557.

Luke, K., Radek, A., Liu, X., Campbell, J., Uzan, M., Haselkorn, R., Kogan, Y. (2002). Microarray analysis of gene expression during bacteriophage T4 infection. *Virology* 299: 182–191.

Orsini, G., Igonet, S., Pène, C., Sclavi, B., Buckle, M., Uzan, M., Kolb, A. (2004). Phage T4 early promoters are resistant to inhibition by the anti-sigma factor AsiA. *Molecular Microbiology* 52: 1013–1028.

Pande, S., Makela, A., Dove, S.L., Nickels, B.E., Hochschild, A., Hinton, D.M. (2002). The bacteriophage T4 transcription activator MotA interacts with the far-C-terminal region of the σ70 subunit of *Escherichia coli* RNA polymerase. *Journal of Bacteriology* 184: 3957–3964.

Rabinovitch, A., Hadas, H., Einav, M., Melamed, Z., Zaritsky, A. (1999). Model for bacteriophage T4 development in *Escherichia coli*. *Journal of Bacteriology* 181: 1677–1683.

Slavcev, R.A., Hayes, S. (2003). Blocking the T4 lysis inhibition phenotype. *Gene* 321: 163–171.

7

Genetics of Other Intemperate Bacteriophages

Bacteriophage T4 is probably the most intensively investigated intemperate virus, but there are many other viruses that have also been the subject of considerable study. In this chapter, descriptions of selected bacteriophages are presented to illustrate the high degree of genetic diversity available to a bacterial geneticist and to provide comparisons among them and with T4. To facilitate these comparisons, the physical properties of each phage discussed in this chapter are summarized in Table 7.1. Yeast viruses are not considered because, like all fungal viruses, they are transmitted only by cell fusion and do not spend any portion of their life cycle in an independent state.

Major topics include:

- Strategies used by viruses to minimize the size of their genome
- Replication of viruses having nontraditional genetic material (not double-strand DNA)

Table 7.1. Properties of some common intemperate bacteriophages.

Phage	Usual Host	Nucleic Acid Type	Size (kb or kbp)	Topology	Virion Morphology	Dimensions (nm)	Related Phages
T1	E. coli	2-DNA[a]	50.7	Linear, circularly permuted, terminally redundant	Polyhedral head, flexible non-contractile tail	55–60, 150 × 7	TLS
T4	E. coli	2-DNA[a]	166	Linear, circularly permuted, terminally redundant	Oblong head, contractile tail	80 × 120, 113 × 20	T2, T6, U5
T5	E. coli	2-DNA	121.3	Linear, unique sequence,[b] terminally redundant	Octahedral head, noncontractile tail	90, 200	BF23
T7	E. coli	2-DNA	39.937	Linear, unique sequence, terminally redundant	Octahedral head, noncontractile tail	60, 15 × 15	T3, φ11

SPO1	B. subtilis	2-DNA	140–145	Linear, unique sequence, terminally redundant	Icosahedral head, contractile tail	87, 140 × 19	SP8, SP82, φe, 2c, H1
φ6	P. syringae	2-RNA	2.948, 4.063, 6.374	Linear	Enveloped sphere	58 (no envelope)	φ7–φ14
φ29	B. subtilis	2-DNA	19.285	Linear, unique sequence	Oblong head, noncontractile tail	32 × 42, 32 × 6	φ15, Nf, GA-1, M2, PZA, B103, BS32
f1	E. coli	1-DNA	6.407	Circular	Filamentous	870 × 5	M13, AE2, HR, Ec9, fd, δA (the Ff group)
φX174	E. coli strain C	1-DNA	5.386	Circular	Icosahedral	27	S13, φR, G4
MS2, M12,	E. coli F+	1-RNA	3.569	Linear, unique sequence	Icosahedral	26	fr, f2, R17, Qβ

[a] The number in front of the nucleic acid refers to the number of strands in the molecule.

[b] As contrasted to a circularly permuted sequence.

- Viruses that are used as the basis for cloning vectors
- Additional methods of regulation discovered by the study of viruses

---

THINKING AHEAD

What are the essential and desirable functions for a virus?

Other Members of the T Series

Delbrück chose seven distinct, lytic bacteriophages isolated in Milislav Demerec's laboratory to be members of the T series of viruses. He believed that focusing attention on a small number of viruses would yield information more quickly than a diffuse effort covering many bacteriophages. Closer examination of the chosen phages showed that all even-numbered phages were similar to one another, but were different from the odd-numbered phages. Electron microscopic studies have shown that all even-numbered phages possess contractile tails and elongated heads, but the T-odds have noncontractile tails with octahedral heads. Phages T3 and T7 are similar, but phages T1 and T5 are unlike T7 or each other.

Bacteriophages T2 and T6

Bacteriophages T2 and T6, the remainder of the T-even series, are homologous to T4. They have similar physiology and genetic composition, although they differ in the way in which their DNA is glucosylated. T2 and T6 have 25% of their hydroxymethylcytosine (HMC) residues unglucosylated, whereas T4 has none. Nearly all glucosylated T2 HMC residues have only one glucose attached in the α-configuration, whereas in the case of T6 phage, the HMC residues are diglucosidic, having first an α-linkage and then a β-linkage.

The degree of genetic homology (i.e., the extent to which DNA sequences are identical) among the T-evens can be readily demonstrated by the technique of heteroduplex mapping, described in Chapter 2. If DNA carrying a deletion is heteroduplexed to nondeleted DNA, the reannealed structure has a characteristic loop of single-strand DNA originating from a point on the nondeleted DNA strand that represents the site of deletion on the corresponding molecule (Fig. 7.1). Regions of nonhomology (i.e., different base sequences)

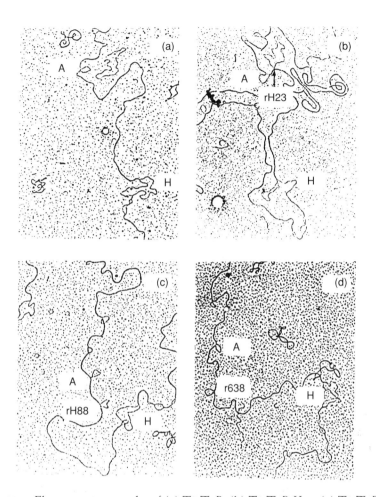

Figure 7.1. Electron micrographs of (**a**) T2/T4B, (**b**) T2/T4B*rH23*, (**c**) T2/T4B*rH88*, and (**d**) T2/T4B*r638* heteroduplex DNA molecules around the *r*II region. In this region, T2 and T4 DNA molecules have a short segment in which their base sequences are drastically different (a substitution) and a short segment is present in T4 DNA but is deleted in T2. These differences give rise to the substitution loop H and deletion loop A that are seen in (**a**). Each of the *r*II mutations is also a deletion, and in (**b–d**) the extra deletion loop observed in the heteroduplex is also labeled. The distance between loops A and H in (**a**) is 11,500 bp. The distances from loop A to the deletion loops *rH23*, *rH88*, and *r638* correspond, respectively, to 400, 1100, and 2800 bp. The small circular DNA molecules seen in the background are double-strand φX174 DNA used as a size standard. (From Kim, N., Davidson, N. [1974]. Electron microscope heteroduplex study of sequence relations of T2, T4, and T6 bacteriophage DNAs. *Virology* 57: 93–111.)

between two-phage DNA molecules result in the formation of two single-strand loops (which may not be of identical size).

By carefully measuring the distance between deletion loops with known genetic map position and unknown loops along the DNA molecule, and knowing the final magnification of the measured molecule, it is possible to determine the distance between various points on a phage DNA molecule, in terms of not recombination units but physical units. These physical distances are often expressed as a percentage of total genome length. This simplifies calculations by eliminating the need to know the exact measurement of the magnification. Kim and Davidson (1974) estimated the overall homology between phages T2 and T4 at 85%, with late genes being more homologous than early ones. Similarity of sequence implies similarity of virion (late) proteins, which accounts for the large amount of immunologic cross reactivity among the T-even phages. Some bacteriophage T2 alleles are specifically excluded during crosses with T4. More recently, Monod et al. (1997) used PCR technology to identify sequence similarities between phage T4 and a large group of relatively uncharacterized bacteriophages. Four members of the group appear to be more distantly related to T4 than T2 or T6 (30–40% sequence similarity). The new phages apparently do not modify their DNA bases as extensively as the classic T-evens because they are susceptible to restriction endonucleases. They also vary in their possession of introns. Sandegren and Sjøberg (2004) have shown that only T4 and U5 phages have all three introns. Most have only one, and their model is that the introns are fundamentally unstable and require constant reinheritance by recombination (see Chapter 15).

Some early work with T2 involved a class of mutations that does not occur in T4. These mutations were designated *h* (**host range**) and affected the tail fibers and their ability to attach to certain bacterial cell walls. *Escherichia coli* can mutate so normal T2 no longer infects the cell owing to a change in the cell wall surface receptor site (a phage-resistant mutant). However, the appropriate kind of tail fiber mutation once again permits attachment of the T2 tail fiber to a mutated cell and a successful infection of either mutant or wild-type bacteria. This *h* phenotype is another example of one that can be scored by plaque morphology. However, unlike the *r* phenotype, it is necessary to use a mixed indicator. When dealing with *h* or *h*+ virions and using a mixture of normal and phage-resistant bacteria as indicators, two kinds of plaque are possible. One kind is perfectly clear, meaning that both sorts of indicator bacteria have lysed and the phage is an *h* mutant. The other kind of plaque is turbid, meaning that the phage-resistant cells have survived the infection and the phage is not an *h* mutant.

The existence of host range mutants has led to the suggestion that an endless cycle of bacterial resistance and phage compensatory mutation is possible. In fact, *h* mutants are best looked on as extended host range mutants: They still infect wild-type cells and are found in only certain viruses. Moreover, in the case of phages T3 and T7 (see later), bacteria that are resistant to host range mutants can be obtained, and the phages do not produce additional compensatory mutations.

Several unusual discoveries have emerged from studies dealing with T2 morphogenesis. Drexler and coworkers have shown that tail fiber protein gp37 is normally processed posttranslationally by removing about 120 amino acids from the carboxy terminus. If a mutation is introduced into gene 37 that deletes 87 amino acids from the region normally removed, protein processing is not possible, and the net effect of deleting 87 amino acids is the production of a finished protein product that is longer than the unmutated one. Tail fibers determine the host range of a T-even virus by recognizing specific receptors on the cell surface. Each of the T-even viruses has its own specific tail fibers and adhesins. In phage T4, the carboxy terminus of gp37 includes the adhesin sequence. However, phage T2 has its adhesin encoded in gene 38, and gp38 binds to the tips of the tail fibers (gp37). The gp38 protein from one phage or mutant can recombine with a heterologous adhesin from another phage (Tetart et al. 1998). This is an example of functional cassettes in bacteriophages.

Bacteriophage T1

Although T1 was one of the first phages used in bacterial genetics, it has not been studied as thoroughly as the T-evens. One reason is its incredible persistence. Once the phage has been brought into a laboratory, it is difficult to eliminate, because it is capable of surviving for years on laboratory surfaces and of forming stable aerosols. These activities are in sharp contrast to those of a phage such as T6, which survives only a few hours in a desiccated state. However, Roberts et al. (2004) have now reported the complete DNA sequence for the virus. Despite earlier suggestions that T1 was in fact a virulent form of phage lambda, DNA sequence comparisons indicate that T1 has little similarity to any of the well-studied bacteriophages. Its circularly permuted, terminally redundant DNA, codes for 71 open reading frames, many of which are quite small and have no similarities to other proteins in the database. There are three classes of mRNA, but none corresponds to middle RNA. Instead there is a group of molecules that are synthesized at all times during the infection (continuous).

Bacteriophage T5

The structure of bacteriophage T5 is similar to that of the T1 virion except that it has four tail fibers (one straight and three curved) instead of one (Fig. 7.2). The T5 virion has a linear DNA molecule of unique sequence (has defined ends) with no unusual bases but with a large terminal redundancy, amounting to 9% of the

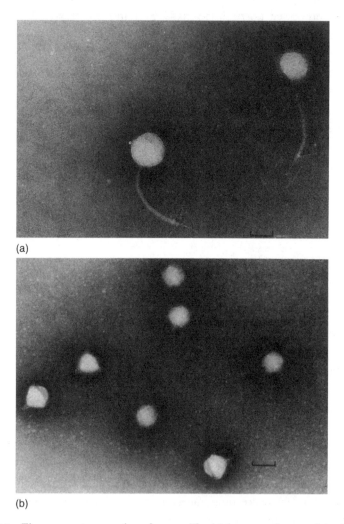

(a)

(b)

Figure 7.2. Electron micrographs of some T-odd bacteriophages: (a) phage T5, (b) phage T7. The length of the bar in each micrograph is 50 nm. (Courtesy of the late Robley C. Williams, Virus Laboratory, University of California, Berkeley.)

total DNA. The DNA also contains a series of four to five **interruptions** (nicks in the phosphodiester backbone of one DNA strand) that can be repaired by the enzyme DNA ligase. These most commonly occur at definite positions on the DNA molecule (although exceptions are known); are all on the same strand of the double helix (Fig. 7.3); and have no known function. The frequency of individual nicks can be increased or decreased by mutation without affecting phage activity. Mutant phages are known to have lost all their nicks and function normally. Therefore, the biological reason for the interruptions is unclear.

After attachment of a phage to an *fhuA* protein, either free or as part of an *E. coli* cell, DNA ejection is triggered. About 8% of the phage genome (shown as the left end in Fig. 7.3) penetrates the cell wall and membrane regardless of the presence of ionophores, the absence of a capsid, or a lack of DNA superhelicity. The only property apparently required is appropriate membrane fluidity. Bonhivers et al. (1998) suggest that naked phage DNA is transported through the FhuA channel that is formed on binding of the phage to the transporter. The DNA entry process then stops at a specific region. The cause of the stoppage is not known, but that region has the potential to form various stem-and-loop structures and also has sequences compatible with *dnaA* binding. Continuation of DNA entry requires the expression of certain "first step transfer" (FST) genes contained within the already injected DNA. The mRNA produced by these FST genes is referred to as pre-early mRNA by analogy to T4. After translation of this mRNA, DNA transfer resumes, apparently only due to the presence of the newly synthesized viral proteins, as removing the head and tail proteins of a phage particle after the occurrence of initial attachment does not prevent DNA entry. One proposed mechanism is progressive binding of DNA to the cell

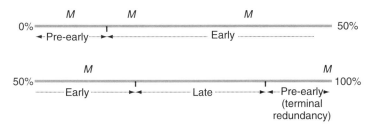

Figure 7.3. Simplified map of the phage T5 genome. The terminal redundancies include all the pre-early genes. There are five interruptions, sites where permanent nicks occur in the DNA, and they are indicated by M. All interruptions occur in the DNA strand that has its 3′-end at the left of the drawing.

membrane, possibly via transmembrane channels opened at *fhuA* sites by viral protein pb2. Under normal conditions, the entire infection process is complete within 1 min.

As the infection proceeds, host RNA polymerase is modified by addition of two pre-early proteins. Further modifications seem possible, as there are three classes of mRNA produced: class I, which codes for the FST synthesis; class II, which codes for viral metabolic proteins; and class III, which codes for proteins needed for continued DNA synthesis, virion assembly, and cell lysis. One of the FST proteins degrades host DNA, and two others antagonize the *Eco*RI restriction endonuclease. Evidence for the latter point comes from experiments that introduced an artifical *Eco*RI site into the FST DNA. The infections aborted. However, naturally occurring *Eco*RI sites elsewhere in the genome have no effect. Assembly of the virions is probably accomplished by a mechanism that is more like the one used by T7 than that used by T4 because, like T7 DNA, T5 DNA has a unique sequence.

Some T5 promoters are remarkably efficient, and it is possible to clone them specifically onto plasmids carrying a selectable marker that lacks its normal promoter. The T5 promoters thereby obtained are so efficient that they can outcompete all normal host cell promoters to give preferential transcription of cloned DNA. Hsu et al. (2003) have investigated some of the biochemical properties of a strong T5 promoter. Plasmids carrying certain T5 promoters initiate 90% of all their transcripts at the T5 site. T5 promoters are distinctive for their 75% (A + T) content and for the in vitro ability of at least some of them to accept 7-methyl guanosyl triphosphate instead of adenosine triphosphate as the initiator base for transcription. This ability allows production of capped mRNA molecules from cloned eukaryotic DNA.

A T5-specific DNA polymerase catalyzes DNA replication. This polymerase is a single molecule that causes strand displacement when initiating synthesis at a nick developing multibranched, concatemeric structures.

Bacteriophages T7 and T3

Bacteriophages T7 and T3 are the smallest of the T phages (see Fig. 7.2) and are best studied after the T-evens. Both infect species of *Escherichia*, *Shigella*, *Salmonella*, *Klebsiella*, and *Yersinia*. The primary member of the group is T7, but results obtained using T3 are discussed whenever the two phages are known to differ. The T7 DNA molecule contains no unusual bases, has a unique sequence, and has a short terminal redundancy of 160 bp (less than 1% of the total DNA). The redundancy for T3 DNA is 230 bp, and sequence analysis suggests that it is related to T7 and to *Klebsiella* phage 11. All three phages seem to have terminal redundancies based on the progenitor sequence TTAAC-CTTGGG.

Although T3 and T7 appear to be morphologically similar, they are in fact discrete entities. Their infections are mutually exclusive, a cell infected with T7 cannot be infected with T3 and vice versa, whereas cells infected with other intemperate phages can generally be superinfected with the same phage. Therefore, it was a surprise when Pajunen et al. (2002) concluded that T3 was the result of a recombination event involving T7. T3 is most similar to the *Yersinia* bacteriophage φYeO3-12, a phage that coinfects cells with T7.

Like T5, T7 DNA enters the cell in stages: first 10%, then 50%, and then the remainder. Once again, there is coupled transcription and entry for stage 2 DNA, but the final 40% of the DNA can enter a cell only if translation is permitted.

A reasonable amount of information is available about T7 DNA replication. The replication process begins at a definite point of origin that is located 17% of the distance from the left end of the map (Fig. 7.4). It uses phage-specific proteins that often share some homology with corresponding *E. coli* proteins. The phage replication proteins include the gene 5 product, a nonprocessive DNA polymerase that becomes processive when combined with host thioredoxin; the gene 4 helicase and gene 1 product that make RNA primers; and a viral protein that acts as a single-strand DNA binding protein. The host contributes DNA ligase and DNA polymerase I. It is not necessary for the phage DNA to circularize so that replication can proceed because at least the first two rounds of replication are bidirectional (if replication were not bidirectional, only part of the linear molecule would be replicated). Concatemeric structures are formed by a mechanism that may be similar to that of T4 with recombinogenic single-strand ends. DNA replication is not a prerequisite for late transcription, unlike T4.

Transcription of T7 DNA always uses the right strand of the DNA helix as a template (preferentially binds poly-UG). This polarity leads to an automatic sequencing of events during infections, as the left end of the DNA (as shown in Fig. 7.4) is always injected first. Early mRNA molecules are produced by host RNA polymerase from three promoters located near the left end of the DNA and are processed by host RNase III to yield up to five smaller mRNA molecules. At 6 min after infection at 30°C, late mRNA synthesis begins using a phage RNA

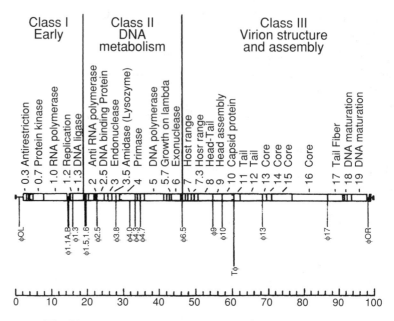

Figure 7.4. The T7 genome. The main genes (*open boxes* drawn to scale) are characterized by their numbers and by a reference to their functions. Promoters for T7 RNA polymerase begin with φ. Tφ is a terminator but is not fully efficient. The small, *filled boxes* at the ends represent the terminal repeats. (From Studier, R.W., Dunn, D.J. [1983]. Organization and expression of bacteriophage T7 DNA. *Cold Spring Harbor Symposia on Quantitative Biology* 47: 999.)

polymerase that consists of a single protein molecule. The molecule functionally differs from the elaborate *E. coli* RNA polymerase complex, as the T3 and T7 polymerases work with either phage DNA but do not transcribe λ, T4, or *E. coli* DNA. The major promoter specificity determinant is located in the region near base −11 and involves binding of nascent RNA to the amino terminus of the phage RNA polymerase. The T7 RNA polymerase is capable of producing three long transcripts per second from its own promoters, a higher rate of synthesis than that of the host RNA polymerase.

Further transcription by the *E. coli* polymerase is prevented by phosphorylation of its β′ subunit catalyzed by the protein kinase from gene 0.7, a threonine–serine protein kinase that also phosphorylates ribosomal protein S1 and initiation factors 1 and 2 (IF1 and IF2). The protein from gene 2.0 provides additional inhibition. RNase III also processes many late transcripts.

Relations between the host cell and phages T3 and T7 are also unusual. T3 but not T7 grows on *E. coli* strains harboring an F plasmid (Hfr, F$^+$, or F'), with the difference between the phages lying at the end of gene 1. Failure to grow is due to an inability of infected cells to synthesize any macromolecules after the beginning of class III mRNA synthesis.

A similar abortive infection can be observed if an *E. coli* strain B cell that is a P1 lysogen (see Chapter 8) is infected with T7. Although phage T3 is not restricted by F$^+$ cells, under starvation conditions it replicates so slowly that the host cell survives to divide and grow, yielding a **pseudolysogen**. A pseudolysogen is a phage-infected cell that grows and divides even though its virus is pursuing a lytic infection. Unlike the case for a true lysogen, an isolation streak of a pseudolysogen can yield uninfected cells because the rate of accumulation of intracellular virus is quite slow. The inhibitory effect of phage T3 on T7 is mediated via suppression of an ochre codon. A ribosomal mutation that increases the fidelity of translation reduces the inhibitory effect, whereas the presence of an ochre suppressor mutation increases it.

Maturation of T7 begins at about 9 min postinfection and proceeds in a manner similar to that of T4. The major unresolved problem in T7 virion assembly is to develop a series of DNA molecules that have a unique sequence rather than a circular permutation as in the case of T4. The conventional headful mechanism does not suffice, as it involves random cuts in the concatemeric DNA. The mechanism is one in which a prohead binds first to a part of the concatemers that corresponds to the right end of the T7 monomer (Fig. 7.5). As in the case of T4, replication occurs within the concatemeric DNA net, and one of the replicated left ends is packaged to yield a fully double-stranded DNA molecule with unique sequence and a terminal redundancy. The burst size is approximately 200.

The ability of T7 RNA polymerase to respond exclusively to its own promoters has led to its frequent use in cloning experiments. T7 promoters are highly conserved in the region from base −17 to base +6 and are thus unlikely to occur by chance. DNA sequences that are cloned downstream from a T7 promoter remain untranscribed until the T7 RNA polymerase is provided. This technique can be used in vitro to generate RNA probes for Northern blotting, although a certain amount of caution is necessary. In the presence of 3'-overhangs generated by restriction enzymes, T7 RNA polymerase can initiate away from its promoter and transcribe incorrect regions or even the incorrect strand of DNA. The problem does not arise if the 3'-overhangs are digested away. The T7 polymerase gene has also been linked with vaccinia viral DNA and is used to

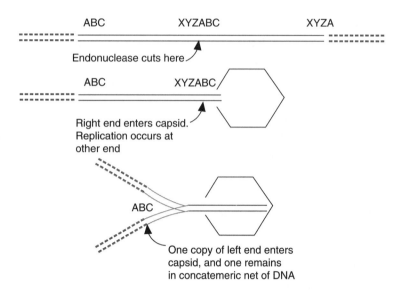

Figure 7.5. Production of unique sequence terminal redundancies by cutting a concatemeric DNA molecule. In this example, the cut occurs at site C. As replication of the catcatemer continues, the segment of DNA enters the phage head, leading to a terminal redundancy.

control the expression of DNA, cloned into other vectors in mammalian cells. Transcription of the desired DNA sequence occurs only after infection by the modified vaccinia virus.

In an interesting use of technology, John Yin and his collaborators have developed a computer simulation of phage T7 infection. The model simulates such processes as phage entry, regulatory processes, and virus assembly. In a recent iteration of the model, You et al. (2002) showed that rapid growth of the *E. coli* host leads to more rapid production of virus.

THINKING AHEAD

What are the ways in which a particular region of DNA could code for more than one protein?

Bacteriophages Containing Single-Strand DNA

Bacteriophages Belonging to the Ff Group

The Ff group is a large one composed of filamentous phages (Fig. 7.6) that contain a circular, single-stranded DNA (1-DNA) molecule of 6408 nucleotides. A typical genetic map is shown in Fig. 7.7. The protein coat consists of about 2700 major subunit protein molecules encoded by gene VIII and minor subunits encoded by genes III, VI, VII, and IX. The rodlike shape of the virion enforces a similar configuration on the DNA molecule, which folds upon itself. The ends of the DNA molecule thus defined can be differentiated by the proteins that are bound to them, the products of genes VII and IX at one end and the products of genes III and VI at the other. A number of morphologic variants are known. Miniphages of less-than-normal length have the replication origin together with a variable amount of the remainder of the genome. Diploid phages occur about 5–6% of the time and are twice the normal length. They contain two complete circular DNA molecules. Longer polyphages contain increased numbers of DNA molecules in proportion to their length.

As usual, infection begins with the specific attachment of virions to the cell. Electron micrographs have shown the phage attached to the tips of F pili (see Fig. 7.6) or I pili. The Ff phages are described as being **male-specific** because these types of pili are produced only by male cells (carrying a conjugative plasmid). Despite their mode of attachment, the viruses do not actually release their DNA into the pili. DNA transfer requires functional *tolQRA* proteins, proteins that form an intrinsic membrane complex in the inner membrane. In the absence of these proteins, phages attach to the pili but do not infect (tolerance). During infection, the DNA is routed to the cytoplasm, and protein VIII, the major capsid protein, enters into the cytoplasmic membrane where it can be recycled into new phages.

Once a virion has attached to a cell, it enters an eclipse phase. Physically, the eclipse means that part of the protein coat opens up, partially releasing the DNA and making it susceptible to nucleolytic attack. The released DNA apparently penetrates through the cell membrane into the cytoplasm of the cell provided certain host membrane proteins are normal. The next steps in the infectious process are analogous to those of phage T5. Gene III protein and *E. coli* RNA polymerase initiate synthesis of a complementary DNA strand, after

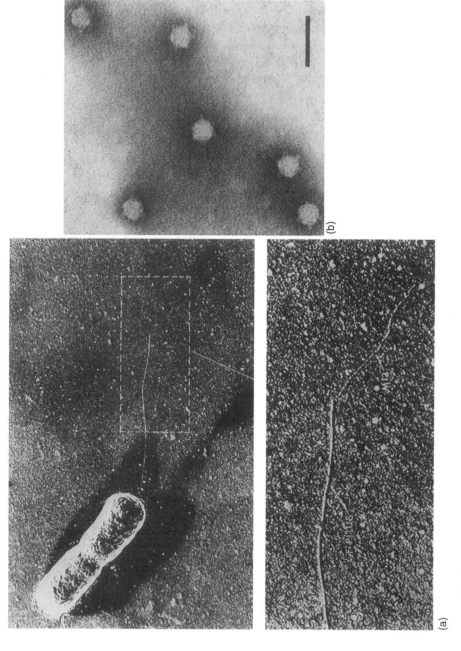

Figure 7.6. (a) Electron micrograph of bacteriophage M13 attached to the tip of an F pilus of *E. coli*. The photograph at the *bottom* is an enlargement of the area indicated by the *dotted lines*. (From Ray, D.S. [1977]. Replication of filamentous bacteriophages, pp. 105–178. In: Fraenkel-Conrat, H., Wagner, R.R. (eds.), *Comprehensive Virology*, vol. 7. New York: Plenum Press.) (b) Electron micrograph of bacteriophage φX174 negatively stained with phosphotungstate. The bar represents a length of 50 nm. (From Denhardt, D.T. [1977]. The isometric single-stranded DNA phages, pp. 1–104, op. cit.)

which the rest of the DNA molecule enters the cell as DNA polymerase III completes the synthesis.

Despite the binding of RNA polymerase to the single-strand viral DNA, all mRNA molecules are synthesized using the new complementary DNA strand as a template. The viral DNA is thus considered plus-strand DNA, as its sequence corresponds to the mRNA, and its complement is minus-strand DNA. Promoters are positioned to allow transcription to begin at genes II, X, V, IX, or IV (see Fig. 7.7). Some processing of the transcripts occurs using host enzymes.

An unusual feature of the genetic map is that gene X is actually the carboxy portion of gene II. They are translated in the same reading frame, and thus the gene X protein is a shortened version of the gene II protein. It is also significant that the IG region is not transcribed at all. Figure 7.8 shows that there is an enormous amount of potential secondary structure within the IG region. There are five major loops that can be predicted from the known base sequence. As outlined below, these loops seem to be associated with both transcription and replication. The lack of transcription can be attributed to loop [A], which functions as a ρ-dependent terminator of gene IV transcription. The loops also prevent binding of single-strand binding protein (SSB) to the IG region, emphasizing the replication and transcription start sites. Although genes II, V, and VII are located in the same transcription unit, there is substantially more synthesis of protein V than the other members of the operon.

The replication mechanism for the Ff phages has to be unusual, as conventional replication models do not generate single DNA strands. DNA replication begins with production of the complementary DNA (minus strand) during

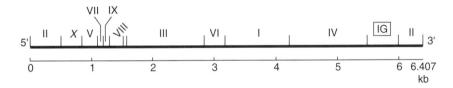

Figure 7.7. Genetic map of Ff phage. The circular genome is presented in linear form as if it had been opened at the unique HindII restriction site. Roman numerals refer to the genes. IG is the intergenic region. X refers to the part of gene II that codes for the X protein. The direction of transcription and translation is from left to right. The *bottom line* is a scale marked by kilobases. (Adapted from Zinder, N.D., Horiuchi, K. [1985]. Multiregulatory element of filamentous bacteriophages. *Microbiological Reviews* 49: 101–106.)

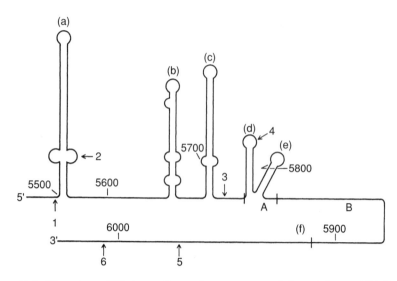

Figure 7.8 IG region of Ff phage. Secondary structure of the viral strand DNA. (a), (b), (c), (d), and (e) represent self-complementary sequence elements. (f) is an approximately 150-base-long, AT-rich sequence without self-complementarity, extending downstream from hairpin (e). Shown by *arrows* are: (1) end of gene IV; (2) rho-dependent termination of gene IV mRNA; (3) initiation site of minus-strand primer synthesis; (4) initiation site of plus-strand synthesis; (5) initiation site of gene II mRNA; (6) start of gene II. Domains A and B of the plus-strand replication origin are indicated. Four-digit numbers represent nucleotide numbers. (From Zinder and Horiuchi, op. cit.)

DNA entry. Only host proteins are necessary for this reaction; they include Ssb, DNA polymerase III, RNA polymerase, DNA ligase, and DNA polymerase I. A 30-base RNA primer is laid down near the base of loop [C] in Fig. 7.8. The primer is extended by DNA polymerase III holoenzyme, and it is then removed and the nick sealed as in conventional replication. The resulting double-strand DNA molecule is nicked in the position indicated by the number 3 in Fig. 7.8, which is where the RNA primer is located. After removal of the primer in the usual way, the covalently closed, circular, relaxed DNA molecule is designated a replicative form (RF) DNA, specifically RF IV DNA (Fig. 7.9). This completes stage I replication. RF IV DNA is converted to RF I (supercoiled) by DNA gyrase.

Although the RF I DNA should in principle be able to replicate in the usual way, it does not. Instead it replicates via a **rolling circle** mechanism that was first proposed by Gilbert and Dressler for concatemer production in phage T4.

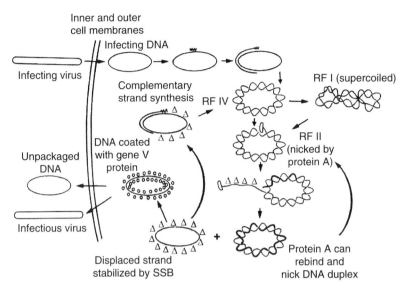

Figure 7.9. Rolling circle model for Ff phage DNA replication. The phage is represented as a long cylinder. The phage DNA enters the cell at the *upper left* as a single-strand, circular molecule. Priming with RNA (*wavy line*) initiates synthesis of a complementary DNA strand to yield an RF IV molecule that is covalently closed and circular. It can be converted to RF I by supercoiling. Both RF I and RF IV can be used to generate RF II by introducing a nick at a specific site. The nick serves to prime stage II DNA synthesis in which the complementary strand is used to synthesize another viral strand (*heavy line*) while the original strand is displaced. The displaced strand is coated with single-strand binding protein (SSB, *triangles*) to stabilize it. Recircularization of the displaced strand is accomplished by intramolecular pairing facilitated by the gene II protein. If a primer displaces SSB, complementary strand synthesis occurs. For stage III replication the displaced strand is coated with gene V protein, which displaces SSB and prevents synthesis of the complementary strand. As the DNA strand migrates through the cell membrane, coat protein subunits replace the gene V protein coating the DNA, and a new virus particle is formed. Some DNA is also extruded without being coated.

This model seems to be adequate to account for all DNA replication in the single-strand coliphages and for some replication in λ and other temperate phages as well as certain plasmids. Ironically, it does not seem to apply to T4.

During stage II replication, RF I replicates to make more RF I. The process begins with a nick introduced into the plus-strand by the gene II protein at site 4 of the IG region (intergenic; see Fig. 7.8) to yield RF II DNA. If both strands are nicked, the resulting linear molecule is designated RF III. The 3'-OH group located at the site of the nick serves as a primer for DNA synthesis by host proteins, DNA polymerase III holoenzyme, *rep*, and SSB. As new bases are laid down, the old plus-strand is displaced (Fig. 7.9) and coated with SSB. Complementary minus-strand synthesis occurs either after the old plus-strand is released in circular form (with the ends originally paired by loop [D] and sealed by gene 2 protein) or in a manner analogous to lagging strand synthesis in conventional DNA replication. The RF IV molecules are then converted to RF I. Most replicating molecules synthesize only a unit length of DNA and then terminate at a site in domain A of Fig. 7.8, presumably because the initiator proteins have not yet been released. Domain A is essential for replication and includes the initiation, termination, and gene II protein binding sites. Domain B functions in a manner similar to a eukaryotic enhancer sequence and is dispensable in phages with a compensatory mutation in gene II.

Stage III replication results in the production of only plus-strand DNA that is suitable for packaging into mature virus. The switch from stage II to III replication is controlled by the concentration of the gene V and X proteins, both of which have an affinity for single-strand DNA. As its concentration builds up, gene V protein may coat the plus-strand DNA instead of SSB, and DNA thus coated is available to be packaged into virions. Stage III replication occurs only after stage II replication has been established but does not completely replace it, although the presence of protein V also inhibits the synthesis of protein II. Therefore, the virus continues to replicate within the cell at all times. Gene V protein is the major regulatory element for phage M13. It is translationally autoregulated and also inhibits translation of proteins I and III.

The Ff phages do not cause cell lysis. Instead, they leak into the medium without significantly damaging the cell membrane. As a result, infection by Ff phages slows the growth of a culture but does not kill it. The "plaques" that are observed with these phages are actually due to the difference in growth rates between infected and uninfected cells and tend to disappear if incubated for prolonged periods. Maturation of the virus occurs as an extrusion process of viral DNA through the cell membrane. The capsid proteins are synthesized as procoat molecules with 23 extra amino acid residues located at the amino terminus. These extra amino acids constitute a **signal peptide** (hydrophobic tail) that causes the protein to be cotranslationally inserted into the cell membrane. During or after the insertion, the signal peptide is cleaved from the procoat molecule by a leader peptidase. As the circular phage DNA molecules are transported across the cell membrane, coat proteins in combination with host *lip* protein attempt to displace the gene V protein from the DNA and form mature virus (Fig. 7.10). However,

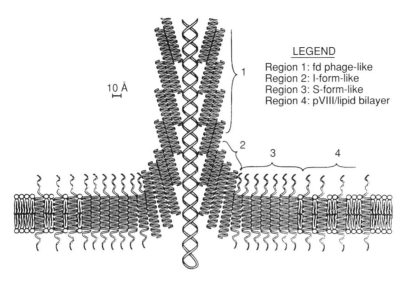

10 Å

LEGEND
Region 1: fd phage-like
Region 2: I-form-like
Region 3: S-form-like
Region 4: pVIII/lipid bilayer

Figure 7.10. Proposed pathway for pVIII during penetration and assembly. pVIII is proposed to pass through several sequential steps during phage penetration and assembly, including: (Region 1) a structure like fd phage; (Region 2) a structure like I-forms; this region might comprise only a few annuli, with the fd I-forms and the I-forms–S-forms conversions occurring one right after the other at the membrane surface; (Region 3) a structure like S-forms with pVIII having a transmembrane helix and nonhelical ends; the loss of helix at the ends is proposed to be due to increased repulsion of like charges as DNA–protein interactions are lost and as the highly charged ends come closer together following the contraction steps; (Region 4) the unusual lipid bilayer containing transmembrane pVIII. (From Dunker, A.K., Ensign, L.P., Arnold, G.E., Roberts, L.M. [1991]. A model for fd phage penetration and assembly. *FEBS Letters* 292: 271–274.)

some viral DNA molecules are also extruded without any coat. Exceptionally high titers of intact phage (circa 10^{12} ml^{-1}) of Ff phage can be routinely obtained.

The Ff phages have assumed an important role in molecular biology thanks to the efforts of Joachim Messing. He constructed a series of DNA cloning vectors based on phage M13, the M13mp series, which numbers some 19 members. The phages carry a portion of the *lacZ* gene to serve as a control marker and a specially constructed DNA insert that contains multiple restriction enzyme sites within a short space (a **polylinker**). This insert is located close to the origin of replication for the minus-strand. Therefore, by using an appropriate primer, it is possible to set up a replication system in vitro that preferentially copies the cloned DNA. The size of the DNA joined to the polylinker is not critical because of the method of virion assembly. The longer conjoint DNA is simply coated with additional coat protein subunits, up to a point.

Bacteriophage φX174

Bacteriophage φX174, as its name implies, was the 174th isolate in group 10 of a large series of bacteriophages. It is the principal representative of a group of phages that are simple icosahedrons with 5 nm spikes extending from all 12 vertices (see Fig. 7.6b). The capsid is composed of 60 molecules of protein F arranged in pentamers with vertices constructed from five molecules of protein G for each pentamer plus one molecule of protein H. Also associated with the capsid are 60 molecules of protein J, one of A* (the carboxy terminal portion of A), and the polyamines spermidine and putrescine. The DNA molecule is again single-stranded and circular. As in the case of the Ff phages, φX174 eclipses outside a cell, protruding some portion of its DNA. Protein H is partially located within the capsid, and ejection requires disrupting its interaction with the DNA. If this DNA is not converted to a double-strand form, the infection is abortive.

Priming of stage I replication occurs via a primosome such as that seen for bacterial chromosome replication. More than one initiation site is apparently possible. Extension of the primer is via the same mechanism as that used by Ff phages. Complementary strand synthesis is initiated at a stem-loop structure and generates an RF molecule.

Stage II replication is initiated by a nicking protein, protein A, which attaches covalently to the nicked DNA strand via a tyrosine residue (reviewed by Novick 1998). About 30 bp of φX174 DNA are necessary and sufficient to initiate rolling circle replication when cloned into a plasmid. The general outline of this rolling circle replication is the same as for the Ff phages. Host *rep* protein is required for its helicase activity. Termination of synthesis of plus-strand DNA is via the attached protein A. As protein C concentration increases, it inhibits the initiation or reinitiation of rolling circle replication by binding to the RF-protein A-Rep complex in the presence of ATP. The new complex is a substrate for the packaging system when proheads and protein J are supplied. Protein A* also assists in the switch to stage III replication.

Unlike members of the Ff group, φX174 infections cause cell lysis, which is dependent on gene E function. E protein integrates into the inner membrane to inhibit the MraY enzyme required for peptidoglycan synthesis, resulting in cell lysis due to catastrophic failure at cell division (Bernhardt et al. 2002). This process requires normal functioning of the host cell *ftsZ* and *ftsA* genes, both of which are involved in the cell division process, as well as the SlyD protein (Bern-hardt et al. 2002). The SlyD protein is a peptidyl-prolyl *cis–trans* isomerase that stabilizes the phage E protein.

There was originally a problem concerning mRNA synthesis and transla-tion in φX174-infected cells due to a peculiarity of the genetic map. Based on the number and size of the proteins resulting from an infection, it has been estimated that 6100 nucleotides are required to code for all the necessary amino acid sequences. The genome of the virus, however, consists of only 5386 nucleotides. This discrepancy was clarified by the discovery that three of the genes are embed-ded in other genes (Fig. 7.11). Gene B lies at the end of gene A, and gene E lies at

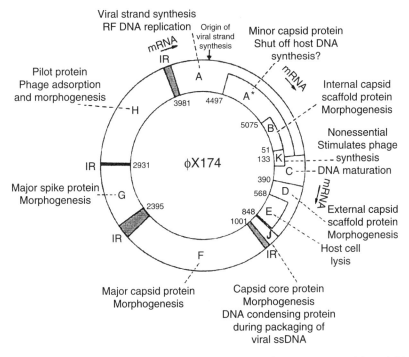

Figure 7.11. Genetic map of φX174. Ten genes are drawn, separated by *solid lines.* The 11th is A*, which is an internal restart (i.e., an alternative ribosomal binding site) within gene A. The intergenic spaces are marked in *gray* and labeled IR. Inter-nal numbers indicate gene positions on the DNA sequence. The origins and direc-tions of transcription are shown. (Adapted from Weisbeeck, P. [1990]. Genetic and restriction map of bacteriophage φX174, p. 1.79. In: O'Brien, S.J. (ed.), *Genetic Maps,* 5th ed., Cold Spring Harbor, NY: Cold Spring Harbor Laboratory.)

the end of gene D. Gene K translation begins at an overlap of the two terminator codons for gene B and spans the last 86 bases of gene A and the first 89 bases of gene C. The proteins produced by the **embedded genes**, unlike the case for the A and A* proteins, do not resemble the proteins encoded by the larger, surrounding genes because the mRNA molecules are translated in different reading frames.

Genetic relatedness is often determined by codon analysis, which is based on the principle that when several codons are possible for a single amino acid, an organism may characteristically use only one or two. In the case of leucine residues in protein G, for example, six codons are possible, but GAT is never used and GAA is used 50% of the time. An analysis of the codons comprising the embedded and nonembedded genes suggests that gene A was formerly shorter but lost its terminator signal and now reads through the B gene in a different reading frame. The amino-terminal portion of A and all of the B gene use codons that are typical for ϕX174, but the region of the A gene that overlaps B uses unusual codons. By contrast, all of gene E uses codons that are rarely used anywhere else in the ϕX174 genome, even though gene D uses normal codons. This finding suggests that the E gene evolved from a preexisting D gene by formation of a new translation start signal, and that D was always as large as its present size. Codon analysis for gene K suggests that its origin is similar to that of E.

Obviously, most mutations in an overlapping region affect both proteins. Nevertheless, it is theoretically possible to mutate one protein without changing the other. This feat has been accomplished using site-directed mutagenesis to make a T→A transversion that created an amber codon with respect to gene K but retained a leucine codon in the gene A reading frame. The mutated phage was still viable but had a reduced burst size.

THINKING AHEAD

What are the problems inherent in replicating an RNA molecule?

RNA-Containing Bacteriophages

Single-Stranded RNA Viruses

The RNA phages are small icosahedral viruses that contain a molecule of linear, single-strand RNA inside a capsid (reviewed by van Duim [2005]). The coliphages can be subdivided into four groups based on criteria such as the immunologic cross reactivity of the coat protein, the buoyant density of the virion, the ratio of adenine to uracil residues in the RNA molecule, and the amino acids that do not occur in the coat protein. Each group also apparently produces a specific enzyme for RNA replication that does not replicate the RNA from any other group. Commonly encountered group members are (I) f2, MS2, R17; (II) GA; (III) Qβ; and (IV) SP, Fl. The capsids of groups I and II consist of 180 subunits of coat protein plus one molecule of maturation protein, while groups III and IV have an additional 12 copies of a "read-through" protein that results from improper termination of the coat translation.

The RNA phages are further examples of male-specific phages. They infect Hfr, F⁺, or F′ *E. coli* cells or cells of any other genus into which an F plasmid has been transferred (e.g., *Salmonella*, *Shigella*, and *Proteus*). The site of phage attachment is the side of the F pilus. Therefore, male cells that have been depiliated by shearing in a blender are resistant to phage infection because they lack receptors. Cells that have been infected or superinfected with a DNA phage are also resistant to RNA phage infection.

When the phage RNA reaches the cytoplasm of the cell, it serves as its own mRNA (i.e., it is a plus-strand virus). It also has guanosine-5′-triphosphate as the initial base. The viral RNA contains sufficient nucleotides to code for at least four proteins: the coat protein, the A (maturation) protein, the replicase protein, and the L (lysis) protein. In Qβ, a fifth protein, A1, has been observed that results from inefficient termination at the end of the A gene and consequent read-through into the first portion of the coat protein gene. This is another example of overlapping genes.

Genetic maps of RNA are difficult to prepare because reversion rates for mutations in RNA phages approach 0.1%, which is much higher than what is generally observed for DNA recombination. In turn, the high reversion rate may be due to lower accuracy requirements for RNA synthesis in contrast to DNA synthesis in an *E. coli* cell. The genetic sequence was finally determined by biochemical techniques in which the RNA molecule was fragmented in a known way, and then fragments were translated using an in vitro protein-synthesizing system. The genetic map for the MS2 group is shown in Fig. 7.12. Note that two reading frames are used and the L protein is the product of an embedded gene.

Fiers and coworkers determined the complete sequence of the RNA molecule from phage MS2. Some in vitro protein-synthesizing systems cannot translate the replicase gene without first translating the coat protein. This phenomenon is known as **translational coupling** and results from the existence of a secondary structure in the RNA molecule. In order to reflect possible intramolecular structure, the base sequence of the RNA can be presented so as to maximize

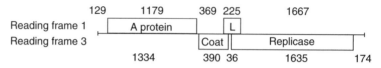

Figure 7.12. Genome of MS2. The four genes are shown as *rectangles;* untranslated regions are indicated by *narrow spaces*. The 5'-end of the RNA molecule is shown at the *left*. Reading frame 1 begins at the first base, and reading frame 3 begins at the third base. The numbers above and below the diagram indicate the number of bases in a particular segment of the reading frame. In each case the initiator codon is taken to be part of the gene, and the terminator codon is taken to be part of the untranslated region. No proteins are known to be read using reading frame 2. (Redrawn from Fiers, W., Contreras, R., Duerinck, F., Haegeman, G., Iserentant, D., Merregaert, J., Min Jou, W., Molemans, F., Raeymaekers, A., Van den Berghe, A., Volckaert, G., Ysebaert, M. [1976]. Complete nucleotide sequence of bacteriophage MS2 RNA: Primary and secondary structure of the replicase gene. *Nature* 260: 500–507.)

the number of intramolecular hydrogen bonds. Figure 7.13 presents such a structure as developed by van Duin and his coworkers. The highly convoluted loops are frequently referred to as "flower structures," and Fiers and coworkers described the entire molecule as a "bouquet." Note that a double-stranded region of the molecule masks the start codon for the replicase protein. This structure should make the start codon unavailable to 30S ribosomal subunits and thus block translation. Confirmation of this idea comes from the work of Mills and coworkers who cloned Qβ replicase on a plasmid and then grew mutated Qβ in the presence of the plasmid. Some mutations in the 3' region of replicase are *cis*-acting (cannot be complemented by the clone) and serve to identify a region whose functions are both coding and structural (stem-formation).

The fact that translation of the coat protein allows translation of the replicase protein provides concrete evidence that a ribosome can disrupt hydrogen bonds in an mRNA molecule once it has attached but cannot disrupt them in order to effect its own attachment. This observation generalizes conclusions drawn from RNA polymerase binding studies, namely, that nucleic acid secondary structure or previously bound proteins prevent initiation of macromolecular synthesis but are no impediment to continued synthesis. Poot et al. (1997) have shown a similar phenomenon with respect to the maturation of A protein. The 5'-end of the MS2 RNA forms a stem-loop structure. Immediately adjacent to this region is a stretch of RNA that can form three adjacent stem-loop structures

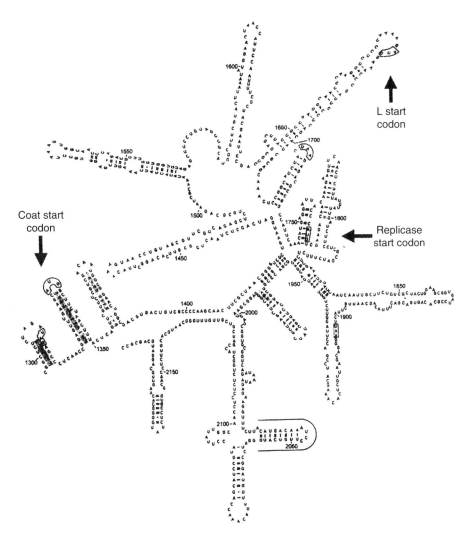

Figure 7.13. Proposed secondary structure for the central region of MS2 RNA (bases 1297–2190). Helices for which phylogenetic evidence exists are drawn with base-pair dashes. Helices conserved in group A have shaded base pairs and the I helix conserved in groups A and B is framed by a *continuous line*. Initiator and terminator codons are *boxed*. (Adapted from Skripkin, E.A., Adhin, M.R., de Smit, M.H, van Duin, J. [1990]. Secondary structure of the central region of bacteriophage MS2 RNA. Conservation and biological significance. *Journal of Molecular Biology* 211: 447–463.)

that interact with one another for some distance. The long distance interaction thus developed renders the GUG start codon inaccessible. Poot et al. suggest that the advantage of blocking the start codon is that it allows the RNA to replicate properly. However, some translation must occur. Formation of the long distance interaction may require as much as 5 min, giving the ribosomes a "window of opportunity" to initiate translation.

Another interesting point about translation of the MS2 RNA molecule is that the ribosome binding site upstream from the region coding for the L protein (and therefore part of the coat protein coding sequence) is normally blocked by a hairpin loop. Again, there is translational coupling, this time between the L and coat proteins. However, an added complication is that the two proteins are in different reading frames (see Fig. 7.12). van Duin and coworkers have demonstrated that ribosomes translating the coat terminate just downstream from the beginning L protein coding sequence. The ribosomes then restart by moving upstream and initiate the synthesis of L protein in a new reading frame. The frequency of restart can be affected by changing the base sequence upstream from the ribosome binding site for coat protein, or by altering the translational fidelity of the ribosome since termination that occurs at a large distance from the Shine–Dalgarno box for L protein renders it effectively invisible to a ribosome. The successful restart is an example of programmed frameshifting.

The presence of a number of regulatory loops in the MS2 RNA molecule raises the question of whether the loops make the RNA susceptible to attack by RNA processing enzymes such as RNaseIII. An MS2 RNA molecule carrying an RNase III stem-loop target sequence in a nonessential region can reproduce in an RNaseIII mutant strain but not in a wild-type strain. Phages that manage to reproduce in the wild-type strain acquire mutations that shorten the stem, introduce mismatches, or introduce short deletions or insertions. The effect of these mutations is to render the RNA inaccessible to RNaseIII attack.

Replication of phage RNA is an involved process (Fig. 7.14). It requires a complex of four proteins designated by Greek letters, three of which are provided by the host and are, surprisingly, part of the RNA translation system. These are ribosomal protein S1 (α) and the two elongation factors for protein synthesis, Tu (γ) and Ts (β). The fourth protein is, of course, the viral RNA replicase (ρ). The complex constitutes an RNA-dependent RNA polymerase. In Qβ, the replicase has an absolute specificity for the sequence CCCA3′ (Tretheway et al. 2001). Full length replication is assured in any RNA molecule with that particular base sequence, although there are some optional binding sites located internally.

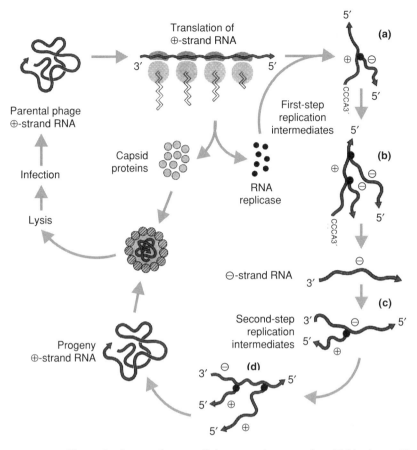

Figure 7.14. General scheme of intracellular reproduction of an RNA phage. Upon infection, the parental RNA plus-strand is released from the capsid and translated to form capsid proteins and the RNA replicase. The replication of the phage RNA can now begin. At the initial stage of the replication process, the single-strand parental plus-strand serves as the template for the synthesis of complementary minus-strands. It results in the formation of first-step replication intermediates (**a**) and (**b**), open structures in which the template plus-strand and the replica minus-strand do not form an RNA double helix. In the next stage of the replication process, the single-strand minus-strand serves as the template for plus-strand synthesis. It results in the formation of second-step replication intermediates (**c**) and (**d**), which are similar in structure to the first-step replication intermediates except that their full-length template is a minus-strand rather than a plus-strand. The replica plus-strands are encapsulated by the capsid proteins to form structurally intact progeny phages, to be released on lysis of the infected cell. (Throughout the scheme shown here all complete plus- and minus-strands have the same length. (Adapted from Stent, G.S., Calendar, R. [1978]. *Molecular Genetics*, 2nd ed., San Francisco CA: W.H. Freeman.)

This protein complex uses the viral plus-strand to produce minus-strands, and the minus-strands then serve as templates for production of more plus-strands. In vivo, few duplex RNA molecules can be detected, a condition that leaves plus-strands free to be translated or packaged into phage particles. Binding of the replicase to the S-site located between bases 2664 and 2788 on the 3680 base molecule (where 3680 is the 3′ terminus) inhibits translation of the coat protein gene.

If an in vitro reaction for Qβ replicase is set up but no template RNA is provided, the enzyme apparently generates RNA molecules spontaneously. Should one of these molecules accidentally prove to be a suitable substrate for further replication, it will reproduce rapidly and account for most of the RNA in the tube. Comparisons between reaction mixtures indicate that each time the reaction is run, different RNA molecules are obtained. In a similar vein, normal phage RNA can be added to an in vitro system under conditions where only the replicase protein must be transcribed and translated for RNA replication to occur. If the reaction mixture is diluted at regular intervals, a premium is placed on minimizing the size of the RNA molecule to be replicated, and self-shortening of the RNA molecule is observed (but always preserving the replicase function). This system therefore provides an in vitro model for a type of evolution.

Maturation proceeds in much the same way as in the DNA phages. Coat protein binds to plus-strand RNA molecules and prevents further replication or translation, thereby serving the incidental function of turning off the expression of the viral replicase. A terminator mutation in the coat protein results in accumulation of greater-than-normal amounts of replicative RNA molecules. The maturation protein (A_2) is necessary for phage infectivity and, in the case of Qβ, provides lytic functions. Bernhardt et al. (2001) have shown that as in the case of φX174, lysis is accomplished by inhibition of cell wall biosynthesis. The function inhibited in this case is MurA, an enzyme belonging to the same biosynthetic pathway as MraY, which in fact suggests a general strategy for phage-induced cell lysis. If a phage carries a terminator mutation in the A gene, phage particles of normal appearance are produced, but they are noninfectious.

Double-Stranded RNA Viruses

There is a small group of bacterial viruses with a genome that is double-stranded RNA, similar to the Reoviruses of animals. Unlike the viruses discussed so far, the genome consists of three separate double-strand RNA molecules of different sizes. These viruses are also unusual in that they have a lipid envelope surrounding the capsid. The best-studied member of this group is φ6, but other, essentially similar viruses have been isolated (Mindich et al. 1999). Sequence analysis shows that they are not necessarily genetically related.

Replication and maturation of φ6 presents some obvious problems, especially the difficulty in ensuring that three different RNA molecules be present in each virion. Butcher et al. (1997) have examined the problem and shown that proteins P1, P2, P4, and P7 form a procapsid. The individual plus-strands enter the procapsid, and minus-strand synthesis occurs within it. However, minus-strand synthesis only occurs when all three segments are present. After minus-strand synthesis has occurred, plus-strands may be produced.

Like other RNA viruses, φ6 can exit the cell without lysis. Its phospholipid membrane contains five viral-specific proteins. Subsequent infections occur by membrane fusion and involve the type IV pilus of *Pseudomonas syringae* or a similar pseudomonad.

Bacteriophages Infecting *Bacillus subtilis*

Bacteriophage SP01

Bacteriophage SP01 (and similar phages listed in Table 7.1) has the same general morphology as T4 but is somewhat larger. Its linear DNA molecule has a unique sequence with 12.4 kb direct terminal repeats and a unique chemical composition. Instead of thymine (5-methyluracil, MdUMP), the DNA contains the modified base 5-hydroxymethyluracil (HMdUMP), which serves as a chemical label to permit restriction enzymes to distinguish between host and viral DNA. The phage codes for enzymes that degrade thymidine and uridine deoxyribonucleotide triphosphates (dTTP and dUTP) to their respective monophosphates and those that convert deoxyuridine monophosphate to HMdUMP and then to HMdUTP, so that it can be used in DNA synthesis. The enzymes that break down dTTP and dUTP are not essential for phage growth, but in their absence up to 20% of the HMU residues are replaced by thymine.

Infection occurs by slow movement of DNA into a cell, a process that requires several minutes to complete. Crude genetic maps of the phage genome can be prepared by the simple expedient of violently shearing the virion and uninserted DNA away from the host cell with a blender. The cells are then superinfected with various defective phages and observed to see if cell lysis occurs. The results are analyzed as for a *cis–trans* test. As expected, genes controlling DNA metabolism are the first to enter.

After injection of the phage DNA, the physiologic pattern is similar to that of T4. Host DNA replication shuts down within 5–7 min postinfection at 37°C, although little or no DNA degradation is observed. The shutdown of replication is independent of the enzymatic breakdown of dTTP. Host rRNA synthesis continues until cell lysis. However, host mRNA synthesis is rapidly replaced by viral RNA synthesis partly due to extensive modifications to the RNA polymerase complex. The SP01 DNA polymerase gene contains a Group I self-splicing intron very similar to the one found in phage T4.

Three proteins are produced that confer different template specificities on the RNA polymerase holoenzyme by replacing the normal sigma factor. Transcription itself is subdivided into the temporal groups early, middle, and late (*e*, *m*, and *l*). Early mRNA is produced by host polymerase (RNA polymerase A) transcribing primarily from the terminal repeats. This region has been described as the "host takeover module," and consists of 24 genes in 12 operons. Most of them are less than 100 codons in length and show little homology to other proteins in the data bank. Synthesis of middle mRNA requires RNA polymerase B, formed by replacement of host σ^{43} by viral gene 28 protein. Late mRNA synthesis requires that gene 33 and 34 proteins replace gene 28 protein to give RNA polymerase C. This cascade of sigma factors is typical of *B. subtilis* regulation (see also Chapter 14). Six subclasses of mRNA have been identified on the basis of time of appearance and time of shutoff measured from the start of infection. They are *e*, 1–5 min; *em*, 1–12 min; *m*, 4–12 min, m_1l, 4 min until lysis; m_2l, 8 min until lysis; and *l*, 12 min until lysis. Type e mRNA is processed by an *E. coli* RNase III-like host enzyme. Transition from early to middle RNA requires the products of genes 44, 50, and 51 (Sampath and Stewart 2004).

DNA replication in SP01-infected cells uses at least two origins that produce concatemeric structures up to 20 genomes in length. The maturation and packaging systems are assumed to be similar to those of phage T5. The onset of sporulation in the host cell effectively blocks further development of SP01 and prevents cell lysis. Consequently it is possible to have endospores that carry viral genomes and release phage particles during outgrowth. The mechanism of inhibition probably involves further modification of the RNA polymerase holoenzyme so it no longer transcribes SP01 DNA efficiently but transcribes the bacterial *spo* genes.

Bacteriophage φ29

Of all the phages discussed in this chapter, bacteriophage φ29 is the smallest double-strand DNA phage, in terms of both head size and DNA size (see Table 7.1). The DNA molecule is unusual in that it is circularized by a protein (product of gene 3) that is covalently bonded via a serine residue to the 5′-ends of the linear DNA, which contain short inverted repeats. Removal of this protein dramatically reduces the efficiency of transfection (see Chapter 10). Infection by φ29 does not significantly affect the level of macromolecular synthesis by the cell prior to the time of lysis.

DNA transcription occurs in the familiar manner. Three major transcripts have been identified: two early transcripts that correspond roughly to the ends of the DNA molecule and one late transcript that corresponds to the middle of the DNA molecule. The early mRNA molecules are transcribed from the L strand, whereas late mRNA transcription uses the H strand (Fig. 7.15). The time lag between infection and start of major mRNA synthesis is 6–8 min, which is longer than for most phages. The shift from host cell to viral transcription occurs as the result of the synthesis of a new polypeptide, p4, which tends to replace the host sigma factor in the RNA polymerase holoenzyme. P4 is an example of a protein that can be both an activator and a repressor. Promoters A2b and A2c have the typical *B. subtilis* σ^A binding sites. They activate promptly upon arrival in the cytoplasm. After p4 is synthesized, a dimer binds to the −35 site of promoter A2b in a reaction facilitated by p6, bending the DNA. The bent

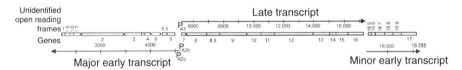

Figure 7.15. Genetic and transcriptional maps of bacteriophage φ29. The *boxes* represent open reading frames, regions that should code for proteins. Numbers below the boxes indicate definite gene assignments. Numbers above the boxes are not yet assigned to genes. *Thin arrows* represent mRNA transcripts, and the numbers associated with them are the base-pairs from the left end. All early transcription occurs on the strand defined as the L or 5′ strand. (Adapted from Pacek, V., Vlcek, C. [1993]. Genetic and restriction maps of *Bacillus* bacteriophage φ29, pp. 1.78–1.82. In: O'Brien, J.S. (ed.), *Genetic Maps*, 6th ed. Book 1 Viruses. Cold Spring Harbor, NY: Cold Spring Harbor Laboratory.)

configuration blocks RNA polymerase access to promoter A2b, but the p4 subunit not involved in the bending binds to the carboxy terminus of the RNA polymerase α subunit in a manner similar to that for CRP and the *lac* operon (see Chapter 4). This anchorage allows the normally inactive promoter A3 to function even though it lacks at good −35 binding site.

Promoter A2c is located 91 bp away from promoter A3 and therefore is not affected by the bend introduced by p4 into the DNA. Instead p4, p6, and RNA polymerase bind so strongly to promoter A2c that the complex is unable to move from the promoter, thus blocking the promoter to further transcription.

APPLICATIONS BOX

What sort of experiment would you conduct to prove that protein P4 binds to RNA polymerase to form a stable complex that blocks promoter A2c?

The DNA replication system for φ29 has been studied extensively in the phage and the cloned phage DNA (Meijer et al. 2001). No RNA primer is needed for φ29-specific replication to be initiated, but protein 3 must be covalently joined specifically to the physical ends of the DNA molecule, making it a terminal protein (TP). Cloned φ29 DNA does not self-initiate unless the plasmid is linearized adjacent to the TP binding sites. Protein 2 catalyzes a reaction between free protein 3 and 5′-dAMP so as to give the initiation complex, because the DNA polymerase requires a base for initial synthesis. The reaction is stimulated by the presence of protein 6 that binds to a specific 24 bp sequence, bending the DNA to make a positive superhelix. The protein–AMP complex acts as a primer and is extended by a phage-encoded DNA polymerase coupled with single- and double-strand DNA binding proteins. Protein p17 is stimulatory at low multiplicities of infection, but dispensable later in the infectious cycle. The initiation process has a sliding-back step in which the adenine residue that was attached to the TP is removed and replaced with the correct base for what is present in the template strand. After synthesis of a short stretch of DNA, the TP dissociates, and the two replication forks (one from either end) proceed toward one another. When they meet, the type I replication intermediate becomes a type II (double strand at one end, single strand at the other). The phage produces two single-strand DNA binding proteins, 5 and 16.7. The former is similar to cellular proteins. The latter has a membrane insertion domain and apparently anchors the replicating DNA to the membrane (Serna-Rico et al. 2002).

Phage proheads copurify with a small (120 bases) packaging RNA molecule (pRNA) whose presence is indispensable for packaging to occur. The promoter for pRNA is A1. Six pRNA molecules bind to the portal vertex of the procapsid and act sequentially to package DNA. There is significant homology between the left end of the DNA molecule and the prohead RNA, which may assist in the actual packaging operation. After packaging is complete, the pRNA is released.

Bacteriophages Infecting the Archaea

The surge of interest in the genetics of the archaea has led to the discovery of several bacteriophages that infect them. The catalog is not very extensive, but phage SNDV, which infects *Sulfolobus*, has some interesting features (Arnold et al. 2000). It is droplet-shaped with many tail fibers at the pointed end. It is an unusual virus in that the only time virions are released from infected cells is during the stationary growth phase. At other times the cells are growing too fast for the viral synthesis to keep up. It is thus another example of a pseudolysogen.

Summary

Bacteriophages are a heterogeneous group of "organisms." They may be rodlike, spherical, or complex (having a head and tail). Their nucleic acids may be single- or double-strand RNA or DNA.

As a general rule, the single-strand DNA molecules are circular, whereas all other viral nucleic acids are linear. The circularity can be attributed to the difficulty of protecting linear single-strand DNA from exonucleolytic attack. Infection begins as the virus attaches to the cell surface. In all cases, at least some of the impetus for the transfer of nucleic acid from the virion into the cell is provided by the host cell itself. Often it takes the form of a required replication or transcription step. Transcription of viral DNA is a highly regulated process, both temporally and spatially. In many instances, the genes are arranged in the order of their use so that transcription provides an automatic sequencing of events. As in the case of phage T4, the host RNA polymerase is often modified so that the promoters for genes needed late are recognized and those for regions that have already been adequately transcribed are not.

Nucleic acid synthesis is as varied as the phages themselves. The large viruses show little dependence on host metabolism and frequently degrade the host

DNA. Such phages protect their own DNA from degradation by the use of unusual bases such as hydroxymethyluracil or hydroxymethylcytosine. The smaller phages have shorter nucleic acids that lack the coding capacity to produce large numbers of polypeptides. As a result, they are more dependent on host cell function. They may also exhibit extensive genetic overlap (embedded genes). The single-strand DNA phages convert their DNA to the double-strand form and then use the rolling circle mode of replication, which can generate either more double-strand DNA or else single-strand DNA for the production of more virus particles.

Assembly of new virions has been studied extensively in only a few cases. The complex phages with circularly permuted, terminally redundant DNA molecules are presumed to function similar to T4. In cases where the DNA is not circularly permuted, the model predicts a site-specific cut in a concatemeric molecule followed by packaging. The single-strand phages seem to assemble by the simple expedient of having single-strand DNA binding protein subunits attach to the nucleic acid as it is synthesized. These proteins are replaced by the coat proteins later.

The bacteriophages have made a number of important contributions to DNA cloning procedures. Phage T5 promoters are exceptionally strong and can be used to generate large amounts of product from the cloned DNA. The T7 RNA polymerase is specific for its late promoters and has found a use in regulating transcription of cloned DNA in both prokaryotic and eukaryotic cells. The DNA remains quiescent until the polymerase is provided, and then it is efficiently transcribed.

Questions for Review and Discussion

1. Describe the strategies used by bacteriophages to sequentially express their genes. Which of these strategies would interfere with infection by (a) a different bacteriophage; (b) the same bacteriophage?
2. What unusual modes of nucleic acid replication are found in bacteriophages? What are the advantages of using those modes instead of conventional Watson–Crick semiconservative replication?
3. How would you identify bacteriophage nucleic acid having the following properties?

Contains interruptions (nicks)
Single-stranded
Circular
Contains an intron

References

General

Ackerman, H.W., Krisch, H.M. (1997). A catalogue of T4-type bacteriophages. *Archives of Virology* 142: 2329–2345.

Campbell, A. (2003). The future of bacteriophage biology. *Nature Reviews Genetics* 4: 471–477.

de Smit, M.H., van Duin, J. (1990). Control of prokaryotic translational initiation by mRNA secondary structure. *Progress in Nucleic Acid Research and Molecular Biology* 38: 1–35.

Henkin, T.M. (2000). Transcription termination control in bacteria. *Current Opinion in Microbiology* 3: 149–153.

Meijer, W.J.J., Horcajadas, J.A., Salas, M. (2001). φ29 family of phages. *Microbiology and Molecular Biology Reviews* 65: 261–287.

Novick, R.P. (1998). Contrasting lifestyles of rolling-circle phages and plasmids. *Trends in Biochemical Sciences* 23: 434–438.

van Duin, J. (2005). Bacteriophages with ssRNA. *Encyclopedia of Life Science.*

Yin, J. (2004). Genome function—A virus world view. Advances in systems biology, pp. 31–46. In: Opresko, L.K., Gephart, J.M., Mann, M.B. (eds.), *Advances in Experimental Medicine and Biology*, vol. 547. New York: Kluwer Academic/Plenum Publishers .

Specialized

Arnold, H.P., Ziese, U., Zillig, W. (2000). SNDV, a novel virus of the extremely thermophilic and acidophilic archaeon *Sulfolobus. Virology* 272: 409–416.

Bernhardt, T.G., Wang, I.-N., Struck, D.K., Young, R. (2001). A protein antibiotic in the phage Qβ virion: Diversity in lysis targets. *Science* 292: 2326–2329.

Bernhardt, T.G., Roof, W.D., Young, R. (2002). The *Escherichia coli* FKBP-type PPIase SlyD is required for the stabilization of the E lysis protein of bacteriophage φX174. *Molecular Microbiology* 45: 99–108.

Butcher, S.J., Dokland, T., Ojala, P.M., Bamford, D.H., Fuller, S.D. (1997). Intermediates in the assembly pathway of the double-stranded RNA virus φ6. *The EMBO Journal* 16: 4477–4487.

Gary, T.P., Colowick, N.E., Mosig, G. (1998). A species barrier between bacteriophages T2 and T4: Exclusion, join-copy and join-cut-copy recombination and mutagenesis in the dCTPase genes. *Genetics* 148: 1461–1473.

Hsu, L.M., Vo, N.V., Kane, C.M., Chamberlin, C.M. (2003). In vitro studies of transcript initiation by *Escherichia coli* RNA polymerase. 1. RNA chain initiation, abortive initiation, and promoter escape at three bacteriophage promoters. *Biochemistry* 42: 3777–3786.

Mindich, L., Qiao, X., Qiao, J., Onodera, S., Romantschuk, M., Hoogstraten, D. (1999). Isolation of additional bacteriophages with genomes of segmented double-stranded RNA. *Journal of Bacteriology* 181: 4505–4508.

Monod, C., Repoila, F., Kutateladze, M., Tetart, F., Krisch, H.M. (1997). The genome of the pseudo T-even bacteriophages, a diverse group that resembles T4. *Journal of Molecular Biology* 267: 237–249.

Pajunen, M.I., Elizondo, M.R., Skurnik, M., Kieleczawa, J., Molineux, I.J. (2002). Complete nucleotide sequence and likely recombinatorial origin of bacteriophage T3. *Journal of Molecular Biology* 319: 1115–1132.

Poot, R.A., Tsareva, N.V., Boni, I.V., van Duin, J. (1997). RNA folding kinetics regulates translation of phage MS2 maturation gene. *Proceedings of the National Academy of Sciences of the USA* 94: 10110–10115.

Roberts, M.D., Martin, N.L., Kropinski, A.M. (2004). The genome and proteome of coliphage T1. *Virology* 318: 245–266.

Sampath, A., Stewart, C.R. (2004). Roles of genes 44, 50, and 51 in regulating gene expression and host takeover during infection of *Bacillus subtilis* by bacteriophage SPO1. *Journal of Bacteriology* 186: 1785–1792.

Sandegren, L., Sjöberg, B.-M. (2004). Distribution, sequence homology, and homing of group I introns among T-even-like bacteriophages: Evidence for recent transfer of old introns. *The Journal of Biological Chemistry* 279: 22218–22227.

Serna-Rico, A., Salas, M., Meijer, W.J.J. (2002). The *Bacillus subtilis* phage φ29 protein p16.7, involved in φ29 DNA replication, is a membrane-localized single-stranded DNA-binding protein. *The Journal of Biological Chemistry* 277: 6733–6742.

Tetart, F., Desplats, C., Krisch, H.M. (1998). Genome plasticity in the distal tail fiber locus of the T-even bacteriophage: Recombination between conserved motifs swaps adhesin specificity. *Journal of Molecular Biology* 282: 543–556.

Tretheway, D.M., Yoshinari, S., Dreher, T.W. (2001). Autonomous role of 3′-terminal CCCA in directing transcription of RNAs by Qβ replicase. *Journal of Virology* 75: 11373–11383.

You, L., Suthers, P.F., Yin, J. (2002). Effects of *Escherichia coli* physiology on growth of phage T7 in vivo and in silico. *Journal of Bacteriology* 184: 1888–1894.

8

Genetics of
Temperate Bacteriophages

For all bacteriophages discussed in the preceding chapters, a successful phage infection always results in the immediate production of progeny virions. However, many bacteriophages are known for which there is an alternative outcome to phage infection. Instead of the customary unrestrained DNA replication and phage assembly, there is a temperate response in which a bacteriophage sets up housekeeping within a bacterial cell and maintains a stable relationship with that cell and all its progeny for many generations. The varied ways in which the temperate response can be accomplished are the subject of this chapter. The physical properties of the temperate bacteriophages discussed in this chapter are summarized in Table 8.1.

Major topics include:

- Events that occur during the basic life cycle of phage lambda
- Regulation using DNA inversion

Table 8.1. Physical properties of various temperate bacteriophages.

Phage	Usual Host	DNA Molecule (kb pairs)	Topology	Morphology	Dimensions (nm)	Related Phages	Prophage DNA
λ	E. coli	48.5	Unique sequence, cohesive ends	Icosahedral head, non-contractile tail	60, 150 × 17	21, φ80, 82, 424, 434	Circular permutation
P22	Salmonella	43	Circularly permuted, terminally redundant	Icosahedral head, six short tail spikes around central core	65, 18	L, LP7, SF6	Circular permutation
P2	E. coli, Shigella, Serratia	33.8	Unique sequence, cohesive ends	Icosahedral head, contractile tail	57, 140 × 17	PK, 186, 299, Wφ	Circular permutation

P4	E. coli, Shigella, Serratia	11.6	Unique sequence, cohesive ends	Icosahedral head, contractile tail	46, 133 × 17		Circular permutation
P1	E. coli, Shigella, Serratia	97	Circularly permuted, terminally redundant	Icosahedral head, contractile tail	87, 226 × 18	P7	Circular, nonintegrated
Mu	E. coli	42	Unique sequence, ends are host DNA	Icosahedral head, contractile tail	56, 130 × 18	D108	Colinear
SSV1	S. solfataricus	15.465		Spindle-shaped			
SSV2	S. solfataricus	14.796		Spindle-shaped			

Note: Terminology is that of Table 7.1 except for "Prophage DNA," which refers to whether the vegetative and prophage genetic maps have the same gene order.

- Integration of one DNA molecule into another
- Replication of nonintegrated prophage
- Types of interactions among phages

General Nature of the Temperate Response

The key characteristic of the temperate response is the modulation of phage growth. The viral DNA replicates at the same rate (on a molecule-for-molecule basis) as the host cell DNA and is distributed to both daughter cells at each cell division. However, despite DNA replication, most of the phage-specific proteins, especially those involved in late functions, are not produced. Because the virion structural proteins are among those not produced, there is no possibility of new phage particles being assembled, and the host cell thus survives the infection.

The survival of the host cell has important implications for the interaction between phage and host, as both temperate and lytic infections begin in the same way. This fact means that any effect the virus has on the host cell during the early stages of infection must either be nondetrimental to cell survival or be reversible. Therefore, activities such as degradation of the nucleoid, which occurs shortly after T4 infection, should not be expected among the temperate viruses. Neither should one expect a temperate virus to carry out wholesale modifications of the bacterial RNA polymerase as part of its regulatory system. Instead, the expected pattern should be one of careful utilization of existing host biochemistry, at least during the potentially reversible portion of a viral life cycle.

A **lysogen** is a cell that carries a temperate bacteriophage, and a **prophage** is the quiescent phage DNA inside a lysogen. Generally, a cell that is a lysogen is immune to superinfection by the same phage (homoimmune), but not by heterologous phages. Accordingly, it is possible for a cell to carry more than one prophage, a state that constitutes multiple lysogeny. Generally, multiple lysogens involve heterologous phages, as superinfection immunity is due to the presence of repressors that bind to the viral DNA and turn off lytic functions in the prophage. The repressors can also act on newly arrived DNA and prevent its expression, thereby conferring immunity against any phage to which the repressor may bind. Conversely, if a prophage is transferred to a cytoplasm that does not contain a repressor, it can reactivate and produce a lytic infection.

The lysogenic state is not always maintained in all cells of a culture. It is possible for a prophage to revert to the vegetative state and further produce a lytic infection. The reasons for spontaneous reversion of a prophage to the lytic state are unknown, but it seems to occur at a rate that yields roughly 10^6 phage particles per milliliter of midexponential phase culture (about 2×10^8 cells/ml). If any lysogenic culture of moderate cell density always contains some cells that have spontaneously induced their prophages, it is impossible to obtain a phage-free culture of a lysogenic bacterium. This fact can be used to identify lysogenic cultures. For many phages, treatments that damage DNA (e.g., ultraviolet radiation [UV], mitomycin C, or other mutagenic agents) result in an increased rate of conversion of a prophage to the vegetative state. This increase is termed **induction**, and viruses that are so stimulated are considered inducible.

Occasionally, a cell derived from a lysogenic culture gives rise to a cell line that does not produce infectious virus particles. This event occurs because a prophage is susceptible to the same genetic processes as the bacterial DNA. In particular, it may undergo mutation. Simple mutations may inactivate a prophage so that it can no longer be induced, either endogenously or exogenously. More complex mutations such as deletions may result in the production of defective phages (e.g., only tails). A completely inactivated prophage that does not produce infectious particles is referred to as a **cryptic prophage**.

Temperate phages can be detected during routine screening because they produce **turbid plaques** (that still contain a thin layer of growing bacterial cells) in lawns of nonlysogenic bacteria. They are true turbid plaques in the sense that some of the infected cells have lysed, and others have formed lysogens. The newly formed lysogens are, of course, immune to superinfection and continue to grow in the region of the plaque. In contrast, the turbid "plaques" produced by the RNA phages discussed in Chapter 7 are not the result of cell lysis but rather are due to the retardation of growth.

THINKING AHEAD

What biochemical functions are essential for a temperate virus?

Bacteriophage Lambda as the Archetypal Temperate Phage

The Paris group headed by the senior Wollmans was the first to recognize that some bacterial cultures are persistently contaminated by bacteriophages and therefore must be lysogens. Later, Andre Lwoff demonstrated that lysogens are stable in the absence of inducing agents and yet are capable of being lysed by the virus that was inside them. The major experimental effort, however, came from François Jacob and Elie Wollman working with *Escherichia coli*. Although they identified a large number of distinct phages, most of their efforts were concentrated on a single phage known as lambda (λ), which had been originally identified by Esther Lederberg. Nearly all the λ phages studied today have descended from a hybrid phage known as $\lambda PaPa$ produced by the Pasadena and Paris labs and lack the tail fibers possessed by the original λ (Hendrix and Duda, 1992).

Figure 8.1 is an electron micrograph of the hybrid phage (λ wild type), which is of an average size (see Table 8.1) and contains a 48.5-kb linear DNA molecule. At each 5'-phosphate end of the DNA molecule there is a short single-strand region of 12 bases that are complementary to the bases at the other end. These special ends of the DNA are described as cohesive because they hydrogen-bond readily in the same manner as restriction fragments and allow the DNA to circularize rapidly after infection to form nicked circles. Phage DNA that lacks one cohesive end cannot circularize and thus cannot replicate. Phages of this type are called *doc* (defective, one cohesive end) and cannot reproduce themselves.

The nicked DNA circles resulting from the presence of the cohesive ends are sometimes called **Hershey circles** after their discoverer and can be permanently sealed by host DNA ligase in the same manner as RF II is converted to RF I in the Ff phages. A careful analysis of λ virions indicates that the right-hand cohesive end of the DNA molecule (as defined in Fig. 8.6) is always located at the head–tail junction. During abortive infection, it is the first portion of the λ DNA that becomes nuclease-sensitive and is therefore always presumed to enter a cell first.

Lytic Life Cycle

A λ phage attaches to its host cell in two stages: first reversibly with the tip of its single tail fiber, then irreversibly with the end of the tail to a membrane structure on the surface of an *E. coli* cell that is a component of the maltose permeation system. Some strains of *E. coli* that are *mal* mutants are deficient in both maltose transport and the λ receptor (see Chapter 14) and consequently are resistant to λ infection. Chemicals that uncouple energy-producing reactions from the membrane proton gradient have no effect on DNA entry, and the DNA

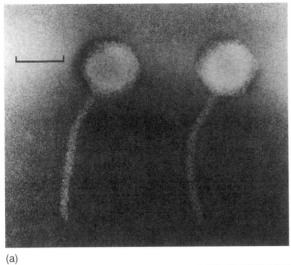

(a)

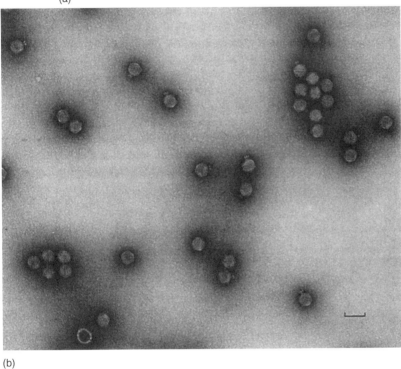

(b)

Figure 8.1. Electron micrographs of some temperate bacteriophages. (a) Lambda phage negatively stained with potassium phosphotungstate. The length of the *bar* is 50 nm. (b) Phage P22, also negatively stained. The length of the *bar* is 100 nm. (From King, J., Casjens, S. [1974]. Catalytic assembling proteins in virus morphogenesis. *Nature* 251: 112–119.) (c) Phages P2 (large) and P4 (small) photographed by minimal beam exposure. The length of the *bar* is 50 nm. (Courtesy of the late Robley C. Williams, Virus Laboratory, University of California, Berkeley.)

267

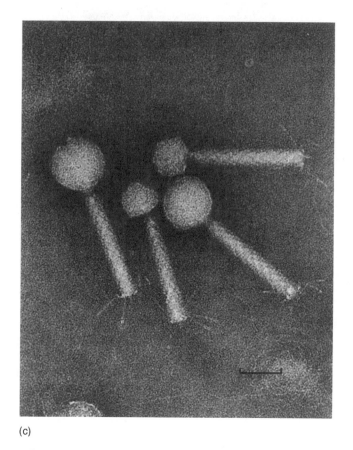

(c)

Figure 8.1. (*Continued*)

may enter by simple diffusion. Consistent with that model is the observation that if artificial lipid bodies (liposomes) carrying Mal protein are prepared, λ DNA enters a liposome until it is full or the phage head is empty. Successful entry depends on normal functioning of a host PTS IIc protein, which is a component of the inner membrane-associated phosphotransferase system for mannose transport. After DNA insertion, the λ genome circularizes as described earlier and is ligated into a supercoiled molecule.

RNA transcription begins promptly and is divided into three temporal classes designated immediate early, delayed early, and late. Immediate early transcription produces two short mRNA molecules that are translated to give only three proteins: N and Ral (restriction alleviation) from one transcript and Cro from the other. Ral interacts with host cell methylases to give enhanced modification of all DNA in the cell, thereby improving the chance that incoming λ DNA will avoid restriction. Discussion on the functions of N and Cro follows.

Delayed early transcription gives rise to substantially longer transcripts that still hybridize to the same DNA sequences as the immediate early mRNA and thus must include the immediately early sequences within them. In fact, they are merely extensions of the immediate early mRNA and represent an additional, well-studied example of antitermination. Careful study of the antitermination function in λ has provided substantial information about RNA polymerase function.

The critical phage protein in antitermination is N, a 107 amino acid basic protein. To function, it must have host RNA polymerase with four additional host proteins, products of the *nusA*, *nusB*, *nusG*, and *rpsJ* genes. The holoenzyme is set up for antitermination using the transcript of two binding sites located just downstream from the promoter. The first of these sites is called *boxA* (sequence 5′-CGCTCTTA), the second is called *boxB*, and the pair of sites is the *nut* (N utilization) site. They are separated by one helical turn of the DNA and constitute an RNA enhancer element. The *boxB* sequence (5′-**AGCCCT**-GAAAA**AGGGCA**) contains a short region of dyad symmetry (boldface) that forms a stem. N protein binds to this stem in the nascent RNA. NusA protein is an elongation factor apparently necessary for normal termination of transcription and binds to the N–*boxB* complex. It interacts with N protein to form an RNA–RNA polymerase complex (Fig. 8.2). Ribosomal protein S10, also known as NusE or RpsJ, binds directly to the RNA polymerase holoenzyme. NusB protein binds first to *boxA* and then to S10 to stabilize the N:NusA:RNA polymerase antitermination complex. It is specifically necessary to allow antitermination, and appropriate NusB mutant proteins can substitute for N and *nusA* function. NusG protein is a transcription elongation factor that interacts with rho protein to promote efficient termination in the absence of N. In terms of biochemistry, without N, RNA polymerase pauses at specific terminators downstream from a *nut* site and terminates transcription. In the presence of N and its associated factors, an RNA polymerase complex speeds up and pauses too briefly to allow termination, even at terminators found in other transcription units that have been joined to λ. The terminators may be ρ-dependent or ρ-independent. Antiterminated transcription eventually stops, but the mechanism is unknown. It may involve multiple, closely linked terminators. Continued antitermination requires constant transcription of N because host *lon* protease rapidly degrades the protein. Antitermination in which the RNA polymerase is stably modified by the addition of proteins is designated as **processive antitermination** (Weisberg and

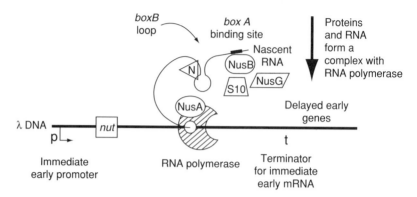

Figure 8.2. A model for the assembly of N antitermination complex. Key protein–protein and protein–RNA interactions inferred from genetic and biochemical evidence are indicated by *arrows*. N protein binds to the RNA transcript of a nut site in the looped region labeled *boxB*. In a reaction mediated by NusA protein, it then binds to the RNA polymerase complex. Proteins NusB, NusG, and ribosomal protein S10 bind to *boxA*. Interactions with proteins bound to *boxA* serve to stabilize the N:NusA:RNA polymerase complex. NusG protein interacts with rho in the absence of N protein to give efficient termination at t, thereby keeping transcripts short until antitermination occurs. (Adapted from Das, A. [1992]. How the phage lambda *N* gene product suppresses transcription termination: Communication of RNA polymerase with regulatory proteins mediated by signals in nascent RNA. *Journal of Bacteriology* 174: 6711–6716.)

Gottesman 1999) because the affected polymerase ignores termination signals and is less sensitive to pause signals.

A similar phenomenon is observed with respect to regulation of late transcription. Protein Q functions at a *qut* site; acts to antiterminate a transcript without host Nus factors; and allows late gene expression. By 10–12 min after infection, late mRNA synthesis has entirely replaced the early mRNA synthesis, which has ceased because of the inhibitory effect of the accumulated Cro protein, a repressor of early mRNA synthesis.

This pattern of antitermination is not unique to λ. It occurs in several other lambdoid phages such as P22 (see later in this chapter). The phages remain unique entities in the sense that the N proteins are not interchangeable between phages. However, a phage mutated in *N* function can be rescued by superinfection with a normal phage that supplies protein N. The protein diffuses through the cytoplasm and causes **transactivation** of the defective DNA (i.e., it activates a DNA molecule other than the one that produced the functional mRNA). Protein Q, on the other hand, functions efficiently only in *cis*, primarily turning on only the DNA molecule that coded for the Q-specific mRNA. The difference in activity is apparently due to the relative binding constants of the proteins to their specific DNA sequences. Protein N binds more strongly and hence is less influenced by the presence of extraneous DNA sequences. The weaker binding Q protein can easily attach to the site located near its gene but loses its specificity in the face of large quantities of DNA as would be seen when acting in *trans*.

The temporal sequence of the events so far described is depicted in Fig. 8.3. The short leftward and rightward transcripts shown in (b) are designated L1 and R1, respectively. Note that in several cases transcription termination is not perfect even in the absence of protein N. For example, R1 transcription beginning at p_R occasionally extends past t_{R_1} and terminates at t_{R2}. After N protein is present, the L1 transcript becomes longer and is now designated as L2. The R1 transcript also becomes longer, and protein Q accumulates. Q acts to antiterminate transcription from promoter p_R' that yields a transcript designated R2. Ko et al. (1998) have demonstrated that σ^{70} that has not yet been released from transcription initiation can recognize a repeat of the promoter -10 element and can cause the RNA polymerase complex to pause. At that time Q attaches to the complex and modifies it for antitermination. The binding of DnaA protein just downstream of the promoter stimulates transcription from p_R.

The increased level of R2 transcription provided by p_Q ensures that structural components for a large number of phages are produced. However, various functions in the L2 region tend to be harmful to the host, and the progeny phage yield is not high unless these functions are repressed after they have served their purpose. The repression is accomplished by the small *cro* gene product (66 amino acids) whose symbol is an acronym for control of repressor and other things. Experiments with fusion proteins have shown that the Cro protein binds to operators at three sites (o_L, o_R, and near p_E) and prevents or reduces the binding of N protein-modified RNA polymerase. The DNA sequence for o_L and o_R is basically the same and includes three sites at which Cro dimers can bind. Cro binds preferentially to the third site and gradually fills in the other sites if its concentration reaches sufficient levels. Figure 8.4 shows that the Cro binding sites physically overlap the promoter region so that, as Cro is sequentially bound, the leftward and rightward promoters are progressively blocked as the DNA bends due to Cro binding. In vitro experiments have shown that host DnaA protein stimulates transcription from p_R. In this case, however, the connection appears to

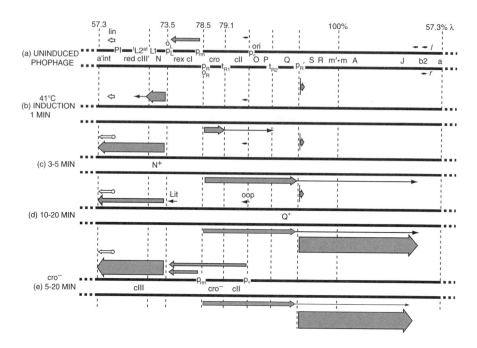

Figure 8.3. Temporal sequence of transcriptional events in prophage λ. The *leftward* transcripts are coded by the *l* strand and drawn above the λ DNA; the *rightward* transcripts are coded by the *r* strand and drawn below the λ DNA. (**a**) Transcription in the uninduced prophage. The *cl-rex* transcripts correspond to 80–90% of the total prophage-specific RNA. (**b**) Immediate early transcription after induction (the same pattern occurs during lytic infection). (**c**) Delayed early transcription. (**d**) Late transcription. (**e**) Decontrolled transcription in an induced *cro⁻* mutant of λ. The prophage maps are not drawn to scale, but, rather, with the immunity region expanded. The numbers in the *top line* indicate the positions of various sites with respect to the left end (0% λ) and the right end (100% λ) of mature λ DNA. The width of the *arrows* is a measure of the rate of transcription. In the case of the 198 nucleotide 6S RNA transcribed early (*arrow* under), it was found that in vitro synthesis provides 10- to 20-fold more of the 5′-proximal 15 nucleotide sequence (represented by the *vertical line* in parts **a–c**) than of the total 6S RNA. Thus, p_R is the strongest λ promoter, but is immediately followed by strong termination signals that can apparently be overcome by the Q product, with resulting synthesis of late RNA. (From Szybalski, W. [1977]. Initiation and regulation of transcription and DNA replication in coliphage lambda. In: Copeland, J.C., Marzluf, G.A. (eds.), *Regulatory Biology* (Ohio State University Biosciences Colloquia, no. 2). Copyright © 1977 by the Ohio State University Press. All rights reserved. Used by permission of the author, the editors, and the publisher.)

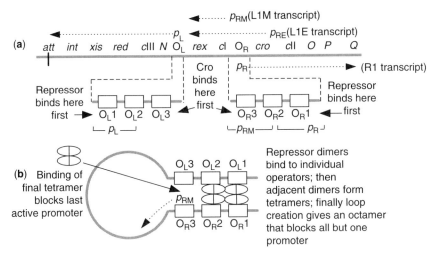

Figure 8.4. Regulation of early λ transcription. (a) Transcription of early λ genes is initiated at p_L and p_R and is subject to repression by repressor at o_L and o_R, respectively. Transcription of the cI (repressor) and rex genes can be initiated either at p_{RM} (the maintenance promoter) or at p_{RE} (the establishment promoter). Transcripts are indicated by *dashed lines*. (b) The expanded diagram illustrates spatial relations between p_L and three repressor binding sites in o_L. The *cro* protein also binds to o_R and o_L to reduce transcription from p_R, p_M, and p_L. Repressor molecules bind to the operator sites as dimers. Adjacent dimers form a tetramer. If the DNA between the two operator regions loops, the repressor molecules form an octamer that stabilizes the loop and physically blocks transcription from p_R and p_L. Binding of a final repressor tetramer would stop transcription from p_{RM}. The effect of Cro binding is similar but in reverse.

be to the β subunit instead of the α subunit. As would be expected, the DnaA binding site is located downstream of the promoter.

DNA replication during vegetative growth of λ phage is a complex process that begins during delayed early mRNA synthesis. A large number of host cell proteins are required for viral DNA replication, some of which are listed in Table 2.1. There are two λ proteins (products of genes O and P) that are essential for replication. The actual initiation site, *ori*, is located within the O gene. As shown in Fig. 8.5, O protein binds specifically to four 19-bp inverted repeats in the *ori* region. The base sequence of this region imparts a curve to the DNA that is accentuated into a loop by the binding of four O protein dimers. P protein binds to host DnaB protein and then adds to λ ori where host proteins (chaperone system DnaK, DnaJ, and GrpE) release the P protein, liberating

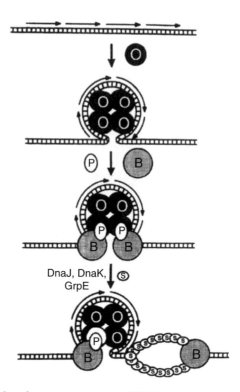

Figure 8.5. Inferred pathway to initiation of DNA replication at *ori*λ. O protein (O) binds to the direct repeats in *ori*λ and self-associates to form the "O-some." P protein (P) binds to O and to DnaB (B), attracting DnaB from its normal position at the *E. coli* replication fork. Host DnaJ, DnaK, and GrpE proteins disrupt the P–B complex and allow DnaB to act as a helicase, unwinding the origin region. The locally unwound DNA serves as a substrate for DnaG primase to initiate leading strand replication. SSB(s) stabilizes the single strands of DNA prior to replication. (Adapted from Echols, H. [1986]. Multiple DNA-protein interactions governing high-precision DNA transactions. *Science* 233: 1050–1056.)

DnaB. Other host proteins follow to generate a typical initiator complex that gives localized unwinding of the DNA helix. Accidental initiation of replication at other sites within the DNA molecule that resemble the normal *ori* is prevented by the requirement for multiple O protein binding sites. The circular mode of replication persists for about 16 min and is then supplanted by a rolling circle mode.

Evidence for rolling circle replication is provided by experiments showing that concatemeric DNA molecules (two to eight times the normal genome length) are formed during replication even in the absence of recombination functions. In vitro, λ DNA replication is unidirectional unless transcription is permitted, and unidirectional replication can serve as a trigger for rolling circle replication. Significant rolling circle activity cannot begin until late mRNA synthesis commences. The problem lies with exonuclease V, which is involved in the host cell recombination pathways (see Chapter 5) and preferentially degrades single-strand DNA. The presumption is that the exonuclease would normally attack the single-strand intermediates emanating from the rolling circle. However, the phage *gam* gene (transcribed as part of the delayed early mRNA) codes for protein gamma that specifically inhibits exonuclease V activity. The actual molecular mechanism for the rolling circle replication is the same as that described in Chapter 7.

DNA gel mobility shift assays performed by Grzegorz Wegrzyn and his collaborators (Fig. 2.17) have shown that DnaA binds extensively to phage λ DNA. The implication is that DnaA is necessary for bidirectional replication from the phage origin of replication. Under this model, as λ DNA accumulates, it binds up all the DnaA protein, thereby preventing bidirectional replication and starting rolling circle replication.

Concatemeric DNA is essential for proper DNA packaging. Concatemers result from rolling circle replication or from recombination. Adjacent to *gam* are two genes that together constitute the λ *red* system, an analog of the *E. coli* general recombination system. Included is an *exo* gene that codes for an exonuclease whose crystal structure is related to certain type II restriction enzymes. It is possible to genetically alter λ so that it is *red gam*. Therefore, it can only grow poorly because of an inability to switch to the rolling circle mode for DNA replication, resulting in a lack of concatemeric DNA molecules. Cells infected with such mutant λ phages produce few progeny and yield only small plaques. However, mutant phages that produce many progeny under these conditions and give rise to large plaques can be found. The phages carry mutations at several sites, all of which are designated **Chi**. Franklin Stahl and coworkers have shown that Chi mutations increase the recombination frequency in their immediate vicinity (10–20 kb), but primarily in one direction only. The increase is seen only when the *recA recBCD* pathway is used, not in the case of *recF* or *red* pathways (see Chapter 5).

Maturation of λ follows a pattern similar to that of phage T4. Tails, proheads, and scaffold proteins are produced in the usual manner. The concatemeric DNA molecule is broken into unit lengths of DNA by offset nicks at the *cos* sites (cohesive ends, Fig. 8.6). The terminase enzyme (a product of the *Nu1* and *A* genes) binds at one site (*cosB*) and in the presence of integration host factor (IHF) or

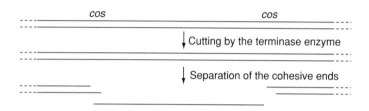

Figure 8.6. Production of the cohesive ends of λ DNA. Unit length λ DNA molecules are cut from concatemers by the terminase enzyme. The cutting takes place at *cos* sites located at the ends of the vegetative DNA molecule. In the diagram, each line represents a single strand of DNA. Only a short portion of the concatemer is shown. The cuts made by the enzyme are offset, generating single-strand DNA tails. Because all cuts occur at identical base sequences on the concatemer, the tails carry complementary sequences.

terminase host factor (THF) makes offset cuts at a 22 bp second site (*cosN*) located about 26 bp to the left on the conventional genetic map. Its behavior thus resembles an asymmetric type II restriction enzyme. Arens et al. (1999) have shown that the symmetric *cosN* site responds in an asymmetric manner to point mutations because those in the left half reduce the burst size while mutations in the right half do not. The actual cut requires the presence of an intact prohead to neutralize the effect of a host cell inhibitor, and the left end of the DNA is inserted into a prohead first. Ion etching experiments similar to those for T4 indicated that the spiral fold model may apply to λ as well, and that the right-hand end of the DNA is located on the outside of the phage DNA mass. As the head fills, it expands slightly.

Following assembly of the progeny virions, the host cell is lysed by the products of the *R* and *S* genes, where R protein is an endolysin that attacks peptidoglycan glycosidic linkages and S protein is a holin. A holin is a protein that produces holes, usually lethal, in a cell membrane. In this case S protein accumulates in the cell membrane (not in the cytoplasm) for about 50 min, and then a triggering event causes multimerization of the S protein to form large pores. The pores allow passage of the R protein, which has no intrinsic signal to cause its export, into the periplasmic space where it can cause cell lysis. The S protein actually comes in two forms, a 105 amino acid active form and a 107 amino acid antagonist of hole formation. The two sizes arise from an alternative translation start at a methionine codon in position 3 of the open reading frame.

Temperate Life Cycle

The first stages of a λ phage infection that ultimately results in a temperate response are the same as those that initiate the lytic cycle. However, when a temperate response occurs, the bulk of the phage-specific RNA synthesis gradually slows to a halt at a somewhat indeterminate time after delayed early mRNA synthesis has begun.

Concomitantly with the reduced mRNA synthesis, the phage DNA physically inserts itself into the bacterial DNA and becomes a prophage. This association can be shown experimentally by genetic mapping studies. The prophage behaves like any *E. coli* genetic element. It possesses a definite map position, and individual viral genes can be mapped like host genes. In many cases, however, the prophage must be induced before the genotype can be ascertained, as most prophage genes are not normally transcribed. Insertion of a prophage increases the genetic distance between the host markers at either end of the prophage (the flanking markers, *gal* and *bio*), resulting in an increased recombination frequency. The amount of increase is what would be expected if a piece of DNA the size of a λ genome had inserted linearly between the flanking markers. In essence, then, the bacterial and phage DNA recombine to form a single integrated molecule.

All the experimental evidence is in accord with a mechanism for this integration proposed by Campbell (Fig. 8.7). This model assumes that the λ DNA has circularized via its cohesive ends. The *E. coli* DNA is already known to be a circular molecule, so one recombination event involving the two DNA molecules generates a single, larger circle. The λ prophage has a definite genetic location within the *E. coli* genome (between the *gal* and *bio* loci) and has a specific orientation (i.e., the ends of the prophage are always the same). In order to account for these observations, it is necessary to assume that the integrative recombination event always occurs at the same site (called *att*) on both the phage and the host DNA. The two sites are designated *attP* and *attB*, respectively, and are represented in Fig. 8.7 as consisting of subsites P and P′, as well as B and B′. Each pair of subsites is connected by a 15-bp segment of DNA, which is the actual region of homology for pairing of the *att* sites. Other regions of partial homology are present in the *E. coli* genome, however, because if the *attB* site is deleted λ DNA can still integrate but at a much reduced frequency and at relatively random positions along the genome.

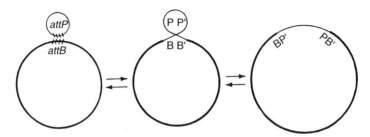

Figure 8.7. Campbell model for λ integration. The double-strand λ and bacterial DNA molecules are represented by the *small* and *large circles*, respectively. Initially, these two circles associate in a region of homology designated *att*. This region can be considered to consist of two halves. A genetic exchange is assumed to occur so that the left half of the bacterial *att* region (B) is linked to the right half of the phage region (P′) and vice versa, generating a figure-eight structure. When this structure unfolds, a larger circular DNA molecule carrying an integrated λ DNA molecule is observed. During excision of the phage DNA, the entire process is reversed.

It might be presumed that the site-specific recombination event between *attP* and *attB* could be catalyzed by any recombination system including the bacterial *rec* or phage *red* systems. However, phages mutant in a gene called *int* fail to form stable lysogens after infection, which indicates that the amount of recombination in the *att* region catalyzed by the *red* or *rec* systems is negligible. Instead, the *int* gene codes for a 140,000 Da **integrase** protein that is specific for the two *att* sites (see Chapter 15).

Thus if all types of recombination are considered, λ phage has three independent systems available to it: *rec*, the bacterial system for generalized recombination; *red*, the phage system for generalized recombination; and *int*, the phage system for recombination at the attachment site. The relative contribution of each system can be assessed using the appropriate mutant strains. Data from one such experiment are presented in Table 8.2. The amount of recombination in the cI–R interval is specific to the *red* or *rec* systems, and the general conclusion is that both systems can catalyze significant amounts of recombination. However, *red* is more efficient than *rec* at catalyzing recombination in the interval. All three systems can catalyze exchanges between *J* and cI, with *int* causing exchanges between two PP′ sites on free phage DNA. Approximately 26% of the exchanges between *J* and cI are observed to be *int* promoted. Note that when all three recombination systems are mutant, no genetic exchanges are detected.

Table 8.2. Comparison of the relative contributions of the *red*, *int*, and *rec* systems to phage recombination.

Phage		Host	Percent Recombination	
red	*int*	*rec*	*J–c*	*c–R*
+	+	+	7.5	3.6
+	−	−	4.1	3.0
−	+	−	2.0	≤0.05
−	−	+	1.3	1.3
+	+	−	7.8	3.1
−	−	−	≤0.05	≤0.05

Note: Average values for recombination were determined in four sets of crosses of the type *susJ* × cl *susR*, carried out in two hosts. The *J+ R+* recombinants were selected and scored visually for cl (turbid or clear). The percentage recombination was determined by the frequency of J+ R+ turbid (crossover in the interval *J–c*) or J+ R^{qd} clear (crossover in the interval *c–R*) × 2 × 100. The interval *J–c* contains *att* and shows generalized as well as specific recombination, whereas the interval *c–R* shows generalized recombination only. (From Signer, E.R., Weil, J. [1968]. Site-specific recombination in bacteriophage λ. *Cold Spring Harbor Symposia on Quantitative Biology* 33: 715–719.)

Yu et al. (2000) have developed a cloning system using the *red* genes that permits highly efficient recombination of linear DNA into the *E. coli* chromosome. For further discussion, see Chapter 16.

Explicit in Campbell's model is the concept that prophage excision (induction) is a simple reversal of the insertion process, which implies that *int* function is necessary and sufficient for excision as well as integration. However, biochemically this is not strictly true. Prophage excision requires two phage functions, *int* and *xis* (pronounced excise). The *xis* gene is small, and its product is absolutely necessary for successful prophage excision. Apparently, integrase can recognize the BP' end of a prophage (see Fig. 8.7) but not the PB' site. The *xis* protein remedies this deficiency. Two host genes, *him* and *hip*, that encode integration host factor (IHF) have also been shown to be necessary for integration and excision. IHF binds to the sequence WATCAA(N)$_4$TTR (where N is any base, R is any purine, and W is A or T) to yield a molecule bent at a 140° angle. The Xis protein has a very limited half-life and is subject to degradation by the Lon and FtsH host proteases.

A protein repressor, the product of a phage gene, maintains the prophage DNA in the integrated state. Its existence can be demonstrated by conjugating a donor strain of *E. coli* that carries a λ prophage to a recipient that is not a lysogen (see Chapter 11). When the prophage travels from one cell to the other, induction is immediate and almost inevitable. The phenomenon is called **zygotic induction** because it occurs after a mating and is due to the lack of repressor protein in the recipient cell.

Geneticists have isolated mutations in λ that affect its ability to produce a functional repressor. These mutations result in the production of clear rather than turbid plaques and hence are called c mutations. Three classes of c mutations have been identified. Mutants for cI function never produce any lysogens, whereas mutants for cII or $cIII$ do produce occasional lysogens that are normally stable. It is reasonable to assume, therefore, that the cI gene must code for the repressor protein, and cII and $cIII$ serve to enhance its expression. Certain mutations that map within cI result in the production of a temperature-sensitive repressor. Lysogens carrying such a mutation grow normally at 30°C, but promptly undergo induction if the culture temperature exceeds 40°C. Mark Ptashne and coworkers finally isolated a 30,000 Da protein and showed that it bound preferentially to λ DNA that includes the regions immediately adjacent to cI, but not to similar regions from heterologous phages such as 434.

The region of DNA that includes the cI repressor binding sites is referred to as the **immunity region** because it determines the type of superinfection immunity conferred by the prophage. It is possible to produce various types of recombinant phages that carry the structural genes of λ, but varying immunity regions from other lambdoid phages (see Table 8.1). The product of a cross between phages λ and 434 would be described as λ *imm434*, etc. A λ *imm434* will superinfect a normal λ lysogen, but cannot grow on a 434 lysogen. The immunity of the lysogen is thus due to the presence of the repressor in the cell that prevents expression of all DNA molecules of the proper immunity type regardless of whether they are integrated.

The action of the repressor is not perfectly effective. It is possible to obtain homoimmune (i.e., identical cI genes), double λ lysogens in which a second λ phage has infected a lysogen and integrated itself by generalized recombination into the middle of the existing prophage (Fig. 8.8). The result is two prophages in a row, but both of them are recombinants. A second general type of double lysogen occurs when a lysogenic cell is infected with a heteroimmune λ such as λ *imm434*. In this case the second phage undergoes integrase-catalyzed insertion at one of the ends (equivalent to a normal *att* site) (see Fig. 8.8) of the original prophage to give a tandem duplication of λ in which each prophage has the same genetic composition with which it began. This type of integration is not possible for a homoimmune phage due to the action of the λ repressor on *int*.

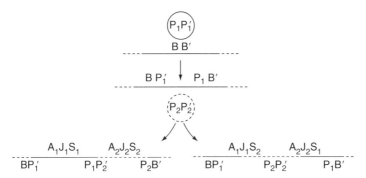

Figure 8.8. Possible mechanisms for insertion of a second phage DNA molecule into a lysogen to form a double lysogen. The *small circles* represent the phage DNA, and the *solid horizontal lines* represent a portion of the bacterial chromosome. Attachment sites are indicated as either PP′ (phage) or BB′ (bacterium). In the first step a phage DNA inserts into a bacterial DNA by an exchange event catalyzed at the attachment sites by phage integrase. The insertion of the second prophage may occur in two ways: the recombination may be catalyzed by integrase again and occur between P_2P_2' and one of the recombinant attachment sites (BP_1' and P_1B') to yield the structure shown at the left; alternatively, the recombination may occur within one of the phage genes such as J, catalyzed by a nonspecific recombination system, as shown at the right. Note that in the first case each of the tandem prophages are intact, but in the second case the first prophage is separated into two pieces by integration of the second prophage. The result is still two prophages arranged in tandem, but each prophage is recombinant.

The repressor protein is the regulator of the temperate response of phage λ. It must shut down the major lytic functions of the phage while allowing the integration activity necessary for prophage formation. The shutdown is accomplished at some of the same promoters used by Cro. Specifically, the repressor protein binds cooperatively to o_L and o_R, but adds to the multiple binding sites in the reverse order, so that it blocks the leftward and rightward promoters first (Fig. 8.4). Thus, if the repressor is available in large quantity, it prevents a lytic infection. The cI gene coding for the repressor can be transcribed from either of two promoters, p_E or p_{RM}. Neither of the promoters is initially active, so they require positive regulatory elements to be functional.

Among the genes contained within the L2 and R1 transcripts are cII and cIII, which have an important role in any temperate response. Their gene products not only act as regulators of λ repressor expression but also stimulate *int* transcription from p_I. Their effect is exerted via the cII protein, which binds to the similar −35 regions on both p_I and p_E in the major groove of the DNA helix (i.e., on the opposite side of the helix from the RNA polymerase) and links to the α subunit of RNA polymerase as seen with other regulatory proteins. Together, the cII and cIII proteins form a regulatory circuit with host FtsH (*hflB*) protein, a protease that degrades cII protein.

The cII protein, as it accumulates, stimulates the appropriate λ promoters, but under normal conditions the stimulation is not great owing to the proteolytic action of FtsH. Mutations in that locus confer a high frequency of lysogeny because the resultant high level of Int protein converts nearly all of the phage DNA into integrated prophages. The normal frequency of lysogeny is maintained by cIII action, which antagonizes the FtsH protein so as to preserve a minimum level of cII activity. High levels of cII are toxic to *E. coli*. Kedzierska et al. (2003) have shown that the effect is due to inhibition of DNA synthesis.

The binding of cII protein to p_E initiates the establishment mode of repressor transcription (the long, thin leftward transcript in Fig. 8.3e). This mode yields relatively larger quantities of repressor protein and results in the backwards transcription of the *cro* region. This reverse transcription antagonizes transcription of *cro* and may facilitate a lysogenic response. Note, however, that transcription from p_E cannot be maintained because binding of the repressor to o_L and o_R eventually prevents further production of the cII and cIII proteins. The maintenance mode of repressor synthesis is from p_{RM}, a promoter that overlaps o_R (see Fig. 8.4) and yields the L1M transcript. As the repressor binds to turn off rightward transcription, it also activates its own promoter (another example of autoregulation). Transcription is maintained from this promoter so long as repressor is present in the cell. Should the level of repressor become too high, binding of repressor to the leftmost portion of o_R prevents further transcription until the repressor concentration is reduced. The p_M transcript includes information for cI and *rexAB*, the latter being genes coding for proteins that prevent phage T4 *rII* mutants from growing. The transcriptional state of the prophage is also shown in Fig. 8.3. The cI mRNA molecule is unusual in that it is "leaderless," meaning that the initial AUG codon forms the 5′-terminus of the molecule and there is no Shine–Dalgarno sequence present.

In molecular terms, the binding of repressor is interesting because of supramolecular interactions. The repressor binds to each operator element as a dimer. Adjacent dimers associate to become tetramers (Dodd et al. 2001), and the DNA forms a looped configuration linking the two sets of operator sites (Fig. 8.4) as was described for the galactose operon (see Chapter 4). This particular repressor looping event forms the largest loop yet described. Binding of a repressor tetramer to the third operator site results in loss of repressor tran-scription. Mutations in this site that prevent repressor binding actually make it more difficult to induce the prophage because of excess repressor concentration in the cytoplasm (Hochschild 2002).

THINKING AHEAD

Enumerate some other consequences of antitermination effects on RNA polymerase.

An interesting type of regulation was observed when transcription of the *int* region from p_L was examined in a *c*II mutant strain (p_I poorly activated). The observation was that transcription is relatively inefficient unless the *b* region is separated from it (as would be the case for an integrated prophage). The ineffi-cient transcription is an example of retroregulation, regulation of a transcript by a downstream sequence. The deleted *b* sequence includes DNA that codes for a stem-loop structure of a particular sort (Fig. 8.9). The region that includes this potential loop is designated *sib* (sitio inhibidor en b). The loop formed by the p_L transcript has been shown to be a signal for RNase III processing both in vivo and in vitro. The cut RNA may be a substrate for further degradation by the RNase II enzyme, as mutants in this function do not exhibit normal retroregula-tion. Transcription from p_I, on the other hand, terminates at a different spot (recall that transcription from p_L involves antitermination) and cannot form the same stem-loop. The differing transcription end points are another example of how RNA polymerase can be programmed for certain termination signals while it is initiating transcription.

Both Cro and repressor recognize the same sequence on the DNA (i.e., TATCACCGCCAGAGGTA). Both proteins fold to give a pair of α-helical regions that fit into the DNA major groove, and both bind as dimers. However, Cro bends the bound DNA more sharply than does repressor. The larger λ repressor protein dimer extends its amino-terminal arms around the helix, thereby protecting more of the DNA structure during chemical alteration exper-iments. It also forms a loop in the operator DNA that is visible by electron microscopy. Counting from the right in the recognition sequence, Cro keys on base 3, and repressor keys on bases 5 and 8. This observation of similarity in binding site emphasizes the major remaining problem in λ regulation: under-standing why certain infected cells form lysogens, whereas other cells in the

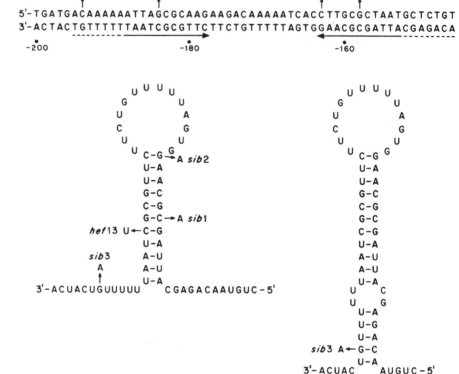

Figure 8.9. The *sib* region. Locations of four *sib* mutations are shown in the DNA sequence (*top*). Possible secondary structures of the RNA from this region are shown below. The RNA structure on the right can form from p_L RNA, but not from p_I RNA because the latter terminates at the last of the six consecutive U bases (position −193). (From Echols, H., Guarneros, G. [1983]. Control of integration and excision, pp. 75–92. In: Hendrix, R.W., Roberts, J.W., Stahl, F.W., Weisberg, R.A. (eds.), *Lambda II*. Cold Spring Harbor, NY: Cold Spring Harbor Laboratory.)

same culture lyse and release phage particles. One possibility is the opposing transcripts in the *cro* region. Minor fluctuations in the efficiency of one transcript may shift the metabolic balance of the cell toward either lysogeny or lysis by emphasizing either the L1 or R1 transcripts. Building on this type of fluctuation, Santillán and Mackey (2004) have developed a physical model of the lysis–lysogeny decision process, suggesting that under normal concentration levels, lysogeny is the only stable state.

Genetic Map of Lambda

There are two genetic maps for lambda: one for the vegetative phage and one for the prophage. This situation arises because the *attP* site used for integration of phage DNA is not located at one of the cohesive ends of the DNA molecule. The result is that integration of phage DNA via the Campbell mechanism (see Fig. 8.7) results in a circular permutation of the genetic map for the vegetative phage.

The vegetative map has been developed from standard phage crosses such as those used for T4. Interpretation of the data is much simpler because λ DNA is neither circularly permuted nor terminally redundant. The prophage map, on the other hand, has been derived from standard bacterial crosses using transduction (see Chapter 9) or conjugation (see Chapter 11). A simplified version of the map is presented in Fig. 8.10. It includes arrows indicating major transcripts together with the major promoters and terminators. There is a distinct clustering of genetic functions. The *A–J* region and *R* and *S* represent the regions of late transcription. Although they appear separated on the vegetative map, they are actually continuous in the circular vegetative DNA molecule. The rest of the functions are arrayed about the *cI* gene, with recombination functions to the left and DNA replication functions to the right. Located next to *cI* are the *rex* genes, which determine the ability of the lysogen to exclude *rII* mutants of phage T4.

It is also possible to develop so-called **physical maps** for λ in which distances are based not on recombination frequencies but on measurements of linear distance using several different techniques. One of the techniques is similar to the heteroduplex analysis discussed in Chapter 7. When the heteroduplexes are examined in an electron microscope, regions of nonhomology appear as "loops" or "bubbles" (see Fig. 7.1). By measuring distances from the ends of the λ DNA to various nonhomologous regions, physical distances can be estimated. Figure 8.11 shows some physical maps comparing various *imm* recombinants. Modern technology has provided a new form of physical map based on restriction digests (see Fig. 16.1). A restriction map positions all known restriction sites on the genetic map of the organism or virus using the size of the restriction fragments to estimate the distance between recognition sequences.

A wide variety of DNA cloning vectors based on phage λ has been prepared. Although not especially useful for DNA sequencing experiments, they can carry relatively large DNA inserts and, of course, offer the advantage of automatic packaging of the cloned DNA. Some examples of λ cloning vectors are discussed in Chapter 16.

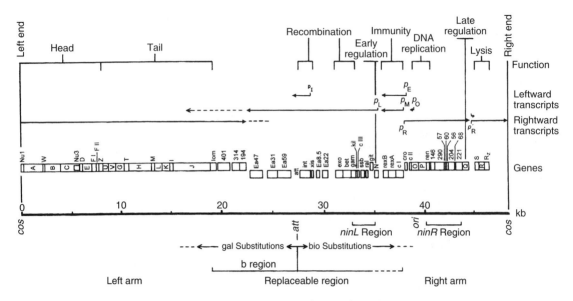

Figure 8.10. Genetic map of phage λ. The *heavy line* is a kilobase scale beginning and ending at a cohesive end site (*cos*), and above it is a scale drawing of the known λ genes. The brackets at the top indicate functional clusters of loci. The *vertical lines* show the positions of the major regulators N and Q. Known promoters are denoted by *p* with a subscript to indicate their unique points of origin. p_I, int protein promoter; p_E, establishment promoter for cI; p_M, maintenance promoter for cI; p_L, major leftward promoter; p_R, major rightward promoter; p_O, oop promoter; $p_{R'}$, late promoter. →, extent and direction of transcription; - - -, read through transcription that can occur when termination fails. Known genes (identified either genetically by mutation analysis or functionally by SDS–gel electrophoresis of protein product, or both) are indicated with letter names. Genes known only as open reading frames (ORFs) in the DNA sequence analysis are given numbers corresponding to the predicted size of the protein. In the area below the map are indicated the major areas of substitution and deletion mutations: *att*, attachment site; *ori*, origin of replication. (Adapted from Daniels, D.L., Schroeder, J.L., Szybalski, W., Sanger, F., Blattner, F.R. [1984]. A molecular map of coliphage lambda, pp. 1–21. In: O'Brien, S.J. (ed.), *Genetic Maps 1984*. Cold Spring Harbor, NY: Cold Spring Harbor Laboratory.)

Other Lambdoid Phages

There are numerous other phages having the general appearance of phage λ. Collectively, they are designated as lambdoid phages. They include some of the

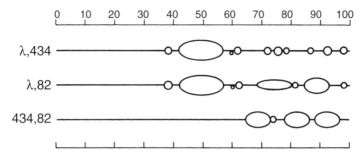

Figure 8.11. Heteroduplex analysis of hybrid phages. The lines represent stylized drawings of heteroduplex DNA molecules as observed in the electron microscope. Each substitution loop (see Fig. 7.1) represents a region of nonhomology. All possible combinations of the three lambdoid phages are shown. The immunity region is roughly between 70 and 80 (Fig. 8.10). (From Campbell, A. [1977]. Defective bacteriophages and incomplete prophages, pp. 259–328. In: Fraenkel-Conrat, H., Wagner, R.R. (eds.), *Comprehensive Virology*, vol. 8. New York: Plenum Press.)

phages noted above (434, 82) as well as the series of HK phages isolated in Hong Kong. Comparative studies of the lambdoid phages have shown that they are all very similar in terms of genetic structure, even if specific proteins such as cI repressors are different. It is as though the lambdoid phages have arisen by recombination of discrete pieces of their genome so that similar functions are always located in the same general area, an idea put forward by David Botstein and further developed by Hendrix and his collaborators (Juhala et al. 2000). For example, phage HK022 has a Q function analagous to the one found in λ that is located in the same relative locus on the genetic map.

Even when functionality is not preserved, the discrete region may be. A case in point is phage HK022, which lacks an *N* gene and has in its place a gene designated *nun* (reviewed by Henkin 2000). Nun protein, whose amino-terminal sequence is interchangeable with that of the N protein, competes with N for binding to *boxB* in phages λ and 434. When Nun binds, it causes inhibition of RNA polymerase translocation at a site located just downstream of *boxB*, thereby preventing the other phages from carrying out lytic infections. The action of Nun can be blocked by overproduction of N or by mutations in *nusA*. HK022 antiterminates its own early transcription by a unique mechanism, RNA structure. In HK022 there are regions called *put*, polymerase utilization, that are located in the analogous position of the *nut* sites in λ. When transcribed, each *put* site can fold to give two stem-loop structures separated by a single base.

The double-loop signals RNA polymerase to antiterminate in the absence of any protein cofactors.

Bacteriophage P22

The general morphology of phage P22 (see Fig. 8.1) is roughly similar to that of phage T3, consisting of a polyhedral head attached to a six-spiked baseplate. The DNA is a linear molecule that is circularly permuted and terminally redundant like that of the T-even phages. The extent of the redundancy is about 2.5% of the genome. Assembly of the P22 coat protein requires normal host *groEL* and *groES* chaperonins.

When P22 infects *Salmonella typhimurium*, its DNA circularizes by recombination within the terminal redundancies, provided the essential recombination function (*erf*) gene product is present. Because it also catalyzes generalized recombination, the *erf* system is analogous to the λ *red* system. A companion gene, *abc2*, is analogous to the λ *gam* function but has different biochemistry (Murphy 2000). It binds to the RecBCD complex but instead of inhibiting the enzyme as Gam does, it makes the enzyme independent of Chi activity. A *cro* gene functions to turn off genes whose products lead to lysogeny. During the lytic cycle, concatemeric DNA molecules arise that are probably produced by the rolling circle mechanism. The concatemers are cut to unit length and the DNA packaged by a headful mechanism similar to that used by phage T4. As is the case with T4, there is a scaffolding protein inside the developing capsid that is subsequently lost.

During the temperate response, P22 forms a prophage inserted between the *proA* and *proC* genes on the *Salmonella* genetic map. The Campbell model suffices to explain the mechanism, and a phage integrase is present to provide the necessary catalyst. The prophage state is maintained by two repressors, the products of the *c2* and *mnt* genes. Mutations in either gene can prevent establishment of lysogeny. The protein product of the *c2* gene functions similarly to the repressor encoded by the *cI* gene of λ, and the two proteins are about 45% homologous with respect to the amino acid sequence of their carboxy halves. The function of the *mnt* gene is, however, unusual. It acts by binding to overlapping promoters for *ant* and itself. When bound, the protein stimulates its own transcription and represses that of *ant*. The *ant* gene codes for an **antirepressor**. In the presence of the antirepressor, the normal repressor protein molecules (from P22) are inactivated, and prophage induction occurs. Interestingly, the antirepressor is nonspecific and inactivates the repressors of other temperate phages, including some that are not normally inducible.

Induction of the P22 prophage and its genetic mapping can be carried out by the same techniques used for λ. The genetic map that has been obtained for P22 is shown in Fig. 8.12. Superimposed on the P22 map is a λ map in order to emphasize the similarity of gene arrangements in P22 and in λ. Although viral taxonomists do not consider P22 to be in the same species as λ, it is obvious that they are genetically closely related. For example, the Q proteins of λ and P22 are interchangeable functionally, and antitermination of early mRNAs follows the

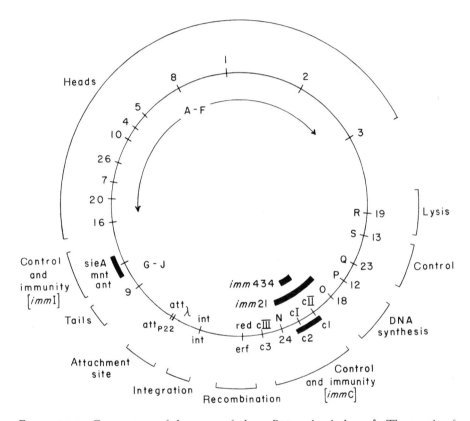

Figure 8.12. Comparison of the maps of phage P22 and coliphage λ. The inside of the circle shows the λ map and the outside shows the P22 map. A *solid line* connecting markers of the two phages indicates substantial similarity in genetic function. The *heavy bars* inside the λ map indicate the extent of material substituted in certain λ variants. The *heavy bars* outside the P22 map indicate the extents of the *immI* and *immC* regions. (From Susskind, M.M., Botstein, D. [1978]. Molecular genetics of bacteriophage P22. *Microbiological Reviews* 42: 385–413.)

same pattern as in λ. This relationship supports the idea of assembly of bacteriophages from discrete genetic pieces as noted for the lambdoid phages.

THINKING AHEAD

How could a phage activate a particular portion of another virus' genome?

Bacteriophages P2 and P4

P2 and P4 constitute an unusual pair of bacteriophages. Morphologically they are similar (see Fig. 8.1), with the exception of their head sizes, a P4 head being only about two-thirds the diameter of a P2 head. Because head size is held to reflect genome size, the size differential suggests that the P4 DNA molecule is smaller, which is in fact the case (see Table 8.1). Upon analysis, P2 and P4 DNA molecules have unique sequences, but with one odd similarity. Heteroduplex analysis shows that the base sequences of the DNA molecules are not similar, except for the cohesive ends, which are identical. A further analysis of the virions themselves shows that they are composed of absolutely identical subunits. The key to this similarity of proteins between the two phages can be found in the nature of their life cycles.

Phage P2 follows much the same life cycle as does λ and can insert its DNA directly into the bacterial DNA in order to form a lysogen. Like λ, P2 lysogens prevent the growth of certain other phages, in this case the T-evens and λ. In contrast to λ, however, P2 prophages have been located in at least ten sites on the *E. coli* genome. The preferred site varies with the strain used (e.g., the phage uses a different site in *E. coli* C than in *E. coli* K-12), but multiple lysogens are possible, with integration occurring at separated locations. A further contrast to λ is that the lysogens cannot be artificially induced to give phage particles, even by zygotic induction, although spontaneous induction does occur at the usual low frequency. The basis for noninducibility is that P2 repressor protein (product of gene C) lacks the specific amino acid pair used to destabilize the λ repressor. When a phage enters the vegetative state, it replicates its DNA solely by the rolling circle mode, using the host cell *dnaB*, *dnaE*, and *dnaG* functions. P2 A protein is equivalent in function to the A protein of phage φX174 (Odegrip and Haggård-Ljungquist 2001). Late transcription requires DNA replication and the product of the P2 *ogr* gene. P2 undergoes maturation and assembly by mechanisms similar to those used by λ except that the substrate is covalently closed (circular) monomeric DNA, instead of concatemers, so only one *cos* site is needed.

P2 carries an *old* gene that specifically prevents it from infecting *recBCD* hosts because the infection is lethal to the host before viral progeny form. The mechanism of lethality is not known. The *old* gene also interferes with the growth of phage λ probably because normal λ produces the Gam protein that inhibits exonuclease V, mimicking a *recBCD* mutant cell. Obviously, λ *red gam* mutants that have a Chi site to stimulate recombination are immune to the effect of P2. Such λ mutants are called *spi⁻*.

Phage P4 can complete its life cycle only if it infects a cell that already has a helper phage such as P2 within it or if a helper phage is supplied later. In the event that no helper phage is available, there are two possible outcomes for P4 infection, and they are not mutually exclusive. In both cases, there are a few rounds of uncommitted DNA replication. After that, the P4 can stabilize as a prophage integrated at a specific site, as a low copy number plasmid or, if suitably mutated so that some DNA replication functions are expressed (i.e., a clear plaque mutant), as a high copy number plasmid (30–50 copies per cell) sometimes called a phasmid or **phagemid**. If repressor function is absent, those phages that attempt to follow the lysogenic pathway kill the host cell. Virulent phage mutants replicate bidirectionally from a single point of origin using the classic replication mechanism of leading and lagging strands.

Normal DNA replication depends on proper functioning of initiator protein α, which binds to the *ori* and *crr* sites to form a loop. The same protein is responsible for helicase and primase activities. The product of the *cnr* gene negatively regulates the number of copies of replicating P4 DNA. It prevents proper binding of α to the DNA. P4 repressor (*cI*) is unusual in that it is not a protein, but rather a small, untranslated RNA molecule. Briani et al. (2000) have shown that it functions by binding to nascent RNA molecules near p_{LE} (see Fig. 8.13) and causing premature termination (equivalent to attenuation).

When a helper phage is supplied, either in the vegetative or in the lysogenic state, P4 progeny are produced. A successful P4 infection requires the helper phage to be completely normal for all functions essential to the structural integrity of the virions and for lysis of the host cell. Taken together with the fact that P4 virions are composed of P2 subunits, it means that P4 is using some P2 genes for its own purposes and is acting as a **satellite phage**, one that parasitizes another virus instead of having its own genes for capsid proteins (Fig. 8.13). However, the differences in replication mechanisms noted above indicate that

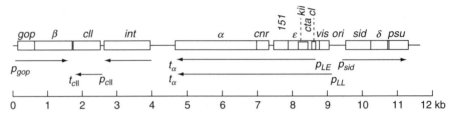

Figure 8.13. Genetic map of bacteriophage P4. Promoter sites are indicated by p and transcription terminators by t. The *cII* gene causes clear plaques, and the *gop* gene blocks multiplication on *E. coli argU* mutants. The α gene is a primase and helicase; β is a negative regulator of gop, and ε derepresses the helper prophage. An *orf* is an open reading frame whose function is unknown. (Adapted from Ziermann, R., Six, E.W., Julien, B., Calendar, R. [1993]. Bacteriophage P4, pp. 1.70–1.74. In: O'Brien, S.J. (ed.), *Genetic Maps*, 6th ed., Book I *Viruses*. Cold Spring Harbor, NY: Cold Spring Harbor Laboratory.)

P4 produces its own replication proteins and is therefore not totally dependent on P2.

The nature of the interaction between P2 and P4 phages is interesting. Integrated P4 DNA does not activate well when the host cell is infected with P2, unless P2 DNA replication is blocked. A helper phage lysogen is not significantly induced during P4 infection. Simultaneous infection with P2 and P4 yields progeny of both phages, but P4 is more abundant. The genetic map of P4 (Fig. 8.13) shows ten identified genes and four open reading frames, but the interaction of P4 and P2 revolves around only two of these, genes ε and δ. Transcriptionally active P4 has two mechanisms for turning on the late functions of P2. The E protein can derepress P2 transcription by acting as an antirepressor (Liu et al. 1998). This activity does not, however, guarantee an abundance of P2 late transcription because P2, like T4, has a requirement for DNA replication as a part of late transcription. The δ protein, on the other hand, performs true transactivation as seen with λ N protein, and derepresses P2 late genes even in the absence of P2 DNA synthesis.

The mechanism by which P2 capsid assembly is modified to produce a smaller-size head for P4 has not been worked out entirely, but it is dependent on *sid* function and normal host ρ protein. The promoter for *sid* is activated by δ protein from P4 and by Pag protein from P2 (Reiter et al. 1998). Another member of that operon is *psu*, whose product is a protein that suppresses polar mutations by preventing normal rho function. Both P2 and P4 process the P2 *N* gene product to give capsid proteins. However, P2 gives mainly one processed product, while P4 yields several major products. The P2 enzyme system can package P4 DNA into the smaller heads, but not a complete P2 genome. In the absence of *sid* activity, normal-sized P2 heads are prepared, and functional P2 results. Some P4 DNA does appear inside phage particles, but it is nearly always two or three complete genomes per particle.

Some investigators have used P4 as a cloning vector. Its behavior as a phage allows it to deliver cloned DNA to a cell easily. At the same time its mode of replication as a multicopy plasmid allows an experimenter to recover large quantities of cloned DNA with relatively little effort. It is more advantageous than λ because large quantities of λ are obtained only after induction, which, of course, kills the cells. P4 has no such effect in its plasmid mode.

THINKING AHEAD

How could a virus have a circular DNA molecule for its genome but a linear genetic map?

Bacteriophage P1

P1 is the largest of the *E. coli* phages and, indeed, one of the largest phages known (see Fig. 8.14). It is similar to phage P7 (which was formerly known as φ-amp because its lysogens are resistant to ampicillin). Heteroduplex analysis indicates that the two phages are 92% homologous, although they are heteroimmune (i.e., they produce different repressor molecules). The terminal redundancies also differ, that for P1 being 9–12% whereas that for P7 is only 1%. Both phages seem to have variable assembly mechanisms, as up to 20% of virions examined have smaller heads than usual and hence are defective. Most defective virions carry only 40% of the phage genome. However, because of the circularly permuted DNA, multiple infection of a single cell by defective viruses can result in progeny virus production as a consequence of complementation between defective phages.

Infection by P1 is basically similar to that of T4. Instead of tryptophan, calcium ions are absolutely required for attachment of virions to the cell. One interesting variation from the T4 pattern is that as the tail sheath contracts, the tail fibers are released in a process found only in P1 and P2. Moreover, P1

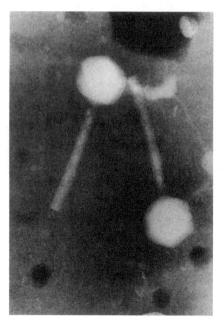

Figure 8.14. Electron micrograph of phage P1. The phage was negatively stained with sodium phosphotungstate. The *bar* indicates a length of 25 nm.

infection does not produce extensive host DNA degradation. Phage DNA must circularize for replication to begin. During lytic infection, the initial replication is via a circular mechanism that later converts to the rolling circle, as in the case of λ. Concatemeric DNA molecules are then packaged by a headful mechanism beginning with a *pac* site. The host range for P1 is controlled by a system of invertible DNA (C segment) that shows strong homology to the system used by phage Mu, as is discussed in the next section.

Promoters for late genes are not recognized by normal host RNA polymerase. Instead, the host protein SspA (stringent starvation protein) and phage protein Lpa allow the polymerase to transcribe late genes (Hansen et al. 2003).

The lysogenic state for P1 is unusual because the prophage is integrated into the host chromosome only rarely, generally remaining as an autonomous plasmid. Therefore, P1 DNA must be capable of two styles of DNA replication: viral when a lytic infection is in progress and a low copy number plasmid at other times. In accord with this idea, experimenters have demonstrated multiple replication origins. Westwater et al. (2002) have constructed a P1 phagemid that behaves similarly to P4.

Given that P1 plasmids are circular DNA molecules, it is surprising that the genetic map for P1 is linear. The linearity is due to extensive recombination catalyzed by the *cre* protein at a site called *loxP* that lies between the "ends" of the genetic map and prevents any genetic linkage between markers situated on either side of it (recall that a recombination frequency of 50% means that it is impossible to tell how far apart the two markers are situated). This recombination is also important in circularizing newly infecting DNA molecules and can cause inversions and deletions under appropriate conditions. Aranda et al. (2001) have examined the properties of mutated Cre protein and have shown that the overexpression of Cre switches from inversions to deletions. When a prophage integrates, it frequently does so at the *loxP* site. A molecular map (Fig. 8.15) based on DNA sequence analysis is circular, confirming the analysis.

Superinfection immunity depends on multiple factors. The *c*1 gene codes for a repressor protein that is necessary, but not sufficient, for immunity and is identical in P1 and P7. However, in combination with the *c*4 protein it confers the usual superinfection immunity in a manner such as that seen in P22. Like P22, P1 also possesses an antirepressor function. Mutant *c*6 proteins can interact with *c*1 protein to make it more sensitive to antirepressor. As in the case of λ, a P1 prophage is inducible by UV radiation.

Bacteriophage Mu

Phage Mu is notable because it is not only a noninducible temperate virus, but also the most efficient transposon known. It has an icosahedral head, contractile tail, and six tail fibers. Upon infection of a host cell, the linear Mu DNA molecule is converted to a circular form by binding a 64 kDa phage-specific protein. Unlike the case with φ29, this protein is not linked covalently to the DNA. Regardless of whether the infection is destined to be lytic or temperate, Mu DNA always integrates into the host chromosome.

The integration mechanism used is not the Campbell model, as with λ, but a simple transposition. The basis for the transposition can be seen in the structure of the phage DNA molecule (Fig. 8.16). The DNA is about 39 kb and always includes some host DNA sequences at both ends. The host DNA at the left, or *c*, end is short (~55–154 bp) and is always in a multiple of 11 bp. At the other end the host DNA is about 1–2 kb. The nature of the transposition mechanism is such that there is no reason to expect any two Mu DNA molecules to have the same host sequences, and hence the presence of host DNA can be

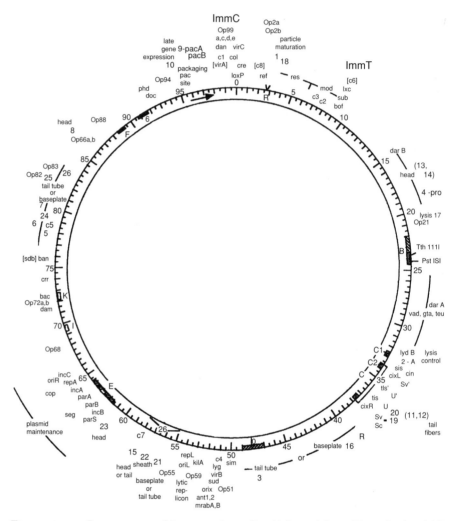

Figure 8.15. Genetic map of bacteriophage P1. (Adapted from Yarmolinsky, M.B. Lobocha, M.B. (1993) Bacteriophage P1, p. 1.50. In: O'Brien, S.J. (ed.), *Genetic maps*, 6th ed., Book I *Viruses*. Cold Spring Harbor, NY: Cold Spring Harbor Laboratory.)

demonstrated readily by heteroduplex analysis. Integrating the phage DNA then requires a conservative transposition from the previous host sequences (still present inside the virion) to some sequence in the new host. The choice of sequence is not entirely random, but it is close to being so, because insertions into the same gene can usually recombine to restore the wild type. Insertion of Mu into a coding region causes the inactivation of that gene and produces a mutation.

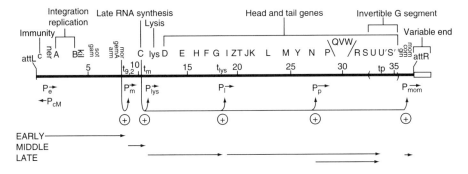

Figure 8.16. Genetic and physical map of phage Mu. The *solid line* represents Mu-specific DNA. The *vertical line* at the *left* end and *open box* at the *right* end represent attached host DNA sequences. Parentheses flank the invertible G segment. The functions of some of the genes are indicated above the map. *Vertical tick marks* above the line represent 5-kb intervals. (The *arrows* at the bottom of the figure show the possible mRNA molecules resulting from a lytic response, with their left ends indicating the promoter sites and the *arrowheads* indicating the terminator sites. Note that transcription of the c (immunity) gene is oppositely oriented and shown above the *other arrows*. The *mor* gene product is a positive regulator of middle transcription, and the C gene product is a positive regulator of late transcription. (Modified with permission from Jiang, Y. [1999]. Mutational analysis of C protein: The late gene activator of bacteriophage Mu, Ph.D. dissertation, The University of Tennessee, Memphis.)

The specific mechanism of transposition is considered in Chapter 15 as part of the discussion on recombination.

As a noninducible phage, Mu is stable following integration and gives rise to stable mutations. This situation is obviously inconvenient for phage geneticists, and therefore, most of the work is done on Mu cts mutants whose repressor (Repc) fails to function at high temperatures. Shortening of the Repc protein by insertion of a nonsense mutation restores temperature stability. Repressor–operator complexes are more stable in the presence of IHF. Some repressor mutations are true virulence mutants with which no lysogens are possible. Some of these mutants are unstable repressor proteins that are rapidly degraded by host proteases. Presence of the unstable repressor also destabilizes normal repressor protein.

Ranquet et al. (2001) have shown a special role for the host tmRNA function in phage Mu physiology. The tmRNA apparently modulates the translation of Repc by controlling whether a full length protein can be made, and, in the absence of tmRNA, the temperature-sensitive repressor functions normally. The mRNA function of tmRNA is not involved in this process because Ranquet et al. did not detect any modified Repc protein.

After induction, Mu replicates not as an autonomous molecule, but by repeated replicative transposition. The transposase protein is the product of gene *A* (see Fig. 8.16), but maximum efficiency is obtained only when the *B* function is also supplied. The B protein is an ATP-dependent DNA binding protein that first binds to DNA and then promotes recombination. Yamauchi and Baker (1998) have proposed that ATP hydrolysis is the switch that controls whether B protein stimulates DNA binding or recombination. IHF is also essential for lytic growth of Mu. As is the case with all transposons, the nature of the DNA in the middle of the molecule is immaterial, but the ends of the phage DNA must contain specific 22-bp sequences, two on the left and one on the right.

The phage life cycle is regulated in the usual way. Middle transcription depends on the presence of the *mor* protein, a product of early transcription. The only essential protein from translation of the middle transcript is the C protein, the activator of late transcription. Basak and Nagaraja (1998) report that C protein binds to specific sites on Mu DNA and induces localized unwinding of the DNA, reorienting promoter elements relative to one another. Among the late functions is one called *mom*, modification of Mu. The *mom* protein alters certain base sequences in the phage DNA, replacing about 15% of the adenine residues with acetamidoadenine. This modification prevents many restriction enzymes from attacking the viral DNA. In order to get transcription of the *mom* region, three GATC sites upstream of *mom* must be methylated by the host *dam* function, which is one of the few examples of regulation of transcription in prokaryotes by methylation.

Mu DNA is packaged by a headful mechanism so that if the length of the DNA is changed by insertion or deletion, the left end (see Fig. 8.16) is held constant and the amount of host DNA varies from its customary 1 to 2 kb. This observation suggests that the packaging mechanism begins at the left. Proper assembly of phage heads requires functional GroEL and GroES, host chaperonins.

When Mu DNA is prepared from an induced lysogen and then heteroduplexed, a bubble indicative of nonhomology is routinely observed in the region labeled G (see Fig. 8.16). Careful analysis shows that the sequence has been inverted with respect to its previous orientation and is flanked by 34-bp **inverted repeat** sequences (i.e., identical sequences rotated 180° with respect to one another). The orientation of the G region correlates with the host range of the phage particle. In one orientation, G(+), the phage is able to infect *E. coli* K-12. In the other orientation, G(−), the phage is able to infect *Citrobacter*, *Serratia*, and *E. coli* strain C. The inversion is under the control of the *gin* gene located in the β region. It requires host *fis* protein for full activity. The rate of inversion is gin-dose-dependent and thus usually low. Inversion rarely occurs during lytic infections, but the large number of cells and longer times involved during lysogenic growth produce substantial numbers of prophages with inverted G segments. Within the G segment are two sets of genes (*S* and *U*, *S'* and *U'*) capable of tail fiber biosynthesis, but these are oriented in opposite directions and have no promoter. The promoter is located near either *S* or *S'*, but outside the invertible region. Inversion then acts to bring various DNA segments into position for transcription.

Essentially, the same phenomenon is seen with phage P1 and its invertible C segment. The central 3 kb of the P1 C segment are homologous to the 3-kb G segment of Mu. Again, two sets of genes and only one promoter are used to regulate host range. Additional examples of this type of regulation are found in *Salmonella* flagellar regulation and *E. coli* fimbria regulation. The mechanism of inversion is considered further in Chapter 15.

SSV1 and SSV2

SSV1 and SSV2 are two spindle-shaped viruses with short tails that infect *Sulfolobus solfataricus*, a member of the Crenarchaeota. Both have completely sequenced genomes (Stedman et al. 2003), and they are approximately 55% similar in base sequence even though they were isolated in Japan and Iceland, respectively. They form lysogens that have integrated prophages in a tRNA gene, with SSV1 using arginine and SSV2 glycine. Few viral proteins are similar to those in the prokaryotic or eukaryotic databases. SSV1 is inducible by ultraviolet radiation, while SSV2 is much less responsive.

The preliminary genetic maps for these viruses are remarkably colinear as in the case of λ and P22. Moreover, nearly all of the promoters are identical between the two viruses. Thus, the general pattern seen in viruses that infect Bacteria continues in the Archaea.

Summary

Temperate bacteriophages are phages that can establish a long-term, stable relationship with a bacterium that prevents host cell lysis and ensures that all progeny cells also carry the virus. The combination of a cell and a virus is called a lysogen, and latent viral DNA within a cell is designated as a prophage. Prophages maintain themselves in a quiescent state by producing a protein repressor that prevents transcription of the DNA coding for vegetative DNA replication and for phage structural proteins. The repressor also confers superinfection immunity on a host cell. Prophages revert to the vegetative state following inactivation of the repressor, which may occur spontaneously or in response to external stimuli (in which case the process is called induction).

Two types of prophage have been observed. Those such as λ, Mu, or P2 are inserted into the bacterial chromosome by either: (1) a single recombination event between the circular phage DNA and the circular bacterial DNA (e.g., Campbell's model, which applies to λ or P2); or (2) transposition (Mu). Prophages such as P1 or P4 (at times) do not insert into the genome but, rather, exist as plasmids, independently replicating DNA molecules within the bacterial cell.

During lytic infections, temperate viruses follow many of the same biochemical pathways as the virulent phages. Concatemeric DNA molecules are often produced by rolling circle replication or recombination. Antitermination is a commonly used strategy for regulating viral late functions. Packaging of DNA can be via a headful mechanism as with P1, Mu, or P22 or else via the use of specific cohesive ends that demarcate the region to be encapsidated as with λ, P2, or P4.

Many of the temperate phages show remarkable similarity in their genetic maps that may be indicative of evolutionary relations within the group. Heteroduplexing is often used as a tool for identifying such relations. Phages P1 and Mu when examined in that way exhibit invertible regions that serve as host range determinants.

Questions for Review and Discussion

1. Prophages are known to occur in two possible states, integrated and plasmid. Discuss the physical nature of each state, the functions that are necessary to maintain them, and the types of phage transcription that would be needed in each prophage state.
2. Control of transcription termination is an important aspect of temperate phage biology. Discuss the types of phage-specific termination and antitermination events that can occur.
3. In what ways can temperate phages interact with one another or with lytic phages?

4. What evidence can you present to support the idea that phages are in fact assemblages of gene groups each of which tends to recombine as a unit? Can you think of any evidence against this idea?

References

General

Botstein, D. (1980). A theory of modular evolution for bacteriophages. *Annals of the New York Academy of Sciences* 354: 484–491.

Friedman, D.I., Court, D.L (2001). Bacteriophage lambda: Alive and well and still doing its thing. *Current Opinion in Microbiology* 4: 201–207.

Hendrix, R.W., Duda, R.L. (1992). Bacteriophage λ*PaPa*: Not the mother of all λ phages. *Science* 258: 1145–1148.

Hochschild, A. (2002). The λ switch: cI closes the gap in autoregulation. *Current Biology* 12: R87–R89.

Howe, M.M. (1998). Bacteriophage Mu, pp. 65–80. In: Busby, S.J.W., Thomas, C.M., Brown, N.L. (eds.), *NATO ASI Series, Vol. H 103 Molecular Microbiology*. Berlin: Springer-Verlag.

Symonds, N., Toussaint, A., van de Putte, P., Howe, M.M. (eds.) (1987). *Phage Mu*. Cold Spring Harbor, NY: Cold Spring Harbor Laboratory.

Weisberg, R.A., Gottesman, M.E. (1999). Processive antitermination. *Journal of Bacteriology* 181: 359–367.

Specialized

Aranda, M., Kanellopoulou, C., Christ, N., Peitz, M., Rajewsky, K., Dröge, P. (2001). Altered directionality in the Cre-loxP site-specific recombination pathway. *Journal of Molecular Biology* 311: 453–459.

Arens, J.S., Hang, Q., Hwang, Y., Tuma, B., Max, S., Feiss, M. (1999). Mutations that extend the specificity of the endonuclease activity of lambda terminase. *Journal of Bacteriology* 181: 218–224.

Basak, S., Nagaraja, V. (1998). Transcriptional activator C protein-mediated unwinding of DNA as a possible mechanism for *mom* gene activation. *Journal of Molecular Biology* 284: 893–902.

Briani, F., Ghisotti, D., Dehò, G. (2000). Antisense RNA-dependent transcription termination sites that modulate lysogenic development of satellite phage P4. *Molecular Microbiology* 36: 1124–1134.

Dodd, I.B., Perkins, A.J., Tsemitsidis, D., Egan, J.B. (2001). Octamerization of λ cI repressor needed for effective repression of P_{RM} and efficient switching from lysogeny. *Genes & Development* 15: 3013–3022.

Hansen, A.-M., Lehnherr, H., Wang, X., Mobley, V., Jin, D.J. (2003). *Escherichia coli* SspA is a transcription activator for bacteriophage P1 late genes. *Molecular Microbiology* 48: 1621–1631.

Juhala, R.J., Ford, M.E., Duda, R.L., Youlton, A., Hatfull, G.F., Hendrix, R.W. (2000). Genomic sequences of bacteriophages HK97 and HK022: Pervasive mosaicism in the lambdoid phages. *Journal of Molecular Biology* 299: 27–51.

Kedzierska, B., Glinkowska, M., Iwanicki, A., Obuchowski, M., Sojka, P., Thomas, M.S., Wegrzyn, G. (2003). Toxicity of the bacteriophage λ cII gene product to *Escherichia coli* arises from inhibition of host cell DNA replication. *Virology* 313: 622–628.

Ko, D.C., Marr, M.T., Guo, T.S., Roberts, J.W. (1998). A surface of *Escherichia coli* σ^{70} required for promoter function and antitermination by phage lambda Q protein. *Genes & Development* 12: 3276–3285.

Liu, T., Renberg, S.K., Haggård-Ljungquist, E. (1998). The E protein of satellite phage P4 acts as an antirepressor by binding to the C protein of helper phage P2. *Molecular Microbiology* 30: 1041–1050.

Murphy, K.C. (2000). Bacteriophage P22 Abc2 protein binds to RecC increases the 50 strand nicking activity of RecBCD and together with λ bet, promotes Chi-independent recombination. *Journal of Molecular Biology* 296: 385–401.

Odegrip, R., Haggård-Ljungquist, E. (2001). The two active-site tyrosine residues of the A protein play non-equivalent roles during initiation of rolling circle replication of bacteriophage P2. *Journal of Molecular Biology* 308: 147–163.

Ranquet, C., Geiselmann, J., Toussaint, A. (2001). The tRNA function of SsrA contributes to controlling repression of bacteriophage Mu prophage. *Proceedings of the National Academy of Sciences of the USA* 98: 10220–10225.

Reiter, K., Lam, H., Young, E., Julien, B., Calendar, R. (1998). A complex control system for transcriptional activation from the *sid* promoter of bacteriphage P4. *Journal of Bacteriology* 180: 5151–5158.

Santillán, M., Mackey, M.C. (2004). Why the lysogenic state of phage is so stable: A mathematical modeling approach. *Biophysical Journal* 86: 75–84.

Stedman, K.M., She, Q., Phan, H., Arnold, H.P., Holz, I., Garrett, R.A., Zillig, W. (2003). Relationships between fuselloviruses infecting the extremely thermophilic archaeon *Sulfolobus*: SSV1 and SSV2. *Research in Microbiology* 154: 295–302.

Yamauchi, M., Baker, T.A. (1998). An ATP–ADP switch in MuB controls progression of the Mu transposition pathway. *The EMBO Journal* 17: 5509–5518.

Westwater, C., Schofield, D.A., Schmidt, M.G., Norris, J.S., Dolan, J.W. (2002). Development of a P1 phagemid system for the delivery of DNA into Gram-negative bacteria. *Microbiology*: 148: 943–950.

Yu, D., Ellis, H.M., Lee, E.C., Jenkins, N.A., Copeland, N.G., Court, D.L. (2000). An efficient recombination system for chromosome engineering in *Escherichia coli*. *Proceedings of the National Academy of Sciences of the USA* 97: 5978–5983.

9

Transduction

Transduction is the term used to designate bacteriophage-mediated transfer of DNA from one cell (a donor) to another cell (a recipient). It was first described by Norton Zinder and Joshua Lederberg for *Salmonella* and phage P22, but has since been shown to occur in many other bacteria and found to involve a variety of bacteriophages. Depending on which virus is involved, the donor cell DNA may or may not be associated with viral DNA inside the capsid of the bacteriophage. However, in all cases of transduction, it is necessary for donor cells to lyse and for virions carrying host DNA (the transducing particles) to be able to infect new host cells. A cell that has acquired a recombinant phenotype by this process is called a transductant.

Transducing particles are by-products of normal phage metabolism and are considered to be of two basic types. **Generalized transducing particles** are associated with the progeny from lytic infections involving numerous virulent or temperate phages. Almost any suitably sized portion of a bacterial genome may be found inside a transducing particle, although not all genes appear with equal frequency. The bacterial DNA is usually not associated with any viral DNA, so these generalized transducing particles are sometimes called pseudovirions. **Specialized transducing particles**, on the other hand, are associated with the progeny from an infection by an integrative temperate phage such as λ. The only bacterial DNA found within these particles is a segment immediately adjacent to one end of the prophage, and is always covalently linked to viral DNA.

This chapter presents various mechanisms for the production of transducing particles and provides examples of the kinds of data obtained.

Major topics include:

- Reasons why some virus particles become transducing particles and others do not
- Types of mutations that can turn conventional phages into better transducing phages
- Use of the information derived from transduction experiments to prepare a genetic map

Bacteriophage Lambda: A Specialized Transducing Phage

Production of Transducing Particles

A number of observations, made over a period of years, have contributed to the development of an appropriate model for the genesis of λ transducing particles. The first fact to emerge was that only two bacterial markers were involved, *gal* and *bio*, and their genetic map positions flank the *att*λ site (Fig. 9.1). As might be expected from the definition of specialized transduction, it later became apparent that any other genes mapping between the attachment site and *gal* or *bio* could also undergo specialized transduction with λ. However, the classic nomenclature has predominated, and all subsequent discussion deals only with *gal* or *bio*. The gene(s) carried by a transducing particle is indicated by appending the genotype symbol to the phage (e.g., λ*gal* or λ*bio*).

The overall frequency for specialized transducing particles among normal virions produced after induction of a lysogen has been shown to be about 10^{-6}. Such a low frequency suggests that a rare event is involved in the production of

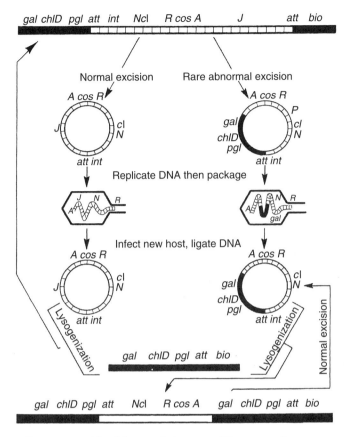

Figure 9.1. Formation of λ*gal* from a λ lysogen. The upper linear structure represents a portion of the genome of a cell that is lysogenic for λ. Each *horizontal line* represents a single DNA strand, and the *short vertical lines* represent hydrogen bonds. Both normal and abnormal excision mechanisms are shown. Note that abnormal excision may take the *bio* region instead of the *gal* region. Lysogenization following normal excision regenerates the structure at the *top* of the figure. Lysogenization following abnormal excision results in the *bottom* structure, which carries a duplication for the portion of the bacterial genome immediately adjacent to the left hand end of the prophage. (From Campbell, A. [1977]. Defective bacteriophages and incomplete prophages, pp. 259–328. In: Fraenkel-Conrat, H., Wagner, R.R. (eds.), *Comprehensive Virology*, vol. 8. New York: Plenum Press.)

transducing particles. For obvious reasons, this type of lysate is referred to as a **low-frequency transducing (LFT) lysate**.

The size of the DNA molecule contained in a λ transducing particle is essentially normal. This observation is not surprising, as studies on λ deletion and insertion mutations have shown that the λ maturation process requires a DNA molecule between 75% and 109% of normal size. Molecules bigger or smaller do not permit proper encapsidation. What the size constancy does mean, however, is that there must be an approximately equal amount of viral DNA lost vis-à-vis the amount of bacterial DNA carried by the transducing particle. A loss of viral DNA implies a loss of viral function, which is indeed the case. The *gal* genes are somewhat farther from the phage attachment site than are the *bio* genes, and consequently λ*gal* particles tend to be more defective than λ*bio* particles. A defective transducing particle is indicated by adding the letter d to the name (e.g., λ*dgal*).

The nature of the defects in a λ transducing particle is correlated with the prophage map. A λ*dgal* phage is defective in tail gene functions, whereas λ*dbio* is defective in recombinational functions. Examination of the prophage map (see Fig. 8.11) shows that the missing viral DNA is always lost from the end of the prophage opposite the newly acquired bacterial DNA. In fact, it is as though the phage excision function occasionally cuts the prophage off-center, and the result is the transducing particle. It should be noted that deletions produced in this way have been extremely useful for elaboration of the λ genetic map.

A molecular mechanism that accounts for all of the above observations can be designed by a slight modification of Campbell's model for integration and excision of λ (see Fig. 9.1; see also Fig. 8.7). In this revised model, excision of the prophage after induction requires that the DNA form a loop structure. If the base of the loop is not centered on the *att* sites at the ends of the prophage when the recombination event occurs, a specialized transducing DNA molecule results. An excision of this type is an example of nonhomologous recombination; other examples are discussed in Chapter 15. The size of the newly excised viral DNA molecule depends on the size of the initial loop.

There are, however, certain limitations to the displacement of the center of the loop from the center of the prophage if a viable phage particle is to be produced. If the *cos* site is lost, a λ*doc* phage results that cannot package its DNA. Furthermore, if the *ori* site happens to be deleted, the DNA would be unable to initiate replication. It is obvious, then, that transducing phages that have lost relatively little viral DNA are less defective and are more likely to be found during a random search. In fact, genetic analysis shows that it is possible to recover λ*gal* phages in which some of the sequences lying between *gal* and the left end of the prophage have been deleted, resulting in less defective phages.

Physiology and Genetic Consequences of Transduction

The physiology of transducing phage infection depends on the genetic alteration that produced the DNA. If the phage is fully functional, events can proceed as described in Chapter 8. This situation is not necessarily as rare as might be expected from the preceding discussion, as the $b2$ region, which lies near the right-hand end of the prophage (Fig. 8.11), can be deleted with no apparent effect on the phage. However, if the phage is defective, a normal helper phage is required for successful lysogeny, but not for initial production of transducing particles or actual infection. Independence of initial production is due to the multiple genome equivalents present in exponentially growing Escherichia *coli* cells that allow intracellular production of both normal and defective phages. In the case of λ*dgal*, the helper also supplies the necessary region of homology for integration of the transducing phage by integrating itself into the bacterial genome first. This process is similar to that diagrammed in Fig. 8.8.

The results of infection of a cell by a transducing particle are variable. In about one-third of cases, new *gal* or *bio* genes are substituted for old by simple recombination, and the transductant does not become a lysogen. In the case of λ*dgal*, nonlysogenic transductants are recovered when cells are infected in the absence of a helper phage. Alternatively, in the presence of a helper about two-thirds of the transductants are observed to be lysogenic and to carry a duplication for the transduced marker (e.g., two sets of *gal* genes) (see Fig. 9.1). In this case, the transductant cells frequently segregate *gal*+ and *gal*- progeny because of recombination involving the flanking *gal* markers that results in loss of the prophage.

Cells that are lysogenic for a specialized transducing phage can be induced by the usual techniques. If the transducing prophage is defective, it is necessary to supply the missing functions by means of a helper phage unless one has integrated with the original transducing phage to give a double lysogen. The progeny virions produced include wild-type (from the helper phage) and specialized transducing particles (from the appropriate prophage). However, the transducing particles may represent as much as 50% of the burst, and this type of lysate is called a **high-frequency transducing (HFT) lysate**. The transducing particles are identical to the original phage that gave rise to the lysogenic transductant, but HFT lysates do apparently differ from LFT lysates. When an HFT lysate is used as the source of phage, the proportion of lysogenic transductants is more than 90%, in contrast to LFT lysates for which the figure is about 70%.

Lambda Phages that Transduce Additional Genetic Markers

In its original form λ transduction is of limited genetic usefulness because of the restricted regions of host DNA that can be transduced. However, there has been a concerted effort to extend the range of markers that are transducible by λ because there are many biochemical advantages to obtaining a specialized transducing phage for the particular gene under study. A major advantage is that, upon induction, all of the prophage genes are turned on (i.e., derepressed), including the extra bacterial DNA. Therefore, the amount of a bacterial gene product in the culture can be greatly increased by attaching the appropriate gene to λ phage DNA and inducing the resulting prophage. As a result, a given amount of product can be obtained from a substantially smaller culture than would otherwise be required (i.e., it is at a higher concentration relative to other host cell products).

One good method for constructing new transducing phages is to make them artificially by DNA splicing. However, prior to the discovery of the linking techniques, a method was developed by Ethan Signer to extend the range of transduction by randomizing the prophage location on the bacterial genome. This method requires that the normal attachment site for λ be deleted from the bacterial genome. This deletion lowers the frequency of lysogens by several 100-fold but does not completely eliminate them. Cells that manage to form lysogens carry their prophages at secondary attachment sites distributed about the genome. Each new site offers the opportunity for obtaining a new type of specialized transducing phage.

THINKING AHEAD

How would you extend Signer's method so that the prophage is inserted at a new site selected by you, and not at random?

Specialized Transducing Phages Other than Lambda

Bacteriophage ϕ80

Phage λ is the best-studied member of a large group of morphologically (and sometimes serologically) similar viruses that infect *E. coli*. Several members of this group have been mentioned in Chapter 8, including phages 21 and 434. Another well-studied member is bacteriophage φ80, isolated in Japan in 1963 by Matsushiro. This phage is heteroimmune to λ and has its own attachment site near *trp* on the *E. coli* genetic map.

The life cycle of φ80 is essentially identical to that of λ, including its ability to produce specialized transducing particles. The usual marker trans- duced by φ80 is *trp* in the form of a φ80*dtrp*. One of the genetic markers lying between the attachment site for φ80 and the *trp* genes is *tonB*, whose protein product is required for infection by phages T1 and φ80, and which can also be carried on a transducing particle. Insertion of any new DNA into the *tonB* gene has two effects: The cell becomes resistant to phage T1, and if it was already a φ80 lysogen, the inserted material could become part of a transducing phage particle.

Gottesman and Beckwith developed a technique called **directed transpo- sition** to exploit the convenient location of the *tonB* gene. The first requirement is that the marker that is to be inserted at *tonB* must be a dominant one. The marker must also be carried on a particular type of F plasmid that is temperature- sensitive for replication, which means that the plasmid DNA is capable of self- replication only at low temperatures. At high temperatures (42°C) the plasmid ceases to replicate itself and is gradually lost from the culture unless it recom- bines and integrates itself into the bacterial chromosome.

The F plasmid carrying the desired marker is transferred by conjugation into a φ80 lysogen that carries a deletion at the site on its genome correspon- ding to the extra marker on the F plasmid. After conjugation, the phenotype of the cell is nonmutant due to the presence of the F plasmid marker. While maintaining selection for this phenotype, the temperature of the culture is raised. The selection forces the cell to keep the plasmid in order to grow, but the high temperature prevents plasmid replication and leads to the segregation of cells lacking the plasmid, and thus renders the cells incapable of growth. However, if the F plasmid integrates into the bacterial genome, it can be repli- cated by the bacterium in the same way as a prophage, and loss of the plasmid is avoided. Homologous recombination is not possible because of the deletion, so integration is at more or less random sites, including *tonB*. Cells carrying the F plasmid integrated at *tonB* can be selected readily from among the survivors of the high-temperature selection by treating the culture with T1 phage. The T1-resistant cells made lysogenic with phage φ80 can be induced with ultraviolet radiation (UV) to give specialized transducing particles of the appropriate type.

Bacteriophage P1

Bacteriophage P1 is normally considered to be a generalized transducing phage because it produces a plasmid-type lysogen. However, P1 does integrate into the bacterial chromosome at a frequency of about 10^{-5}. Utilizing various combinations of bacterial and phage mutants, experimenters have shown that integration of P1 may be catalyzed either by the host *rec* system or by phage Cre protein.

Upon induction, a cell carrying an integrated P1 gives rise to the usual specialized transducing particles. Although integrated P1 lysogens are difficult to detect, the LFT lysate they produce can be demonstrated in a straightforward manner using a strain of *Shigella* as a recipient. Members of the genus *Shigella*, although nearly identical to *E. coli*, do not use the sugar lactose as a sole carbon source and have no DNA corresponding to the *E. coli lac* region. When such a strain is treated with P1 from *E. coli*, Lac$^+$ cells can be obtained. Because there is no homology, the *lac* DNA is not integrated into the *Shigella* genome but is part of a self-replicating P1 plasmid. As predicted by this model, Lac$^+$ transductants are generally immune to P1 superinfection and occasionally segregate Lac$^-$ cells.

Lac$^+$ transductants can also be induced with UV to produce HFT lysates, and in many cases the presence of a helper phage is not required during induction. This independence is due to the long terminal redundancy in the P1 DNA, which allows insertion of relatively large pieces of extraneous DNA into a DNA molecule of constant size with no corresponding loss of function. The practical limitation is that enough terminal redundancy must remain to allow the DNA to circularize after infection.

Bacteriophage P22

As a temperate phage with a life cycle similar to that of λ, P22 can form the same type of specialized transducing phage particles as does λ. It is, however, more versatile than λ in the sense that during a normal infection it occasionally integrates at secondary sites on the *Salmonella* genome, providing a broader range of transducible markers than does λ.

A second type of specialized transducing particle has been observed for P22. It consists of a P22 genome into which DNA coding for resistance to one or more antibiotics has been inserted by a transposon. The origin of the resistance DNA presumably is one of the R plasmids (see Chapter 12) that can coexist with P22. The DNA of P22 is terminally redundant, although not to the same extent as that of P1, and therefore, the same strictures concerning the size of the inserted DNA that apply to P1 also apply to P22. If the size of the P22 DNA plus the insertion becomes too large to fit into a head, each virion is able to carry only a partial genome. However, an infection is still possible if the multiplicity of infection is more than 1.0. There is sufficient randomness in the way in which circularly permuted P22 DNA is produced so that two defective phage particles generally have a complete set of P22 genes if their DNA molecules are joined by recombination (Fig. 9.2).

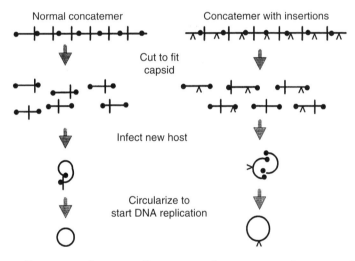

Figure 9.2. Formation of an intact P22 genome from two partial genomes. Each *horizontal line* represents a portion of a double-strand, concatemeric DNA molecule. The *vertical lines* indicate genome lengths, and the ^ symbol indicates where the extra DNA has been inserted. The normal packaging mechanism is shown at the *left* of the figure. It generates circularly permuted terminally redundant molecules by cutting the DNA at points indicated by the *dots*. The effect of a large insertion on the products of the same cutting mechanisms is shown at the *right* of the figure. The molecules produced are the same length as normal packaging, but they no longer include an entire genome. A complete genome can only form when two different DNA molecules recombine.

Generalized Transduction

Bacteriophage P22

Despite the fact that P22 can integrate itself into a bacterial genome and produce specialized transducing particles, it is able to form generalized transducing
particles as well. Formation of these particles occurs during packaging of the
DNA into capsids, not at the time of excision of the prophage. Production of
generalized transducing particles can occur during a wholly lytic infection and
in the absence of any recombination functions. Therefore, P22 DNA apparently
is not required to participate in the packaging process.

The headful mechanism of packaging phage DNA generates circularly
permuted DNA molecules, which implies that cutting of the P22 DNA can occur
at many sites. Ozeki proposed that generalized transduction occurs by a **wrapping choice** mechanism in which sites on the bacterial chromosome are recognized by P22 enzymes and used to package bacterial DNA. Because more P22
DNA is packaged than host DNA, the P22 must have a greater number of sites,
more efficient ones, or both, for the cutting enzyme to act on.

The site on the DNA molecule where the first cut occurs is called a *pac*
site. The cutting action is a property of the products of genes 2 and 3. The *pac*
sites for P22 tend to be clustered in one region of the genome, as heteroduplex
analysis shows that the end sequences of the phage DNA molecules are fundamentally similar. This situation occurs in cases where there is an initial cut within
a limited region followed by sequential encapsidation of terminally redundant
DNA from a concatemer of finite length (Fig. 9.3). Competition between the
process of continued encapsidation and the process of reinitiation of packaging
at a new *pac* site limits the amount of DNA packaged in any one series of reactions. Available data suggest that the average series length is about 5 phage
equivalents, and that the initial cut is somewhat variable in position, as all four
bases can be found as the last base at the *pac* terminus. There is a possibility that
the nuclease has only limited precision.

If there were only one *pac* site on the bacterial chromosome, all generalized transducing DNA fragments would be homogeneous in the sense that a
given marker would appear on one and only one kind of fragment (Fig. 9.4). On
the other hand, if fragments were generated from a variety of sites, many of the
transducing DNA molecules would be heterogeneous, meaning that a given bacterial marker might appear on more than one kind of fragment. Genetic analysis

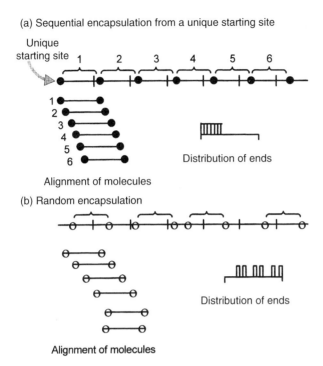

(a) Sequential encapsulation from a unique starting site

Unique starting site

Alignment of molecules

Distribution of ends

(b) Random encapsulation

Distribution of ends

Alignment of molecules

Figure 9.3. Comparison of unique site-sequential encapsidation and random encapsidation. (a) If the concatemer is long enough for only a small number of headfuls, sequential encapsidation at a unique starting site results in a restricted distribution of ends (i.e., restricted permutation). (b) Random encapsidation results in a random distribution of ends (i.e., random permutation). (From Susskind and Botstein [1978].)

of cotransduction (i.e., simultaneous inheritance of two markers) can be used to distinguish between the two possibilities. If one gene can be cotransduced with either of the other two genes, but all three genes cannot be cotransduced, at least two types of transducing DNA fragment are necessary. John Roth and Paul Hartman demonstrated this for the *Salmonella ilv* region.

Independent estimates suggest that *Salmonella* carries at least five or six *pac* sites. The existence of multiple cutting sites for the production of transducing DNA fragments also serves to explain another set of observations. Early experimental results showed that deletion mutations can affect the **cotransduction frequency** of nearby markers (the frequency with which two markers are co-inherited). If the deletion occurs between two markers, the case obviously is trivial and the frequency of cotransduction increases. However, there are instances when the deletion occurs to one side of a pair of markers and yet changes the

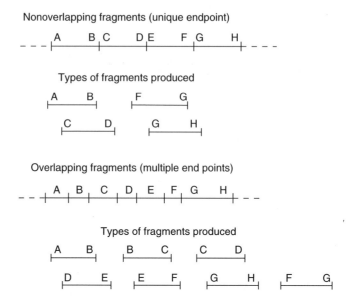

Figure 9.4. Production of transducing DNA by the headful method. The *horizontal lines* represent double-strand DNA, and the *vertical lines* indicate potential starting points for DNA packaging. Note that if nonoverlapping fragments are produced, genes B and C or D and E or F and G lie close to one another, but can never cotransduce, because they will not be on the same DNA fragment. However, if overlapping fragments are produced, all nearby markers are cotransducible.

cotransduction frequency. In one instance, the deletion occurred at such a distance from the pair of markers that it was not included in the same transducing DNA fragment, still the cotransduction frequency was altered. If the deletion mutation either removes a potential site for the initiation of the headful packaging mechanism or else brings the markers closer to such a site, it alters the kinds of expected DNA transducing fragment. The appearance of a new type of fragment carrying only one of a pair of markers would result in a lower cotransduction frequency, whereas the disappearance of such a fragment would result in an increased frequency.

One might assume that after formation of the transducing particles, the balance of the transduction process would be straightforward, but that is not the case for P22. At least 90% of all generalized transducing DNA introduced into a host cell fails to recombine and remains as a persistent, circular, nonreplicating DNA fragment. The genes in these fragments may be expressed, resulting in the production of **abortive transductants**. These cells are phenotypically recombinant, but fail to produce daughter cells with identical phenotypes. Instead, one daughter cell has the DNA fragment (and the recombinant phenotype), whereas the other daughter cell does not.

If a nutritional selection is applied to the culture, abortive transductants produce a microcolony composed of the nonrecombinant daughter cells and a single dividing cell, which carries the transducing DNA. Benson and Roth (1997) have reported a class of P22 mutants that cannot produce abortive transductants, but the mechanism is not certain.

Roughly 2–5% of all generalized transducing particles do give rise to complete transductants (true recombinants). The recombination process involves replacement of both DNA strands in the recipient duplex (see Chapter 15). The amount of DNA replaced ranges from 2×10^6 to 2×10^7 Da (1.3–6.6 kb).

Bacteriophage P1

Phage P1 is similar to P22 in many respects, forming both specialized and generalized transducing particles. However, it is not usually convenient to exploit P1 as a specialized transducing phage because any recombinants that have become P1 lysogens would also acquire the P1 restriction and modification system. The consequence of this acquisition would be that any further genetic manipulation using the recombinant cell as a recipient would require that the donated DNA come from another P1 lysogen. In order to alleviate this problem, P1 stocks used for transductions are usually mutants in the *virB* gene that causes overproduction of an antirepressor protein, which allows lytic replication at all times. Such phages cannot produce lysogens even though they introduce and circularize their DNA in the same fashion as wild-type P1. By sacrificing specialized transducing ability, an extremely valuable generalized transducing phage can be obtained.

There is evidence to indicate that the mechanism by which generalized P1 transducing particles are produced is the same as that used by P22 (i.e., a mistake by the headful packaging system). However, because the size of the P1 phage head is larger, P1 can carry 2.5 times as much DNA as P22, and thus, transduction experiments in *E. coli* generally involve segments that are more than 2% of the genome, whereas similar experiments in *Salmonella* involve about 1% of the genome.

The efficiency of transduction of selected markers varies over a 25-fold range, depending on the marker used. This variability is attributed to fluctuations in the efficiency of the recombination system, but it is assumed that the same phenomenon does not affect cotransduction frequencies for the unselected markers. Despite the differences in efficiency, all regions of the *E. coli* chromosome have been linked by cotransduction to yield the genetic map.

There is a technical note worth mentioning at this point. Although a generalized transductant survives the original phage infection since, in fact, it receives no viral DNA, it would not survive a subsequent P1 infection. Because at least 99.9% of the P1 virions in any given lysate are functional virulent phages, as soon as the infected cells lyse, large quantities of P1 are released in the culture. These P1 particles normally would infect the transductants and lyse them. However, if calcium ions, which are required for phage adsorption, are removed from the medium before any cells can lyse, but after the transducing particles have attached, the transductants are protected against superinfection by P1 virions.

THINKING AHEAD

What properties of other viruses that are discussed in this book would have to be changed to make them capable of generalized transduction?

Other Phages

Even phage λ can be shown to carry out generalized transduction, albeit under unusual conditions. Sternberg (1987) has shown that if the λ exonuclease function is inactivated and if host cell lysis is prevented by a mutation in the *S* gene, generalized transducing particles do accumulate in the host cell after 60–90 min. The length of time required for production of generalized transducing particles is longer than a normal infection, and thus, cells that lyse normally release only specialized transducing phages.

Dhillon et al. (1998) looked for transducing phages that would infect *E. coli* cells that were resistant to P22 and λ. They found four new phages that would carry out generalized transduction and resemble P22. Their most successful isolate was HK620, which does not form plaques on K-12 strains at all, but only on strain H.

Many scientists are interested in establishing genetic systems for less-studied prokaryotes. The initial task is often daunting, given that there is no guarantee that conjugative plasmids will be available. Sander and Schmieger (2001) have described a technique that identifies generalized transducing phages in environmental samples without knowing the original host strain. They isolate phage particles from culture fluid, extract the DNA, and use PCR to amplify any ribosomal RNA genes carried by transducing phages. By sequencing the amplified DNA, they can use standard taxonomic tools to identify the host organism.

Analysis of Transductional Data

Generalized Transduction

Examples could be taken from a large number of organisms, but for convenience *E. coli* is used because its genetic map is the most detailed and therefore the most amenable to transductional analysis. The phage used is P1, but the analytic principles hold true for any generalized transducing phage.

For any given transduction, four parameters can be measured: number of phage particles added per milliliter; number of recipient cells per milliliter; number of transductants per milliliter; and number of cotransductants. The cotransduction frequency is determined by replica-plating the selected transductants to check for the presence of unselected markers from the donor strain. The other parameters can be determined by the usual colony or plaque counts. Table 9.1 presents an example of the data obtained.

Unselected marker analysis can determine relative map positions of the three genetic markers used in the experiment. The *pdxJ*20 allele is coinherited with *purI* 46% of the time, whereas the *nadB* allele is coinherited only 26% of the time. Because the multiplicity of infection was roughly 0.17, the probability of cells infected by more than one phage particle as calculated by the Poisson approximation (see Appendix 1) is only 1%. It is thus reasonable to assume that coinheritance of two markers means that they were on the same transducing fragment. It implies, however, that the *pdxJ* locus lies closer to *purI* than does *nadB* because there were more DNA molecules carrying *purI* and *pdxJ*20 than molecules carrying *purI* and *nadB*. In other words, a higher cotransduction frequency means less recombination and implies that the markers are closer together.

Table 9.1. Data from a typical P1 transduction.

Donor genotype	*purI*+	*naB*+	*pdxJ20*	
Recipient genotype	*purI66*	*nadB4*	*pdxJ*+	
Selected marker	*purI*+			
Results of unselected marker analysis	*nadB*+	*pdxJ*+		3 colonies
	nadB+	*pdxJ20*		10 colonies
	nadB4	*pdxJ*+		24 colonies
	nadB4	*pdxJ20*		13 colonies
	Total			50
Cotransduction frequencies:				
	purI+	*nadB*+		13/50 (0.26)
	purI+	*pdxJ20*		23/50 (0.46)

Note: The multiplicity of infection was 0.17. (Data from Apostolakos, D., Birge, E.A. [1979]. A thermosensitive *pdxJ* mutation affecting vitamin B$_6$ biosynthesis in *E. coli* K-12. *Current Microbiology* 2: 39–42.)

The foregoing analysis is consistent with either of two gene arrangements: *purI–pdxJ–nadB* or *pdxJ–purI–nadB*. Note that, because there is no way to determine which end of the transducing fragment contains which gene, it is impossible to distinguish between mirror images of the arrangements (i.e., *nadB–purI–pdxJ* is indistinguishable from *pdxJ–purI–nadB*). All that remains, then, is to decide whether the unselected markers are on the same side of *purI* or on opposite sides.

Figure 9.5 presents two diagrams showing how crossovers would have to occur for each possible map order to generate the observed recombinants. Note that the left-hand diagram includes one class of recombinants that requires four crossovers whereas all of the recombinants in the right-hand diagram require only two crossovers. Assuming a random distribution of genetic exchanges, four events in a given space should occur less often than two events, and therefore that class of recombinant should be recovered infrequently.

Another way of looking at the problem is to consider what happens whenever a transductant receives both the selected marker and the more distal of the two unselected markers. The unselected marker closer to the selected marker will be inherited more often than expected whenever it lies on the same side of the selected marker as the other unselected marker. Conversely, if the unselected marker lies on the opposite side of the selected marker, it will be inherited

Donor gene arrangement

$purI^+$ $pdxJ$ $nadB^+$ $pdxJ$ $purI^+$ $nadB^+$
$purI$ $pdxJ^+$ $nadB$ or $pdxJ^+$ $purI$ $nadB$

Recipient gene arrangement

Transductant genotype

$purI^+$ $nadB^+$ $pdxJ^+$

$purI^+$ $nadB^+$ $pdxJ$

$purI^+$ $nadB$ $pdxJ^+$

$purI^+$ $nadB$ $pdxJ$

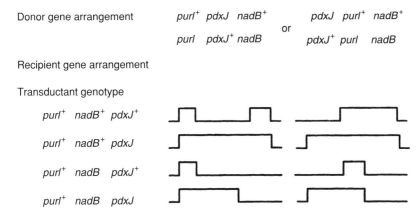

Figure 9.5. Recombination analysis of the data in Table 9.1. The *lines* indicate how the recombinant DNA molecule would have to be constructed to give the observed results. Each step represents one genetic exchange. Because the donor DNA is only a short fragment, an even number of exchanges is always necessary to regenerate a viable genome. Note that only one arrangement of the genes would require a quadruple crossover in order to produce some of the observed recombinants.

less often. Because the $purI^-nadB^-pdxJ$ recombinants were indeed the rarest type and because 10/13 (77%) of $purI^+$ $nadB$ transductants are also $pdxJ20$ (see Table 9.1), the correct map order must be $purI–pdxJ–nadB$.

For purposes of comparison, Table 9.2 presents data obtained when the unselected markers lie on opposite sides of the selected marker. In this case, two genetic exchanges suffice to produce any type of recombinant (Fig. 9.5), and no single class of recombinants can be considered to be significantly rarer than any of the others. Moreover, of the $pdxJ$ transductants, only 22% were also $glyA8$, confirming that the unselected markers are arrayed on opposite sides of the selected marker and are not frequently coinherited.

The *E. coli* genetic map has the minute as its unit of distance, where 1 minute represents the average amount of DNA transferred from a typical Hfr cell to an F$^-$ cell in 1 minute. It is possible to convert cotransduction frequencies into minutes by use of an appropriate formula. For *E. coli*, Barbara Bachmann and her collaborators adapted the formula of Wu to establish all of the genetic map (see Fig. 11.3):

$$\text{Contransduction frequency} = \left(1 - \frac{d}{L}\right)^3$$

Table 9.2. Data from a P1 transduction in which the unselected markers flank the selected marker.

Donor genotype	$purI^+$	$pdxJ^+$	$glyA8$	
Recipient genotype	$purI66$	$pdxJ20$	$glyA^+$	
Selected marker	$purI^+$			
Results of unselected marker analysis	$pdxJ^+$	$glyA8$		5 colonies
	$pdxJ^+$	$glyA^+$		18 colonies
	$pdxJ20$	$glyA8$		19 colonies
	$pdxj20'$	$glyA^+$		8 colonies
		Total		50
Cotransduction frequencies:				
	$purI^+$	$pdxJ^+$		23/50 (0.46)
	$purI^+$	$glyA8$		24/50 (0.48)

Note: The multiplicity of infection was 0.17. (Data from Apostolakos, D., Birge, E.A. [1979]. A thermosensitive $pdxJ$ mutation affecting vitamin B_6 biosynthesis in E. coli K-12. Current Microbiology 2: 39–42.)

where d represents the distance between the selected and unselected markers in minutes, and L represents the size of the transducing DNA fragment in minutes. Although P1 can theoretically carry 2.2 minutes worth of DNA, in practice only a value of 2.0 is used because genetic exchanges are difficult near the ends of DNA molecules. Using the data of Tables 9.1 and 9.2, it is possible to estimate the distance between $purI$ and $pdxJ$ as 0.46 minute, the distance between $purI$ and $nadB$ as 0.72 minute, and the distance between $purI$ and $glyA$ as 0.43 minute.

Specialized Transduction

Nonlysogenic transductants obtained via specialized transduction can be analyzed in the same manner as transductants obtained via a generalized process. However, lysogenic transductants present difficulties for analysis that preclude their use in genetic mapping. As noted earlier in the chapter, lysogenic transductants actually represent a gene duplication rather than recombination. As such they do not readily yield information regarding the normal gene arrangement.

Summary

Transduction is a genetic process in which DNA is removed from a cell by a virus, carried through the culture medium inside a virion, and introduced into a recipient cell in the same manner as viral DNA. Bacteriophages that do not integrate themselves into the bacterial DNA and do not degrade the host DNA are responsible for generalized transduction, a process by which any bacterial gene may be transported. Phages that do integrate themselves are responsible for specialized transduction, a process by which only DNA located near the ends of the prophage is transported.

Both types of transducing phage particle result from mistakes by the enzymatic systems responsible for excising and packaging viral DNA. Generalized transducing phages package their DNA by the headful mechanism. If the enzymes responsible for packaging accidentally use bacterial DNA instead of viral DNA, generalized transducing particles result. Each particle contains only bacterial DNA. Specialized transducing phages occur when the enzymes that are excising a prophage cut out a proper-sized piece of DNA that is not wholly prophage. The result is a DNA molecule that contains both viral and bacterial DNA. As might be expected, specialized transducing particles are usually defective in one or more functions.

Examples of phages that can carry out generalized transduction are P22 and P1. Analysis of the cotransduction frequencies of unselected markers from generalized transductants permits ordering of genes on a genetic map. It is even possible to convert cotransduction frequencies into normal map units by formulas specific for the bacterium and the phage.

An example of a specialized transducing phage is λ. The range of markers it can transduce can be increased by forcing λ to integrate at abnormal sites around the genome. Although not generally used to construct genetic maps, specialized transductants that are also lysogens can be useful biochemically. Induction of such a lysogen in the presence of a helper phage results in production of large numbers of virus particles carrying a particular segment of the bacterial genome.

Questions for Review and Discussion

1. Why are some phages capable of either specialized or generalized transduction (not both of them at the same time), while others are incapable of any transduction?

2. What factors influence cotransduction frequencies?
3. Suppose a donor strain has the genotype *proA⁺ lacZ⁻ argF⁺ hemB⁻* and a recipient strain is *proA⁻ lacZ⁺ argF⁻ hemB⁺*. In an experiment using a generalized transducing phage, *argF⁺* transductants are selected. Scoring of the unselected markers gives the following results:

proA⁺ lacZ⁺ hemB⁺	29
proA⁺ lacZ⁺ hemB⁻	1
proA⁺ lacZ⁻ hemB⁺	1
proA⁺ lacZ⁻ hemB⁻	9
proA⁻ lacZ⁺ hemB⁺	39
proA⁻ lacZ⁺ hemB⁻	1
proA⁻ lacZ⁻ hemB⁺	3
proA⁻ lacZ⁻ hemB⁻	17

What is the order of the genes on a genetic map?

References

General

Campbell, A. (1977). Defective bacteriophages and incomplete prophages, pp. 259–328. In: Fraenkel-Conrat, H., Wagner, R.R. (eds.), *Comprehensive Virology*, vol. 8. New York: Plenum Press.

Nordeen, R.O., Currier, T.C. (1983). Generalized transduction in the phytopathogen *Pseudomonas syringae*. *Applied and Environmental Microbiology* 45: 1884–1889.

Susskind, M.M., Botstein, D. (1978). Molecular genetics of bacteriophage P22. *Microbiological Reviews* 42: 385–413.

Specialized

Adams, M.B., Hayden, M., Casjens, S. (1983). On the sequential packaging of bacteriophage P22 DNA. *Journal of Virology* 46: 673–677.

Backhaus, H. (1985). DNA packaging initiation of *Salmonella* bacteriophage P22: Determination of cut sites within the DNA sequence coding for gene 3. *Journal of Virology* 55: 458–465.

Benson, N.R., Roth, J. (1997). A *Salmonella* phage-P22 mutant defective in abortive transduction. *Genetics* 145: 17–27.

Blahova, J., Kralikova, K., Krcmery, V., Mikovicova, A., Bartonikova, N. (1998). Two high-frequency-transduction phage isolates from lysogenic strains of *Pseudomonas aeruginosa* transducing antibiotic resistance. *Acta Virologica* 42: 175–179.

Dhillon, T.S., Poon, A.P.W., Chan, D., Clark, A.J. (1998). General transducing phages like *Salmonella* phage P22 isolated using smooth strain of *Escherichia coli* as host. *FEMS Microbiology Letters* 161: 129–133.

Iida, S., Hiestand-Nauer, R., Sandmeier, H., Lehnherr, H., Arber, W. (1998). Accessory genes in the *darA* operon of bacteriophage P1 affect antirestriction function, generalized transduction, head morphogenesis, and host cell lysis. *Virology* 251: 49–58.

Sander, M., Schmieger, H. (2001). Method for host-independent detection of generalized transducing bacteriophages in natural habitats. *Applied and Environmental Microbiology* 67: 1490–1493.

Sternberg, N. (1987). The production of generalized transducing phage by bacteriophage lambda. *Gene* 50: 69–85.

10

Genetic Transformation

This chapter introduces the first major bacterial genetic transfer process to be discovered—genetic transformation. Initially, the mechanism appears to be improbable. Donor cells release large DNA fragments (as heavy as several million daltons), and the fragments diffuse through the culture medium to recipient cells. They are then transported across the cell wall and cell membrane into the cytoplasm where recombination occurs. The process is distinct from another biological phenomenon also denoted transformation, the conversion of normal mammalian cells into tumor cells. To emphasize this difference, the bacterial process has always been described as genetic transformation in this book.

The natural susceptibility of DNA to degradation by nucleases prompted many workers to question whether genetic transformation is only a laboratory artifact. However, natural genetic transformation systems occur in over 40 species, including members of the genera *Achromobacter, Azotobacter, Bacillus, Butyrivibrio, Campylobacter, Clostridium, Hemophilus, Micrococcus, Mycobacterium,*

Neisseria, Pseudomonas, Streptococcus, Streptomyces, and *Synechococcus,* and it is reasonable to believe that such a general phenomenon has some genetic significance. Moreover, studies from Wilfried Wackernagel's laboratory have shown that plasmid or chromosomal DNA added to nonsterile soil is slowly degraded and can genetically transform *Pseudomonas stutzeri* even after several days. It is thus probable that genetic transformation has significance beyond the boundaries of the research laboratory, even if the extent of the naturally occurring process is difficult to assess.

In this chapter, the basic process of genetic transformation and some of its variations are discussed. Only a few of the many transformable genera are considered, primarily *Bacillus, Streptococcus,* and *Hemophilus,* which are the best-studied ones.

Major topics include:

- Binding and intake of DNA by cells
- Conditions under which cells bind DNA
- Types of nucleic acids used in transformation
- Use of transformation experiments to obtain information about the genetic map of an organism

Standard Genetic Transformation Systems

Discovery of Genetic Transformation

Frederick Griffith made the original observations on transformation in 1928 during the course of a series of investigations into the mode of infection of Pneumococcus. Griffith noted that there were two colony morphologies present in isolates of these bacteria. One colonial type was smooth and glistening owing to the presence of a polysaccharide capsule surrounding each cell. The other colonial type was much rougher in appearance owing to the absence of a capsule. Colonial morphologies were stably inherited and thus genetically determined.

The presence or absence of a capsule had a profound influence on bacterial virulence. Only encapsulated bacteria could produce a fatal septicemia (infection of the blood) within a mouse. In the absence of a capsule, the mouse's immune system soon destroyed the invading bacteria. As might be expected, Griffith showed that encapsulated cells could be heat-killed prior to infection and no septicemia would result.

However, Griffith also showed that if he mixed live, nonencapsulated bacteria with heat-killed encapsulated ones and injected them into a mouse, the mouse soon died from a septicemia. When he cultured the bacteria that had killed the mouse, he found that all of them had capsules, even though the only encapsulated bacteria that had been injected were already dead. Griffith said that bacteria without capsules had been transformed into bacteria with capsules, but he was never able to show how it was accomplished. That discovery was left to microbial chemists who painstakingly purified the "transforming principle" and identified it.

Finally, Oswald T. Avery, Colin MacLeod, and Maclyn McCarty showed that genetic transformation was caused by an obscure chemical of unknown function found in all cells and named deoxyribonucleic acid (DNA). Their results were the first concrete evidence that DNA was in fact the genetic material of a cell, and so revolutionary was the idea that it did not take hold until 8 years later when the experiments of Hershey and Chase (see Chapter 6) led to the same conclusion. Soon after that, Watson and Crick proposed a molecular structure for DNA and launched the era of molecular biology.

Pneumococcus is still studied extensively; however, over the years it has undergone several name changes at the hands of the bacterial systematists. Presently, they have classified it as *Streptococcus pneumoniae*, but other names still in use are Pneumococcus or *Diplococcus pneumoniae*. For the rest of this book, it is referred to as *S. pneumoniae*.

Another bacterium discussed in this section is *Bacillus subtilis*, but it should be pointed out that not all members of *B. subtilis* are transformable. Indeed, nearly all work has been done with certain special strains, variously denoted "168" or "Marburg," which are derived from that originally used by John Spizzizen and coworkers. The differences between these strains and other members of the species have not been elucidated.

Hemophilus influenzae is a Gram-negative organism that is naturally transformable. Although there are some specific differences in detail between the mechanism used by *Hemophilus* and Gram-positive organisms, the fundamental mechanism of genetic transformation is the same.

THINKING AHEAD

How would you identify and locate the genes coding for the competence-specific proteins whose synthesis is induced by binding of the competence factor?

Competent Cells

Among the naturally transformable bacteria (in contrast to those that can undergo genetic transformation only after special laboratory treatments), the fundamental processing of entering DNA is very similar. However, nearly all vary in the timing and duration of their **competence** as well as their ability to bind DNA and protect it from external nucleases. For this reason, most recent experimental studies have focused on regulation of competence by diverse bacteria. Only *Neisseria gonorrhoeae* seems to lack the ability to regulate its competence. For the remainder, competence is a definite physiologic state inducible by certain growth patterns, often a "shift-down" (mimicking the transition from the exponential phase of growth into the stationary phase) in which cells are transferred from a relatively nutrient-rich medium to a nutrient-poor one. In the case of *B. subtilis*, even a shift from 42°C to 37°C can induce competence. A *Hemophilus* culture generally requires a nutritional shift-down to produce competent cells and therefore is similar to *B. subtilis*. Moreover, an appropriate treatment that blocks nucleic acid synthesis but allows continued protein synthesis leads to essentially 100% of the cells in a culture becoming competent. The competent state can persist for a reasonable period of time so long as the cells are not allowed to grow. Transfer into a rich medium rapidly eliminates competence.

For Gram-positive cells, the regulatory system involved is a set of proteins that activates transcription of flagellar and chemotaxis genes on the one hand or competence genes on the other (Lui and Zuber 1998). Involved in the process is a ComA or a similar protein (Fig. 10.1), a kinase/phosphatase whose enzymatic activity can be stimulated or inhibited by additional factors. Alternatively, *Staphylococcus aureus* can have competence induced by a bacteriophage and *S. pneumoniae* becomes competent only during the exponential growth phase. The number of competent cells in a culture is variable, ranging from 15% for *B. subtilis* to 100% for *S. pneumoniae*. The process of competence formation depends on appropriate cell density. Therefore, in terms of process regulation, it is a quorum sensing function.

The development of competence may require only a few minutes, but at least for organisms that become competent during stationary phase, cultures with competent cells can be maintained for some time even though the competent cells do not divide or replicate their DNA. Chemostat experiments, in which cells are grown in a culture vessel to which fresh medium is constantly added and depleted medium removed, show that competence in *B. subtilis* occurs

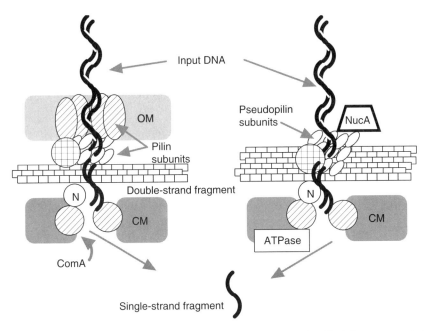

Figure 10.1. Entrance of transforming DNA into bacterial cells. Movement of the DNA is from top to bottom in the figure through a competence pseudopilus. The complex on the *left* is from a Gram-negative cell possessing Type IV pili (*N. gonorrhoeae*), and the complex on the *right* is from a Gram-positive cell that lacks pili (*B. subtilis*). There are two nuclease activities involved. The first (NucA) trims the double-strand DNA to smaller fragments of uniform size. The second (N) converts double-strand DNA to the single-strand DNA that enters the cytoplasm and undergoes recombination. Abbreviations: CM, cell membrane; OM, outer membrane. The OM proteins are encoded by *pilQ*. The various geometric figures represent other proteins involved in the transport of DNA. Similarity in shading indicates genetic similarity in the proteins. (Adapted from Claverys and Martin [2003].)

at doubling times of 150 and 390 min. Because the chemostat cultures did not sporulate, these experiments also demonstrated that competence is not necessarily linked to sporulation. The latter conclusion is reinforced by experiments showing that mutations blocking the earliest steps in sporulation have no effect on genetic transformation ability.

Competent *B. subtilis* cells are more buoyant than noncompetent cells, and they will separate when centrifuged in a medium such as Renografin that forms a density gradient from top to bottom of the centrifuge tube. They are small and have one nonreplicating genome. Membranes of competent cells contain three- to fourfold more poly-β-hydroxybutyric acid (PHB), which is normally produced as a storage product than do noncompetent cells. However, the possible role played by the polymer in genetic transformation is unclear, as many nontransformable genera also accumulate it. One hypothesis is that PHB assists in forming a transmembrane channel with a lipophilic exterior and a calcium-lined interior channel.

Immunologic tests have shown that competent *S. pneumoniae* cells have a new protein antigen (molecular weight approximately 10,000 Da) on their surfaces. This competence factor is released into the medium and can be used to make noncompetent cells competent by binding to the *comD* histidine kinase and triggering the response. In other words, it is behaving like a pheromone (Claverys and Martin 2003). When competence factor binds, the kinase stimulates the ComX sigma factor that is necessary for production of late competence proteins. The overall process requires more than 100 genes. Moscoso and Claverys (2004) have shown that when *S. pneumoniae* cells become competent they also begin to release DNA into the medium. The process becomes progressively more extensive and continues into the stationary phase, although the cells are no longer competent by that time. Unlike the case for *S. pneumoniae*, there does not seem to be a *Haemophilus* competence protein that is released from the cells.

Claverys and Martin (2003) consider the issue of whether finding a DNA sequence that resembles a competence protein means that the organism is naturally transformable. They point out that the competence factors in *B. subtilis* and *S. pneumoniae* do not appear to be genetically related. However, the DNA transport systems bear a resemblance to either Type IV pili or Type 2 secretion systems (Chen and Dubnau 2004). Indeed, some Pil proteins form the channel through which the DNA travels (Fig. 10.1). Therefore, the DNA transport system is often called the **competence pseudopilus**.

The slow release of competence factor is used to account for the relatively high cell density ($\sim 10^8$/ml) required for the production of competent cells. Mutations in the *com* loci prevent expression of the competence factor. The changes induced by competence constitute a sort of differentiation process in which the cell wall becomes more porous; there is an increase in autolytic enzyme activity; and the average length of chains of cells increases as much as eightfold. Most of the changes occur along the cell's equator, the area where new growth is normally found and where DNA uptake occurs. The *comK* gene of *B. subtilis* encodes a competence transcription factor that activates itself, as well as at least six other genes. It acts by bending the DNA so as to stabilize binding by RNA polymerase (Susanna et al. 2004).

DNA Uptake and Entry

Competent and noncompetent cells bind some double-strand DNA to their surfaces; however, only competent cells bind it so that the binding is not readily reversible by simple washing procedures. The addition of ethylenediaminetetraacetic acid (EDTA), a chelator (complexer) of divalent cations, to the medium allows binding of DNA but prevents any further processing. Measurements of the amount of bound DNA indicate that there are approximately 50 DNA binding sites per competent *B. subtilis* cell. Bound DNA is still sensitive to exogenous nucleases or hydrodynamic shear (as in a blender). Any type of DNA can bind, and foreign DNA (e.g., salmon sperm DNA) is often used in competition experiments to prevent further binding of specific bacterial DNA. Competition experiments have shown that the minimum size for effective DNA binding is about 500 bp and that single-strand DNA, glucosylated DNA, double-strand RNA, or RNA:DNA hybrids rarely, if ever, bind.

Competent *Hemophilus* cells exhibit more selectivity in the DNA they transport efficiently when compared with Gram-positive organisms. Although *H. influenzae* takes up DNA from other *Hemophilus* species, it rejects DNA from disparate organisms such as *Escherichia coli* or *Xenopus laevis* (South African clawed toad), whereas *Bacillus* and *Streptococcus* do not. The basis for the specificity can be determined by competing total unlabeled *Hemophilus* DNA with pieces of radioactive *Hemophilus* DNA for cell surface binding sites. The DNA uptake sequence is 5'-AAGTGCGGT-3' and can be found roughly every 4000 bp (about 600 per genome). Some DNA fragments lacking the DNA uptake sequence are nonetheless taken up by competent cells. *N. gonorrhoeae* recognizes 10 bp target sequences that occur about once every 1000 bp.

After binding, transforming DNA must enter the cell (Fig. 10.1), beginning at its 3'-end. Prior to entry, the DNA must be cut to size by a major endonuclease located near the DNA binding site. This enzyme makes essentially random double-strand cuts to generate DNA fragments no larger than about 15 kb, smaller than the size of DNA fragments carried by most transducing phages. It requires magnesium or calcium for its function and thus accounts for the sensitivity of DNA uptake to EDTA.

If a cell is disrupted at this point, the double-strand donor DNA fragments that are released can still be used to transform another cell. However, the next stage is for the DNA to form an **eclipse complex** with a 19.5 kDa competence-specific protein. In *S. pneumoniae* about 68% of the donor DNA enters the eclipse complex, and about 25% of that DNA actually becomes part of the recipient DNA. When the eclipse complex forms, the DNA becomes single-stranded and completely protected from external nuclease activity. The conversion occurs in a linear fashion beginning at one or several points. There does not appear to be any specificity as to which strand is degraded, and bases from the degraded strand are released into the medium for possible reuse. Because single-strand DNA is virtually nonfunctional in transformation, disruption of a cell at this point does not release any donor DNA that can be used to transform a new recipient. In this sense, the eclipse phase for transformation closely parallels the eclipse phase for bacteriophage infection. The half-time for a single gene to exit the eclipse phase is about 5 min.

Entry of the transforming DNA into *H. influenzae* is different from the previously described mechanism. In competent cells, a series of membranous extrusions appear on the cell surface. The double-strand transforming DNA is cut on the surface of the cell and then transported not into the cytoplasm but into the membranous extrusions that subsequently bud off to form **transformasomes**, independent double-strand DNA-containing vacuoles within the cytoplasm (Fig. 10.2). The binding data fit a model in which there is a single DNA receptor site per transformasome.

Redfield (1993) has proposed that DNA degradation and resultant energy production are the driving evolutionary forces behind the cells' ability to become competent. Genetic transformation itself does not appear to be able to confer sufficient immediate advantage to a cell growing in a natural environment to justify having that cell maintain the genes for the competence state. However, nutritional benefit coupled with long-term genetic benefit might be a sufficient explanation.

Electron microscopic studies have demonstrated the close proximity of cell wall and cell membrane in competent cells in a manner reminiscent of the zones of adhesion used during phage T4 infection. The bacterial chromosome is also membrane-attached, and thus the transforming DNA can be delivered right to its target. A normal proton gradient must be present across the cell membrane for DNA transport to occur in *B. subtilis*, but ATP is the driving force in *S. pneumoniae*. Part of competence induction is production of RecA protein and an associated protein called a colligrin (Masure et al. 1998). The colligrin is a membrane protein that acts as an organizer of RecA in the membrane. The assumption is that DNA passes directly through the membrane and immediately enters recombination. Support for this model comes from the observation that electropora-

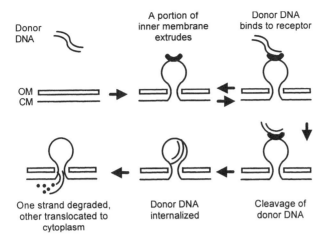

Donor
DNA

A portion of
inner membrane
extrudes

Donor DNA
binds to receptor

OM
CM

One strand degraded,
other translocated to
cytoplasm

Donor DNA
internalized

Cleavage of
donor DNA

Figure 10.2. Hypothetical model for DNA entry into *H. influenzae* cells. DNA entry into *H. influenzae* constitutes the initial step leading to the genetic transformation of these bacteria by DNA present in the medium. The model assumes that small vesicles of inner or cell membrane (CM) protrude through the outer membrane (OM) of the Gram-negative cell wall. DNA first binds reversible to specific receptors present at the surface of the transformasome and is then irreversibly internalized into the vesicle entity with concomitant degradation of the complementary strand. Recombination is assumed to occur elsewhere. DNA uptake is defined as the summation of the binding and internalization step. (Adapted from Barouki, R., Smith, H.O. [1986]. Initial steps in *Haemophilus influenzae* transformation: Donor DNA binding in the *com10* mutant. *Journal of Biological Chemistry* 261: 8617–8623.)

tion (see later) cannot successfully give transformation of chromosomal markers in *S. pneumoniae* (Lefrancois et al. 1998).

Establishment of Donor DNA in the Recipient Cell

Most competent cells bind both homologous and heterologous DNA without discrimination. They also process both types of DNA to single-strand fragments. Because a fundamental principle of genetics is that genetic exchange occurs only between similar organisms, there must be some sort of discriminator mechanism to prevent the establishment of heterologous DNA within a cell. In the cases of *Streptococcus* and *Bacillus*, the discrimination occurs at the level of recombination.

THINKING AHEAD

How would you experimentally demonstrate that heterologous DNA does not enter the recombination process as opposed to failing to complete the process?

Both homologous and heterologous single-strand DNA fragments form an association with the cell membrane and the nucleoid that Gerard Venema and his collaborators referred to as the donor–recipient complex. It is possible to demonstrate that there is an actual pairing of the recipient DNA and the single-strand fragments if the donor DNA is radioactive. The donor–recipient complex can be extracted from lysed cells and treated with S1 nuclease. S1 nuclease attacks only single-strand nucleic acid and does not harm the correctly hydrogen-bonded strands. When the donor–recipient complex from a homologous genetic transformation is tested, it is relatively stable to S1 nuclease, unlike a heterologous complex. Failure to correctly hydrogen-bond would prevent the normal recombination process from functioning (see Chapter 5) and thus prevent assimilation of heterologous DNA. This model predicts that recombination-deficient mutants would also be defective in transformation, which is true. Similarly, treatments that enhance recombination (e.g., introduction of nicks into the recipient DNA by means of x-irradiation) also improve transformation efficiency. Despite the seeming complexity of the genetic transformation process, it is rapid, and the time involved is only about 15 min before donor DNA can be recovered as a double-strand molecule.

The single strand of DNA that recombines is literally substituted for one of the two host DNA strands. This substitution leads to the production of a transient DNA heteroduplex (Fig. 10.3). To the extent that DNA repair enzymes manage to repair the mismatched bases, potential transformants may be lost. In *S. pneumoniae*, but not in *B. subtilis*, there is an enzyme system coded for or controlled by the *hex* gene, which seems to bind to various specific sites on the genome and to correct any mismatched bases in the immediate vicinity. As a consequence, in *hex*$^+$ strains markers that recombine near one of the Hex enzyme-binding sites tend to be lost owing to the correction process, whereas markers that recombine further away are not lost. The former type of marker is called low efficiency (LE), and the latter type is called high efficiency (HE). The Hex system is limited by the number of enzyme molecules available, and thus a method of bypassing its effects is to overload the cell with transforming DNA.

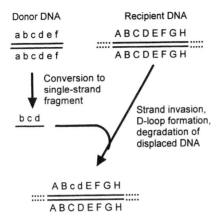

Figure 10.3. Formation of transient heteroduplexes during transformation. Each line represents a single DNA strand, and each letter represents a different genetic marker. Methods for the nick and degradation steps that result in the insertion of the new piece of DNA are discussed in Chapter 5.

One might think that the requirement for recombination as an integral part of genetic transformation is applicable only to chromosomal and not plasmid DNA. After all, by definition plasmids are self-replicating entities, and thus simple arrival in the cytoplasm ought to be sufficient to yield a transformant carrying a plasmid. However, that analysis overlooks the mechanism of transformation. The plasmid DNA, although a small molecule, is linearized by the nucleases at the cell surface. Moreover, it arrives in the cytoplasm as a single-strand entity, whereas it must be a double-strand circle for successful replication. In the light of this analysis, it is not surprising that genetic transformation of a cell by a single plasmid molecule is rare, and when successful plasmid transformation is observed, extra DNA has often been acquired.

The extra DNA may be a second copy of the plasmid (i.e., a concatemer) or the incoming DNA may be homologous either to the bacterial chromosome or to another plasmid that is already resident in the cell. The process is also recombination-dependent. The model based on these observations is that if two plasmid copies are available, there are regions of homology that can be used for circularization (Fig. 10.4). Similarly, regions of homology with another replicon can anchor the DNA and allow circularization to proceed. In general, the larger the region of homology, the greater is the number of transformants obtained. Thus, when homologous DNA is linked with a vector and transformation is attempted, the result is a high frequency of genetic transformation. However, if

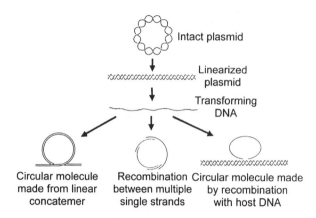

Figure 10.4. Possible mechanisms for successful genetic transformation by plasmid DNA. A plasmid is linearized by cutting at one site. During transformation the strands are separated so that only single-strand DNA arrives in the cytoplasm. Regeneration of circular DNA can occur by synthesis of a complementary DNA strand followed by self-recombination; pairing of multiple single strands of DNA followed by gap-filling DNA synthesis; or recombination with the host cell chromosome.

eukaryotic DNA is linked into the same vector, the transformation frequency may be low unless the vector itself provides homology or there is some accidental homology between the DNA insert and the host cell. Experiments in which a resident plasmid is used to improve the recovery of transformants are often denoted marker rescue experiments.

A comparison of the efficiency of genetic transformation by chromosomal DNA and plasmid DNA in *Hemophilus* indicates that the latter is inefficient. Apparently, only linear DNA exits the transformasome normally, and a linear plasmid has difficulty recircularizing, as was the case with Gram-positive cells (each transformasome would normally have only a single plasmid molecule). Plasmid establishment by double-strand DNA escaping from the transformasome has been reported, but the frequency is low: 10^{-5} to 10^{-7}. The same considerations shown in Fig. 10.4 should apply to plasmid transformation of *Hemophilus*.

Despite the double-strand nature of the transformasome DNA, the recombinant DNA molecules are still heteroduplexes. Therefore, only a single DNA strand is exchanged, as is the case with Gram-positive cells. Electron microscopic experiments have shown that during the acquisition of competence, single-strand gaps appear in the chromosomal DNA. A recombination deficiency mutation that prevents transformation also prevents gap formation. The transforming DNA is processed to yield short, single-strand tails that are presumed to serve as initiators of the recombination process beginning at the gaps.

Other Transformation Systems

Escherichia coli

Although a Gram-negative organism, *E. coli* is unlike *Hemophilus* because it does not undergo transformation spontaneously without exposure to conditions drastically different from its normal environment. For example, **spheroplasts** (cells that have lost most of their cell wall) can take up DNA and be genetically transformed by it. At present, most genetic transformation in *E. coli* is based on the original observation of Mandel and Higa that intact cells shocked by exposure to high concentrations of calcium ions become competent. The most elaborate study on the conditions leading to competence is that of Hanahan (1983), who found that in addition to calcium ion, treatment should include cobalt chloride and dimethylsulfoxide. All cell manipulations must be done on ice, and the transforming DNA be added as a chilled solution. Even so, the treatment may kill as many as 95% of the cells. After allowing time for adsorption, a temperature shift from ice to 37–40°C is used to stimulate uptake and entry of the DNA.

Like *B. subtilis*, competent *E. coli* cells accumulate PHB in their membranes, and the amount of PHB correlates well with the transformability of the cell. The addition of PHB (and possibly other substances) results in a change in the lipid structure of the cell membrane. All lipids have a characteristic phase transition temperature at which they go from being relatively solid to being relatively more fluid. As the intensity of the *E. coli* phase transition increases, so does the transformability of the cells. One suggestion is that the role of calcium in genetic transformation is to activate the biochemical pathway leading to the synthesis of PHB, making competence in *E. coli* also a regulated phenomenon.

A wide variety of plasmids readily transform *E. coli*, which is one reason it has been so widely used as a host for genetic cloning experiments. Unlike the naturally transformable organisms, *E. coli* does not seem to linearize plasmids during transformation, a property very important to scientists cloning DNA in plasmid vectors. If the experimenter linearizes the plasmids prior to transformation, low levels of transformants are obtained. The rare successfully transformed linear plasmids usually contain deletions that include the site of linearization.

Sequence analysis indicates that recombination occurred between 4 and 10 bp directly repeated sequences.

The enzyme responsible for the failure of linearized plasmid transformation is exonuclease V, the same enzyme that must be inactivated to allow rolling circle replication. Because it attacks only linear duplexes, circular plasmid DNA, even if nicked, is not a substrate, and transformation can proceed normally. Many strains of *E. coli* that are used for transformation are *recB recC* (to inactivate exonuclease V) and *sbcB* (to suppress the recombination-deficient phenotype). *E. coli* thus possesses two natural barriers to transformation—lack of competent cells and presence of exonuclease V—both of which can be overcome in the laboratory. A comparison of the transformation frequencies for the various organisms that have been discussed is given in Table 10.1.

Electroporation

The invention of electroporation has greatly extended the concept of genetic transformation to include **electrotransformation**. During this process, a high-voltage electric current is passed through a bacterial suspension for several milliseconds (Fig. 10.5). The effect of the current is to cause holes in the cell walls and membranes through which substances can enter or leave the cells according to their respective concentration gradients. Small plasmids and artificial genetic constructs can readily enter cells during this period. After entry, the same processes discussed earlier are applicable to the newly arrived DNA. Significant cell death does occur, but particularly with microbial populations the number of treated cells is so large that there are sufficient survivors for most experimental needs.

The real strength of the procedure lies in its potential to open pores in any cell, regardless of whether that cell normally undergoes genetic transforma-

Table 10.1. Representative transformation efficiencies in various bacteria using homologous donor DNA.

Organism	Transformants/Viable Cell
B. subtilis	0.001
S. pneumoniae	0.05
H. influenzae	0.01
E. coli	0.05

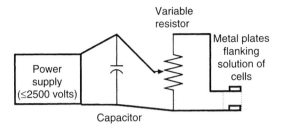

Figure 10.5. Simple electroporation apparatus. The power supply charges a capacitor. When the capacitor is fully charged, it discharges through a circuit that includes a variable resistor and two metal plates inside a container holding a solution of bacterial cells. The solution is usually something like high-percentage aqueous glycerol so that it is not very conductive. The duration of the current pulse is controlled by the variable resistor. More resistance means a longer pulse to the cells.

tion. It is not a universal panacea, but it does work well in many instances, especially when a strong selection is available to detect rare events. Factors working against electrotransformation include the presence of an S layer on the cell surface and the tetrad mode of cell growth. The procedure is simple enough that it is quicker to introduce DNA into *E. coli* by electroporation than by the laborious process of inducing competence, particularly given that both techniques kill a majority of the treated cells.

THINKING AHEAD

What would happen if you tried to use viral DNA for genetic transformation? Would the size of the virus have any effect?

Transfection

An unusual variant of transformation is transfection, a process in which the source of donor DNA is not another bacterial cell but a bacteriophage. The result of successful uptake and entry of DNA into a recipient cell is an infected cell (infectious center), and the success of the procedure can be measured by a standard plaque assay. By varying the amount of phage DNA added, experimenters have shown that the number of infectious centers obtained is proportional to the amount of DNA added to the cells.

Transfection has proved to be a powerful tool for investigating the transformation process as well as other genetic problems. Its biggest advantage is that, unlike DNA extracted from bacterial donors, carefully prepared phage DNA is a homogeneous population of molecules. Therefore, it is much easier to detect the action of various bacterial enzymes on newly introduced DNA, as all the molecules were originally identical. An important use of transfection is to study the DNA recombination and repair process by adding artificially created heteroduplex phage DNA molecules to cells and observing the pattern of inheritance of the heteroduplex region.

E. coli as a Recipient

Oddly enough, the initial observation of transfection did not take place in the *Bacillus* or *Streptococcus* systems but, rather, in *E. coli*. In 1960, Dale Kaiser and David Hogness observed that purified DNA from λ*dgal* high-frequency transducing lysates could cause the production of *gal*⁺ recombinants, provided the recipient cell was simultaneously infected with a nontransducing λ helper phage to provide certain unspecified functions.

Further examination of the **helped transfection** system showed that it had several interesting properties: The helper phages must not be ultraviolet (UV)-inactivated, and they must contain functional DNA. In general, the best multiplicity of infection for the helper phage was high, around 5–15 per cell. Under these conditions, the efficiency of transfection was increased from 10^{-7} to 10^{-3}, making the λ transfection system one of the most efficient known.

Certain strictures also applied to the λ DNA molecules used in the experiment. If less-than-intact genomes are used, each transfecting fragment must contain a *cos* site. Additionally, the *cos* sites must be identical, or nearly so, to the *cos* sites of the helper phage. Rolf Benzinger interpreted these results to mean that as a helper DNA molecule is introduced into the bacterial cell in the usual manner, its *cos* site attaches to the *cos* site of a transfecting DNA molecule, thereby pulling in the transfecting fragment. Hence, fragments without *cos* cannot enter a cell.

Helped transfection is of more general use than might be expected because markers other than *gal* can be used. In fact, if the helper phage carries a mutation that prevents its replication in the recipient cell, the results of phage recombination can be studied. Phages other than λ, such as φ186 or P2, also participate in transfection. Their *cos* DNA sequences are distinct from those of λ, so they cannot act as helpers for λ transfection. However, they do help each other, as 17 of the 19 bases comprising their *cos* sites are identical.

Nonhelped transfection also occurs in *E. coli*. In this case, phage DNA enters a cell by a more usual uptake and entry mechanism such as the artificially competent cells described earlier. Alternatively, the experimenter can remove most of the cell wall to give a spheroplast, most commonly by treating the culture with a mixture of lysozyme and EDTA. The cells remain viable provided that they are contained in a medium of sufficient osmotic strength. New phage infections due to released progeny virions cannot occur because attachment sites for intact phage are not present on spheroplasts.

B. subtilis as a Recipient

Transfection in this organism is much simpler than that in *E. coli* because normally competent cells serve as recipients without any special treatment. The rate of entry of phage DNA seems to be much slower than that of chromosomal DNA. Conversion of phage DNA to a nuclease-insensitive state may require an hour or more, depending on the temperature. DNA uptake and entry by these cells is cooperative. Trautner and coworkers demonstrated that there are various levels of cooperativity, with smaller phages such as φ29, whose DNA tends to aggregate, showing first-order kinetics (Fig. 10.6). Kinetic order is an indication of the number of individual molecules or aggregates that must come together to give the observed result, in this case an infected cell. The presumed explanation for these observations is fragmentation of phage genomes as they attempt to enter the cell and the necessity for two single strands of opposite polarity to produce a replicating DNA molecule. This situation results in several fragments being necessary to provide the equivalent of an intact phage DNA. After transformation, the infectious process proceeds normally, although the yield of phage is lower.

There are some indications that transfection as a process differs from transformation. One strain of *B. subtilis* carries a mutation making it nontransformable owing to an inability to bind DNA (less than 1% normal activity). Nevertheless, it undergoes transfection with DNA from φ29 but not with DNA from SP01. The difference between these two phages is that SP01 requires recombination for DNA circularization, whereas φ29 is circularized by a protein linker. Apparently, phage DNA can attach to competent mutant cells even though bacterial DNA cannot, which suggests the existence of at least two mechanisms for

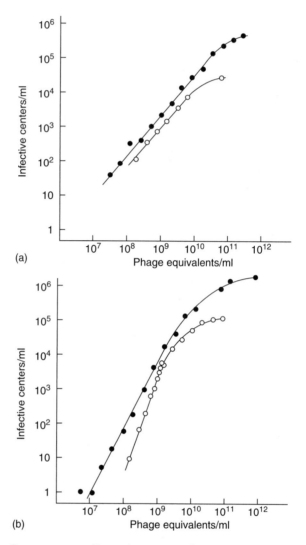

Figure 10.6. Cooperativity effects during transfection. The number of infectious centers observed is plotted as a function of the number of genome equivalents of phage DNA added to the cells. For first-order kinetics, a tenfold increase in phage equivalents would result in a tenfold increase in infectious centers. For second-order kinetics, a tenfold increase in phage equivalents would result in a 100-fold increase in infectious centers, and so on. The *curves* in (**a**) approximate first-order kinetics, whereas the *curves* in (**b**) approximate second- and third-order kinetics. (**a**) The phage DNA molecules used were φ29 (*closed circles*) and SP02 (*open circles*); (**b**) SPP1 (*closed circles*) and SP50 (*open circles*). (From Trautner, T.A., Spatz, H.C. [1973]. Transfection in B. subtilis. *Current Topics in Microbiology and Immunology* 62: 61–88.)

DNA uptake. The lack of transfection by SP01 suggests that the strain in question is also recombination-deficient, which seems to be borne out by other evidence.

Genetic Mapping Using Transformation

Analysis of Transformational Data

Conceptually, the analysis of transformational data is similar to that of transductional data. In both cases one marker is selected, and other unselected markers are then checked to determine cotransfer frequencies. Also, in both the cases, the deduced map order is only relative, as there is no way to unambiguously distinguish left from right on the donor DNA molecule.

Measurable parameters during a transformation experiment include the total number of viable recipient cells, the amount of donor DNA added, the number of transformants obtained, and the frequency with which the unselected markers were obtained. A sample of such data is given in Table 10.2. As with transduction, the cotransfer frequencies are inversely proportional to the genetic distance between the selected and unselected markers. The ordering of genes on the genetic map can be determined using the techniques discussed in Chapter 9.

Genetic Map for *B. subtilis*

The genetic map for *B. subtilis* has been developed piecemeal using both transformation and transduction. An abbreviated map is presented in Fig. 10.7. One problem that slowed mapping in *B. subtilis* was that transformation or transduction can transfer only small regions of the genome at one time, especially as compared to the average amount of DNA transferred by conjugation in *E. coli*.

The largest *Bacillus* transducing phage, SP01, can transfer only some 10% of the genome at one time, and even using gently lysed protoplasts in genetic transformation allows for transfer of only 2.3% of the genome. On the other hand, during conjugation, *E. coli* can easily transfer 30% or even 50% of its genome. Therefore, basic relations between widely separated markers are more difficult to establish in *Bacillus*. An indication of the difficulty is that until 1978 the total number of *Bacillus* markers was not sufficient to permit every known

Table 10.2. Typical data from a *B. subtilis* transformation experiment.

Donor strain	*gua*⁺	*pac-4*	*dnaH*⁺
Recipient strain	*gua-1*	*pac*⁺	*dnaH151*
Selected marker	*gua*⁺		
Distribution of unselected markers	*pac-4*	*dnaH*⁺	6 colonies
	pac-4	*dnaH151*	1 colony
	pac⁺	*dnaH*⁺	44 colonies
	pac⁺	*dnaH151*	56 colonies
Total transformants tested			107
Cotransformation frequencies	*gua*⁺	*pac-4*	6.5%
	gua⁺	*dnaH*⁺	47.0%
Selected marker	*dnaH*⁺		
Distribution of unselected markers	*gua*⁺	*pac-4*	6 colonies
	gua⁺	*pac*⁺	30 colonies
	gua-1	*pac-4*	5 colonies
	gua-1	*pac*⁺	55 colonies
Total transformants tested			96
Cotransformation frequencies	*dnaH*⁺	*pac-4*	11%
	dnaH⁺	*gua*⁺	38%
Deduced map order	*gua-dnaH-pac*		

Note: Abbreviations are listed in Table 1.2, with the exception of *pac*, which refers to resistance to the antibiotic pactamycin. (The data are from Trowsdale, J., Chen, S.M.H., Hoch, J.A. [1979]. Genetic analysis of a class of polymyxin resistant partial revertants of stage 0 sporulation mutants of *B. subtilis*. *Molecular and General Genetics* 173: 61–70.)

marker to be cotransformed or cotransduced with at least two others. It was thus impossible to verify genome circularity prior to that time.

The *Bacillus* genetic map does not show any obvious similarities to genetic maps for other bacteria (compare with Figs. 11.3, 12.1, and 12.2). It is interesting to note that the sporulation genes seem to be scattered about the genome, rather than being clustered in one area. The problem inherent in regulating such a system is considered in Chapter 14.

Summary

Genetic transformation was the first genetic exchange process to be observed in bacteria; despite its comparative antiquity, the molecular details are only now

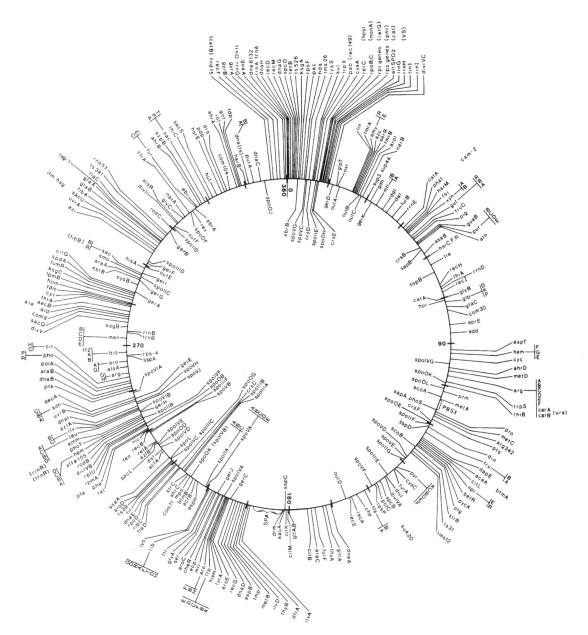

Figure 10.7. A genetic map of the *B. subtilis* chromosome. The *outer circle* presents the complete genome, and the *inner circle* includes only those functions that affect sporulation. Many of the genotype abbreviations used in this map are presented as part of Table 1.2. (Additional information can be found in O'Brien, S.J. [1993]. *Genetic Maps.* Cold Spring Harbor, NY: Cold Spring Harbor Laboratory; or in Moszer [1998]). (From Piggott and Hoch [1985].)

being clarified. The process requires that the recipient cells be in a state of competence for binding and transport of double-strand DNA. In most organisms, competence is regulated in some manner, usually by growth rate. In many cases, stationary phase cells are competent and exponentially growing cells are incompetent, but that is not universally true. Competence is often associated with the appearance of poly-β-hydroxybutyrate in the cell membrane. Competence-specific proteins appear on the cell surface and within the membrane and cell wall. They are related to or comprise pilus proteins. In some cases, certain protein pheromones can be purified and used to induce competence in other cells.

DNA bound to competent cells is processed by nucleolytic attack to yield shorter double-strand fragments that penetrate a cell. Entering DNA is linear and may involve either a single strand (*Bacillus, Streptococcus*) or a double strand (*Hemophilus*). The DNA appears directly in the cytoplasm in the case of Gram-positive bacteria, but in vesicular transformasomes in the case of *Hemophilus*. DNA normally exiting from a transformasome is linear. In all cases, recombination is necessary for completion of genetic transformation, and only a single donor strand is inserted into the recipient chromosome.

Genetic transformation by plasmid DNA presents some difficulties. *E. coli* is unusual in allowing high-efficiency plasmid transformation, for in most organisms the processing of incoming DNA results in the formation of a linear molecule that must be recircularized before it can self-replicate. The recircularization is fundamentally a recombination process and is facilitated by the presence of plasmid concatemers, multiple transforming DNA molecules, or homology to DNA sequences already present in the host.

In most cases, recipient cells exhibit little selectivity with respect to the origin of the DNA they take up. *E. coli*, for example, can be transformed by a wide variety of plasmids, as well as by its own DNA. The exception is *H. influenzae*, whose DNA contains about 600 copies of a 9-bp sequence that is required for normal uptake of DNA. The organisms lacking specificity of uptake compensate for it by a stringent recombination process after the DNA arrives in the cytoplasm. Nonhomologous DNA is rapidly degraded.

All genetically transformable organisms also undergo transfection in which purified phage DNA molecules are transported into the cell, establishing a normal infection. Transfection systems have proved to be particularly advantageous for analyzing the biochemistry of various transformation, recombination, or repair processes, as they use a homogeneous population of molecules.

Questions for Review and Discussion

1. What evidence can you provide that the process of genetic transformation is fundamentally similar in Gram-positive and Gram-negative bacteria? What specific differences can you cite?
2. Explain why plasmids present special problems during genetic transformation. How have experimenters worked around these problems?
3. Give three examples of naturally transformable bacteria. How and when is competence expressed in your examples?
4. Why is it that although the input donor DNA is very large, the amount of donor DNA that appears in transformants is small—smaller even than the amount of DNA involved in most generalized transduction?
5. Why should organisms bother to become competent? What are the short-term and long-term advantages? What are the disadvantages?

References

General

Chen, I., Dubnau, D. (2004). DNA uptake during bacterial transformation. *Nature Reviews Microbiology* 2: 241–249.

Claverys, J.P., Martin, B. (2003). Bacterial 'competence' genes: Signatures of active transformation, or only remnants? *Trends in Microbiology* 11: 161–165.

McCarty, M. (1985). *The Transforming Principle: Discovering that Genes Are Made of DNA*. New York: Norton. (Scientific history written by one of the participants.)

Moszer, I. (1998). The complete genome of *Bacillus subtilis*: From sequence annotation to data management and analysis. *FEBS Letters* 430: 28–36.

Piggott, P.J., Hoch, J.A. (1985). Revised genetic linkage map of *Bacillus subtilis*. *Microbiological Reviews* 49: 158–179.

Specialized

Campbell, E.A., Choi, S.Y., Masure, H.R. (1998). A competence regulon in *Streptococcus pneumoniae* revealed by genomic analysis. *Molecular Microbiology* 27: 929–939.

Chung, Y.S., Dubnau, D. (1998). All seven *comG* open reading frames are required for DNA binding during transformation of competent *Bacillus subtilis*. *Journal of Bacteriology* 180: 41–45.

Hanahan, D. (1983). Studies on transformation of *Escherichia coli* with plasmids. *Journal of Molecular Biology* 166: 557–580.

Lefrancois, J., Samrakandi, M.M., Sicard, A.M. (1998). Electrotransformation and natural transformation of *Streptococcus pneumoniae*: Requirement of DNA processing for recombination. *Microbiology* 144: 3061–3068.

Lui, J., Zuber, P. (1998). A molecular switch controlling competence and motility: Competence regulatory factors *comS*, *mecA*, and *comK* control σ^D-dependent gene expression in *Bacillus subtilis*. *Journal of Bacteriology* 180: 4243–4251.

Masure, H.R., Pearce, B.J., Shio, H., Spellerberg, B. (1998). Membrane targeting of RecA during genetic transformation. *Molecular Microbiology* 27: 845–852.

Moscoso, M., Claverys, J.-P. (2004). Release of DNA into the medium by competent *Streptococcus pneumoniae*: Kinetics, mechanism and stability of the liberated DNA. *Molecular Microbiology* 54: 783–794.

Sikorski, J., Graupner, S., Lorenz, M.G., Wackernagel, W. (1998). Natural genetic transformation of *Pseudomonas stutzeri* in a nonsterile soil. *Microbiology* 144: 569–576.

Susanna, K.A., van der Werff, A.F., den Hengst, C.D., Calles, B., Salas, M., Venema, G., Hamoen, L.W., Kuipers, O.P. (2004). Mechanism of transcription activation at the comG promoter by the competence transcription factor ComK of *Bacillus subtilis*. *Journal of Bacteriology* 186: 1120–1128.

11
Conjugation and the *Escherichia coli* Paradigm

The discovery of the process of conjugation in prokaryotes was due to one of the most fortuitous experimental designs in recent scientific history. Many other scientists had tried to demonstrate conjugation and had failed. Nevertheless, Joshua Lederberg and Edward Tatum, working at Yale University, chose to make another attempt using a common laboratory strain of bacteria, *Escherichia coli* K-12, as their experimental system. They also decided to test only a few isolates of the strain. By chance, they chose not only one of the few organisms that readily undergoes conjugation, but also a fertile strain of that organism. By doing so, they started an extremely fruitful branch of bacterial genetic research. *E. coli* remains the standard conjugation system to which all others are initially compared. This chapter presents the *E. coli* conjugation system. Several alternative conjugation systems are discussed in Chapter 12.

Major topics include:

- Types of cells capable of conjugation
- Structure of DNA transferred between cells
- Lack of reciprocity in the process of conjugation
- Biochemical events occurring during conjugation
- Use of conjugation experiments to obtain information about the genetic map of an organism

Basic Properties of the *E. coli* Conjugation System

Discovery of Conjugation

Lederberg and Tatum developed an experimental design that was both simple and elegant. They realized that to observe a rare event it is necessary to have a powerful method of selection. Yet most biochemical mutations revert to wild type with a frequency of 10^{-6} to 10^{-8}, a rate that may well be greater than that of the desired event. In order to circumvent this problem, they decided to use strains that were multiply **auxotrophic** (required more than one biochemical added to their medium) and to look for complete conversion to **prototrophy** (no biochemicals required for growth). This design permitted the system to be as sensitive as necessary, if they assumed that simultaneous transfer of all markers needed for prototrophy was possible. For example, if a cell were doubly auxotrophic and each mutation reverted at a frequency of 10^{-7}, the probability of a double revertant, assuming independence of reversion events, would be 10^{-14}.

Pairwise tests of various *E. coli* strains gave results indicating that mixtures of some strains would produce prototrophs at a rate of 10^{-6} instead of the expected 10^{-14}. It soon became apparent that there were in fact two types of *E. coli* in the culture collection: those that were fertile when mixed with any other *E. coli* strain, and those that were fertile only with certain strains. The former type was designated F^+ (possessing fertility) and the latter F^-. The various pairings gave the following results:

$$F^+ \times F^+ \rightarrow \text{prototrophs}$$
$$F^+ \times F^- \rightarrow \text{prototrophs}$$
$$F^- \times F^- \rightarrow \text{no prototrophs}$$

The observed genetic transfer appeared to be a generalized phenomenon, as all types of auxotrophs tested tended to give rise to prototrophs at the same low frequency. Interestingly, the prototrophs were invariably F$^+$, but their frequency (10^{-6}) was substantially lower than the conversion of F$^-$ cells to F$^+$ cells in the culture as a whole (70% in a 1-h mating).

They quickly realized that the F$^+$ quality was unique. Not only was it infectious, it was independent of the other genetic properties of the cell. It became customary to speak of the "F factor" as a discrete entity. In fact, the F factor was a plasmid and is now known as the F plasmid.

In order to demonstrate that the genetic transfer process involving F$^+$ cells was unlike any other, Bernard Davis performed an experiment using a U-shaped tube. He observed that if one auxotrophic strain of *E. coli* (either F$^+$ or F$^-$) was placed in one arm of the U and an auxotrophic F$^+$ strain was placed in the other, prototrophs were produced. However, if the bottom of the U tube was closed off by a microporous fritted glass filter that allowed soluble factors such as DNA molecules or viruses to pass through but blocked the passage of intact cells, no prototrophs were observed. From these results, Davis concluded that cell-to-cell contact was necessary for transfer. Lederberg and Tatum were inclined to interpret these results as demonstrating homothallic cell fusion (in contrast to the heterothallic fusion described for *Saccharomyces*), but their opinion was not accepted universally.

Discovery of Efficient Donor Strains and Partial Transfer

William Hayes was one scientist who disagreed with the idea of homothallic cell fusion and decided to inquire about obvious differences in the roles played by the F$^+$ and F$^-$ strains. He used two pairs of strains with each pair carrying identical auxotrophic markers, but one member of the pair was F$^-$ and the other was F$^+$. Hayes mated the F$^+$ strain from one pair with the F$^-$ strain from the other pair and selected for prototrophic recombinants. He then examined the unselected markers of the prototrophs for vitamin production, ability to degrade sugars, and streptomycin resistance. The data presented in Table 11.1 indicate that 87–99% of the prototrophs carried *mal* and *rpsL* markers appropriate to the F$^-$ cell. The conclusion to be drawn from the Hayes and Davis experiments is that the process is not one of cell fusion, but of conjugation (direct transfer of DNA from cell to cell), and that one cell plays the role of a donor of genetic material and the other the role of a recipient just as seen in

Table 11.1. Effect of reversal of F polarity on the genetic constitution of recombinants from otherwise similar crosses.

Strain A = *thr⁺, leu⁺, met⁻, mal⁺, rpsL⁻*
        B = *thr⁻, leu⁻, met⁺, mal⁻, rpsL⁺*
Select: Met⁺, Thr⁺, Leu⁺
Examine: 300 colonies for unselected markers (*rpsL* amd *mal*)

Strains Mated		Percent of Transconjugants		
F⁺	F⁻	Mal⁺RpsL⁻	Mal⁻RpsL⁺	Other Combinations
A	B	0	99	1
B	A	86.8	8.3	4.9

Note: Adapted from Hayes, W. (1953). The mechanism of genetic recombination in *E. coli. Cold Spring Harbor Symposia on Quantitative Biology* 18: 75–93.

transduction or genetic transformation. Transfer is thus one-way rather than reciprocal. By obvious analogy, donor cells are sometimes referred to as male cells and recipient cells as female cells. The recombinant progeny are **transconjugants**.

Not much information could be gained about a system in which genetic transfer took place so inefficiently. Therefore, a thorough understanding of the conjugation process had to await the discovery of some method that would increase the frequency with which genetic exchange took place. Two donor strains of *E. coli* representing a new sort of fertility were isolated independently from F⁺ cultures by Luigi Cavalli-Sforza and William Hayes. These strains gave rise to prototrophic transconjugants at a much higher frequency than did F⁺ cultures (i.e., 10^{-1} instead of 10^{-6}). For this reason, they were designated high-frequency-of-recombination strains, or **Hfr strains**. In honor of their discoverers, they are still referred to as HfrC and HfrH.

The type of genetic transfer carried out by Hfr cells appeared to be different from that carried out by F⁺ cells. Although it was true that certain types of prototrophs arose with a high frequency, other types of prototrophs were produced rarely, if at all. Moreover, the two Hfr strains differed with respect to the types of transconjugant that appeared at high frequency after direct selection (Table 11.2), and the transconjugants that did arise were nearly always F⁻. These data provided further evidence that something more complex than simple cell fusion was occurring.

Table 11.2. Results of Hfr matings using two donors.

Donor Used	Transconjugant cells/ml		
	Arg⁺RpsL⁻	Leu⁺RpsL⁻	Lac⁺RpsL⁻
Similar to HfrH	0	1.9×10^6	8.7×10^5
Similar to HfrC	3.2×10^3	2.8×10^5	2.4×10^6

Note: Recipient strain genotype = *leu, arg, lac, rpsL*.

THINKING AHEAD

If a genetic marker is not inherited after a conjugation, how would you decide whether the failure was in the transfer process or in recombination?

Nature of the Transfer Process

François Jacob and Elie Wollman at the Pasteur Institute in Paris soon used the Hayes Hfr for an important series of experiments. They conjugated a multiply marked F⁻ strain to HfrH and then observed the distribution of unselected markers. Data from a similar series of experiments are presented in Table 11.3. In cross 1, there is apparently a gradient of inheritance, or polarity gradient, of the unselected markers among the Leu⁺ transconjugants. Some markers are coinherited with varying frequency, whereas others are never inherited. The data are arranged to emphasize this gradient.

Table 11.3. Gradient of transfer in Hfr × F⁻ cross.

Cross No.	Selected Marker	Transconjugants/ml	Percent of Transconjugants Inheriting the Hfr Allele for					
			Leu	Pro	Lac	Trp	His	Thy
1	Leu⁺	3.5×10^6	100	42	32	1	0	0
2	Trp⁺	4.1×10^4	71	60	52	100	7	0

Note: Matings were carried out at 37°C using a 10:1 ratio of F⁻ to Hfr cells. After 40 min, aliquots of the culture were placed on agar selective for the indicated markers. When the transconjugant colonies had grown, they were tested for the unselected markers by replica plating.

However, the data from cross 2 show a strikingly different pattern. There are basically two classes of unselected markers: those that are coinherited with the selected marker frequently and those that are rarely or never coinherited. There seems to be little middle ground. The explanation for such results is not obvious. One possible explanation that occurred to Jacob and Wollman was that transfer might be not only unidirectional, but also linear. In such a case, if DNA were transferred from the donor such that the selected marker arrived early (**proximal marker**), the unselected markers need not be inherited by the recipient cell. However, if the selected marker were transferred late (**distal marker**), then necessarily the recipient cell had to receive all of the proximal markers as well (Fig. 11.1).

Such an interpretation also suggests, but does not prove, the possibility that only part of the donor chromosome is transferred to the recipient. One way to ascertain if partial transfer does occur is to take advantage of the phenomenon of zygotic induction (see Chapter 8). Because induction of the transferred prophage was postulated to be due to lack of the λ repressor in the F⁻ cytoplasm, any transfer of the prophage should result in a phage infection. According to the hypothesis of partial transfer, zygotic induction should be observed only when the transferred fragment includes the prophage. Data from an experiment demonstrating the proof of this proposition are given in Table 11.4.

Crosses 1 and 2 represent positive controls showing the distribution of unselected markers in the presence of a λ prophage in both conjugating strains. Crosses 3 and 4 are additional controls designed to show that basically two DNA

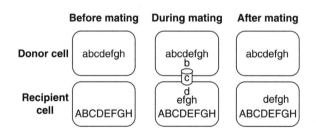

Figure 11.1. Hypothesis of unidirectional transfer. The *rectangles* represent donor and recipient cells and their opposed surfaces are the points at which the cells conjugate. Each genome is represented in linear form. The model assumes that transfer is preceded by replication so the donor cell is not depleted of DNA. Note that transfer of *e* is impossible unless *f*, *g*, and *b* precede it. After mating, the recipient cell is shown as a merodiploid. The diagram assumes that any untransferred DNA in the donor cell is lost.

Table 11.4. Effect of prophage transfer on unselected marker inheritance.

Cross No.	λ Prophage Present in Hfr	λ Prophage Present in F⁻	Selected Marker	Transconjugants/ml	Percent of Transconjugants Inheriting the Hfr Allele for Leu	Lac	Gal	Bio
1	Yes	Yes	Leu⁺	7.1×10^5	100	29	18	13
2	Yes	Yes	Gal⁺	8.4×10^4	25	27	100	93
3	No	Yes	Leu⁺	7.3×10^5	100	19	11	9
4	No	Yes	Gal⁺	9.3×10^4	9	13	100	99
5	Yes	No	Leu⁺	1.1×10^4	100	20	5	1
6	Yes	No	Gal⁺	1.2×10^3	36	30	100	75

Note: Mating conditions were as described for Table 11.3.

molecules can recombine normally, even though one molecule lacks a segment of DNA that is present in the other. In this case the deficiency is the lack of a prophage in the Hfr strain, and inheritance of Hfr DNA means loss of prophage DNA. Induction does not occur in this case, as the λ repressor is present at all times until the prophage DNA is degraded or lost by cell division.

The crucial test of the hypothesis of partial transfer occurs in crosses 5 and 6 in which zygotic induction is possible. The data show that the total number of transconjugants recovered is reduced dramatically. When selection is made for Leu⁺, there is a specific loss of Gal⁺ recombinants. This result is expected from the hypothesis because of the location of the λ prophage next to *gal*. Cells that would have become Gal⁺ transconjugants usually also acquire the λ prophage and lyse. Note that some Gal⁺ recombinants are recovered and that they are generally Bio⁻, thereby confirming that the λ prophage is located between *gal* and *bio*. The occasional inheritance of *bio* can happen if the transferred prophage manages to produce sufficient repressor to turn itself off prior to cell lysis. The same phenomenon, of course, can be observed during normal phage infection of a cell. Note also that direct selection for Gal⁺ requires that the prophage be successfully turned off to obtain the transconjugants in the first place. Therefore, the unselected marker frequencies are essentially normal, although the number of transconjugants recovered is low.

The partial transfer of donor DNA might result from mechanical forces in the culture disrupting the mating aggregates. However, if the mixture of cells to be mated is immobilized on membrane filters, the polarity gradient seen in Table 11.3 is still observed. Therefore, the property of partial genetic transfer is inherent in the conjugation system itself and is not a function of culture agitation, and so on.

The conclusions drawn from the data of Tables 11.3 and 11.4 lead to a model for conjugation in which transfer of DNA begins at a specific point on the genome of the Hfr cell and proceeds in a linear manner into the F⁻ cell. Apparently, this transfer is spontaneously and randomly stopped, or interrupted, in some fashion during the normal course of events. Jacob and Wollman reasoned that if this model were correct what could be done by nature could also be done artificially. They developed an experimental design for an **interrupted mating** in which the conjugating bacterial cells would be broken apart and their further transfer prevented by dilution or immobilization of the cells in an agar layer.

By taking samples from a mating culture at various times and disrupting the cells, it should be possible to observe a specific **time of entry** for any particular genetic marker. Jacob and Wollman, as well as subsequent workers, tried a variety of methods for interruption. The most successful methods have been the following: (1) lysis from without, using T6 phage and phage-sensitive Hfr cells but phage-resistant F⁻ cells in a manner analogous to premature lysis experiments; (2) use of a blending device that applies intense hydrodynamic shear to the mating bacteria causing them to separate and breaking any DNA connecting them; (3) introduction of the chemical nalidixic acid, which instantaneously stops transfer by stopping DNA replication. Because DNA replication is required for transfer (see discussion later in this chapter), when replication halts, so does DNA transfer. Regardless of the method of interruption used, the same type of graph is obtained; a typical one is shown in Fig. 11.2. It is obvious that each type of transconjugant has a specific time of entry, which is obtained by extrapolation of the early time points back to zero. The extrapolated time of entry is expressed in terms of minutes elapsed since the Hfr and F⁻ cells were mixed together. This confirmation of the prediction of the model was the final piece of evidence in the development of a basic theory of conjugation. By means of such an experiment, it is also possible to verify that gal transfers before λ prophage, the time differential being about 0.5 min.

Recombination Following Conjugation

Although only partial transfer usually occurs from an Hfr cell, recombination is still necessary for the production of transconjugants. This statement can be proved by using a *recA* recipient strain, which gives no recombinant progeny.

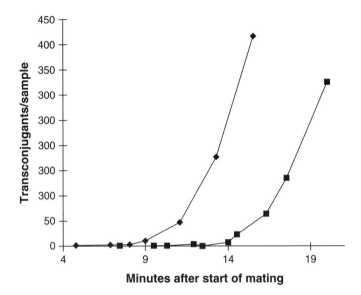

Figure 11.2. Typical interrupted mating curves. The Hfr (donor) and F⁻ (recipient) cells were mixed at time zero in a ratio of 1:100. The mixed culture was incubated at 37°C. Samples of 0.05 ml were removed at various times, blended to break apart the mating aggregates, and then plated on selective medium. The diamonds are Leu⁺ transconjugants, and the squares are ProA⁺ transconjugants.

Lanzov et al. (2003) have examined the coinheritance of unselected markers following conjugation. They find that the transfer gradient shown in Table 11.3 can be affected by the state of LexA induction and the stability of the RecA protein. Inactivation of LexA (SOS induction) decreased the coinheritance fourfold while a RecA protein that is constitutive for its "protease" activity decreases coinheritance sevenfold. The effects of the single mutants are multiplied when in the same strain and increase recombination (decrease coinheritance) 17-fold. The authors suggest that perhaps inhibition of complementary strand synthesis following transfer (see later) may leave more single-strand regions that are recombinogenic.

Construction of the First Genetic Map

Interrupted mating is a powerful tool for the unambiguous determination of gene order, so an obvious next step was to use data obtained from such experiments to construct a genetic map. Most genetic maps have as their unit of distance some function of the number of recombinants observed in various crosses. However, the interrupted mating experiments permitted easy determination of times of entry for different markers. Therefore, the basic unit of the *E. coli* genetic map became the minute, the amount of DNA transferred by an average Hfr cell in 1 min. For example, Fig. 11.2 indicates that *proA* and *leu* are about 5 minutes apart on the genetic map. However, note that there is no inherent directionality to the genetic map (i.e., there is no structure equivalent to the centromere in eukaryotic cells, only the Hfr origin itself).

Any given Hfr strain does not transfer all markers efficiently (see Table 11.2), and thus other Hfr strains had to be found in order to complete the genetic map. F⁺ cultures were tested systematically for the presence of Hfr cells and even treated with ultraviolet radiation (UV), x-rays, or nitrogen mustard to induce DNA error-prone repair and see if new types of Hfr strains could be obtained. It soon became apparent that the number of types of Hfr strains was not only finite but actually quite small. Approximately 22 discrete types of naturally occurring Hfr strains have been observed. After adoption of a few conventions, the bulk of the data obtained could be presented in a simplified genetic map such as that shown in Fig. 11.3. The necessary conventions are the following:

1. After taking into consideration the overlapping regions of DNA transferred by various Hfr strains, the length of the genetic map is taken to be 100 minutes (early estimates were 90 minutes).
2. An arbitrary zero point is picked to lie at *thr*.
3. The direction of transfer of HfrH is considered to be clockwise and is represented by an arrowhead pointing in a counterclockwise direction, where the arrowhead should be viewed as leading in the DNA.

Figure 11.3 presents a simplified *E. coli* K-12 map together with the points of origin and directions of transfer of some commonly used Hfr strains. Note that, in confirmation of the physical studies described in Chapter 2, the genetic map is circular rather than linear. This finding was a surprise at the time, although circular maps have since been found to be the rule rather than the exception among prokaryotes. To enter the data obtained from P1 transduction, it is necessary to use a conversion factor. Bachmann assumed that P1 carries a piece of DNA equivalent to 2 minutes on the genetic map and used the formula of Wu (see Eq. 9.1). A more complete *E. coli* genetic map is available on the World Wide Web (see Appendix 2).

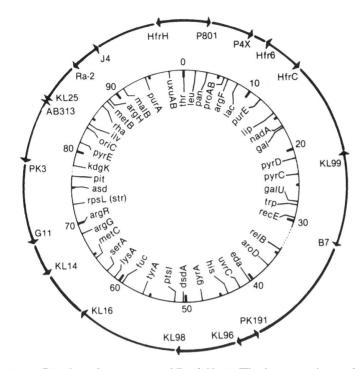

Figure 11.3. Circular reference map of *E. coli* K-12. The large numbers refer to map position in minutes, relative to the *thr* locus. From the complete linkage map available on the Internet (see Appendix 2), 43 loci were chosen on the basis of greatest accuracy of map location, utility in further mapping studies, or familiarity as longstanding landmarks of the *E. coli* K-12 genetic map. Outside the circle, the orientations and leading transfer regions of a number of Hfr strains are indicated. The tip of the *arrow* marks the first DNA transferred in each case. (Adapted from Bachmann, B.J., Low, K.B. [1980]. Linkage map of *E. coli* K-12, edition 6. *Microbiological Reviews* 44: 1–56.)

Note the correlation between map positions determined by interrupted matings and the gradient of transfer seen in Table 11.3. It is consistent with the model of partial unidirectional transfer. If the distance between the various markers is known accurately and is not too great, it is possible to determine the probability of a spontaneous interruption in the DNA transfer per minute of DNA transferred. In this particular case, it works out to about 12% of all DNA molecules being interrupted for each minute of DNA passed along to the recipient cell.

Some of the same genetic oddities seen in viral genetic maps have also been detected in bacterial genetic maps. There are several examples of overlapping genes known, usually of the type where the overlap is in the same reading frame.

An example is the *cheA* gene that codes for two chemotaxis proteins. The *dnaQ* and RNase H genes do not directly overlap, but are transcribed from divergent, overlapping promoters.

A genetic map should not be thought of as a static entity. Rather, mutagenic and evolutionary processes are constantly changing it. Sometimes these changes can be dramatic. Two variant strains of *E. coli* K-12 carrying substantial inversions have been identified. W3110 and its descendants carry an inversion of the DNA lying between *rrnD* and *rrnE*, whereas 1485IN carries an inversion of 35% of the chromosome (roughly *gal* to *his*).

Interactions of the F Plasmid with the Bacterial Chromosome

The F plasmid can be isolated as an independent entity from F$^+$ cells, but not from Hfr cells. Nevertheless, blotting and heteroduplex electron microscopy experiments clearly demonstrate the presence of F plasmid sequences in both types of cell. The difference between the two types of cell is that in Hfr cells the plasmid is integrated into the bacterial DNA in a manner analogous to a λ prophage. Any plasmid that can exist in either the autonomous or the integrated state is said to be an **episome**.

Even when replicating autonomously, the F plasmid is nonetheless in close spatial association with its host cell chromosome. As was noted in Chapter 2, the bacterial chromosome exists in a series of giant supercoiled loops. The covalently closed, circular F DNA is also supercoiled and resembles one of the chromosomal loops. When F$^+$ *E. coli* cells are gently lysed with a detergent and their DNA examined, the F plasmid is generally found to be entangled in the fibers of the extracted bacterial chromosome. This association between the F supercoil and the bacterial chromosome is not strand-specific. The association is presumably useful to F because such an arrangement would automatically segregate F DNA to the daughter cells along with the host cell chromosomal DNA.

Integration of the F Plasmid

The interaction of F DNA with bacterial DNA to effect integration is similar to the situation when λ phage integrates itself. In both cases, circular molecules are formed subsequent to entry into the cell. Therefore, the Campbell model for integration of λ should function equally well for F integration. The subsequent discussion is based on this assumption.

Campbell's model also predicts that F⁺ and Hfr cells are in equilibrium and that aberrant excisions of the F plasmid from the bacterial chromosome should be seen. In λ, there are size constraints on DNA that is incorporated into specialized transducing particles. However, an F plasmid does not have to be packaged, and so the size of the bacterial DNA excised along with an F plasmid can range from a fraction of a minute up to 10–15% of the genome. However, because larger plasmids have a greater retarding effect on cell growth, there is a strong natural selection for shorter F plasmids within a culture carrying a large plasmid. K. Brooks Low found that this shortening was due to recombination events, and that it could be prevented or greatly reduced by using *recA* strains as hosts for F plasmids carrying bacterial DNA.

To distinguish an F plasmid carrying bacterial DNA sequences from a normal F plasmid, the former is designated **F-prime** (F′). F′ molecules are usually stable, but can be isolated in various sizes and may carry any portion of the bacterial DNA (Fig. 11.4). They can also exist in the integrated form, but are difficult to maintain and distinguish from normal Hfr cells. Each unique F′ molecule is designated either by a number or by the gene symbol for the bacterial markers it carries. Thus F42 is also referred to as F′ *lac*.

In addition to the naturally occurring Hfr strains, a number of Hfr strains have been produced by essentially the same process of directed transposition as that used by Jon Beckwith for phage ϕ80. The Hfr process was also developed by Beckwith and coworkers and involves taking an F′ plasmid that carries a temperature-sensitive mutation for DNA replication and putting it into an F⁻ strain that carries a deletion for the genetic material carried on the F′. When the temperature of the culture is raised, the plasmid stops replicating. However, if selection is applied to the culture for the gene carried on the F′ plasmids the bacteria need to avoid the loss of the F′ DNA if they are to survive. They can do it by **integrative suppression**, which is integrating the F′ into the bacterial genome where it can be replicated by bacterial enzymes. The integration would normally occur in the bacterial region corresponding to the bacterial DNA carried by the F′. However, because this material has been deleted, there is no extensive region of homology between the F DNA and the bacterial DNA, and the integration occurs more or less at random. If the integration occurs in the middle of a bacterial gene, the gene is inactivated. Therefore, the site of F integration can be "directed" by an appropriate selection that incidentally results in transposition of the bacterial DNA carried by the F′ from its normal location to a new site on the genome.

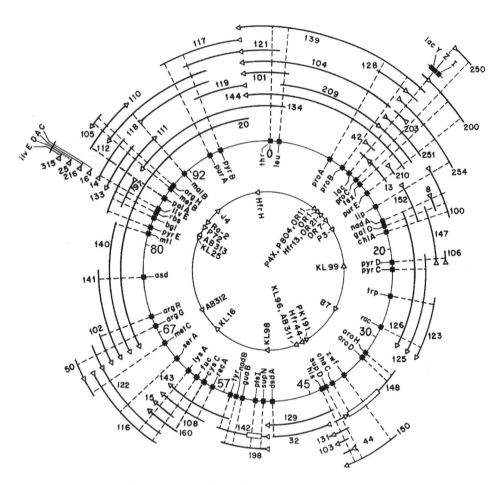

Figure 11.4. Genetic map of *E. coli* K-12 showing approximate chromosomal regions carried by selected F′ plasmids. Each F′ is represented by an *arc* that has an *arrowhead* drawn to show the point of origin of the ancestral Hfr strain (see *inner circle*). The *dashed lines* that extend radially from the genetic markers on the *outer circle* indicate the approximate termini of the F′ plasmids as far as they are known. Those deletions that are known to be present are indicated by narrow *rectangles* (e.g., F142 is deleted for *ptsI*). (Adapted from Low, K.B. [1972]. *E. coli* K-12 F-prime factors, old and new. *Bacteriological Reviews* 36: 587–607.)

Although it seems that an F plasmid should establish the same sort of relationship with its host cell each time it integrates, that does not seem to be the case. Some Hfr strains (e.g., HfrC) are extremely stable. Only rarely does the F plasmid manage to excise itself and convert a cell from Hfr to F⁺. Most Hfr strains are somewhat less stable, segregating F⁺ cells at a perceptible but not inconvenient rate. However, certain Hfr strains are extremely unstable, segregating F⁺ cells at a high frequency. In practice, these unstable Hfr strains must be purified by isolation streaking every few weeks, or they come to consist almost entirely of F⁺ cells. Therefore, when thinking of Campbell's model as it applies to F, the process of integration should be looked on as a dynamic one in which the equilibrium state may lie anywhere along the continuum from 100% Hfr to 100% F⁺ cells.

All observations discussed thus far suggest that, unlike the situation for λ, there are multiple specific sites on the bacterial genome at which F can insert. Originally these sites were called sex factor affinity (*sfa*) and were considered to be unique. However, electron microscopic studies (see discussion later in this chapter) have indicated that the *sfa* sites are merely examples of special discrete DNA sequences known as insertion sequences (IS), which are found on both F plasmid and bacterial DNA. Integration can thus be considered a normal recombination event between homologous DNA regions. However, no F plasmid enzyme system analogous to the λ *int* protein has been identified. Instead, the normal cellular recombination system seems to catalyze most of the integration events. Thus in a *recA* strain, which lacks normal recombination ability, the amount of F integration is reduced 100- to 10,000-fold.

THINKING AHEAD

What are the potential consequences of having an integrated plasmid excise improperly from the bacterial chromosome?

Excision of the F Plasmid

As noted earlier, it is possible for an integrated F plasmid to excise itself and convert an Hfr cell to an F⁺ cell. Campbell's model also accounts for this process as it did for λ excision. There are, however, major differences between F excision and λ excision. Because F is not a virus, the excision event does not affect the host cell significantly. For example, it does not result in cell lysis, and F pili (see discussion later in this chapter) continue to be produced. Because there is no size limit to F′ DNA, there is also no restriction as to the location of the aberrant excision event.

John Scaife proposed that F′ plasmids be categorized according to the original position of the bacterial DNA they carry (Fig. 11.5). Type IA F′ plasmids carry bacterial DNA sequences located near the origin of the ancestral Hfr, whereas type IB plasmids carry bacterial DNA sequences located near the terminus of the ancestral Hfr. Type II plasmids carry both proximal and distal bacterial sequences and are the most prevalent type of F′ isolated, although the relative amounts of proximal versus distal bacterial DNA may be disproportionate.

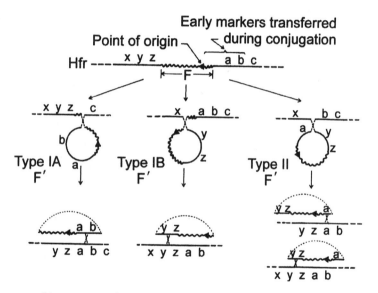

Figure 11.5. Variations in the topology of F′ formation and chromosome mobilization. The line in the top portion of the figure represents part of the chromosome of a hypothetical Hfr strain that transfers the genetic markers a, b, and c early and x, y, and z late in conjugation. The middle part of the figure indicates the relative orientation of the F plasmid and chromosomal markers during the formation of the three types of F′ plasmids shown. The bottom portion of the figure indicates regions of homologous pairing and crossover with the chromosome when the various types of F′ plasmids are in secondary F′ strains (i.e., transferred into new F⁻ hosts). The *dotted lines* symbolize the circularity of the F′. (Adapted from Low [1972].)

Physiology of Conjugation

Formation of Mating Aggregates

Hfr cells form aggregates with F⁻ cells, and the outer membrane protein (Omp A) makes them more stable. Walmsley used an electronic particle counter to show that as mating aggregates form after the Hfr and F⁻ cultures are mixed, there is a net decrease in the total number of particles counted. Blending the cultures breaks apart the aggregates and restores the original number of particles. Achtman et al. have shown that mating aggregates with Hfr cells are more stable than those with F⁺ or F′ cells. They proposed that the stability difference might be due to the amount of DNA that can be transferred.

At one time it was assumed that the mating aggregates were formed on a 1:1 basis, but this assumption is not valid. Achtman et al. used a particle counter to demonstrate that mating aggregates usually consist of either 2–4 cells or 8–13 cells. Such an analysis, however, does not indicate the relative proportion of Hfr and F⁻ cells within the aggregate. Conjugation experiments are customarily performed with an Hfr/F⁻ ratio of approximately 1:10. Skurray and Reeves (1973) have shown that when the ratio is reversed, the F⁻ cells suffer severe membrane damage and die, a phenomenon they called **lethal zygosis**. An unexplained and extremely puzzling observation is that the phenomenon is seen only with Hfr cells. All other types of donor are harmless, and even Hfr cells cause no damage if the culture is not aerated during mating. Recombination-deficient cells seem to be particularly sensitive to lethal zygosis, even in nonaerated cultures.

Mating aggregate formation depends on the presence of donor-specific pili or **F pili**, which are long filamentous structures visible on electron micrographs of the cell surface (Fig. 11.6). They have a vague similarity to eukaryotic flagella but are not used for locomotion and have a different biochemical structure. Pilin (the F pilus protein) is encoded on the F plasmid DNA, and cells have one to three pili whenever F⁺, F′, or Hfr cells are grown at temperatures above 33°C. The pili disappear when cells are grown at lower temperatures, enter the stationary growth phase, or are cured of their F plasmids. Donor cells that lack F pili behave like recipient cells during conjugation experiments and are thus referred to as **F⁻ phenocopies**. Other surface structures (fimbriae or common pili) may be present at all times. The simplest assay system to detect the presence of F pili involves testing for sensitivity to certain male-specific phages (e.g., f1, MS2, or M13) that use the pili as attachment sites for initiating infections.

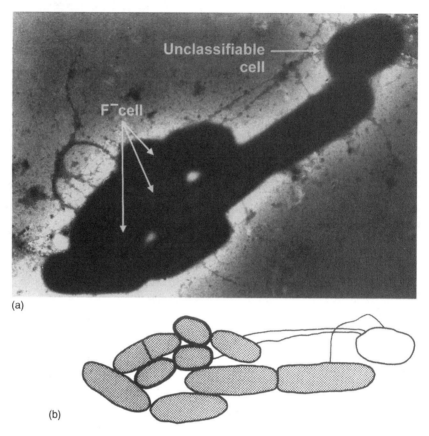

(a)

(b)

Figure 11.6. (a) Hfr × F⁻ mating aggregates. Formaldehyde-fixed cells were negatively stained with phosphotungstic acid and examined under the electron microscope. (b) The cells are differentiated on the basis of size, with the longer cells being Hfr cells, and the three thick-walled cells being F⁻. The long, thin structures connecting various cells are F pili. (Adapted from Achtman, M., Morelli, G., Schwuchow, S. [1978]. Cell–cell interactions in conjugating *E. coli*: Role of F pili and fate of mating aggregates. *Journal of Bacteriology* 135: 1053–1061.)

It is the tips of the F pili that are important for conjugation. Jonathan Ou has demonstrated that MS2 phage particles, which can be shown by electron microscopy to bind to the sides of the F pili, have little effect on mating aggregate formation, whereas f1 phage particles, which bind at the tip of the pili, rapidly prevent mating aggregate formation. Electron micrographs such as that shown in Fig. 11.6 seem to show a connection between cells via the F pilus, and one proposal suggests that the function of the pili may be to overcome the mutual repulsion of the negative charge normally found on bacterial cell surfaces. Although it has been suggested that DNA might be transferred through the F pilus, Achtman et al. demonstrated that using the detergent sodium dodecyl sulfate (SDS) to disaggregate the F pili after mating aggregate formation does not necessarily disrupt DNA transfer. Figure 11.7 presents a model in which the F pilus brings the cells together so that membrane fusion can occur, and DNA is directly transferred from one cytoplasm to the other.

The cells in the mating aggregate must be stabilized, otherwise conjugation is less efficient by orders of magnitude. Both the recipient and donor cells participate in the stabilization process. The recipient cell must have normal lipopolysaccharide attached to its outer membrane and the OmpA protein inserted into the membrane. The donor cell must have both TraN and TraG

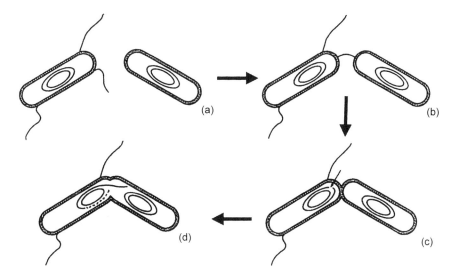

Figure 11.7. Simplified conjugation model. (a) Two cells of opposite mating type are shown. The donor cell carries specific sex pili. (b) Initial mating contact is shown. The F pilus plays an essential role. (c) A stable mating pair. The cells are tightly connected; the F plasmid DNA has been nicked; and conjugative replication is beginning. (d) Transfer of donor DNA into the recipient cell is occurring. New DNA is being synthesized (*dotted line*), and old DNA is being transferred to the recipient cell. Note that although large areas of the cytoplasm are shown as continuous, it cannot be the case because mating interruption by blending does not seem to harm the cells. (Adapted from Hoekstra, W.P.M., Havekes, A.M. [1979]. On the role of the recipient cell during conjugation in *E. coli*. *Antonie van Leeuwenhoek Journal of Microbiology* 45: 3–18.)

proteins present. TraN interacts specifically with OmpA to stabilize the conjugating cells (Klimke and Frost 1998).

Cells carrying an F plasmid do not normally form mating aggregates with other F+ cells. This phenomenon is called **surface exclusion** and depends on the presence of two F-specific proteins, TraT and TraS. TraT is located in the outer membrane and TraS in the inner membrane.

Transfer DNA Synthesis

In order for DNA transfer to take place, both recipient and donor cells must be active metabolically. Starvation of recipient cells or uncoupling of their oxidative phosphorylation results in failure of DNA transfer. The metabolic necessity is, among other things, for a signal of unknown nature to be transmitted from the recipient to the donor cell, which initiates a special round of **transfer DNA synthesis**. The only long-term biosynthetic activity in the donor that is essential for transfer is this DNA synthesis, as streptomycin-sensitive donor cells replicated onto streptomycin-containing agar still form mating aggregates and transfer some of their DNA.

Transfer DNA synthesis is a special process that has been analyzed by Sarathy and Siddiqi (1973) (Fig. 11.8). They used density shift experiments similar to those of Meselson and Stahl that established semiconservative DNA replication to demonstrate that when transfer DNA synthesis is initiated, the normal replicating forks stop and a new round of replication begins. The new synthesis results in the premature appearance of light–light DNA (DNA that carries no density label in either strand), but at a rate that is one-half normal. No such effect is seen if two donor cell cultures are mixed. If DNA synthesis is prevented by expression of a conditional *dnaB* mutation, transfer occurs regardless, but the Hfr cells rapidly become nonviable. Transfer DNA synthesis is less accurate than normal replication because the transferred DNA tends to accumulate mutations at known hotspots.

Transfer DNA synthesis is unique in yet another way. Its product is a single strand of DNA that is transferred without any significant amount of Hfr cytoplasm into the F− cell. To prove the latter point, an experiment was performed in which an F+ cell that was synthesizing the enzyme β-galactosidase was mated to an F− cell that could not produce the enzyme. After DNA transfer had occurred, the F+ cells were lysed by temperature induction of a λ prophage, and the amount of β-galactosidase associated with the F− cells was measured (F+ proteins

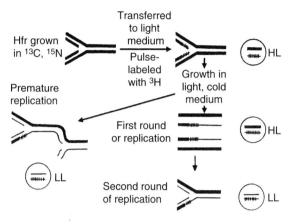

Figure 11.8. Replication of a segment of DNA pulse-labeled with ³H. The experiment is a variation on the classic experiment of Meselson and Stahl. The DNA is initially labeled with heavy isotopes of carbon and nitrogen (*thick lines*, H) and then shifted to a medium containing the normal light isotopes (*thin lines*, L). Each line represents a single strand of DNA. At the time of the shift to light medium, ³H-thymidine is added to the medium for a brief period and incorporated into the DNA (short vertical lines). The density of the resulting ³H-labeled DNA is shown in the *upper inset*. The right-hand side of the figure shows the normal course of events, whereas the left-hand side shows the effects of a premature initiation of DNA replication. The density of the ³H-labeled DNA at each step is shown by the *insets*. Note that premature initiation results in premature conversion of the ³H-labeled DNA to the light–light (LL) density. (Adapted from Sarathy, P.V., Siddiqi, O. [1973]. DNA synthesis during bacterial conjugation. *Journal of Molecular Biology* 78: 427–441.)

adsorbed to cell surfaces had been eliminated with the enzyme chymotrypsin). Within the limits of error of the techniques used, no transfer of β-galactosidase was observed. In terms of the conjugation model shown in Fig. 11.7, these results imply that the connection between the cells is substantially smaller than is diagrammed. In a different conjugation system, that of CollB (see Chapter 12), transfer of donor proteins into recipient cells has been demonstrated.

The proof that only a single strand of DNA is transferred during conjugation was more complex and was achieved independently in the laboratories of Dean Rupp and of J. Tomizawa. One set of experiments took advantage of zygotic induction. An Hfr (λ⁺) culture whose DNA was labeled with ³²P was mated to an F⁻(λ⁻) culture in the absence of ³²P. After lysis of the transconjugant, the progeny phage were collected and the DNA extracted. Barring degradation and resynthesis, any ³²P counts in this preparation represented DNA inherited directly from DNA transferred from the Hfr. The strands of λ DNA were separated by poly(UG) binding and checked for radioactivity. Because the direction of transfer of the Hfr and the orientation of the prophage DNA were known, the experimenters determined that only a single DNA strand was transferred, beginning with the 5′-phosphate end (Fig. 11.9). The logical assumption is that the single strand of DNA transferred is produced by a rolling circle mechanism, which allows the Hfr cell to have a complete set of genetic information at all times. The molecular mechanism involved in plasmid transfer is discussed in Chapter 13. Trgovcevic et al. (1998) have shown that in at least one Hfr strain functional exonuclease V is required for conjugation even through the F plasmid itself can transfer without it.

Hopper et al. (1989) have addressed the problem of what happens to DNA synthesis in an Hfr cell that has its transfer DNA synthesis stopped by

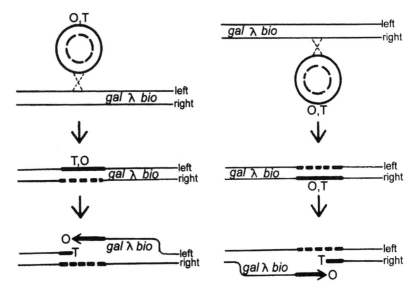

Figure 11.9. Insertion of the F plasmid and transfer of one strand. The *thick lines* represent the complementary strands of the plasmid DNA and the *thin lines* a portion of the bacterial DNA. During mating, one strand of the F plasmid is broken specifically between O (the origin) and T (the terminus), with O being transferred first and T last. Attachment of either strand, *left* or *right*, to the same plasmid strand is the consequence of inserting the plasmid in opposite orientations. (Adapted from Rupp, W.D., Ihler, G. [1968]. Strand selection during bacterial mating. *Cold Spring Harbor Symposia on Quantitative Biology* 33: 647–650.)

DNA gyrase inhibitors. After nalidixic acid treatment (enzyme subunit A inhibited) interrupted mating curves show that transfer resumes from the original spot. However, after coumermycin treatment (enzyme subunit B inhibited), transfer resumes from the point of interruption. Their hypothesis from these results is that nalidixic acid disrupts the DNA replication complex so that continuation is impossible, but coumermycin does not.

The process of DNA transfer is not entirely synchronous. For example, two cultures can be mixed and allowed to form mating aggregates for 5 min before dilution prevents additional aggregates from developing. If interrupted mating curves are prepared for the culture, entry of the selected marker is spread over a period of about 20 min, even though all mating aggregates perforce developed over a 5-min interval. Therefore, some cells experienced a **transfer delay** of 15 min. Other experiments have shown that once transfer begins the rate of transfer is essentially constant over the temperature range 30–40°C but is slower at all other temperatures. Therefore, transfer delay occurs prior to the actual moment at which the first DNA arrives in the recipient.

THINKING AHEAD

How could a scientist determine the ratio of the number of molecules of a plasmid to the number of molecules of chromosomal DNA?

Analysis of the F Plasmid

General Structure

One of the first questions to arise in the study of the F plasmid was that of copy number, the number of plasmid copies per bacterial genome equivalent. There are always some problems in determining the copy number for a particular plasmid due primarily to the difficulty of extracting DNA from a cell without having any of it fragment into smaller pieces. Any fragmentation of the DNA being measured gives a copy number that is too high if both fragments react with the detection system. The most satisfactory results have been obtained either by using radioactive probes to make heteroduplex molecules with the F plasmid or by comparing the rate of reversion of a marker carried on an F′ plasmid to the rate of reversion of the same marker carried on the bacterial chromosome. In the latter case, the ratio of the reversion rates is then the copy number. There is general agreement that the copy number for F is low, perhaps on the order of two copies per genome in cells that are growing slowly. In rapidly growing cells, the number must of course be larger to allow for the shorter time between cell divisions.

The first technique of physical analysis to be applied to the F plasmid was that of heteroduplex mapping. The work was begun in the laboratory of Norman Davidson and continued primarily by Richard Deonier and E. Ohtsubo, his collaborators. When analyzing a circular DNA molecule, the first necessity is to establish reference points for use when measuring distances along the DNA molecule. Unless at least two such points are available, it is impossible to distinguish the clockwise from the counterclockwise direction. The first reference point, an arbitrarily chosen zero point, was obtained by examining a heteroduplex molecule composed of one strand of F DNA and one strand of F100 DNA. The point at which the insertion loop (representing the bacterial DNA of the F') diverges from the double-strand DNA was taken as the origin of the F physical map. Because F DNA is 94.5 kb in length and is circular, the zero point can also be called 94.5 and is usually designated 94.5/0 F. In order to define the clockwise direction, an F plasmid [FΔ(33–43)] that has a deletion of some 10 kb was used as a reference, and the deleted segment was arbitrarily assigned to the interval 33 F–43 F.

As more F and F' DNA molecules were examined by the heteroduplex technique, other differences were found. Small insertions or deletions were observed. Also, certain base sequences were seemingly present several times in any one F DNA strand and could, under proper conditions, anneal with one another. There seemed to be several of these sequences, and they were designated by paired Greek letters: $\alpha\beta$, $\gamma\delta$, $\epsilon\zeta$. In order to differentiate between repeats of the same sequences, they were given subscripts: α_2 β_2, and so forth. Complementary sequences were designated $\beta_1'\alpha_1'$, $\beta_2'\alpha_2'$, and so forth. It soon became obvious that the junction point between F DNA and bacterial DNA in an F' tended to occur at one of these sequences. The same appeared to be true for F integrated to form an Hfr. The natural conclusion was that these sequences represented the actual points of recombination for F and corresponded functionally to the *att* site on lambda.

Other work has shown that some of the repeated sequences are the same as certain insertion sequences that seem to be able to move themselves about the chromosome and that may form the ends of transposons. These sequences are discussed in greater detail in Chapter 15, but a brief description is necessary here. Each unique sequence has been given a number (e.g., IS$_1$, IS$_2$), and it can be shown that several copies of the IS DNA may be found on any genome. Hybridization analysis has shown that $\alpha\beta$ is the same as IS$_3$, and $\varepsilon\zeta$ is the same as IS$_2$, which means that homologous DNA sequences exist on F DNA and bacterial DNA that might be used for integrative recombination. The current model for F integration takes these findings into account and is shown in Fig. 11.10. Different Hfr strains seem to use different repeated sequences for integration, which results in the integrated F plasmids being circular permutations of one another. All Hfr strains have *oriT* (transfer DNA origin of replication) located away from the ends of the plasmid, which means that part of the plasmid can transfer only after the entire bacterial chromosome, a fact that accounts for the lack of Hfr transconjugants.

F′ plasmids are assumed to result when pairing takes place either between two repeat sequences inverted with respect to each other and flanking the integrated F plasmid or between one such sequence and one of the normal F termini. The possible genetic consequences of recombination events of this sort are discussed in Chapter 15.

Genetic Analysis of the F Plasmid

The major barrier to genetic analysis of the F plasmid was the existence of **incompatibility**, the inability of two plasmids to occupy the same cell at the same time, which precluded the use of conventional *cis–trans* tests to assign mutations to genes. Achtman et al. (1972) overcame this problem by means of an ingenious experimental design.

The basic plan of action was to take advantage of the fact that two strains carrying F′ plasmids can be mated if one of them has been F$^-$ phenocopied, which gives rise to phenocopy cells carrying two different F′ plasmids. These cells would, of course, soon exclude one of the F′ plasmids (such exclusion could be prevented by using *inc* plasmids, defective for incompatibility, but they were not discovered until later). However, prior to its exclusion, the genetic information the plasmid carried would be expressed. The result of that expression during the transient merodiploid state would be a potential phenotypic change that would permit a *cis–trans* test, provided the trait could be assayed before exclusion had time to operate. One easily assayed trait is transfer ability itself, as plasmids are ready to retransfer within 50 min of their arrival in a new cell. Therefore, if the two plasmids were both *tra*, and the merodiploid cell were mated to an F$^-$ tester strain before exclusion could become effective, transfer of one or the other of the F′ plasmids into the tester strain would indicate a positive *cis–trans* test.

Achtman et al. began by isolating a series of *tra* mutations in an F′ *lac* plasmid. From among these mutant strains, they selected *tra* mutants that could be suppressed by a nonsense suppressor and moved the appropriate suppressors into strains carrying the plasmids, giving them a set of donor strains that were genotypically *tra* but phenotypically Tra⁺. A second set of F′-carrying strains that had no suppressors (and were thus genotypically and phenotypically *tra*) were prepared for use as recipients.

The actual experiment consisted of mating a suppressed donor F′ strain (Tra⁺) to an F⁻ phenocopied recipient F′ strain (Tra⁻). After about 45 min of mating, the donor strain was destroyed by lysis from without by phage T6. The recipient strain, which was resistant to phage T6, was unaffected. The culture was then diluted to prevent any retransfer of the plasmid between cells, and samples were removed and tested for donor ability. The tester F⁻ strain was resistant to T6 (*tsx*), resistant to the antibiotic spectinomycin (*rpsE*), and unable to use lactose as a sole carbon source (*lac*). If resistant cells appeared that could ferment lactose, they must have received an F′ *lac* from the initial recipient strain, which would imply that either complementation or recombination had occurred between the two F′ plasmids. The two alternatives were easily distinguished, because complementation would cause transfer of a *tra* plasmid that would be incapable of further transfer, whereas recombination would cause transfer of a *tra⁺* plasmid that could be retransferred indefinitely.

The experiments were so successful that the same workers devised a way to test *tra* mutations that could not be suppressed. Instead of mating the first donor strain to a phenocopied F′ strain, they grew P1 phage on it. They then used transduction to carry portions of the F′ plasmid into other plasmid cells. Again, the donor DNA was expressed, and complementation was possible and could be detected in the usual way. However, the efficiency of transfer was, of course, considerably lower, and thus the transductional method lacked the sensitivity of the direct mating.

Paul Broda and coworkers devised an interesting variation of the same experimental scheme in which the primary donor was an Hfr strain rather than an F′ strain. They were investigating the prediction that only part of the F plasmid transfers at the beginning of the Hfr DNA, a prediction based on the observation that transconjugants from an Hfr × F⁻ mating are generally F⁻ and not Hfr. If the prediction were true, a given Hfr should be able to complement only certain *tra* mutations during a mating in which only short pieces of DNA are transferred. In this instance, experiment exactly corresponded to theory, and complementation was obtained. As indicated in Fig. 11.10, subsequent work confirmed that not all Hfr strains transfer equal amounts of F DNA during initiation of conjugation owing to the manner in which the F plasmid integrated.

It is obvious that the experimental method discussed above does not suffice to study the genetics of F plasmid traits that do not affect transfer ability. The electron microscopic technique of physical mapping via heteroduplex DNA molecules has been applied to F with considerable success, as described earlier. This technique has been augmented by the use of restriction enzymes to cut F DNA into fragments that can be individually analyzed or linked with other DNA molecules for analysis, as discussed in Chapter 2. In fact, the *tra* complementation experiments can be done much more readily if cloned DNA fragments are used in vectors that can coexist easily with F.

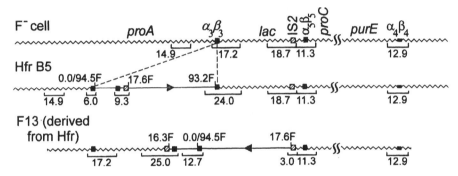

Figure 11.10. Summary of the physical structures of an Hfr and an F'. A physical map of a portion of the bacterial chromosome is shown at the top of the figure with the approximate locations of selected genes and IS elements indicated. A portion of the bacterial DNA between *proC* and *purE* has been omitted for clarity. Below this area the sequence arrangements of Hfr strain B5 and F' F13 are shown. *Sawtooth lines* represent bacterial DNA, and *smooth lines* represent the integrated F sequences. Selected F coordinates are indicated by the numbers above the lines (see also Fig. 11.11). The *small arrowheads* indicate the point of origin and direction of transfer. The various IS3 elements are designated in the $\alpha\beta$ notation with subscripts used to identify the individual components of each IS3 sequence. They are represented by the *solid boxes*. The IS2 elements are represented by the *hatched boxes*. Note that in the Hfr the F plasmid is integrated between the two halves of $\alpha_3\beta_3$ in a left-to-right orientation. In F13, the orientation is reversed, and the site of the F plasmid insertion is an IS2 element. (Adapted from Hadley, R.G., Deonier, R.C. [1979]. Specificity in the formation of type II F' plasmids. *Journal of Bacteriology* 139: 961–976.)

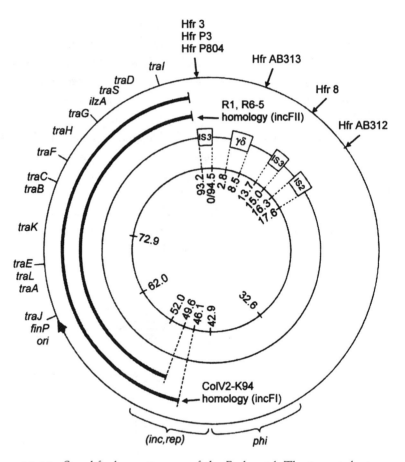

Figure 11.11. Simplified genetic map of the F plasmid. The *inner circle* gives some physical coordinates in kilobase units. The next circle indicates the locations of identified insertion elements. The two *arcs* show regions of extensive homology with ColV2-K94, R1, and R605 plasmids. The *outer circle* indicates the locations of certain genetic loci involved in phage inhibition (*phi*), incompatibility (*inc*), replication (*rep*), transfer (*tra*), fertility inhibition (*fin*), and immunity to lethal zygosis (*ilz*). The origin of transfer replication (*ori*) is also indicated. The positions where insertion elements on F recombine with the bacterial chromosome to form Hfrs are also indicated. A detailed map of the *inc* and *rep* region is presented in Fig. 13.2 and a detailed map of the *tra* region in Fig. 13.5. (Adapted from Shapiro, J.A. [1977]. F, the *E. coli* sex factor, p. 671. In: Bukhari, A.I., Shapiro, J.A., Adhya, S.L. [eds.], *DNA Insertion Elements: Plasmids and Episomes.* Cold Spring Harbor, NY: Cold Spring Harbor Laboratory.)

The result of these investigations is a map for F, shown in Fig. 11.11. Although not as detailed as some genetic maps, it nonetheless represents a large amount of work using a difficult experimental system. Indicated on the map are both the known genes and the known repeated sequences. Additional information about the molecular biology of the F plasmid can be obtained from Chapter 3.

THINKING AHEAD

Suppose you could do conjugation experiments with diploid instead of haploid *E. coli* cells. How would that change affect the recombinant analysis discussed above?

Something Completely Different— Possible Diploidy

Gratia and Thiry (2003) have examined an unusual phenomenon that occurs in certain rare strains of *E. coli*. These cells lack an F plasmid, yet in an experiment similar to that of Lederberg and Tatum prototrophs still form, even if the mutations are scattered all over the chromosome. The process, which they call spontaneous zygogenesis, occurs even in *recA* mutant cells. When they labeled DNA in one culture with 5-bromodeoxyuridine and used an immunofluorescence procedure to detect the label after conjugation, up to 98% of all cells in a mating mixture had the label. They suggest that this phenomenon represents a spontaneous fusion of two haploid cells to produce a diploid cell. Unlike diploid cells reported in *Bacillus subtilis*, both chromosomes apparently remain active.

Summary

The conjugation process was first observed in *E. coli* and has proved to be a powerful tool for rapid elucidation of its genetic map. The basic process is one of linear, ordered transfer of DNA from a donor to a recipient. The length of DNA transferred is variable. Donor ability is conferred by the presence of the

F plasmid, and cell-to-cell contact is required for DNA transfer to occur. Mating aggregates form as a result of the presence of a specific F pilus on the cell surface. DNA is transferred from either Hfr or F′ cells. The former carry an integrated F plasmid, and the latter carry an autonomous plasmid that has incorporated some bacterial DNA in an aberrant excision process. DNA transfer in Hfr cells has been observed from specific points of origin on the bacterial chromosome and requires that a special transfer DNA synthesis occur. The DNA is transferred as a single-strand entity beginning with the 5′-end.

The genetic map of *E. coli* is based on conjugative data and has the minute as its unit of linear measure. This unit represents the average amount of DNA transferred by a donor cell in 1 min under standard physiologic conditions. The quickest method of determining map order involves the use of interrupted matings.

The F plasmid is 94.5 kb in size and includes several insertion sequences within its DNA. These sequences serve as regions of homology for integration of the F plasmid during Hfr formation. Excision of the F plasmid may result in accurate regeneration of the F plasmid or in production of an F′ plasmid carrying all of the F DNA as well as some bacterial sequences. F plasmid genetics have been investigated using transfer defective mutants and various types of suppressor.

Questions for Review and Discussion

1. Why are there only a limited number of Hfr types in *E. coli*? How would you try to force an F plasmid to integrate into a new position?
2. Why is it that the *E. coli* genetic map is now based on transductional data instead of interrupted mating data?
3. How would you determine whether the F plasmid has genes in common with other plasmids and with the host bacterium?
4. Most of the discussion in this chapter about F plasmid genetics has focused on the *tra* genes. What other kinds of mutations could be expected in the F plasmid?
5. What sort of selection would you use to identify transconjugants that have probably received an entire Hfr genome?

References

General

Berlyn, M.K.B. (1998). Linkage map of *Escherichia coli* K-12, edition 10. *Microbiology and Molecular Biology Reviews* 62: 814–984.

Clewell, D.B. (ed.) (1993). *Bacterial Conjugation*. New York: Plenum Press. (The first two articles deal specifically with the F plasmid, and the third article discusses plasmid R100.)

Jacob, F., Wollman, E.L. (1961). *Sexuality and the Genetics of Bacteria*. New York: Academic Press.

Lederberg, J. (1986). Forty years of genetic recombination in bacteria: A fortieth anniversary reminiscence. *Nature* 324: 627–628.

Rudd, K.E. (1998). Linkage map of *Escherichia coli* K-12, edition 10: The physical map. *Microbiology and Molecular Biology Reviews* 62: 985–1019.

Willetts, N., Wilkins, B. (1984). Processing of plasmid DNA during bacterial conjugation. *Microbiological Reviews* 48: 24–41.

Wollman, E.L., Jacob, F., Hayes, W. (1956). Conjugation and genetic recombination in *Escherichia coli* K-12. *Cold Spring Harbor Symposia on Quantitative Biology* 21: 141–162.

Specialized

Achtman, M., Morelli, G., Schwuchow, S. (1978). Cell–cell interactions in conjugating *E. coli*: Role of F pili and fate of mating aggregates. *Journal of Bacteriology* 135: 1053–1061.

Achtman, M., Willetts, N.S., Clark, A.J. (1972). Conjugational complementation analysis of transfer-deficient mutants of Flac in *E. coli*. *Journal of Bacteriology* 110: 831–842.

Gratia, J.-P., Thiry, M. (2003). Spontaneous zygogenesis in *Escherichia coli*, a form of true sexuality in prokaryotes. *Microbiology* 149: 2571–2584.

Hadley, R.G., Deonier, R.C. (1979). Specificity in formation of type II F' plasmids. *Journal of Bacteriology* 139: 961–976.

Hayes, W. (1953). The mechanism of genetic recombination in *E. coli*. *Cold Spring Harbor Symposia on Quantitative Biology* 18: 75–93.

Hopper, D.C., Wolfson, J.S., Tung, C., Souza, K.S., Swartz, M.N. (1989). Effects of inhibition of the B subunit of DNA gyrase on conjugation in *Escherichia coli*. *Journal of Bacteriology* 171: 2235–2237.

Klimke, W.A., Frost, L.S. (1998). Genetic analysis of the role of the transfer gene, *traN*, of the F and R100-1 plasmids in mating pair stabilization during conjugation. *Journal of Bacteriology* 180: 4036–4043.

Kunz, B.A., Glickman, B.W. (1983). The infidelity of conjugal DNA transfer in *Escherichia coli*. *Genetics* 105: 489–500.

Lanzov, V.A., Bakhlanova, I.V., Clark, A.J. (2003). Conjugational hyperrecombination achieved by derepressing the LexA regulon, alter the properties of RecA protein, and inactivating mismatch repair in *Escherichia coli* K-12. *Genetics* 163: 1243–1254.

Singleton, P. (1983). Zeta potential: A determinative factor in F-type mating. *FEMS Microbiology Letters* 20: 151–153.

Skurray, R.A., Reeves, P. (1973). Characterization of lethal zygosis associated with conjugation in *Escherichia coli* K-12. *Journal of Bacteriology* 113: 58–70.

Trgovcevic, Z., Salaj-Smic, E., Ivancic, I., Dermic, D. (1998). Conjugation in *Escherichia coli*: Difference in the molecular mechanism of F self-transmission and Hfr transfer. *Periodicum Biologorum* 100: 273–275.

Xia, X.-M., Enomoto, M. (1986). A naturally occurring large chromosomal inversion in *Escherichia coli* K-12. *Molecular and General Genetics* 205: 376–379.

12

Plasmids and Conjugation Systems Other than F

Not all plasmids are conjugative, and those that are do not necessarily have chromosome-mobilizing ability (i.e., the ability to promote transfer of chromosomal markers to a recipient cell). Nevertheless, these plasmids can have major economic and genetic importance. The diversity of plasmids that have been identified is staggering, and they are ubiquitous. In one study, 34 of 87 hospital isolates of enteric bacteria or *Pseudomonas* carried at least one plasmid. *Bacillus megaterium* routinely has eight or more plasmids in its cytoplasm. All of these plasmids have only two things in common. They can be identified in cell lysates as autonomous DNA molecules, and they are capable of self-replication. If the presence or absence of a plasmid has no observable effect on the cell phenotype, it is a **cryptic plasmid**. Researchers often subdivide plasmids according to their capacity for self-transfer from one host cell to another using the F plasmid as the standard for comparison.

This chapter is intended to serve as an introduction to a variety of commonly encountered plasmids. It is important to note, however, that *Escherichia coli* alone is reported to have more than 270 naturally occurring plasmids. The number of plasmids that can be discussed in a text such as this one is obviously highly restricted, and thus the choices are frequently arbitrary. As in Chapter 11, the emphasis here is on the physiology and classic genetics of these systems. The molecular biology of plasmids is discussed in Chapter 13. The first section here deals with chromosome-mobilizing plasmids, the second with a particular set of *E. coli* plasmids called colicinogenic plasmids, and the last with a general group of plasmids, the R plasmids.

Major topics include:

- Types of conjugative systems in bacteria
- Differences between some conjugative systems and the F plasmid systems
- Reasons behind many plasmids apparently having identical sets of genes
- The ways in which the study of DNA sequences could provide information about potential functions of unknown genes

Major Chromosome-Mobilizing Plasmids

E. coli may have gotten a great deal of publicity, but there are well-documented conjugation systems in many other bacterial genera. Their number is such that it is not possible to even list them all. Thus this chapter considers only selected examples (Table 12.1).

Salmonella

The enteric bacteria, of which *E. coli* is a member, are closely related, and thus it is not surprising that several of them have conjugation systems. The best-studied genus other than *Escherichia* is *Salmonella*, whose genetic apparatus seems to have many of the attributes of the *E. coli* system. Hfr-type cells can be produced by transfer of an F plasmid from *E. coli* to *Salmonella enterica serovar* Typhimurium (formerly *Salmonella typhimurium*). Using such Hfr strains, it is possible to transfer DNA from *Salmonella* to *Escherichia*, or vice versa, and to obtain transconjugants within the limits of DNA homology.

Table 12.1. Some plasmids with chromosome-mobilizing ability.

Organism	Plasmid
S. typhimurium	F
P. aeruginosa PAO	FP2
	FP39
	FP110
	R91-5 (transposon-mediated)
E. faecalis	pAD1
S. coelicolor A3(2)	SCP1
	SCP2*

Interestingly, Miroslav Radman and coworkers have shown that the transfer of single-strand DNA via intraspecies conjugation does not trigger SOS repair, but interspecies conjugation does. Matic et al. (2000) report that mutations in the mismatch repair system increase interspecies recombination 2000-fold.

A unique feature of S. typhimurium is that the length of its genetic map is longer than that for E. coli. A map constructed in the same manner as that for E. coli has a length of 135 minutes instead of 100 minutes, even though the relative order and spacing of most markers is the same. The coordinators of the respective maps conferred and decided that this stretching of the map was due to slower transfer of DNA by Salmonella Hfr strains. Accordingly, they recalibrated the Salmonella map to 100 minutes. After recalibration, extensive areas of homology appeared in the two genetic maps.

Nevertheless, there is one striking difference between the two maps. In addition to small insertions/deletions, which account for generic differences, there is a large inversion of about 12% of the genome that is roughly bounded by purB and aroD (25–37 minutes). There is no obvious phenotypic effect due to this inversion, and the reason for its stabilization within the population remains unclear. Some effort has been made to map the end points of the inversion, but there is interstrain heterogeneity present that prevents an accurate determination.

The small insertions/deletions observed when the maps of Escherichia and Salmonella are compared indicate one of the ways in which evolution of the bacterial genome occurs. Various insertion elements such as plasmids or transposons may cause the addition or deletion of genetic material in the course of their normal insertion/excision processes. Possible mechanisms by which such events might occur are discussed in Chapter 15, and some evolutionary considerations are presented in Chapter 17.

Pseudomonas

The genus *Pseudomonas* is one of the most nutritionally diverse of all the prokaryotes, a fact due in large measure to the presence of an extensive number of plasmids. Among this large body of plasmids are some that have chromosome-mobilizing ability superficially similar to that of F in *E. coli*, and conjugation has been reported in *Pseudomonas aeruginosa*, *Pseudomonas putida*, *Pseudomonas fluorescens*, *Pseudomonas glycinea*, *Pseudomonas. syringae* and *Pseudomonas morsprunorum*. Conjugative plasmids found in *P. aeruginosa* include FP2, FP5, FP110, and R65. R65 is representative of a group of plasmids called IncP1 that can transfer among many Gram-negative bacteria. Transducing phages are also known in four of the species.

Most conjugation in *Pseudomonas* takes place by an unknown mechanism that is obviously different from that discussed in connection with *E. coli*. Rarely is any sort of Hfr strain observed and then only with special plasmids. The FP plasmids and others commonly used for conjugation experiments do not integrate themselves stably into the bacterial chromosome, although they do transfer from one or two specific points of origin. Other plasmids can be forced to integrate by preparing transposon-carrying derivatives whose integration is then mediated by the transposon itself. The most successful conjugative plasmid is R68.45, a variant of R68. It is 100,000-fold more successful in mobilizing the host chromosome than FP2. This change results from the addition of an insertion sequence, IS21. This insertion sequence pairs with a similar one located adjacent to the new one. Together they constitute a transposon that can catalyze the insertion of R68.45 into a variety of organisms including species of *Pseudomonas*, *Agrobacterium*, *Azospirillum*, *Erwinia*, *Escherichia*, *Rhizobium*, *Rhodopseudomonas*, and *Zymomonas* (reviewed by Holloway [1998]).

Most of the *P. aeruginosa* PAO map derives from experiments with a mutated form of the R68 plasmid, pMO514. This plasmid carries a Tn2521 transposon that catalyzes insertion into a variety of sites and is temperature-sensitive for replication. It is, therefore, very easy to select cells carrying an integrated plasmid (i.e., Hfr cells).

Unlike the F plasmid, the FP plasmids do not code for pili, and thus the process for mating aggregate formation must also be different. Moreover, some 30–80% of the transconjugants inherit the conjugative plasmid, whereas it is rare for an *E. coli* Hfr donor to give a plasmid-carrying transconjugant. Presumably, this phenomenon results from the transient association of the plasmid with the chromosome.

The map developed for *P. aeruginosa* strain PAO is shown in Fig. 12.1. Like *E. coli*, the genetic map is circular, but the total genome size of 5900 kb is somewhat larger than that for *E. coli* (4500 kb). There are no obvious genetic homologies with the genetic map for *E. coli*. One particularly interesting observation is that the clustering of similar biosynthetic functions in *E. coli* and *S. enterica* serovar Typhimurium (e.g., the *trp* or *his* genes) is not seen in *P. aeruginosa*, where *trp* genes are located at four discrete sites. Appendix 2 provides an Internet address for the *Pseudomonas* Genome Database. Another difference between the Enterobacteriaceae

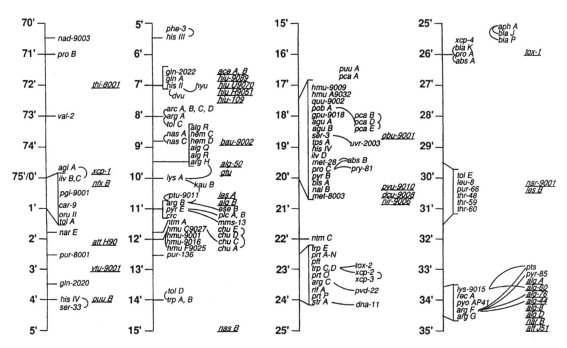

Figure 12.1. Genetic map of *P. aeruginosa* PAO. The origin (0 minute) has been arbitrarily designated at the *ilvB/C* locus. The following notation has been used. (i) Markers whose location is indicated by a *bar* joining the locus designation to the map have been located by interrupted matings using FP2, pM0514, or IncP-1 insert donors. (ii) *Curved lines* indicate that the markers so joined are cotransducible using one or more of the bacteriophages F116, F116L, G101, and E79*tv*-2. (iii) The marker designation is underlined for those cases where there is evidence to locate the marker in the area in which the symbol is placed, but the relationships to flanking markers have not been determined. (From Holloway, B.W., Carey, E. [1993], pp. 2.98–2.105. In: O'Brien, S.J. [ed.], *Genetic Maps: Locus Maps of Complex Genomes*. Cold Spring Harbor, NY: Cold Spring Harbor Press.)

and *Pseudomonas* is that all *P. aeruginosa* isolates appear to carry at least one prophage.

Enterococcus

The study of conjugation among members of the genus *Enterococcus* (formerly *Streptococcus*) has uncovered several unusual phenomena. In *Enterococcus faecalis*, there are conjugative plasmids known with chromosome-mobilizing ability, one of which is pAD1. Cells lacking pAD1 produce a specific **pheromone**, a compound that elicits enhanced mating aggregate formation from the plasmid-carrying cell. The pheromone is designated cAD1 and is a short polypeptide with the sequence:

$$HOOC\text{--}Leu\text{--}Phe\text{--}Ser\text{--}Leu\text{--}Val\text{--}Leu\text{--}Ala\text{--}Gly\text{--}NH_2$$

Analysis of DNA sequence databases has shown that the cAD1 peptide is actually synthesized as part of the signal sequence for a lipoprotein. After the signal sequence is removed during protein export, it is trimmed to release the pheromone.

Cells carrying pAD1 respond to cAD1 by synthesizing an aggregation substance that results in mating aggregate formation to such an extent that the cells of the mixed culture visibly clump. Presence of the pheromone improves plasmid transfer 200- to 300-fold, and not just because of enhanced aggregation. If cultures carrying distinguishable derivatives of pAD1 are mixed, only the culture that has been treated with pheromone carries out significant transfer of its plasmid. If both cultures received pheromone, the transfer is reciprocal.

Plasmid-carrying cells in a culture do not clump with one another because the plasmid chemically modifies the cAD1 molecule so that it is no longer functional. They also produce a protein inhibitor, iAD1, whose structure is:

$$HOOC\text{--}Leu\text{--}Phe\text{--}Val\text{--}Val\text{--}Thr\text{--}Leu\text{--}Val\text{--}Gly\text{--}NH_2$$

The marked similarity of sequence to that of cAD1 has led to the assumption that iAD1 presumably acts as a competitive inhibitor of the binding of cAD1 to the cell surface. Similar results have been obtained for another pheromone and its inhibitor, cPD1 and iPD1, and the genetic nomenclature is also similar.

Enterococci that produce as many as five distinct pheromones have been found, and all operate on the same basic plan independent of one another.

Pheromone production in all cases depends on the lack of a corresponding plasmid, so cells with multiple plasmids synthesize few or no pheromones.

Another strikingly different mechanism of conjugation observed in some Enterococci is conjugative transposition. The process occurs during **filter matings** in which the donor and recipient cells are not mixed in liquid culture but are trapped on the surface of a membrane filter. After an appropriate period of time has elapsed, the filter is transferred to a selective medium, and the transconjugants are allowed to grow. The process meets all the criteria for conjugation. It is insensitive to the presence of DNase, and it requires cell-to-cell contact. A culture filtrate is not sufficient. Nevertheless, it is not possible to identify any sort of plasmid molecule in the donor cells. The markers transferred in this manner are generally antibiotic markers (most often *tet*M), and it can be shown that within their original cell they behave like transposons. They can move from bacterial chromosome to plasmid or from one site to a different site on the same DNA molecule. Such a process may seem strange, but Lagido et al. (2003) have modeled conjugation with bacteria on solid surfaces and conclude that their model can be used to make predictions about the extent of conjugation in the biosphere.

Perhaps the best-studied member of the conjugative transposons is the one designated Tn916 and originally observed in *E. faecalis*. It transposes with a frequency of 10^{-5} to 10^{-9}/cell/generation. The transconjugants have Tn916 inserted into a variety of sites on their chromosomes and may receive more than one copy of the transposon. The latter phenomenon may be the result of an effect similar to that of zygotic induction where the transposition machinery is normally kept turned off by a repressor that must accumulate to appropriate levels within the cell. If the repressor is slow to form, additional transpositions can occur within the recipient. There is no evidence that presence of a transposon in a cell prevents entry of additional copies of the transposon.

DNA isolation experiments have demonstrated that Tn916-sized circles of DNA occur in the cytoplasm during transposition. Protoplast transformation experiments have shown that those circles are sufficient to produce strains capable of further conjugative transposition. The proteins involved in transposition show strong similarities to λ *int* and *xis* proteins (Marra and Scott 1999), and the assumption is that the process is fundamentally similar to λ integration. However, Connolly et al. (2002) have shown that there is no role for the factor for inversion stimulation (Fis) or for integration host factor (IHF).

Excision of the circular transposon is necessary for expression of the *tra* functions. The *int* and *xis* functions normally are at such low levels that excision is not possible. The regulatory system is based on attenuation (Celli and Trieu-Cuot 1998), and the presence of tetracycline allows synthesis of longer transcripts. After excision, the joining of the ends of Tn916 allows transcription into the *tra* region, and conjugation occurs. During conjugation, one strand of the circle transfers to the recipient cell.

The two conjugation methods are not mutually exclusive, for the *E. faecalis* plasmid pCF10 seems to embody them both. It is a 58-kb plasmid conferring tetracycline resistance that carries a 25-kb TRA region coding for the usual pheromone responses and a separate 16-kb region that includes the antibiotic resistance. The 16-kb region behaves as a conjugative transposon and has been designated Tn925. In addition, conjugative transposition is not restricted to *E. faecalis*. Transposon Tn1545 has been shown to transfer into a variety of Gram-positive bacteria including *Bacillus subtilis*.

Conjugative transposons are also present in certain Gram-negative anaerobic bacteria, particularly *Bacteroides*. They have no marked genetic similarity to Tn916. Similar phenomena occur among the Enteric Bacteria (reviewed by Pembroke et al. [2002]).

Streptomyces

Most of the original genetic work with *Streptomyces*, an industrially important genus, was done with *Streptomyces coelicolor* A3(2) and *Streptomyces lividans*. Hopwood and coworkers have identified two principal conjugative plasmids in *S. coelicolor* A3(2). The first is SCP1, a giant plasmid 350 kb in size that is linear rather than circular. There are 80-kb terminal inverted repeats and protein molecules bound to each 5′-end of the linear duplex. It codes for the antibiotic methylenomycin. The second is SCP2, a 31-kb cryptic, circular plasmid. Both of these plasmids appear to have low copy numbers, but there are other plasmids known in *Streptomyces* that may have copy numbers as high as 800. Both SCP1 and SCP2 as well as at least seven other plasmids can mobilize the chromosome of *S. lividans*. Most *Streptomyces* plasmids cause a **pock formation** in the cultures that carry them that is the result of an inhibition or delay in the production of aerial mycelia and spores. In a lawn resulting from a mixture of cells, some of which are carrying a plasmid, the plasmid-containing clones can be readily identified by their lower profile.

Numerous crosses have been made with SCP1 in *S. coelicolor* A3(2). Unlike the case for *E. coli*, SCP1 promotes recombination between two plasmid-carrying strains (NF strains) as well as between a plasmid-free strain and an NF strain. The efficiency of plasmid transfer itself approaches 100%, but marker exchange is a more conventional 3×10^{-6} to 5×10^{-6}. What is particularly unusual about the recombinants is that donor markers are inherited bidirectionally from a point centered on the 9 o'clock position of the genetic map (Fig. 12.2). The diffusible agarase gene located at this site is deleted in all NF strains along with the right terminal inverted repeat of SCP1. Integration apparently involves two IS _466_ insertion elements, one located near the junction of the main plasmid DNA and the right terminal repeat, and the other located near the agarase gene on the bacterial chromosome.

SCP2 plasmids do not normally promote genetic exchange efficiently, even in the absence of SCP1. Spontaneous variants denoted SCP2* have been obtained whose transfer is somewhat more efficient. The complete DNA sequence for the plasmid (Haug et al. 2003) is known, and the plasmid includes code for a member of the λ integrase family and Tn_5417_, which is most similar to the terminal repeats on the _Streptomyces_ chromosome. Artificially prepared SCP2*-primes transfer their cloned DNA efficiently and have been used for complementation studies. The added homology of the SCP2*-primes apparently promotes their chromosomal integration because strains carrying these plasmids become bidirectional donors in a manner similar to SCP1-carrying strains.

There are far too many other plasmids within the genus _Streptomyces_ to discuss here. One unusual plasmid that deserves mention, however, is the temperate _Streptomyces_ phage φSF1, whose prophage is also a conjugative plasmid. Another is SLP1 that occurs as an integrated element in _S. coelicolor_ A3(2) and gives a low level of transfer (10^{-8}) in cells lacking SCP1 and SCP2.

The development of a genetic and physical map for _S. coelicolor_ is a relatively recent occurrence (Redenbach et al. 1996) and illustrates the power of modern molecular techniques. Redenbach et al. took the entire 8-Mb chromosome and partially digested it with the restriction enzyme _Sau_3A1 (four base recognition sequence; Table 2.2). The DNA fragments thus obtained were in the range of 40–60 kb. These fragments were cloned in a λ phage vector (a cosmid; see Chapter 14) that can accommodate fragments 31–48 kb in size. In a separate procedure using restriction enzymes that cut _S. coelicolor_ DNA only rarely, they obtained 17 fragments in one digestion and 8 fragments in another (lettered arcs in Fig. 12.2). The restriction fragments were used as probes in Southern blots to identify which cosmid clones had sequences that hybridized. RNA synthesized from the ends of the cosmid clones was used to probe other cosmids to establish which cosmids had overlapping ends. Eventually they identified a set of 319 cosmids accounting for 7962 kb of DNA that could be used for further genetic analysis. Individual genes were localized onto these clones by hybridization of the appropriate RNA or DNA sequence or by complementation tests using cosmid transfection.

APPLICATIONS BOX

Can you think of a reliable method to determine whether a bacterial chromosome is linear or circular?

Historically, the genetic map for *Streptomyces* is circular, while the physical map (Fig. 12.2) is linear. Wang et al. (1999) have analyzed the reason for the discrepancy and propose that it is an artifact of a strong bias for an even number of crossovers during recombination. The terminal inverted repeats shown in the figure are seen in many other streptomycetes and are at least 61 kb long. Included in the repeats is the *catA* gene (catalase), which is therefore a gene duplication. The terminal inverted repeats represent an area of instability in the chromosome. Recombination in this region can create a circular chromosome, so the question is why the linear form in the first place? Chen et al. (2002a) review the evidence in favor of a model in which integration of a linear plasmid (known to exist in *Streptomyces*) may result in linearization of the chromosome.

Agrobacterium

One conjugation system found in members of the genus *Agrobacterium* is unique. It promotes transfer not from one bacterial cell to another, but from a bacterial cell to a plant cell. The best-studied case is that of *Agrobacterium tumefaciens*, and the observation is that if any one of a variety of dicotyledonous plants is infected by the bacterium near the crown of the plant, a gall or tumor develops. If there is no infection, a tumor is not produced. If the bacterium does not carry a Ti (tumor-inducing) plasmid, there is no tumor. Blotting experiments using DNA extracted from tumor cells have shown that a portion of the DNA from the tumor cell binds to a Ti plasmid probe. By using various restriction fragments of the Ti DNA, it is possible to demonstrate that only a relatively small region of the Ti plasmid, the 25 kb T-DNA, is actually incorporated into a plant

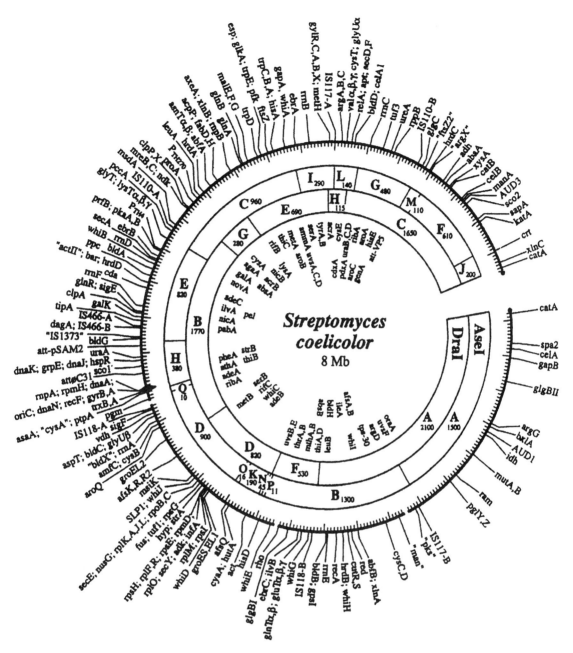

Figure 12.2. Combined genetic and physical map of the *S. coelicolor* M145 chromosome. The positions of markers on the outside of the circle come from their locations on the cosmid clones, assuming average equal lengths for each clone (unique or overlapping segment). Markers inside the circle have been mapped only genetically; their approximate positions relative to the physically mapped markers were determined by interpolation. Sizes of the *Ase*I and *Dra*I fragments are in Kilobase pairs. (•) Chromosomal "telomeres"; (♦) *oriC*. (From Redenbach et al. [1998].)

chromosome. Transfer also occurs into some monocotyledenous plants, but tumors do not develop.

Ti plasmids are large (140–250 kb) conjugative plasmids that code for the ability to degrade the opines nopaline or octopine (amino acid derivatives). The T-DNA segment encodes the ability to produce the opines, and thus a symbiosis between plant and bacterium develops.

The same two sets of transfer genes also mediate transfer of the RSF1010 plasmid between Agrobacteria. The efficiency of this transfer is greatly improved if both donor and recipient carry the Ti plasmid (Bohne et al. 1998), but only a few *vir* genes are necessary in the host to observe the phenomenon. The mechanism of transfer appears to be similar to that of the F plasmid in the sense that there is a nicking event and transfer of a single strand of DNA; however, the Ti plasmids most closely resemble those of the IncP incompatibility group (see Chapter 13).

The opines also act like pheromones to induce conjugational ability in the Ti plasmid, some of which can also mobilize the bacterial chromosome. Transfer only occurs at high cell density, an effect mediated by the TraR protein. TraR is a **quorum sensing** protein, meaning that it detects when a sufficient density of cells is present. In this case, it apparently forms complexes with the *Agrobacterium* autoinducer, an acylhomoserine lactone (*N*-(3-oxooctanoyl)- L-homoserine lactone) synthesized by the product of the *traI* gene. Complexed TraR protein activates transcription of the other *tra* genes to promote transfer. The original bacterial conjugative system was known as VirB, but Chen et al. (2002b) report the presence of a second conjugal system, AvhB, based on the type IV secretion system as in the case of the F plasmid.

Transfer of T-DNA requires the interaction of bacterial, plant, and plasmid proteins. The necessary bacterial genes are *chvA* and *chvB* that code for plant attachment functions. The plant cells to which the bacteria are attached produce phenolic compounds that act as inducers of the virulence genes. Plasmid *virA* and *virG* functions combine to form a two-component regulatory system for the *vir* operons. VirA protein is a membrane spanning protein that binds plant compounds that are released during wound healing. When one of these compounds binds to VirA, the cytoplasmic portion of the molecule acts as a protein kinase to first autophosphorylate itself. The phosphate group is then transferred to the VirG protein, which activates the *vir* operons. In all, there are six essential and two nonessential vir operons including approximately 25 genes. The *virD1* and *virD2* proteins promote excision of a single strand of T-DNA from the Ti plasmid.

DNA sequencing experiments have shown that T-DNA is bounded by two nearly perfect 23-bp direct repeats. If the right-hand border segment is deleted, transfer of T-DNA is abolished, but the left-hand border does not appear to be necessary. A 24-bp sequence called overdrive is required for right-hand border function and is located adjacent to T-DNA. The DNA that is transferred into the plant cell is single-stranded and linear, as in the case of the *E. coli* conjugation system. Transfer occurs through a pore formed by the VirB proteins. The amount of DNA to be transferred is determined only by the positions of the boundary sequences.

Insertions of transposons into T-DNA have no effect on DNA transfer unless they accidentally inactivate the *vir* functions. For example, integration of the transferred DNA into the plant chromosome requires *virE* function. Sometimes T-DNA can serve to carry a virus into a plant cell, a technique known as **agroinfection**. Virus production depends on normal T-DNA functions, so agroinfection is a more direct way of testing for DNA transfer than waiting for tumor formation or probing plant DNA by Southern blotting techniques.

The T-DNA system is the first confirmed example of the transfer of prokaryotic DNA into eukaryotic cells. As such, it is an exception to the genetic principle that genetic exchange occurs only between closely related individuals. From the point of view of the bacterium, the relationship is an unusual form of parasitism. The transformed plant cells produce opines that can be used as carbon sources by *Agrobacterium* cells carrying the Ti plasmid. Some Agrobacteria carry addition plasmids that increase their degradative capabilities. From the point of view of the geneticist, the Ti plasmid system represents a remarkable opportunity to carry out genetic engineering experiments with plants.

Sulfolobus

A large number of different conjugative plasmids have been isolated from the Archaeon *Sulfolobus islandicus* found in solfataras (volcanic fissures emitting sulfur-containing gases) in Iceland. The conjugation process is much slower than in *E. coli*, requiring about 5 h or one generation time, and does not seem to require pili. Stedman et al. (2000) have sequenced the plasmid pNOB8 and a series of pING plasmids, all of which are conjugative. By comparing sequences found to be common, they have identified open reading frames that are necessary (or not) in the conjugation process.

Bacteriocins

General Properties

Microbiologists have long known that mixtures of certain bacteria are incompatible because one of the organisms so combined dies within a few hours. André Gratia first observed that one type of *E. coli* released diffusible substances into the medium that were lethal to some other bacteria, including *E. coli* that did not produce the virulence (V) factor. Fredericq, in collaboration with Gratia, showed that the V factor was proteinaceous in nature (i.e., sensitive to the proteolytic enzyme trypsin) and was merely one example of a large number of antibiotics produced by various types of bacteria. Some are even commercially important. The lactic acid bacteria produce one bacteriocin, nisin, that is approved as a spoilage inhibitor in certain foods, and more than 50 others are known in *Lactobacillus*.

The general name given to substances of this type is **bacteriocin**, and individual bacteriocins are named according to the species of organism that originally produced them. Thus there are colicins from *E. coli*, subtilisins from *B. subtilis*, influenzacins from *Hemophilus influenzae*, pyocins from *P. aeruginosa* (formerly *Pseudomonas pyocyanea*), and halocins from *Halobacterium*. Different protein molecules within each type of bacteriocin are identified by letters and/or numbers. In all cases, cells of the producing culture are immune to their own bacteriocin, but are sensitive to different bacteriocins from the same species. However, the same bacteriocin may be produced in slightly different forms by different strains or species of bacteria. Therefore, each bacteriocin designation also includes the strain number of the producing culture. In the case of Gratia's colicin V, for example, a typical designation might be colicin V-K357.

Many bacteria and archaea produce these toxins, known respectively as bacteriocins and archaeocins, and their genetic analysis has shown that each type is encoded by certain specific DNA sequences, either chromosomal or plasmid. All bacteriocins can be subdivided into two general categories. One type, exemplified by colicin V, occurs either as a pure protein molecule or as a protein complexed to a portion of the outer membrane of the producer cell. The colicin V protein is generally found complexed to O antigen from the *E. coli* host. The other type of bacteriocin, best exemplified by pyocin R from *P. aeruginosa*, closely resembles all or part of a bacteriophage, except that there is no DNA associated with it (Fig. 12.3). DNA located within the bacterial chromosome is always the source for the latter type of bacteriocin and has been

Figure 12.3. Electron micrograph of pyocin R. These structures were obtained by treating a culture of *P. aeruginosa* with mitomycin C. Structures resembling both contracted and noncontracted phage tails can be seen. The *bar* indicates a length of 25 nm. (From Bradley, D.E. [1967]. Ultrastructure of bacteriophages and bacteriocins. *Bacteriological Reviews* 31: 230–314.)

assumed to be a defective prophage. In the case of pyocin R, sequence analysis (Nakayama et al. 2000) shows that it is derived from the same ancestor as phage P2. Moreover, it is not simply a defective prophage because no other viral genes are present at that location. The former type of bacteriocin is associated with the presence of a plasmid within the cell, and the following discussion considers only this type.

Although bacteriocins are present in producing cultures at all times, the amount of bacteriocin present can be greatly increased by the same treatments that cause induction of a λ prophage (turning on of SOS repair). In recognition of this similarity, the process is called bacteriocin induction. Bacteriocin production parallels induction of a prophage in that a producing cell is killed as a result of the induction of lysis genes, whereas nonproducing cells survive. The sacrificed producing cell kills neighboring sensitive bacteria to the presumed betterment of the remainder of the colicinogenic cells. The ecological implications of this process have been reviewed by Riley and Wertz (2002).

The number of producing cells in a culture can be measured using a variation of the plaque assay technique. A diluted cell suspension of a bacteriocin producer is mixed with indicator bacteria in soft agar. When spread on a plate, the presence of bacteriocinogenic cells results in the development of **lacunae**, or holes, in the lawn of indicator bacteria. Each lacuna represents a single producer cell and has in its center a small colony resulting from that cell.

The bacteriocin released by the producing cells binds to specific receptors on the surface of a cell. Elimination of the receptors by mutation results in a bacteriocin-resistant cell. However, protoplasts or spheroplasts made from these resistant cells are found to be once again sensitive to the bacteriocin, which means that the cell wall must protect the cytoplasmic membrane from the colicin. The inference is that the receptor facilitates penetration of the bacteriocin through the cell wall. In contrast to the resistant type of mutant is the tolerant mutant that still possesses receptors on the surface of the bacterial cell, but changes have occurred in the cytoplasmic membrane so that the bacteriocin no longer has any effect. Operationally, it is frequently difficult to distinguish between resistance and tolerance mutations.

Colicins: The Best-Studied Bacteriocins

As might be expected, the *E. coli* bacteriocins have had the most intensive study. One colicin, produced by strain 15, has been identified as a defective prophage, but the remainder are plasmid-derived. A colicin-producing plasmid is frequently denoted as Col followed by the appropriate letter. A short list of the nonphage colicins is presented in Table 12.2, along with the modes of action and receptors for the protein molecules, if known. Note that in every case examined, the colicin receptor functions in the transport of some other substance into the cell, and the colicin kills with first-order kinetics. The relation between this substance and colicin, if any, is not known. The cross-resistance grouping refers to the fact that certain *E. coli* mutations confer resistance to all members of a particular group of

Table 12.2. Properties of some colicins produced by plasmids.

Colicin	Cross-resistance Group	Mode of Action	Normal Function of Receptor	E. coli Gene Coding for Receptor
A	A	Pore formation	Vitamin B$_{12}$ transport	btuB
B	B	Pore formation	Ferric enterobactin receptor	fepA
D	B	Cleaves arginine tRNA near anticodon loop	Ferric enterobactin receptor	fepA
E1	A	Pore formation	Vitamin B$_{12}$ transport	btuB
E2	A	Dnase	Vitamin B$_{12}$ transport	btuB
E3	A	RNase, cleaves 16S rRNA	Vitamin B$_{12}$ transport	btuB
Ia/Ib	B	Pore formation	Chelated iron transport?	cir
K	A	Pore formation	Nucleoside porin	tsx
M	B	Inhibition of murein and LPS synthesis	Ferrichrome receptor	fhuA
V	B			

Note: The colicins are divided into two cross-resistance groups, and any cell that becomes resistant to the action of one colicin is also resistant to all other members of the group.

colicins. Cells that are *tolA* are tolerant to all group A members, whereas cells that are either *fhuA* (phage T1 resistant), *exbB*, or *exbD* (enterochelin uptake defective) are tolerant to group B.

All colicin proteins are modular in their function with three functional domains. There is a receptor binding domain, a lethal activity domain (carboxy terminus), and a domain that interacts with some translocation apparatus in the target cell to move the colicin into the cell (N terminus). Comparison of protein sequences indicates that there are striking similarities in the carboxy termini of each cross-resistance grouping.

The plasmids that code for colicin production may be either conjugative or nonconjugative. They are heterogeneous in size and copy number. Plasmids with a copy number near one are said to be stringently regulated, whereas those with copy numbers near ten or more are considered relaxedly regulated.

Conjugative Colicin Plasmids

One important conjugative plasmid is that which produces colicin I or, more correctly, the two closely related plasmids that produce colicins Ia and Ib. The proteins they produce have molecular weights of about 80,000 Da and also have similar modes of action. However, the producing strains are not cross-immune, and the colicins are thus given separate designations.

The colicin I plasmids produce their own pili as part of the conjugal apparatus, and these pili are thinner and more flexible than F pili. They confer sensitivity to a different group of male-specific phages than do the F pili, and they are immunologically distinct from F pili. In cells having compatible plasmids coding for F and I pili, each plasmid seems to use its own pili preferentially. The colicin I plasmid promotes transfer of bacterial DNA but does not give rise to stable Hfr strains.

The receptor for colicin I (see Table 12.2) passes the protein through the outer membrane (lipoprotein–lipopolysaccharide layer of the cell wall) and into the periplasmic space. Here the TonB system provides transport across the cytoplasmic membrane. Most of the TonB protein is located in the periplasm, with ExbD and ExbB embedded in the nearby membrane (reviewed by Lazdunski et al. [1998]). Colicin I creates channels within the membrane that allow potassium and magnesium ions as well as small phosphorylated molecules to leak out, resulting in membrane depolarization. The depolarization prevents further ATP synthesis, which results in cessation of all macromolecular synthesis and cell death. Cells mutant at the *tol* locus are specifically immune to I colicins, but also suffer from many membrane defects that block active transport and electron transport.

Colicin B and D plasmids code for similar proteins, but the producing strains are not cross-immune. The plasmids also code for F pili, but do not show any ability to instigate the transfer of bacterial DNA. The protein products are roughly the same size (about 89,000 Da) and cleave tRNA. Strains that are *tolB*, tolerate colicins B, A, and K. No tolerance mutations are known that are specific for only colicins B or D. The immunity protein that protects producing cells resembles part of a tRNA (Graille et al. 2004).

The role of colicin V is even more mysterious than that of colicin B/D, as neither its receptor nor its mode of action are known. It is an enhancer of virulence that is found only in *E. coli*. No specific tol mutations are known. On the other hand, colicin V plasmids form excellent Hfr strains possessing F pili, and several of these Hfr strains have been widely used for genetic crosses, after appropriate mutations were introduced to prevent colicin production. The over-all sequence similarity between ColV and F is greater than 50%.

Nonconjugative Colicin Plasmids

An important group of nonconjugative plasmids is the colicin E group. Colicin A (originally found in *Citrobacter freundii*) probably also belongs to this group. The original colicin E designation has been subdivided into three distinct groups, El, E2, and E3, each having its own unique killing mechanism. Colicin El is a 56,000-Da protein whose mode of action is similar to that of colicin I. No specific tolerance mutations are known. Colicin E2 is slightly larger at 62,000 Da and is an endonuclease that makes single- or double-strand breaks in DNA. The apparent direct effect is to activate exonuclease I, which causes DNA degradation. Colicin E2-specific tolerance loci include *tolD* or *tolE*. Colicin E3 is intermediate in size (60,000 Da) and has a well-defined mode of action. It causes the specific cleavage of 49 bases from the 3'-end of the 16S ribosomal RNA molecule in either 30S ribosomal subunits or 70S ribosomes. The *tolD* and *tolE* loci also confer tolerance to colicin E3.

The mechanism of immunity within a producing cell of the colicin E group has been identified. Colicin E2- and E3-producing cells also synthesize a small, acidic protein of about 10,000 Da that acts as an immunity substance. Complexes of colicin and immunity protein have been identified in which the stoichiometric ratio is 1:1. Removal of the protein causes an increase in the in vitro lethality of the colicin. Just as the modes of action of colicins E2 and E3 are different, so are the immunity proteins.

Although the colicin E plasmids are not self-transmissible, they and many other nonconjugative plasmids can transfer into recipient cells if they are in the presence of a compatible self-transmissible plasmid. The mechanism of this mobilization is not certain, but at least in the case of ColE3 the mobilization does not require F functions other than those necessary to form mating aggregates. Research has shown that a colicin E1 derivative, pBR322 (which is often used as a cloning vector), can also be mobilized by F. Colicin E1 can be mobilized in *trans* because it carries *oriT*, the replication origin for transfer DNA synthesis, and appropriate *mob* (mobilization genes). All that is required for transfer is the *tra* functions from a normal F plasmid.

Two other nonconjugative colicins have been tested. Colicin K has a mode of action like that of colicin I and a size of about 43,000 Da. It is sometimes found complexed to the O antigen. Colicin M has been obtained as a complex of protein and phosphatidylethanolamine with a molecular weight of 27,000 Da. It is an inhibitor of murein and lipopolysaccharide biosynthesis, resulting in the loss of cell wall and eventual cell lysis.

Resistance Plasmids

General Properties

Resistance plasmids were originally identified in Japanese clinical isolates of *Shigella* strains during the 1950s. Researchers noted that the *Shigella* strains were simultaneously resistant to several antibiotics and that the combination of resistances tended to be the same. In other words, it did not appear that random mutations to individual antibiotic resistance were occurring. Further testing showed that the drug resistances were inherited by an infectious process just as in the case of the F plasmid. By 1971, 70–80% of all *Shigella* clinical isolates in Japan were found to carry multiple drug resistances. Moreover, the same phenomenon was being observed in other bacteria of clinical importance. Bacteria growing in areas polluted with heavy metals were also found to have acquired a kind of resistance that was inherited in the same manner as the antibiotic resistance.

Soon after the original observations, investigators showed that the DNA coding for the antibiotic resistances was plasmid DNA. The plasmids were called **R plasmids** (originally R factors), and many types of R plasmids were identified. Various combinations of resistance were found, and the nature of the resistance itself was shown to be unusual. Consider, for example, the case of resistance to the antibiotic streptomycin. Mutations on the bacterial genome can alter the 30S ribosomal subunit so that the cell becomes resistant to a high level of streptomycin (as much as 1 mg/ml), but these mutations occur infrequently (10^{-9}). By comparison, R plasmids may code for enzymes that chemically modify the streptomycin by adding adenylyl, phosphate, or acetyl groups to the molecule. This type of resistance can be overcome if the external concentration of the antibiotic is sufficiently high.

Classification of R plasmids is done using a system of incompatibility tests with other R plasmids as well as the F plasmid and various colicin plasmids. The name of an incompatibility group is assigned according to the predominant plasmid member. Thus, F is the primary member of the IncF group of plasmids. Sometimes, it is necessary to subdivide an incompatibility group later, which is done using Roman numerals appended to the group designation. In the case of the IncF group, for example, the F plasmid is now assigned to group IncFI, and the R100 plasmid is a member of IncFII. At present more than 30 incompatibility groups are known for *E. coli*, and *Pseudomonas* has eight more, five of which represent plasmids that do not transfer to *E. coli*. *Staphylococcus aureus* plasmids have been assigned to seven incompatibility groups. *B. subtilis* has no naturally occurring R plasmids but propagates many staphylococcal plasmids transferred to it by transformation or transduction. Additional, artificially constructed plasmids are self-transmissible among some *Bacillus* species.

All of the originally observed R plasmids were self-transmissible, but extensive investigation has shown that nonconjugative R plasmids also exist. Conjugative R plasmids using the FI transfer systems are known, but incompatibility groups N and P have been found to code for their own unique conjugation systems, complete with pili. IncN plasmids can mate only on solid surfaces (filter matings) or in liquids that are foaming. If more than one kind of conjugative plasmid is present in a cell, each system preferentially uses its own pili. The various conjugation systems also recognize different cell surface structures as part of mating aggregate formation. For example, F' *lac* and an R1 derivative (both IncF) transfer well to *lps* mutants, but an R64 derivative (IncIa) does not.

Despite the large number of incompatibility groups, it is rare to find bacteria carrying more than one or two R plasmids, probably because of the retarding effect on cell growth that results from the extra DNA. In a book of this size it is impossible to discuss even a representative number of R plasmids, and thus only one such plasmid is considered as an introduction to this large group.

Plasmid R100

Plasmid R100 is a comparatively simple R plasmid that was isolated from *Shigella flexneri* in Tokyo. As originally isolated, it conferred resistance to 100 μg/ml streptomycin, 200 μg/ml chloramphenicol, 125 μg/ml tetracycline, and 200 μg/ml sulfonamide. The tetracycline resistance is due to changes in cell permeability that reduce influx of the antibiotic and promote its efflux, as well as due to the presence of an intracellular inhibitor. In contrast, the chloramphenicol resistance is due to the presence of a chloramphenicol-acetylating enzyme, and the streptomycin resistance is due to the presence of a streptomycin-adenylating enzyme. The sulfonamide resistance is due to production of a drug-resistant version of the enzyme dihydropteroate synthase, which is required for folate biosynthesis.

Although the plasmid is self-transmissible and produces F pili, it belongs to the IncFII incompatibility group and therefore can coexist with the F plasmid, which is IncFI. R100 can transfer between *Escherichia*, *Klebsiella*, *Proteus*, *Salmonella*, and *Shigella*, but not into *Pseudomonas*. Two synonyms for the plasmid are NR1 and 222. In an attempt to understand the way in which this plasmid behaves in a cell, Rownd et al. transferred it into a *Proteus mirabilis* strain. There were two principal advantages of working with this particular plasmid:host combination. First, the plasmid DNA was observed to be less associated with the bacterial chromosome than it was in *E. coli*, with about 20–25% of all plasmid molecules completely autonomous. This situation greatly facilitated preparation of pure plasmid DNA molecules. The second advantage was that the densities of the plasmid and bacterial DNA molecules were different.

THINKING AHEAD

Which experimental technique would allow you to separate DNA molecules based on their density?

The easiest way to separate molecules of different densities is by means of **density gradients**. DNA density gradients are conveniently prepared by adding a dense chemical, usually cesium chloride or cesium sulfate, to a tube containing DNA. The powdered chemical is added until the density of the solution is approximately the same as that of the DNA, in this case about 1.7 g/cm^3. The tube is then placed in an ultracentrifuge rotor and spun at high speeds for periods ranging from 6 h to several days until equilibrium is reached. At equilibrium, some of the heavy metal ions have migrated to the bottom of the tube as a result of the forces generated by the spinning rotor. The shift in the position of the metal ions results in generation of a concentration gradient for the metal ions (and therefore a gradient of density as well) that is maintained so long as the rotor is spinning. When the centrifuge is stopped, the process of diffusion gradually destroys the gradient, but it occurs so slowly that it does not influence the normal experiment. The concentration difference between the top and the bottom of the tube (the steepness of the gradient) is a function of the rate of rotation in the centrifuge. The faster the rotor spins, the steeper is the gradient.

Density gradients can be analyzed in one of two ways. The centrifuge can be stopped and the tube removed. The liquid in the tube can be carefully extracted in small aliquots and each aliquot assayed for the presence of DNA.

Alternatively, a special centrifuge and rotor can be used that allow a picture to be taken of the liquid sample while it is spinning. Because DNA absorbs ultraviolet (UV) radiation, it is possible to use UV illumination to obtain a picture that is a representation of the amount of DNA present as a function of position within the gradient. By scanning the photographic negative with a densitometer, it is possible to obtain a curve that is essentially a plot of the UV absorbance of the DNA (i.e., the concentration) as a function of position along the gradient. Rownd et al. used the latter technique. A sample set of densitometer scans can be seen in Fig. 12.4. The larger peak with the lower density is the bacterial DNA, and the material of higher density is the R plasmid.

When scans such as these were originally examined, the experimenters noted that there tended to be two peaks of plasmid DNA. Comparison of the profiles obtained with various R plasmids showed that tetracycline resistance and self-transfer functions were usually associated with the less-dense DNA, and that the other antibiotic resistance genes tended to be associated with the more-dense DNA. These observations, made in several laboratories, led to the suggestion that R plasmids consisted of two parts: a resistance transfer factor (RTF), which coded for all the transfer functions and tetracycline resistance, and an r-determinant (r-det), which coded only for antibiotic resistance. However, one unexplained fact remained. In the presence of an appropriate antibiotic such as chloramphenicol, the density of the plasmid DNA increased significantly beyond that normally observed. After removal of the antibiotic, the density gradually returned to its original value (see Fig. 12.4). This phenomenon was observed only in *Proteus*.

In order to explain the changes observed in R100, Rownd et al. postulated the existence of a process they called transitioning that was essentially a model for gene amplification. The basic assumptions of the model were that a large plasmid is more of a retarding force on cell growth than a small plasmid, and that the amount of protein that can be produced by a cell is proportional to the number of copies of the appropriate DNA sequence present in the cell. These two assumptions lead to a contradiction for the cell. Maximum antibiotic resistance is incompatible with rapid cell growth. Therefore, cells that have grown for a long period in the absence of antibiotic will be selected for the presence of a small-sized plasmid. When antibiotic is added, r-determinant DNA is duplicated to provide extra copies of the DNA, increased transcription, and more mRNA for translation. The observations of Fig. 12.4 can be explained if the r-determinant DNA is denser than the RTF. Extra r-determinant copies cause the density of the plasmid to shift to a heavier value. Genetic studies have provided an explanation for how transitioning occurs.

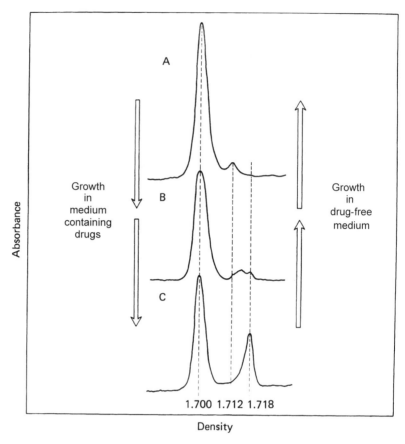

Figure 12.4. Systematic changes in the DNA density profile of an R plasmid. R plasmid R100 DNA was extracted from *P. mirabilis* cells grown in either drug-free or drug-containing medium. Profiles A, B, and C were obtained by sampling at different times after inoculation of the medium. The large peak of DNA on the left at a density of 1.700 is bacterial DNA and the smaller peaks to the right are various forms of R plasmid DNA. (From Rownd, R.H., Perlman, D., Goto, N. [1974]. Structure and replication of R-factor DNA in *Proteus mirabilis*, pp. 76–94. In: Schlessinger, D. [ed.], *Microbiology 1974*. Washington, DC: American Society for Microbiology.)

Genetic Analysis of Plasmid R100

Genetic and physical maps of R100, prepared in a manner similar to that used for the F plasmid, provide evidence to account for the transition phenomenon. The genetic map is shown in Fig. 12.5 and should be compared with the F plas-

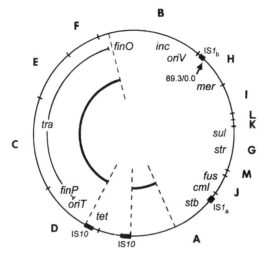

Figure 12.5. Genetic map of the R100 plasmid. The *circle* represents the R100 DNA molecule and the positions of known insertion elements (*black rectangles*). The total length of the DNA is 89.3 kb, with the coordinate system beginning to the right of IS1$_b$. The letters surrounding the map designate *Eco*RI restriction fragments (see Chapter 2). The abbreviations within the circle indicate various genes that have been identified. Resistances to mercuric ion, sulfonamides, streptomycin/spectino-mycin, fusidic acid, chloramphenicol, and tetracycline are indicated by *mer, sul, str, fus, cml,* and *tet*, respectively. The other gene symbols are as follows: *oriT*, origin of transfer; *oriV*, origin of vegetative (normal) replication; *finO* and *finP*, fertility inhibi-tion; *stb*, stability of the plasmid within a host cell; *inc*, incompatibility with other plasmids of the same type; and *tra*, conjugal transfer. Because of the increasingly large number of *tra* genes that have been mapped, no attempt has been made to indi-cate their individual positions. Instead, the *tra* region is indicated by an arc. The innermost *heavy line* represents the region that is approximately 90% homologous with plasmid F. The interruption in the region of F homology is caused by the pres-ence of the Tn*10* transposon. (Redrawn from Dempsey, W.B., Willetts, N.S. [1976]. Plasmid cointegrates of prophage lambda and R factor R100. *Journal of Bacteriology* 126: 166-176; and from Dempsey, W.B., McIntire, S.A. [1979]. Lambda transducing phages derived from a *finO⁻* R100::λ cointegrate plasmid: Proteins encoded by the R100 replication/incompatibility region and the antibiotic resistance determinant. *Molecular and General Genetics* 176: 219-334. Figure furnished by Drs. S. McIntire and W. Dempsey.)

mid map from Fig. 11.11. The *tra* genes are found in the same order and relative orientation on both maps. Note also that each region of R100 that codes for antibiotic resistance is flanked by paired insertion elements. This configuration is characteristic of transposons, and, in fact, the tetracycline resistance domain was quickly shown to be transposon Tn*10*, which is flanked by IS*10* elements in an inverted orientation. Later, Werner Arber and coworkers demonstrated that the r-determinant region is also a transposon, Tn*2671*, which is bounded by IS*1* elements in a direct repeat configuration.

One possible interpretation of transitioning is that addition of antibiotic selects for spontaneous variants in which the Tn*2671* has transposed itself within the plasmid to give extra copies of the r-determinant. Removal of the antibiotic then selects for the opposite phenomenon, transposon-catalyzed deletion of r-determinants. However, removal of the IS*1* flanking elements of Tn*2671* and replacement with artificially prepared direct repeats did not prevent transitioning. The conclusion seems to be that the r-determinant is indeed a transposon, but the transitioning function may be catalyzed by a special enzyme unique to the *Proteus* system. This enzyme requires only a direct repeat for its activity and can use either the normal IS*1* repeats or any suitably-sized, cloned DNA.

This conclusion is reinforced by observations made with different host cells. As mentioned earlier, *E. coli* cells yield R100 DNA with only the composite density. On the other hand, although the entire R100 will transfer into *S.*, in the absence of continued antibiotic selection the complete r-determinant is rapidly lost, leaving only a tetracycline-resistant cell. The process depends on normal functioning of the *recA* gene (major recombination pathway) and may again be related to the direct repeats provided by the IS*1* elements.

Arber and coworkers have shown that it is possible to have the Tn*2671* transposon hop to phage P7 plasmid DNA. Amplification of the r-determinant occurs along with inversion of the P7 DNA. They have also created an artificial r-determinant from the antibiotic resistance markers found in plasmid pBR325 using IS*1* elements in the inverted repeat configuration and cloned it into phage P1 plasmid. This construct also amplifies itself and in the process duplicates various regions of plasmid DNA.

The transposon nature of the antibiotic resistances provides a possible explanation for the origin of R plasmids. Hughes and Datta examined a large number of bacteria that were collected over the years 1917–1924, before the advent of antibiotics, and maintained in a lyophilized state until 1982. The strains showed little antibiotic or mercury resistance, but 14 were colicinogenic. Some 24 of the strains carried plasmids capable of mobilizing a normally non-conjugative plasmid that could not be mobilized by F. The authors concluded that many plasmids were available to act as hosts to transposons and that the present R plasmids may have arisen by transposition of antibiotic markers.

The remaining question, of course, concerns the origin of the antibiotic resistance genes. Julian Davies and Raul Benveniste suggested that these genes must be present in the original producing strains as protection for the producing cell, and there is some evidence to support the notion that they have been transposed onto preexisting plasmids to create new R plasmids. For example, DNA sequence analysis has revealed significant homologies between the aminoglycoside phosphotransferase genes of Tn5 and Tn903 and the corresponding genes in *Streptomyces fradiae* and *Bacillus circulans*.

The evidence for the evolutionary origin of R plasmids is suggestive. Evidence for the ability of transposons coding for antibiotic resistance factors to move rapidly through a population is direct. In the specific case of R100, Nancy Kleckner and coworkers followed transposon Tn10 through a series of genetic exchanges beginning in *S. typhimurium*. Tn10 moved from R100 onto the DNA of phage P22 using at least 20 integration sites. From P22, Tn10 then hopped to the *Salmonella* genome itself, integrating into 100 or more sites. Tn10 was next acquired by phage λ (appropriately deleted to make room for it), yielding five λtet phages. The λ phages then transported Tn10 to *E. coli*, where an additional 20 integration sites were identified. It is not possible for Tn10 to have extensive sequence homology with so many sites, and thus Tn10 is behaving similarly to phage Mu. Some models for transposition are discussed in Chapter 15.

Conjugal Plasmid Interactions

As comparative studies of F and R plasmids continued, it soon became apparent that F transfer was, in general, more efficient. The ability of most R+ cells to transfer their plasmids was found to be less than 1% that of F+ cells. There were exceptions among the R plasmids, however, whose transfer was as efficient as that of F. Moreover, even those plasmids that normally transferred themselves inefficiently were found to be efficient at transfer, provided they had only recently transferred into their current host cell. The simplest explanation for these observations was that most R plasmids produced a substance that gradually repressed the *tra* genes. Thus, a cell carrying a newly arrived plasmid would have only low levels of repressor and therefore good expression of the *tra* genes. However, a cell line that had had the same plasmid for many generations would have maximal amounts of repressor and therefore little *tra* gene expression.

For the experiments used to establish the various incompatibility groups for the R and F plasmids, many combinations of F and R plasmids in the same cell were prepared. Frequently, the R plasmid seemed to have no effect on the F plasmid, but certain R plasmids were observed to reduce the efficiency of transfer of the F plasmid to levels approximating their own inefficient transfer. Such plasmids were said to have the property of fertility inhibition (fi$^+$), which in terms of the repression model presented earlier means that the repressor produced by the R plasmid also affects the F plasmid.

The fi$^-$ R plasmids—those that had no effect on the transfer ability of F— were of two types. One type was observed to be derepressed for its own transfer and therefore presumably did not make a functional repressor. The other was observed to code for I-type pili. As noted earlier, I pili indicate the presence of an entirely different transfer system, and thus it is not surprising that there should be no interaction between the two.

Two F plasmid loci, *finO* and *finP*, determine the fi phenotype. The *finO* gene codes for a protein, but *finP* is the antisense RNA complementary to the leader region of *traJ*. When FinP RNA binds to the promoter region of *traJ*, it blocks transcription of *traJ*. Paranchych and coworkers have suggested that the function of *finO* is to bind to FinP RNA and protect it from degradation. Without TraJ protein, most of the other *tra* genes cannot be expressed because they are coordinately regulated as one huge operon. Without the *tra* functions, self-transfer of the plasmid is impossible. The F plasmid has been shown to be a spontaneous *finO* mutant due to insertion of an IS3 element, which accounts for its high level of fertility. Expression of *traJ* also requires host cpx gene product. The TraJ protein and host SfrA protein then combine to activate the rest of the *tra* genes.

The general model for fertility inhibition is shown in Fig. 12.6. Initial transfer of the R plasmid is presumed to occur from a rare cell that has expressed its *tra* functions. When the first transfer has occurred, further transfer is rapid like an epidemic initiated by the newly created R$^+$ cells (which are derepressed for several generations). Although plasmids tend to spread through a population of cells of their own accord, the process is enhanced considerably if antibiotic selection is applied. This finding constitutes a strong argument against the indiscriminate use of antibiotics.

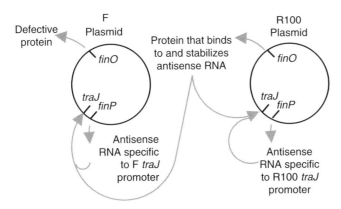

Figure 12.6. Model for the inhibition of F plasmid fertility by R100 as conceived by Willetts. The *finO* gene is a regulatory element for expression of the fertility functions and is common to both the plasmids. It allows expression of these functions unless two products are simultaneously bound to it. These products (one protein molecule and one RNA molecule) are coded by the *finO* and *finP* genes, respectively. In the F plasmid the *finO* gene product is defective so no fertility inhibition is possible. However, R100 produces a functional protein from its *finO* that works with either plasmid. Therefore, the F plasmid, in the presence of R100, represses its fertility functions.

Summary

Bacteria carry many plasmids. The plasmids can be conjugative or nonconjugative, from the archaea or bacteria, linear or circular, but all fulfill the usual plasmid definition. Among the conjugative plasmids are several that frequently integrate into the host cell chromosome to give high-frequency donors. Inserting transposons or bacterial DNA into the plasmid can often enhance this integration. The conjugation systems that have been well studied show many variations on the central theme. Like *E. coli*, both *Pseudomonas* and *Enterococcus* have conjugative systems that require the formation of mating aggregates, but they use different mechanisms. No plasmid-specific pili have been observed in the case of *Pseudomonas*, and *Enterococcus* uses a system of pheromones to cause release of an aggregation factor that acts to clump the cells. In *Enterococcus*, there is also an unusual system of conjugative transposition that does not involve plasmids at all. Conjugal DNA transfer in *Streptomyces* is plasmid-dependent but often occurs bidirectionally from a single point of origin as the plasmid inserts in opposite orientations. Less well-studied conjugative systems have been identified among the colicin and R plasmids.

The *Agrobacterium* conjugative system based on the Ti plasmid serves to transfer DNA not only to another bacterium, but also into a plant cell. The plasmid serves as the basis for an unusual form of parasitism and for genetic engineering of plants. A portion of the Ti plasmid, the T-DNA, is inserted into a plant chromosome, and other DNA cloned into the T-DNA sequence can be carried along.

Evolutionary change in plasmids is apparently often caused by the insertion and deletion of transposons. Similarities and differences among plasmids can be identified by the usual hybridization and mapping techniques as well as by the fact that some plasmids can coexist within a cell while others cannot. The incompatibility groups thereby defined serve to subdivide most plasmids into manageable families. Even plasmids that are not incompatible may nevertheless interact. A good example is the F plasmid whose transfer can be inhibited by closely related plasmids. The inhibition is mediated through the *fin* system.

Questions for Review and Discussion

1. Give at least one possible explanation for why R plasmids carry so many transposons.
2. What do you think is more surprising—that some plasmids transfer only among a few bacteria or that some plasmids transfer among a great many bacteria? Give reasons.
3. What are the criteria that must be met for a genetic transfer process to be conjugation as opposed to transduction or genetic transformation?
4. What are the ways in which a plasmid could mobilize a bacterial chromosome and a nonconjugative plasmid?
5. What are the similarities and differences between the genetic maps for the various bacteria discussed in this chapter?

References

General

Chen, C.W., Huang, C.-H., Lee, H.-H., Tsai, H.-H., Kirby, R. (2002a). Once the circle has been broken: Dynamics and evolution of *Streptomyces* chromosomes. *Trends in Genetics* 18: 522–529.

Clewell, D.B. (1999). Sex pheromone systems in enterococci, pp. 47–65. In: Dunny, G.M., Winans, S.C. (eds.), *Cell–Cell Signaling in Bacteria*. Washington, DC: ASM Press.

Holloway, B.W. (1998). The less traveled road in microbial genetics. *Microbiology* 144: 3243–3248.

Kado, C.I. (1998). Origin and evolution of plasmids. *Antonie van Leeuwenhoek* 73: 117–126. (Includes a good discussion of the Ti plasmid.)

Lazdunski, C.J., Bouveret, E., Rigal, A., Journet, L., Lloubès, R., Bénédetti (1998). Colicin import into *Escherichia coli* cells. *Journal of Bacteriology* 180: 4993–5002.

Meijer, W.J.J., Wisman, G.B.A., Terpstra, P., Thorsted, P.B., Thomas, C.M., Holsappel, S., Venema, G., Bron, S. (1998). Rolling-circle plasmids from *Bacillus subtilis*: Complete nucleotide sequences and analyses of genes of pTA1015, pTA1040, pTA1050 and pTA1060, and comparisons with related plasmids from gram-positive bacteria. *FEMS Microbiology Reviews* 21: 337–368.

O'Brien, S.J. (1993). *Genetic Maps*, 6th ed. Cold Spring Harbor, NY: Cold Spring Harbor Laboratory.

Pembroke, J.T., MacMahon, C., McGrath, B. (2002). The role of conjugative transposons in the *Enterobacteriaceae*. *Cellular and Molecular Life Sciences* 59: 2055–2064.

Riley, M.A., Wertz, J.E. (2002). Bacteriocins: Evolution, ecology, and application. *Annual Review of Microbiology* 56: 117–137.

Zatyka, M., Thomas, C.M. (1998). Control of genes for conjugative transfer of plasmids and other mobile elements. *FEMS Microbiology Reviews* 21: 291–319.

Specialized

Bohne, J., Yim, A., Binns, A.N. (1998). The Ti plasmid increases the efficiency of *Agrobacterium tumefaciens* as a recipient in *virB*-mediated conjugal transfer of an IncQ plasmid. *Proceedings of the National Academy of Sciences of the USA* 95: 7057–7062.

Celli, J., Trieu-Cuot, P. (1998). Circularization of Tn916 is required for expression of the transposon-encoded transfer functions: Characterization of long tetracycline-inducible transcripts reading through the attachment site. *Molecular Microbiology* 28: 103–117.

Chen, L., Chen, Y., Wood, D.W., Nester, E.W. (2002b). A new Type IV secretion system promotes conjugal transfer in *Agrobacterium tumefaciens*. *Journal of Bacteriology* 184: 4838–4845.

Connolly, K.M., Iwahara, M., Clubb, R.T. (2002). Xis protein binding to the left arm stimulates excision of conjugative transposon Tn916. *Journal of Bacteriology* 184: 2088–2099.

Francia, M.V., Haas, W., Wirth, R., Samberger, E., Muscholl-Silberhorn, A., Gilmore, M.S., Ike, Y., Weaver, K.E., An, F.Y., Clewell, D.B. (2001). Completion of the nucleotide sequence of the *Enterococcus faecalis* conjugative virulence plasmid pAD1 and identification of a second transfer origin. *Plasmid* 46: 117–127.

Graille, M., Mora, L., Buckingham, R.H., van Tilbeurgh, H., de Zamaroczy, M. (2004). Structural inhibition of the colicin D tRNase by the tRNA-mimicking immunity protein. *The EMBO Journal* 23: 1474–1482.

Haug, I., Weissenborn, A., Brolle, D., Bentley, S., Kieser, T., Altenbuchner, J. (2003). *Streptomyces coelicolor* A3(2) plasmid SCP2*: Deductions from the complete sequence. *Microbiology* 149: 505–513.

Lagido, C., Wilson I.J., Glover, L.A., Prosser, J.I. (2003). A model for bacterial conjugal gene transfer on solid surfaces. *FEMS Microbiology Ecology* 44: 67–78.

Marra, D., Scott, J.R. (1999). Regulation of excision of the conjugative transposon Tn916. *Molecular Microbiology* 31: 609–621.

Matic, I., Taddei, F., Radman, M. (2000). No genetic barriers between *Salmonella enterica serovar* Typhimurium and *Escherichia coli* in SOS-induced mismatch repair-deficient cells. *Journal of Bacteriology* 182: 5922–5924.

Nakayama, K., Takashima, K., Ishihara, H., Shinomiya, T., Kageyama, M., Kanaya, S., Ohnishi, M., Murata, T., Mori, H., Hayashi, T. (2000). The

R-type pyocin of *Pseudomonas aeruginosa* is related to P2 phage, and the F-type is related to lambda phage. *Molecular Microbiology* 38: 213–231.

Redenbach, M., Kieser, H.M., Denapaite, D., Eichner, A., Cullum, J., Kinashi, H., Hopwood, D.A. (1996). A set of ordered cosmids and a detailed genetic and physical map for the 8 Mb *Streptomyces coelicolor* A3(2) chromosome. *Molecular Microbiology* 21: 77–96.

Stedman, K.M., She, Q., Phan, H., Holz, I., Singh, H., Prangishvili, D., Garrett, R., Zillig, W. (2000). pING family of conjugative plasmids from the extremely thermophilic Archaeon *Sulfolobus islandicus*: Insights into recombination and conjugation in Crenarchaeota. *Journal of Bacteriology* 182: 7014–7020.

Wang, S.-J., Chang, H.-M., Lin, Y.-S., Huang, C.-H., Chen, C.W. (1999). Streptomyces genomes: Circular genetic maps from the linear chromosomes. *Microbiology* 145: 2209–2220.

13

Plasmid Molecular Biology

Plasmids face many of the same problems as bacterial chromosomes: they must replicate themselves, synthesize appropriate gene products, and segregate at least one copy of the plasmid into each daughter cell at cell division. However, plasmids also maintain incompatibility functions that prevent other, similar plasmids from taking up residence in the same cell and in many cases can transfer their DNA to other cells. They exist in a characteristic low or high copy number relative to the bacterial chromosome. This chapter considers the various methods by which plasmids accomplish these tasks. Thomas (2000) has reviewed the basic properties of all plasmids.

Major topics include:

- DNA replication strategies that are common to plasmids
- Means by which plasmids regulate their copy number

- Means by which plasmids ensure that all daughter cells inherit at least one plasmid copy
- Genetic and biochemical mechanisms that mediate incompatibility

Plasmid DNA Replication

DNA replication functions can be subdivided into two parts: those that are concerned with initiating a new round of replication and those that are concerned with the actual synthesis of new DNA strands. Temperature-sensitive (ts) mutations affecting these processes likewise have two general phenotypes. If at high temperature the mutation blocks the initiation process, the replication forks that have already begun continue until the replication is completed, but no new forks are added. With respect to the bacterial chromosome, this type of mutation is called a slow stop mutation. The other type of mutation blocks the elongation process at high temperature and is thus designated a rapid stop mutation. Plasmids interact with the host DNA replication system in varying ways, but in general, the smaller plasmids use more host functions, and the larger plasmids provide more of their own replication functions. Regardless of plasmid type, all fulfill the criteria set down in Table 13.1.

THINKING AHEAD

What happens to the process of DNA replication when a plasmid with its origin of replication integrates into a bacterial chromosome away from the normal chromosomal origin of replication?

Interactions Between Plasmid and Host

It is fairly easy to demonstrate that at least some plasmids supply portions of their replication machinery. Consider the case of integrative suppression by an F plasmid as discussed in Chapter 11. A *dnaB* ts mutation has blocked initiation of new rounds of chromosome replication. The formation of an Hfr strain restores the ability of the chromosome to replicate, but the origin of replication is now within the F plasmid rather than at *oriC*. This indicates that initiation of

Table 13.1. General characteristics of plasmid replication.

There is autonomous replication independent of chromosomal replication and cell division cycles.

There is a wide spread in time intervals between replication events.

Replication occurs throughout the cell cycle.

Characteristic copy numbers are defined by plasmid, host, and growth conditions.

Accidental deviations from normal copy number are adjusted.

Copy mutants exist; some are dominant and some are recessive to the wild type.

Cloned DNA may cause inhibition of plasmid replication or incompatibility.

Note: Adapted from Nordström (1985), pp. 189–214. In: Helinski, D.R. et al. (eds.), *Plasmids in Bacteria*. New York: Plenum Press.

replication by F is independent of *dnaB* function. In a related example, the intercalating dye acridine orange prevents replication of the F plasmid but leaves chromosomal DNA replication intact. If a culture is treated with the dye for a sufficient period of time, curing, or loss of the plasmid, results. A similar curing effect is seen if the culture is incubated above 42°C, but below about 46°C (the maximum temperature for sustained growth of *Escherichia coli*). Even large plasmids require some host cell functions for initiation of DNA replication. Both phage P1 (in the plasmid form) and the F plasmid require normal *dnaA* function in order to replicate. By way of contrast, ColE1 plasmids do not code for any proteins required for their replication.

Most plasmids as well as the bacterial chromosome require protein synthesis as a part of the initiation process. The presumption is that a particular unstable protein must be present in sufficient quantity to allow replication to begin. If an antibiotic that blocks protein synthesis, such as chloramphenicol, is added to the culture, DNA synthesis comes to a gradual stop. Such a requirement is not, however, universal. In the case of the ColE3 plasmid and its derivatives, the addition of chloramphenicol has exactly the opposite effect. The plasmids do not require protein synthesis for initiation and can use the existing replication enzymes. Relieved of the competition for resources, the plasmids continue to replicate for many hours, increasing the copy number to extraordinarily high values (e.g., 1000 per cell). This phenomenon of amplification has been a great boon to scientists working with cloned DNA, and many standard cloning vectors have been based on ColE3 plasmids.

For low copy number plasmids and the bacterial chromosome itself there is a characteristic DNA/cell mass ratio that seems to be invariant when averaged over an entire culture, which implies that DNA replication is always triggered at approximately the same point in the growth cycle (Koppes 1992). The control is apparently much more precise for bacterial chromosomes than for plasmids. In a strain in which the R1 plasmid has been inserted into *oriC*, the origin of chromosome replication, the timing of the replication event becomes nearly random with respect to the rest of the cell cycle.

It is possible to compare an *E. coli* F$^+$ culture and a similar Hfr culture resulting from integrative suppression of a *dnaA* ts mutation. At a high temperature, the DNA/cell mass ratio is lower, indicating that the F plasmid replication system recognizes a signal different from that of the bacterial replicon. A related experiment showed that the F plasmid origin of replication may be functional in Hfr strains, depending on the actual site of insertion. An Hfr strain with transfer origin near *trp* (close to *terC*, the replication terminus) had a slight excess of copies of the *trp* DNA over what was observed for an Hfr strain with an insertion near *oriC*. This observation means that the relatively low number of F plasmid copies near *terC* occasionally triggers the plasmid replication origin, whereas an F plasmid inserted near *oriC* never achieves the correct DNA/cell mass ratio to activate its replication origin.

The study of plasmid replication has been greatly facilitated by DNA splicing technology. Many experiments are now carried out on miniplasmids, constructs that have been prepared by shortening naturally occurring plasmids so as to eliminate nonessential functions. Generally speaking, the miniplasmids contain just enough DNA to self-replicate and to provide markers (usually antibiotic resistance) for detection of the plasmid. They also allow the experimenter to rearrange components from different systems to study their interaction. The essential components appear to be an origin of replication and the necessary sequences for the synthesis of the RNA primer.

DNA Replication Processes

Nonviral plasmids exhibit a variety of replication mechanisms that parallel those of viruses. In structure they may be covalently closed and circular or they may be linear and circularized by proteins. With respect to the nominally circular plasmids, it is possible to distinguish between two basic modes of replication—theta and sigma. Theta replication is exemplified by DNA molecules like bacterial chromosomes or the F plasmid, where the structure of the replicating molecules is similar to that seen by Cairns (Fig. 2.4) and resembles the Greek letter theta. This structure is usually taken as replicating bidirectionally, but that is not necessarily the case. The R100 plasmid as well as ColE1 replicates unidirectionally rather than bidirectionally, but otherwise resembles F. Some plasmids replicate in a sigma mode, which is similar to the rolling circle replication used by φX174 in that it generates unit-length, single-strand circles. In *Bacillus subtilis* and *Staphylococccus aureus*, there are several small R plasmids, typified by pC194, that can be isolated in either a single or a double-strand form. The single-strand form is a replication intermediate that has not yet synthesized its lagging strand, and thus the replication process is considered to be asymmetric. Sigma replication is distinct from rolling circle replication, which generates linear concatemers of DNA.

Linear plasmids can also use theta replication, in part. DNA replication initiates bidirectionally from an internal *ori* and proceeds to the ends. Obviously, the last Okazki fragment cannot be primed in the normal fashion, but the protein bound to the 3′-end of the newly synthesized strand can serve to prime the final fragment.

The rolling circle process is similar in phages and plasmids but with some notable differences owing to their respective life styles (reviewed by Novick [1998]). One obvious point of difference is the initiation of replication. Phages like φX174 can initiate DNA replication and allow it to proceed unchecked. Plasmids, on the other hand, usually need to limit their replication. The point of control is the initiation event. Phages initiate once and then continue to recycle the initiation point. Plasmids must have a new initiation event each time, and the initiation protein is often involved in the termination event that releases a new plasmid molecule. Plasmids that normally replicate in the theta or sigma mode will shift to making linear concatemers in cells that are *recBCD* and *sbc*. Examples include F and ColE1.

Meijer et al. (1998) have examined four small, cryptic plasmids isolated from *B. subtilis* strains other than the transformable strain 168. Each plasmid represented a different plasmid group based on size and restriction enzyme digestion patterns. Like the phages, such plasmids often contain modular elements, in this case the replication initiation gene and its double-strand origin of replication, the single-strand origin of replication, and a *mob* gene for conjugation. All the plasmids had a similar origin of replication, one that is common among plasmids of other Gram-positive bacteria. While there was sequence hetero-geneity in the single-strand origin, there was a highly conserved (nearly identical sequence) region of about 200 bp common to all. The single-strand origins of replication are essential for plasmid replication, and four different ones are known. All require host RNA polymerase for primer synthesis.

Recently, Bentley et al. (2004) sequenced and annotated the linear SCP1 plasmid. This type of analysis is becoming ever more common and ever more fruitful as the genome databases enlarge. Annotation consists of computer comparisons of the sequence with known sequences from other systems to predict genes and their function. They found that the plasmid probably codes for its own sigma factors, antibiotic production, and two sets of partition genes (see later). The gene coding for the terminal proteins that bind to the telomeres was not identified.

THINKING AHEAD

How would you measure the number of copies of plasmid DNA relative to the number of copies of the bacterial chromosome in a cell?

Control of Copy Number

The copy number of plasmids is obviously a function of the rate at which new rounds of replication are initiated. Plasmids with relatively high copy numbers demonstrate little, if any, regulation of initiation of DNA replication. Low copy number plasmids, on the other hand, are highly regulated and have the additional requirement that their initiation step must be coordinated with the cell division cycle. If it were not, cytokinesis might occur at a time when there were insufficient copies of the plasmid to supply at least one complete plasmid DNA molecule to each daughter cell.

Measurement of copy number was not originally a trivial procedure. Simple extraction of plasmid DNA from cells in order to measure its relative amount is complicated by the entanglement of plasmid and bacterial DNA. If any plasmid DNA breaks off during extraction, the copy number for the plasmid is underestimated. The most satisfactory methods for estimating copy number involve either using radioactive probes and Southern blots of gently extracted DNA or looking at the relative rates of reversion of identical mutant alleles located on the plasmid or the bacterial chromosome. In the latter case, the assumption is that the physical location of the gene does not affect its probability of reverting, and thus the number of plasmid revertants divided by the number of chromosomal revertants should equal the copy number.

Typical low copy number plasmids are the IncF plasmids including the F plasmid (IncFI subgroup) and plasmids R100 (NR1) and R1 (IncFII subgroup). Plasmids ColE1 and the *S. aureus* plasmid pT181 behave similarly to the IncFII group. There are three different strategies used by these plasmids for controlling their copy number. The first two of these strategies involve antisense RNA and have been reviewed by del Solar and Espinosa (2000).

Plasmid R1 (R100) codes for a RepA protein whose gene lies near the origin of replication and that is absolutely required for DNA replication. The number of initiations of replication is proportional to the RepA protein concentration. It is mainly a *cis*-acting protein, meaning that it preferentially acts on the DNA molecule from which it was produced. Translation of the *repA* gene is coupled to translation of the leader peptide *tap* (Fig. 13.1).

In this region of the plasmid DNA, there are promoters on both strands; therefore, transcription is possible in either direction. The smaller, leftward transcript is designated *copA* RNA and is of course complementary (antisense) to the leader portion of the CopT transcript. When it binds, it prevents translation of *tap* and consequently *repA*. The CopB protein acts as a repressor to inhibit synthesis of *copT* RNA (regulation at the level of transcription), but the CopB promoter is unregulated. Translation of the 7K *tap* eliminates the hairpin loop and prevents binding of *copA* RNA, so *repA* is more readily translated. When the R1 plasmid transfers to a new cell, neither *copA* RNA nor CopB protein is available in quantity, and the plasmid initially replicates rapidly. This overreplication facilitates establishment of the plasmid in the cell line. The larger number of plasmid molecules yields significant quantities of the two regulators, and the copy number gradually reaches its normal value. That the two regulators do not carry equal regulatory weight is shown by mutagenesis experiments. If *copB* is inactivated, the copy number increases up to tenfold. If *copA* synthesis is prevented, runaway replication occurs in which the plasmid replicates out of control and rapidly fills the cytoplasm with plasmid DNA molecules. Given these differences, the CopB protein is considered accessory.

However, there are plasmids in which the protein is an integral part of the system. An example is the pIP501 plasmid of *Enterococcus*. As shown in Fig. 13.1, the CopR protein is both a negative regulator of RepR and a positive regulator

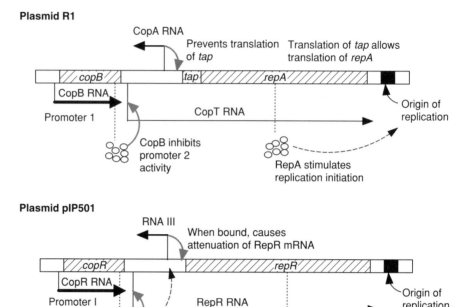

Figure 13.1. Replication control of RNA-regulated plasmids. The *bar* represents the replication region of the plasmid. The *solid lines* above or below the bar are RNA transcripts; the direction of transcription is indicated by the *arrowheads*. The *hatched regions* within the bars are protein-coding regions. The *curved lines* delineate the site(s) of action for each protein. A *gray arrow* indicates an inhibitory effect, and a *dotted arrow* indicates a stimulatory effect.

of the antisense RNA. Binding of antisense RNA produces a loop in the nascent mRNA that causes attenuation.

In both these examples, if the copy number falls, the degree of inhibition is reduced, and replication initiation occurs more frequently. Conversely, under high copy number conditions, inhibition becomes more intense.

Just as the foregoing plasmids have their Rep proteins, so the F plasmid has an essential protein for its bidirectional replication, the RepE protein, a product of the *repE* gene. Like the tryptophan repressor, this protein is autoregulated in that high concentrations of the protein seem to inhibit further transcription of its gene. Unlike R1, the IncFI F plasmid (and P1 prophage replicating as a plasmid) display a copy number control system that does not seem to involve regulating the synthesis of the essential replication protein. Copy number mutants of mini-F plasmids have been isolated and they map to two positions. The *copA* mutations map to the *repE* gene, and the *copB* mutations map to the region near the origin of replication and the region coding for the carboxy-terminus of RepE (Fig. 13.2). DNA sequence analysis of this region has shown that there are five 19–22-bp direct repeats called **iterons**.

By deleting two or more of these iterons, it has been possible to demonstrate that the copy number of the F or P1 plasmid is inversely proportional to the number of repeats. A sequence comparison of the iterons with that of the start region for protein E transcription reveals that there are some striking similarities. RepE is an autoregulated protein and must bind near its own promoter; thus, it seems likely that RepE also binds to the iterons. Because RepE is essential for F replication, it must be present at a reasonable concentration. To the extent that the direct repeats are present, they bind up RepE and prevent it from starting a new replication fork (accidental high copy number means no new replication initiations). If fewer repeats are present, more RepE molecules are available to initiate replication and there is less repression of *repE*, so the copy number increases. Moreover, RepE monomers bound to iterons can dimerize with RepE bound on a different DNA molecule, making replication initiation

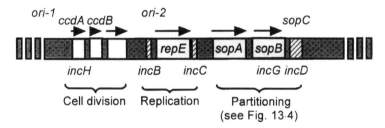

Figure 13.2. Incompatibility and copy number control regions on the F plasmid. Shown here is a functional map of a fragment of the F plasmid. The *sop* region is essential for plasmid partitioning, and thus is also denoted as *par*. Note the extensive overlap of the incompatibility functions with the other regulatory regions. The *open areas* of the map show the locations of protein-coding regions. *Hatched areas* are direct repeats, 19 or 22 bp in *incB* and *incC* and 43 bp in *incD*. The *arrows* show the direction of transcription of the proteins. (Adapted from Hiraga, S., Ogura, T., Mori, H., Tanaka, M. [1985]. Mechanisms essential for stable inheritance of mini-F plasmid, pp. 469–487. In: Helinski, D.R., Cohen, S.N., Clewell, D.B., Jackson, D.A., Hollaender, A. [eds.], *Plasmids in Bacteria*. New York: Plenum Press.)

more difficult because of steric hindrance. This reaction has been called hand-cuffing (Chattoraj 2000). In this manner, replication is controlled without the use of an inhibitory protein or RNA molecule. The ability to delete the iterons without loss of replication ability indicates that they are involved only in control and not in initiation of replication.

The sequences to which RepE binds are not identical. Binding to iterons is a function of RepE monomers. Binding to the operator of the *repE* gene is a function of RepE dimers. Left to itself, RepE is primarily dimeric. Initiation of replication is possible because the host chaperonins DnaK, DnaJ, and GrpE dissociate the dimers to monomers. The functional regions of RepE that determine dimerization are known (Matsunaga et al. 1997).

Partitioning

Partitioning is the act of segregating DNA molecules of each individual type so that there is at least one DNA molecule for each daughter cell. It is the prokaryotic equivalent to the movement of chromosomes during anaphase of mitosis. Pogliano (2002) reviews fluorescence localization experiments showing the position of plasmids at any given time. Basically, these experiments use plasmids engineered to contain many copies of *lacO*. The plasmid also carries *lacI* fused to the green fluorescent protein (GFP). Induction of *lacI* causes fluorescence at the intracellular location of the plasmid(s). By analogy to eukaryotic systems, the DNA site that is presumed to physically control partitioning has been referred to as a centromere. High copy number plasmids do not require any special mechanism for partitioning. Unless the plane of cell division is grossly off center, it is almost impossible that at least one plasmid DNA molecule would not be correctly situated.

The situation is different when the low copy number plasmids in the IncFI and IncFII groups are considered. If the copy number is only one or two, precise control must be exercised to prevent all the plasmid DNA molecules from being inherited by only one daughter cell. For example, a potential problem for a low copy number plasmid can arise if two plasmid DNA molecules undergo a single recombination event to generate a larger circle of DNA. In such an instance, it is conceivable that all of the plasmid DNA might end up in only one daughter cell. The R46 plasmid circumvents this problem by coding for a plasmid-encoded recombinase (*per*) function. The recombinase is an enzyme that catalyzes site-specific recombination of plasmid DNA somewhat like the *loxP* function of phage P1. The effect of the enzyme is to take a large DNA molecule containing two identical sites and convert it to two smaller ones (Fig. 13.3), making it a resolvase. Linear multimers of ColE1 DNA have been reported in cells that carry the *recB* and *recC* mutations. The method of partitioning is different for plasmids R1 and F as opposed to the *E. coli* chromosome. Gordon et al. (1997) used GFP labeling and found that *oriC* was located at the pole of the cell or in the center near the site of cell cleavage (the new cell pole). The plasmids, on the other hand, localized to different regions. Therefore, different mechanisms must operate to move the chromosome and the plasmids.

The partitioning function (type I) used by F is the stability of plasmid (*sop*) function and is essentially the same as the *par* functions of P1 plasmid (Rodionov and Yarmolinsky 2004). Three discrete sites have been identified in the F plasmid: *sopA*, *sopB*, and *sopC* (Fig. 13.4). The corresponding functions in P1 are *parA*, *parB*, and *parC*. The SCP1 plasmid uses a modified type I system in which there is no *parC*. The *sopC* site is a *cis*-acting region that can help to properly segregate an otherwise unstable miniplasmid. It is apparently the location on the DNA that is partitioned and as such is the functional equivalent of a centromere. The *sopA*

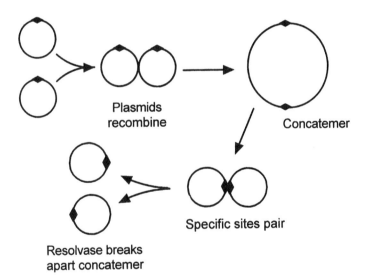

Figure 13.3. Effect of a resolvase enzyme on a concatemeric DNA molecule. The concatemer is assumed to form by a generalized recombination mechanism. Resolvases recognize a particular site (*diamond*) on a given DNA molecule but catalyze an exchange of strands only when two such sites are present. The reactions may be intermolecular or intramolecular. The latter reaction generates two circular DNA molecules where there had been only one concatemer, as in this example.

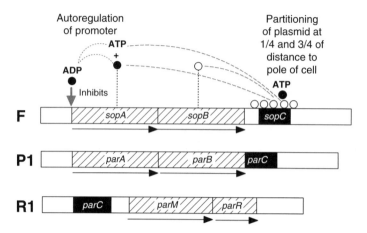

Figure 13.4. Overview of the control of partitioning by various plasmids. The genetic organization of two regions coding for homologous proteins is shown. The "A" or "M" proteins are ATPases characteristic of the partitioning system. The *dark boxes* represent binding sites (centromeres) on the DNA at which the "A" or "R" proteins stimulate partitioning. *Filled circles* are SopA protein molecules, and *open circles* are SopB. *Arrows* under the map indicate the direction of transcription. SopA includes an ATPase activity that presumably supplies the energy necessary to move the F plasmid from its position in the center of the cell to a location about halfway nearer to the pole.

gene product is a modified Walker-type ATPase (normally found in ATP-dependent proteases), and the *sopB* gene codes for a protein that binds specifically to *sopC* and spreads to either side of it. A SopA:ATP complex binds to SopC only when SopB is present and binds to the host chromosome when it is not. Properly bound SopB and SopA cause the plasmid DNA to localize at the 1/4 and 3/4 positions along the cell length, the positions at which the next but one cell division will occur. Thus the binding may be to the developing annular ring required for cytokinesis.

DNA sequence analysis indicates that the *sopC* (*parC*) region is a series of eleven or twelve 43-bp direct repeats with no intervening spaces. Each repeated sequence contains internal 7-bp inverted repeats. Only two of the repeated sequences are required to stabilize the F miniplasmid. Interestingly, there is a small plasmid found in yeast, the 2μ plasmid (named for its size), that displays a similar structure of multiple 62–63-bp direct repeats. It is thus possible that the F partitioning system is widely applicable.

The type II partitioning system involves a different ATPase, *parM*, a member of the actin family of ATPases. An example of a plasmid using this system is R1. The basic pattern of activity is the same, but the arrangement of genes is different (Fig. 13.4). Møller-Jensen et al. (2003) propose that the ParM protein polymerizes as microfilaments, using the energy of ATP to push the plasmid DNA all the way to the poles of the cell.

Logically, there will be times when partitioning fails for one reason or another. One straightforward approach for resolving such problems has been observed with the R1 plasmid. There is a region on the R1 plasmid designated *parB* that can be subdivided in two parts. One portion codes for the *hok* (host killing) function, and the other for an antisense RNA called *sok* (suppressor of *hok*). Together these are known as an **addiction module**. Hok product is routinely present in the cytoplasm. However, the effects of Hok are blocked by Sok in a manner analogous to the way in which a colicin-producing cell is protected against toxic effects. The Sok RNA binds to Hok mRNA, forming a duplex that is sensitive to RNase III cleavage. This cleavage converts the normally stable Hok mRNA into a short-lived molecule, and Hok never accumulates to a lethal concentration. Therefore, clones carrying both the *hok* and *sok* regions do not harm the cell, but a clone of just the *hok* region is lethal.

The P1 plasmid uses a different approach with its *phd* and *doc* genes. This pair of genes triggers a host cell addiction module, *mazEF*, that results in cell death (Hazan et al. 2002). Apparently, the normal function of the host module is apoptotic (programmed cell death), with *mazF* being the toxin.

A related approach is employed by the F plasmid via its coupled cell division (*ccd*) function. Two proteins are encoded with this region. CcdB inhibits cell division by inhibiting DNA gyrase (topoisomerase II). CcdA binds to CcdB protein, and the combination acts as a repressor to autoregulate their operon and does not inhibit gyrase activity. CcdA is an unstable protein because it is readily attacked by the Lon protease, and therefore a daughter cell without an F plasmid will soon be poisoned by its residual CcdB activity. When the number of copies of F DNA in a cell drops below a critical threshold, the CcdB function prevents further cell division until the plasmid number increases and the CcdA/B complex is again predominant. By this means, production of a plasmid-free cell is prevented.

Ron Skurray and coworkers demonstrated that there is a second *cis*-acting partition element in F now designated as *flmA*. The *flmA* gene is nearly identical to *hok*, is transferred early during conjugation, and can improve maintenance of unstable plasmids. The equivalent of *sok* in the F plasmid is *flmB*. There is a third gene involved, *flmC*. Both FlmB and FlmC RNA can bind to FlmA mRNA. The FlmA/C combination is stable and translatable, while the addition of FlmB makes the combination unstable and readily degraded. FlmB RNA is itself unstable and must be continuously supplied. Therefore, a daughter cell lacking an F plasmid would soon be killed by the accumulated FlmA product.

Incompatibility

One of the intriguing problems of plasmid biology has been how a resident plasmid can prevent one of its close relatives from establishing itself in the cell. As might be expected, the control of another plasmid seems to be exercised via the replication system. The same systems that control copy number can also be used for incompatibility functions.

The case of the incompatibility functions for the F plasmid has been particularly well studied. Genetically, two F plasmid-carrying strains can be mated provided one is phenocopied.

If an F′ culture is mated to an Hfr, the progeny are generally like the original Hfr. The occasional exception is a cell carrying two plasmids integrated at different sites on the genome. Such a cell is called a double male, and, oddly enough, some of them are stable. If an F′ culture is mated to a culture carrying a different F′, the resultant cells may carry one or the other of the original F′ plasmids or else may have acquired a single giant plasmid that seems to have resulted from a fusion event. Studies of cloned F plasmid DNA have shown that there are two regions capable of serving as origins of replication. The bidirectional origin is *ori-1* or *oriV*, and the unidirectional origin is *ori-2* or *oriS*.

Experiments with mini-F have identified four incompatibility loci, *incB*, *incC*, *incD*, and *incE*, located near *ori-2*. The *incB* function is to exclude other F plasmids, for if a second plasmid carries a different *incB* locus there is no incompatibility. The DNA sequence for this region includes the four 14-bp direct repeats that constitute the *copB* locus. Acridine orange inhibits replication via the *incC* locus, which is apparently a binding site that controls replication and consists of a series of direct repeats. There are 8 bp within the repeated sequences that could bind RepE, based on a comparison with the regulatory region for protein E. The *incD* locus is the same as *sopC* and thus affects the partitioning function. The *incE* gene seems to prevent plasmid replication via a second replication origin.

The various control loci are summarized in Fig. 13.2. The model for IncFI incompatibility then becomes twofold. The RepE protein is required for initiation of replication; and under normal circumstances all of it is tied up by the resident plasmid. A newly arrived plasmid, therefore, is not able to initiate its replication and establish itself. Similarly, the extra plasmid is not able to bind the proteins necessary for its partitioning and cannot establish itself properly. On the other hand, integrated plasmids are replicated and segregated by the host system and therefore are not incompatible.

Conjugal Functions

Although the F plasmid is certainly not the only conjugative plasmid, it is one of the best studied and can mobilize a wide variety of other normally non-conjugative plasmids. Therefore, this section focuses primarily on F and its known functions.

Pilus Production

Cells carrying the F plasmid have two types of filamentous structure on their surfaces. The fimbriae, or common pili, are the products of *pil* genes found on the bacterial chromosome. They are attached firmly to the cell and difficult to break apart, even in the presence of a strong detergent such as sodium dodecyl sulfate (SDS). By contrast, the F, or sex, pili are longer, thinner, and more loosely attached to the cell. They can be removed by blending the culture and are easily disrupted by SDS. Because they are required for mating pair formation, any plasmid that is pilus-defective is also transfer-defective (*tra*). Table 13.2 summarizes many of the known *tra* functions. All but two *tra* genes are transcribed as a single large mRNA from the *traY* promoter that is under the combined control of *traJ* protein and the same integration host factor normally involved in phage λ activity. Host *arcA* function is also required (Taki et al. 1998). Mutants in *traJ* fail to express the other *tra* functions.

The structural gene for F-pilin, the protein component of pili, is *traA* and is located in a cluster with the other *tra* functions (Fig. 13.5). The DNA sequence codes for the 121 amino acids of the pilin plus 51 extra amino acids at the amino-terminus. Processing the protein to normal size requires *traQ* activity. Physically, the unassembled pilin is associated with the inner membrane of the cell, and presumably this pool of 100,000 molecules makes possible the rapid assembly of pili. The location of the pili does not appear to be random but is associated with

Table 13.2. Genes of the transfer region.

A. Translated genes
  1. Transfer genes
    a. Pilin synthesis
      traA   (prepropillin)
      traQ   (prepropillin hydrolase)
      traX   (pilin *N*-acetylase)
    b. Pilus assembly
      traB, traC, traE, traF, traG (N-terminus), *traH, traK, traL*
      traU, traV, traW, trbC
    c. Mating pair stabilization
      traG (C-terminus), *traN*
    d. DNA transactions
      traD   (DNA transport protein)
      traI   (relaxase/helicase)
      traY   (oriT-nuclease)
  2. Transfer-related genes
    a. Surface exclusion
      traS, traT
    b. Regulation
      finP   (plasmid-specific antisense RNA)
      traJ   (positive regulator of *traY*)
      traM   (binds to *oriT*, allows *tra* expression)
      traY   (allows expression of *traM*)
B. Nontranslated genes
  1. Transfer gene (DNA transaction)
    oriT
  2. Transfer-related genes
    a. Transcriptional promoters
      traIp, traJp, traMp, traTp?, traYp
    b. Repressor binding site
      traJo

zones of adhesion where inner and outer cell membranes touch. Table 13.3 describes the major functions of the pilin protein (TraA protein) as identified by mutational analysis (Manchak et al. 2002). The pili "retract" by disassembly, thereby facilitating the conjugal cell contacts and incidentally providing a route

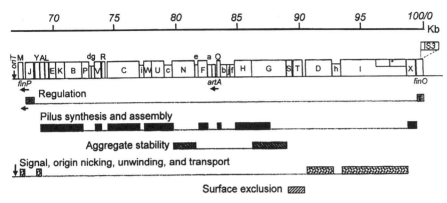

Figure 13.5. Genetic map of the F plasmid transfer region. *Boxes* indicate the position and size of F transfer region genes, transcribed from left to right. Capital letters indicate *tra* genes; small letters indicate *trb* genes. The *finP* RNA product and the *artA* product are encoded by the anti-*tra* strand. Wherever known, functional groups are indicated by the character of the *shaded areas* below the map. A *line* above the map gives F plasmid coordinates from the 100-kb map. (Adapted from Ippen-Ihler and Skurray [1993].)

of infection for the male-specific bacteriophages f1, f2, Qβ, MS2, and M13. Although the pili must be present for male-specific phage infection, a female-specific phage such as φII cannot infect male cells even in the absence of pili.

The early stage of mating pair formation is inhibited if the other cell has both the TraT outer membrane protein and the TraS inner membrane protein;

Table 13.3. Major functions of TraA (pilin) protein as identified by Manchak et al. (2002).

Class of Mutation	Functional Alteration in Mutant Cell
I	Stability of propilin (unprocessed translated protein)
II	Propilin cannot be processed to pilin
III	Prevents phage M13 transduction and conjugation (tip of pilus does not assemble correctly)
IV	Prevent phage R17 attachment (side of pilus)
V	Prevent R17 eclipse (pore formation? pilus retraction?)
VI	Prevent DNA transfer (pore formation? pilus retraction?)

this phenomenon has been called surface exclusion. When cells are pheno-copied, these proteins disappear. The mating aggregates that form are stabilized by the products of the *traN* and *traG* genes. TraN interacts directly with outer membrane protein OmpA and lipopolysaccharide. Generally speaking, those conjugative systems that have long flexible sex pili are capable of mating pair formation in liquid culture. Those that do not are limited to matings on solid surfaces such as membrane filters.

Transfer DNA Replication

Because only a single strand of donor DNA is transferred, only one strand needs to be nicked and replicated. The site of the nick is a separate origin of replication, *oriT*. It is a 373-bp fragment that when cloned converts a nonmobilizable plasmid into a high-frequency donor. Four *oriT* sequences are known among the IncF plasmids. Neil Willetts and coworkers examined a number of other *oriT* regions. That for IncQ is basically the same as ColE1, whereas that from IncN has 13 direct repeats of 11 bp, three pairs of 10-bp inverted repeats, and an (A + T)-rich region.

In the case of the F plasmid, normal *traY* sequence is required for the nicking activity. The *traY* gene codes for a small membrane protein that, in combination with the amino terminus of TraI protein (relaxase), makes a nick at *oriT*. The TraI protein also serves to unwind the DNA duplex during initiation (Matson et al. 2001), a necessary prelude because its nicking activity only functions on single-strand DNA. It is highly sequence specific because the relaxase activity for plasmid R100 will not function with F (Harley and Schildbach 2003).

The experiments discussed in Chapter 11 have revealed that it is the preexisting donor DNA that is transferred to the recipient, and the DNA synthesis that occurs during transfer replaces it. The transfer process is independent of strand replacement, although lack of replacement is lethal to the donor cell. It has been suggested but not proved that the TraM protein may be the signal to begin conjugation.

Physical transfer of the donor DNA requires normal helicase I (*traI*) activity. Mutant cells form mating aggregates, but no transfer can occur. A model for this activity that was proposed by Willetts and Wilkins is presented in Fig. 13.6. The driving force for the transfer is assumed to be the action of the helicase portion of TraI as it unwinds the donor duplex. Under normal conditions, there is concomitant DNA synthesis in both donor and recipient cells. Experiments using the yeast two-hybrid system (Harris et al. 2001) show that three of the *tra*

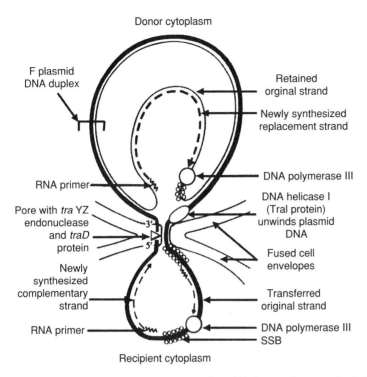

Donor cytoplasm

F plasmid
DNA duplex

Retained
orginal strand

Newly synthesized
replacement strand

DNA polymerase III

RNA primer

DNA helicase I
(TraI protein)
unwinds plasmid
DNA

Pore with *tra* YZ
endonuclease
and *traD*
protein

3'

5'

Fused cell
envelopes

Newly
synthesized
complementary
strand

Transferred
original strand

RNA primer

DNA polymerase III
SSB

Recipient cytoplasm

Figure 13.6. Model for the conjugative transfer of F. A specific strand of the plasmid (*thick line*) is nicked at *oriT* by the *traYZ* endonuclease and transferred in the 5' to 3' direction through a pore in the cell membrane, perhaps involving the *traD* protein, formed between the juxtaposed donor and the recipient cell envelopes. The plasmid strand retained in the donor cell is shown by a *thin line*. The termini of the transferred strand are attached to the cell membrane by a complex that includes the endonuclease. DNA helicase I (*traI* product) migrates on the strand undergoing transfer to unwind the plasmid duplex DNA; if the helicase is in turn bound to the membrane complex during conjugation, the concomitant ATP hydrolysis might provide the motive force to displace the transferred strand into the recipient cell. DNA transfer is associated with the synthesis of a replacement strand in the donor and of a complementary strand in the recipient cell (*broken lines*); both processes require de novo primer synthesis and the activity of DNA polymerase III holoenzyme. The model assumes that a single-strand binding protein coats DNA to aid conjugative DNA synthesis; depending on the nature of the pore, this protein might even be transferred from donor to recipient cell, bound to the DNA. (Adapted from Willetts, N, Wilkins, B. [1984]. Processing of plasmid DNA during bacterial conjugation. *Microbiological Reviews* 48: 24–41.)

proteins (B, K, V) interact to form a linkage that spans the cell wall with TraB in the cell membrane, TraK in the periplasm, and TraV in the outer membrane. Presumably, this structure represents the point of DNA transfer.

The rolling circle model for DNA replication is generally taken as appropriate for strand replacement, and it predicts that the nicked strand will serve as its own primer for synthesis. At least in *dnaB* mutant cells that is not the case. There the replication process is sensitive to rifampin, implying the necessity for RNA synthesis. Single-strand DNA is normally stabilized by single-strand binding protein (SSB), and F plasmid DNA is no exception. The F plasmid itself codes for its own SSB activity in addition to the standard *E. coli* protein.

Mobilization is the process of producing a single-strand DNA molecule for transfer. In the recipient cell, there is a corresponding process of repliconation in which the plasmid DNA is prepared for replication by discontinuous synthesis of a complementary strand via Okazaki fragments followed by circularization. The nature of the priming reaction used by the F plasmid is uncertain, but the *sog* protein from CollB-P9 serves as the primer for complementary strand synthesis in that system. The protein is attached to the transferred conjugative DNA, and in that case radioactive labeling experiments have shown that as much as 0.9% of protein extracted from recipient cells originates in the donor. In the recipient cell, Sog acts as a DNA primase to initiate synthesis of a complementary strand.

Circularization of F cannot occur by recombination of concatemers if each round of rolling circle replication must be separately primed. The model of Fig. 13.6 suggests that the 5'-end of the transferred DNA is held bound to a membrane protein that also recognizes the 3'-end when it arrives. As the two ends are held in close proximity, they are joined to make a covalently closed circle. Such a reaction is known to be possible in the case of phage φX174 using the gene A protein.

The newly circularized plasmid is now ready to follow the usual pattern of replication and partition described earlier. The various events necessary for conjugal plasmid transfer are summarized in Fig. 13.7.

Broad Host Range Plasmids

Careful examination of conjugative plasmids has shown that many of them can transfer their DNA across species or even genus boundaries. However, the transferred DNA may not be able to fulfill all necessary functions in its new host. A plasmid that is defective in replication, copy number control, or partitioning

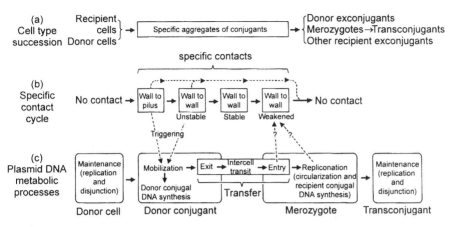

Figure 13.7. Summary of conjugal events that lead to plasmid transmission at the cellular and two subcellular levels. (**a**) Succession of cell types. An exconjugant is any cell that has been part of a mating aggregate, but that has not undergone any genetic change, whereas a merozygote is a recipient cell that has received DNA from a donor cell. (**b**, **c**) Surface and DNA metabolic events underlying the succession. Any of the indicated four types of specific contacts may be present while donor and recipient cells are in mating aggregates. The main pathways are shown by *bold arrows* and alternative pathways by *dashed lines*. DNA metabolic events are shown for only four of the cell types shown in (**a**). Surface events are related to DNA metabolic events by a process called triggering, which may occur following the formation of either of the first two types of specific contacts. (Adapted from Clark, A.J., Warren, G.J. [1979]. Conjugal transmission of plasmids. *Annual Review of Genetics* 13: 99–1253.)

will not persist in its new host. Broad host range plasmids are those that are both conjugative and able to stabilize themselves in their new host. By this definition, the F plasmid has a narrow host range, transferring only among the closely related enterobacteria. Most of the experimentation in this area has dealt with members of the IncP, IncN, or IncW groups.

The IncP group is the most investigated and includes the commonly studied and essentially identical plasmids RP1, RP4, RK2, R18, and R68. They can all transfer better on solid medium than in liquid, and can transfer (but not necessarily maintain themselves) into *Mycobacterium*, *Bacteroides*, some cyanobacteria, and yeast. They transfer and maintain themselves well in a wide variety of Gram-negative bacteria including *Pseudomonas*, *Escherichia*, *Proteus*, *Salmonella*, and purple bacteria. Given such a diverse set of hosts, it is clear that there cannot be any specific cell surface receptors involved.

In RK2/RP4, there are two sets of *tra* genes. Set 1 is involved with conjugal DNA and carries *oriT*. Set 2 handles surface exclusion and/or pilus synthesis. The genes are organized into more than one operon. Notable sequence homologies to the Ti plasmid of *Agrobacterium* are present.

The Incl8 group includes the plasmid pIP501, originally found in *Enterococcus faecalis* but transmissible to *E. coli* and *Streptomyces lividans* (Kurenbach et al. 2003). Protein predictions based on DNA sequence analysis indicate the presence of proteins similar to a type IV secretion system.

Summary

Plasmids that are not integrated into the bacterial chromosome must carry out a number of tasks to ensure their survival. They must replicate themselves at least as often as the chromosome itself. They must arrange for their equal distribution throughout the cytoplasm so that when cell division occurs each daughter cell receives one or more copies of the plasmid. Some plasmids basically overreplicate their DNA and, in essence, saturate the cell with plasmid DNA molecules, thereby ensuring their inheritance by daughter cells. Such high copy number plasmids represent a significant drain on cell resources, and the larger plasmids must regulate themselves or leave their host cells at a significant disadvantage in the competition for nutrients. Plasmids also display incompatibility functions designed to prevent closely related plasmids from displacing a resident plasmid.

Replication control, copy number, and incompatibility are interrelated phenomena. Most plasmids produce one or more proteins that trigger the start of replication. Synthesis of this protein can be controlled at two levels: transcription and translation. Transcriptional control can involve the binding of a protein repressor (which may be the product of translation of that mRNA) so as to block RNA polymerase function. There is also regulation by synthesis of antisense RNA (transcribed from the noncoding strand of DNA). When this RNA binds to the normal mRNA, it makes a complex that is susceptible to nuclease attack, and the mRNA is rapidly degraded. Small antisense RNA molecules also exercise translational control. When bound to the mRNA, they block the ribosome binding site and prevent translation. Incompatibility is basically determined by the same genetic loci as the copy number. The regulatory elements prevent the newly arrived plasmid from significant replication and thus from establishing itself.

Conjugative plasmids must synthesize their surface pili in order to form mating aggregates. When the cell surfaces touch, a nick is introduced into the plasmid DNA at *oriT*, and a single strand of DNA is displaced by the action of a DNA helicase enzyme and transferred into the recipient cell. Donor and recipient cells then synthesize complementary strands to restore the plasmid DNA to its normal state. Plasmids incapable of using pili may still mate on solid surfaces like membrane filters.

Questions for Review and Discussion

1. What are the different ways in which antisense RNA can serve as a regulator?
2. How would you detect regulatory antisense RNA?
3. What are the strategies used by plasmids to control their copy number? How can the same strategies be used to effect incompatibility?
4. What are the strategies used by plasmids to ensure that all daughter cells carry plasmids?
5. What are the processes that occur during DNA transfer?

References

General

Bentley, S.D., Brown, S., Murphy, L.D., Harris, D.E., Quail, M.A., Parkhill, J., Barrell, B.G., McCormick, J.R., Santamaria, R.I., Losick, R., Yamasaki, M., Kinashi, H., Chen, C.W., Chandra, G., Jakimowicz, D., Kieser, H.M., Kieser, T., Chater, K.F. (2004). SCP1, a 356,023 bp linear plasmid adapted to the ecology and developmental biology of its host, *Streptomyces coelicolor* A3(2). *Molecular Microbiology* 51: 1615–1628.

Chattoraj, D.K. (2000). Control of plasmid DNA replication by iterons: No longer paradoxical. *Molecular Microbiology* 37: 467–476.

del Solar, G., Espinosa, M. (2000). Plasmid copy number control: An ever-growing story. *Molecular Microbiology* 37: 492–500.

Gerdes, K., Mùller-Jensen, J., Jensen, R.B. (2000). Plasmid and chromosome partitioning: Surprises from phylogeny. *Molecular Microbiology* 37: 455–466.

Ippen-Ihler, K.A., Skurray, R.A. (1993). Genetic organization of transfer-related determinants on the sex factor F and related plasmids, pp. 23–52. In: Clewell, D.B. (ed.), *Bacterial Conjugation*. New York: Plenum Press.

Khan, S.A. (1997). Rolling-circle replication of bacterial plasmids. *Microbiology and Molecular Biology Reviews* 67: 442–455.

Novick, R.P. (1998). Contrasting lifestyles of rolling-circle phages and plasmids. *Trends in Biochemical Sciences* 23: 434–438.

Pogliano, J. (2002). Dynamic cellular location of bacterial plasmids. *Current Opinion in Microbiology* 5: 586–590.

Thomas, C.M. (2000). Paradigms of plasmid organization. *Molecular Microbiology* 37: 485–491.

Wilkins, B., Lanka, E. (1993). DNA processing and replication during plasmid transfer between Gram-negative bacteria, pp. 105–136. In: Clewell, D.B. (ed.), *Bacterial Conjugation*. New York: Plenum Press.

Specialized

Erdmann, N., Petroff, T., Funnell, B.E. (1999). Intracellular localization of P1 ParB protein depends on ParA and parS. *Proceedings of the National Academy of Sciences of the USA* 96: 14905–14910.

Gordon, G.S., Sitnikov, D., Webb, C.D., Teleman, A., Straight, A., Losick, R., Murray, A.W., Wright, A. (1997). Chromosome and low copy plasmid segregation in *E. coli*: Visual evidence for distinct mechanisms. *Cell* 90: 1113–1121.

Harley, M.J., Schildbach, J.F. (2003). Swapping single-stranded DNA sequence specificities of relaxases from conjugative plasmids F and R100. *Proceedings of the National Academy of Sciences of the USA* 100: 11243–11248.

Harris, R.L., Hombs, V., Silverman, P.M. (2001). Evidence that F-plasmid proteins TraV, TraK and TraB assemble into an envelope-spanning structure in *Escherichia coli*. *Molecular Microbiology* 42: 757–766.

Hazan, R., Sat, B., Reches, M., Engelberg-Kulka, H. (2001). Postsegregational killing mediated by the P1 phage "addiction module" *phd-doc* requires the *Escherichia coli* programmed cell death system mazEF. *Journal of Bacteriology* 183: 2046–2050.

Koppes, L.J.H. (1992). Nonrandom F-plasmid replication in *Escherichia coli* K-12. *Journal of Bacteriology* 174: 2121–2123. (Density shift experiments show that F plasmids replicate at specific times during the cell cycle.)

Kurenbach, B., Bohn, C., Prabhu, J., Abudukerim, M., Szewzyk, U., Grohmann, E. (2003). Intergeneric transfer of the *Enterococcus faecalis* plasmid pIP501 to *Escherichia coli* and *Streptomyces lividans* and sequence analysis of its *tra* region. *Plasmid* 50: 86–93.

Manchak, J., Anthony, K.G., Frost, L.S. (2002). Mutational analysis of F-pilin reveals domains for pilus assembly, phage infection and DNA transfer. *Molecular Microbiology* 43: 195–205.

Matson, S.W., Sampson, J.K., Byrd, D.R.N. (2001). F plasmid conjugative DNA transfer: The TraI helicase activity is essential for DNA strand transfer. *The Journal of Biological Chemistry* 276: 2372–2379.

Møller-Jensen, J., Borch, J., Dam, M., Jensen, R.B., Roepstorff, P., Gerdes, K. (2003). Bacterial mitosis: ParM of plasmid R1 moves plasmid DNA by an actin-like insertional polymerization mechanism. *Molecular Cell* 12: 1477–1487.

Rodionov, O., Yarmolinsky, M. (2004). Plasmid partitioning and the spreading of P1 partition protein ParB. *Molecular Microbiology* 52: 1215–1223.

Taki, K., Abo, T., Ohtsubo, E. (1998). Regulatory mechanisms in expression of the *traY-I* operon of sex factor plasmid R100: Involvement of *traJ* and *traY* gene products. *Genes to Cells* 3: 331–345.

14

Advanced Regulatory Topics

Much of the discussion to this point has only been on a relatively simple type of regulation. There has been a single operon, one or two promoters, and a single repressor or activator complex. In many instances, cells need to be able to coordinate and/or to sequence the activities of a wide variety of diverse functions in order to achieve an appropriate response to an environmental stimulus. This chapter deals primarily with methods used by Bacteria to accomplish these goals, although some Archaea are also included.

Major topics include:

- Strategies used by cells to coordinately regulate functions that are physically located at widely separated sites
- Strategies used by cells to sequence steps in a complex process so that the needed functions are provided at the correct time
- Types of unusual gene structures that serve to prevent protein function prior to appropriate induction

Regulons

A **regulon** is a combination of two or more operons that is regulated coordinately. It is distinguished from the global regulatory networks discussed later in the sense that the operons constituting a regulon are obviously related in function.

THINKING AHEAD

How would you identify members of a regulon?

Maltose

The genes for maltose and maltodextrin degradation and utilization provide a good example of a regulon. The mechanisms used in regulation are not new, but the structural organization of the elements to be regulated is complex.

The maltose regulon of *Escherichia coli* (reviewed by Boos and Shuman [1998]) has been defined by two general methods. The historically older and more straightforward method consists of isolating mutants defective in maltose metabolism. This method defined two widely separated operons that were originally designated as the *malA* and *malB* genes (located at 75 and 91 minutes on the *E. coli* genetic map). However, fine structure mapping by Maurice Hoffnung and Maxime Schwartz and their collaborators has shown that each so-called gene, in fact, is composed of two operons, so the terms MalA and MalB are now used to refer to the groups of operons.

The second method for defining members of the maltose regulon is gene fusions. Winfried Boos and coworkers have used Mu d1 (Ap*lac*) phage to create ampicillin-resistant cells carrying *lacZ* fused to random genes. Induction of the maltose regulon stimulates β-galactosidase production when the phage has inserted into a maltose-regulatable gene. The experimenters can then determine the location of the new gene by mapping the ampicillin-resistance gene carried by the phage. Five additional operons have been identified by this and other methods. Figure 14.1 shows the basic genetic map for these operons.

The MalA region codes for a regulatory protein and for glucose-releasing enzymes. The product of the *malQ* gene is amylomaltase, which acts to join pairs of maltose or maltotriose and longer maltodextrin chains to yield glucose and an

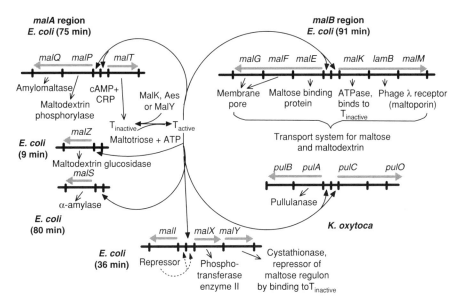

Figure 14.1. Regulatory interactions in the maltose regulon. *Gray lines* indicate the extent and direction of transcription. *Solid arrows* indicate stimulatory transactions, and *dotted lines* indicate inhibitory transactions. (Adapted from Schwartz [1987], pp. 1482–1502. In: Neidhardt, F.C. et al. [eds.], *Escherichia coli and Salmonella typhimurium: Cellular and Molecular Biology.* Washington, DC: American Society for Microbiology.)

even longer polymer. A phophorylase encoded by the *malP* gene hydrolyzes the polymer to release glucose-1-phosphate from the nonreducing end of the molecule. The *malT* gene codes for a protein that acts as a positive regulator (i.e., promotes transcription) for all maltose operons by binding to a specific "maltose box" in the promoters. The consensus maltose box sequence is 5'GGA(T/G)GA. Transcription of *malT* is under the control of the same cyclic AMP/CRP regulatory system as the *lac* operon. Therefore, indirectly, the presence of glucose represses the maltose regulon. MalT protein is synthesized in inactive form and is activated by binding to maltotriose and ATP. The MalK protein binds to the inactive form of MalT and prevents further transcription (Joly et al. 2004). Although the MalQ enzyme can synthesize maltotriose, it is not required for maltotriose production, and *malQ* mutant strains are still inducible. The endogenous sugars are thought to maintain basal levels of function.

The MalB region contains information coding for transport of maltose into a cell and consists of six genes organized into two operons. The first operon consists of the *malGFE* genes. MalE is a high-affinity maltose binding protein found in the periplasm. MalF and MalG form a complex in the cytoplasmic membrane that behaves as a stereospecific pore. MalG has the additional property of anchoring the MalK protein to the inner surface of the cytoplasmic membrane. This combination of a high-affinity periplasmic binding protein, two proteins forming a translocation pore, and a membrane-bound cytoplasmic ATPase is known as an **ABC transporter**, where ABC stands for ATP binding cassette. ABC transporters are ubiquitous in prokaryotes and eukaryotes and may be evolutionarily ancient. Horlacher et al. (1998) have shown that *Thermococcus littoralis*, a hyperthermophilic archaeon, has an ABC transporter system for maltose that includes a *malEFG* cluster very similar to that of *E. coli*. In fact, the MalF and MalG protein sequences are 57% identical.

The second operon within MalB consists of the *malK*, *lamB*, and *malM* genes. The LamB protein is found on the surface of the outer membrane where it serves as a maltoporin and also as the receptor for phage λ. MalM protein has an unknown function in the periplasm, and MalK protein complexes with MalFG proteins to energize the pores in the inner membrane. It also binds to the inactive form of the MalT protein to prevent induction unless maltose is being transported. Other genes shown in Fig. 14.1 are involved in the maltose regulon, but defects in most of these genes do not have the effect of preventing maltose degradation.

The properties of cells carrying *mal* mutations are given in Table 14.1. Note that MalT function is absolutely required for expression of all *mal* genes in wild-type cells. MalT-independent promoters for the MalA operon do exist. However, such mutations have not been observed in MalB. Neither *malS*, which codes for a periplasmic amylase, nor *malX*, whose product is an alternative phosphotransferase enzyme II, are essential for maltose metabolism. Moreover, because LamB protein is both a transport protein and a phage receptor, certain maltose mutants are lambda-resistant as well as defective in sugar metabolism. The MalI protein is a repressor of its own and *malXY* transcription and exhibits about 25% sequence similarity to the LacI and GalR proteins. Constitutive expression of *malY* clones results in lack of induction of the maltose genes. The cystathionase activity of the MalY protein can replace *metC* activity in the synthesis of methionine. What this enzyme activity has to do with its ability to repress maltose induction is unclear. The *malZ* protein is a cytoplasmic enzyme with maltodextrin glucosidase activity whose physiological role is also unclear.

Table 14.1. Phenotypes of cells carrying various *mal* mutations.

Genotype	Maltose	Lambda
Wild type	+	S
lamB	− (Maltoporin missing)	R
malE, malF, malG, or *malM*	− (Maltose transport defective)	S
malI	+ (Defective repressor for *malX*)	S
malP or *malQ*	− (Defective enzymes for maltose degradation)	S
malS	+ (Defective amylase)	S
malT	− (Defective positive regulator)	R
malX	+ (Defective enzyme for inducer degradation)	S
malY	+ (Cystathionase, repressor of maltose regulon expression)	S
malZ	+ (Malodextrin phosphorylase)	S

Note: +, ability to utilize maltose; −, inability to utilize maltose; S, sensitivity to phage λ; R, resistance to phage λ.

MalK protein has three discrete functions that are separable genetically. The first is its function as a transport system component. The second aspect of MalK in its role as a membrane-associated protein is as a target for regulation by the glucose phosphotransferase system. It interacts with a dephosphorylated component of that system to prevent further transport of maltose, an interaction that would only occur in the presence of glucose. Finally, in addition to its role at the inner membrane, it negatively regulates the *mal* regulon because MalK mutants are derepressed for *mal* regulon function. The mechanism of its repression involves binding of the MalK protein directly to MalT (Joly et al. 2004). This binding couples transport of maltose to expression of the *mal* regulon because when maltose is present, MalK will be hydrolyzing ATP and not binding to MalT.

What is particularly intriguing about the maltose regulon of *E. coli* is that it has strong sequence homology to the maltose regulon in *Klebsiella pneumoniae*. However, in *Klebsiella* there is an additional gene, *pulA*, that codes for the enzyme pullulanase. Pullulanase degrades pullulan, an oddly branched glucose polymer, so that it yields maltodextrins. Its absence in *E. coli* probably reflects differences in their lifestyles.

THINKING AHEAD

How would you design a regulatory system so that it responds to potentially lethal damage and then turns itself off when the damage is repaired?

Heat Shock Proteins

All organisms examined, ranging from vertebrates to *E. coli*, show a definite response pattern when shifted to temperatures above their normal growth range. The **heat shock response** is the stimulation of synthesis of a particular subset of cellular proteins, some of which are normally present at low levels and others are synthesized de novo. There are at least 18 known heat shock genes in *E. coli*, including the chaperonins *dnaK*, *dnaJ*, *groEL*, *groES*, and *grpE*, as well as the proteases *clpAP*, *lon*, and *ftsH*. The chaperonins function to attempt to refold the denatured proteins using ATP as a source of energy. The proteases specifically attack misfolded proteins. Peak protein synthetic activity is expressed some 5–10 min after the heat shock. DnaK together with ClpB and the smaller heat shock proteins IbpAB work to prevent clumping of denatured proteins. Deletion mutants of *ibpAB* require greater concentrations of DnaK for normal growth (Mogk et al. 2000). In cases of extreme heat shock, protein Hsp31 comes into play, holding misfolded proteins until conditions return to normal (Mujacic et al., 2004).

Expression of some heat shock proteins is controlled by *rpoH*, which is a new σ factor for the RNA polymerase holoenzyme complex (reviewed by Wösten [1998]). Normal σ protein has a molecular weight of 70,000 Da and is designated σ^{70}. The new protein is substantially smaller and is designated σ^{32}. Frederick Neidhardt and his collaborators theorized that each of the heat shock proteins have two promoters, a low-efficiency one that responds to σ^{70} and a high-efficiency one that responds to σ^{32}. Under normal conditions there is little σ^{32} present in the cytoplasm. The protein has a short half-life owing to its interaction with DnaK. The combination is degraded rapidly by the FtsH protease (Blaszczak et al. 1999). After heat shock, many proteins are denatured, and DnaK binds preferentially to them, leaving σ^{32} free to bind to RNA polymerase and trigger transcription.

A second group of heat shock proteins is controlled by σ^E, a sigma factor triggered by extracytoplasmic factors. In other words, it is part of a signal transduction system. An example of a stimulus for σ^E would be a misfolded outer membrane protein. Activation of *rpoH* requires both σ^{70} and σ^E. At temperatures over 50°C, only σ^E functions to activate transcription of *rpoH*. Klein et al. (2003) have isolated temperature-sensitive mutants with a reduced basal level of σ^E and found that they mapped to a tyrosine phosphatase gene and to a membrane protein RseA that is part of the signal transduction apparatus. Phosphorylated forms of the sigma factors lost much of their normal activity.

The term "heat shock" is actually something of a misnomer in the sense that many varied treatments besides temperature can induce synthesis of heat shock proteins. These treatments include accumulation of normally secreted proteins, carbon or phosphate starvation, exposure to heavy metals, ethanol, ultraviolet (UV) radiation, and inhibition of topoisomerases, or λ infection. Heat shock proteins automatically return to their normal rate of synthesis after a brief interval because the *rpoD* gene codes for σ^{70} and is stimulated by heat shock. Therefore, it gradually displaces σ^{32}, and the situation returns to normal. Optimal functioning of heat shock promoters requires the presence of the HtpY protein, and its overproduction raises the basal level of heat shock proteins.

There are strong similarities between the heat shock proteins found in *Drosophila* and those found in *E. coli*, suggesting a certain generality of function. DnaK protein is a molecular chaperone, a protein that serves to maintain the physical state of another protein in the cytoplasm and facilitate appropriate protein:protein interactions. It uses ATP to disrupt hydrophobic protein aggregates that might form after partial denaturation of a protein. It plays a similar role during normal λ infection in catalyzing the release of protein P from the λ O protein–host DNA helicase complex (see Fig. 8.5). DnaK has about 50% sequence similarity with eukaryotic proteins. Heat shock in *Bacillus subtilis* is less well studied, but again many of the proteins seem to antigenically match those observed in *E. coli*. Oddly enough, even an extreme thermophile like *Haloferax volcanii* has heat shock genes. The archaeans as a group have the equivalent of the small heat shock proteins, GroEL and DnaK (reviewed by Laksanalamai et al. [2004]). The mesophilic Archaea have more heat shock proteins similar to the Bacteria.

Negative regulation of heat shock proteins is found in bacteria other than *E. coli*. There is an inverted repeat element near the promoters for *groE* genes in more than 40 bacteria. It has the consensus sequence TTAGCACTC-N_9-GAGTGCTAA and is known as controlling inverted repeat of chaperone expression (CIRCE). The repressor that binds to CIRCE is the product of the *hcrA* gene, and it requires GroE binding for activity. The system is thus analogous to the situation with DnaK in *E. coli* where the presence of denatured proteins will titrate away the repressor.

Global Regulatory Networks

The distinction between regulons and **global regulatory networks** is that networks are collections of operons that are not obviously related by function. Nevertheless, members of a network are coordinately expressed or repressed. The Bacteria have several well-studied examples of global regulatory networks that demonstrate some unusual regulatory mechanisms. The SOS DNA repair pathway that is discussed in Chapter 5 is also a notable member of this group.

Nitrogen Regulation

A relatively small group of bacteria, both photosynthetic and nonphotosynthetic, is capable of reducing atmospheric nitrogen and using it for metabolism. Some typical genera that include species known to fix nitrogen are *Rhizobium*, *Azotobacter*, *Klebsiella*, *Clostridium*, *Rhodopseudomonas*, and *Anabaena*. The nitrogenase enzyme is extremely sensitive to oxygen and has three protein subunits that are encoded by *nifHDK*, whose base sequence is highly conserved from organism to organism. Cloning of these genes into *E. coli* yields an organism that fixes nitrogen.

The nitrogen-fixing genes are the best-known example of a genetic system regulated by yet another sigma factor, σ^{54} (for a review, see Martinez-Argudo et al. [2004]). Unlike the sigma factors involved in heat shock, this sigma factor is structurally and functionally separate from σ^{70} and is the product of the *rpoN* gene. Promoters responding to σ^{54} are unusual in the sense that they always require binding of an activator protein to an enhancer site located at least 100 bp upstream. The DNA must bend in order to bring the enhancer close to the promoter site.

The best-studied genetic system is that found in *K. pneumoniae*, which has 20 *nif* genes organized into six transcription units along a 25-kb fragment of DNA. The regulatory scheme for *nif* is diagrammed in Fig. 14.2 as a part of a global regulatory network. The network regulates not only nitrogen fixation, but also histidine and proline utilization and glutamine synthetase, all of which aid cells in assimilating nitrogen. Overriding control of the network is vested in the

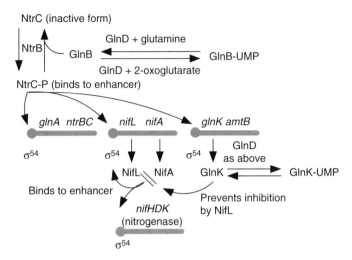

Figure 14.2. Regulation of *nif* gene transcription. The Ntr system regulates multiple functions including *nif* transcription. The sensory transduction is via GlnD removing the uridylyl group from GlnB. When the appropriate signal is transduced, NtrB autophosphorylates and transfers the phosphate to NtrC, activating it. NtrC is one of the enhancer-binding proteins that can help σ^{54} to start transcription. One of the operons that NtrC activates is the NifLA operon. NifL normally antagonizes the action of NifA unless GlnK intervenes after removal of its uridyl group. NifA is an enhancer-binding protein for its own promoter as for well as for the nitrogenase operon (*nifHDK*).

ntr system, which is a typical two-component regulatory system, the focus of which is the enhancer-binding protein NifA. NifA is a member of the AAA+ superfamily of ATPases. Another member of that family is the SopA protein involved in plasmid partitioning.

The NtrB protein (also known as NR$_{II}$) is a kinase/phosphatase that phosphorylates/dephosphorylates NtrC, which is needed for activation of a gene originally called *ntrA*. NtrA must interact with all of the necessary promoters, including that for *nifA*, to turn on transcription; therefore it is a sigma factor, and, in fact it is the same as *rpoN*. The sensory transduction system for nitrogen levels is GlnD operating on the PII-like signaling proteins GlnB and GlnK. In low nitrogen, GlnD adds uridylate groups to GlnB or GlnK. GlnK plays a significant role during nitrogen starvation, operating in the second level of regulation. In *E. coli* and several other Bacteria, the *glnK* gene codes for a protein 67% identical to GlnB.

The NifL protein is a flavoprotein that is activated by oxygen or high ammonium levels to tie up NifA protein and prevent activation of the *nifHDK* operon. Mutation studies have shown that the nitrogen- and oxygen-sensing activities are independent within the protein and that the GlnK protein functions as an allosteric effector to relieve inhibition by NifL. Cells must maintain a stoichiometric relationship between NifL and NifA proteins or regulation will not be possible. *E. coli* accomplishes this feat via translational coupling. The first codon of NifA overlaps the terminator codon of NifL, with the ribosomes that terminate from *nifL* reinitiating at the Shine–Dalgarno sequence for *nifA*. NifL does not antagonize enhancer-binding proteins generally but only NifA. It appears to inhibit the binding of ATP and σ^{54}.

Considerable attention has been given to the function of NtrC protein, especially because it is another enhancer-binding protein. Phosphorylated NtrC binds to a closed complex of promoter and RNA polymerase, hydrolyzes ATP, and converts the closed complex into an open one that is ready for transcription. For this reason, researchers attempting to understand the steps involved in RNA transcription often work with σ^{54} rather than σ^{70}. All enhancer-binding proteins have three general domains within their three-dimensional structure. The central domain is the ATPase function and the region that allows NtrC dimers to form. It, however, cannot function without phosphorylation of the amino terminus, its regulatory region. The carboxy terminus of the protein contains a three-dimensional structure characterized as a helix-turn-helix that is characteristic of DNA binding proteins. Most enhancer-binding proteins are totally modular, and genetic constructs that mix the three domains provide mixed phenotypes. However, NtrC is unusual in that its ATPase and DNA binding domains are not totally independent.

Nitrogen-fixing systems in other organisms are organized along similar lines, although not necessarily as compactly as in *Klebsiella*. One point of difference is seen in *Anabaena*, a cyanobacterium that carries out oxygenic photosynthesis. The sensitivity of the nitrogenase enzyme to oxygen means that a cell cannot carry out both normal photosynthesis and nitrogen fixation. Instead, certain cells within a filament of cells differentiate to form thick-walled heterocysts where nitrogen fixation can occur. During the differentiation process, an 11-kb segment of DNA known as the *nifD* element is recombined from the middle of the *nifD* cistron to yield a normal *nifD*. The presence of the element prevents undifferentiated cells from engaging in futile attempts to fix nitrogen, and it is common among the heterocystous cyanobacteria (Thiel et al. 1997). Excision of the *nifD* element is under the control of *xisA*, which is located within the element. When cloned into *E. coli*, the XisA protein is sufficient to catalyze the *nifD* rearrangement.

Endospore Formation in *Bacillus*

The process of endospore formation is complex and highly regulated. It was recently reviewed by Hilbert and Piggott (2004), who summarized the results of several studies using DNA microarrays to identify active genes. Cells pass through the morphologic stages listed in Table 14.2. Basically, the process is one in which a septum forms asymmetrically within a cell. The smaller cellular compartment is the prespore, and the larger is the mother cell. The mother cell then engulfs the prespore, which forms first a protoplast and later a forespore within the cytoplasm of the mother cell. Eventually, after the mother cell has deposited the various layers constituting the outer surface of the spore, the mother cell lyses and the spore is free. Starvation of a culture for carbon or nitrogen sources triggers the sporulation process, and the process is reversible until the cells reach stage IV. Genes affecting sporulation are designated by "*spo*" followed by a Roman numeral indicating the affected stage and then by the allele designator.

Endospore formation presents a considerable regulatory problem. The *spo* loci are not grouped but are scattered about the genome (see Fig. 8.6). Some loci

Table 14.2. Stages in sporulation.

Stage	Activity	Sigma	Gene
0	Normal vegetative cell	A, H (minor)	*spooH*
I	Axial filament of DNA stretches from pole to pole		
IIi	Septation occurs off center		
IIii	Mother cell engulfs forespore		
III	Prespore protoplast inside mother cell	F (prespore), E (mother cell)	*spoIIA*, *spoIIG*
IV	Spore cortex forms	G (forespore), K (mother cell)	*spoIIIG*, *spoIIIC*
V	Spore coat forms		
VI	Spore matures, resistance to temperature and chemicals		
VII	Mother cell lyses, spore is released		

are active even in vegetative cells, but others turn on only when needed for sporulation. The complete system thus represents an exceptionally large global regulatory network that once again is triggered by a phosphorylation event, this time with Spo0A. Primarily, *Bacillus* uses the same regulatory strategy that is found in some viruses, namely, the use of various σ factors to sequentially modify RNA polymerase function. For a review of the structure and function of sigma factors, see Gruber and Gross (2003). The transcriptional changes produced in sporulating bacilli must be more gradual than those that occur in a system such as phage T4, because many of the vegetative genes must continue to be transcribed in order to provide necessary functions for the developing spore. The use of multiple σ factors provides for the necessary timing of the various stages of sporulation. It also permits the developing spore and the mother cell to have entirely different metabolic processes during the irreversible stages of sporulation. In fact, σ^G is active only in the prespore, and σ^K is active only in the mother cell. Gene fusions to *lacZ* have proved useful in determining the timing of various sporulation processes (Fig. 14.3).

The initial event for sporulation is phosphorylation of Spo0A by a phosphorelay system that includes five kinases, Spo0F, and Spo0B. From that point on, the biosynthesis of most sigma factors is a straightforward cascade process in which the presence of one sigma factor turns on the next one. However, that for σ^K involves some unusual regulatory features (Fig. 14.4). Genetic mapping experiments have shown that the gene for σ^K is actually a split gene consisting of what were originally considered as the separate genes *spoIIIC* and *spoIVCB*. The intervening region is known as sigma K intervening element (*skin*) and includes the gene *spoIVCA*. The SpoIVCA gene product is a recombinase that catalyzes *skin* excision, rejoining the two halves of what is then known as the *sigK* gene. The presumption is that the *skin* DNA represents the relic of a former prophage or transposable element whose insertion was catalyzed by the recombinase. The phenomenon does not occur in *Bacillus thuringiensis*, *Bacillus megaterium*, or *Clostridium acetobutylicum*.

Expression of *sigK* requires σ^E as well as the SpoIIID protein. The protein product is an inactive prosigma K protein that must be cleaved, possibly by the SpoIVFB protein, to yield mature σ^K. The cleavage reaction is subject to inhibition by a variety of factors, including σ^G from the prespore.

Note that σ^K is part of an unusual negative feedback regulatory loop. During the early stages of sporulation, an unstable sigma factor, σ^E, is synthesized. It is apparently stabilized by its interaction with RNA polymerase and stimulates synthesis of a new regulatory protein, SpoIIID, in the mother cell. In

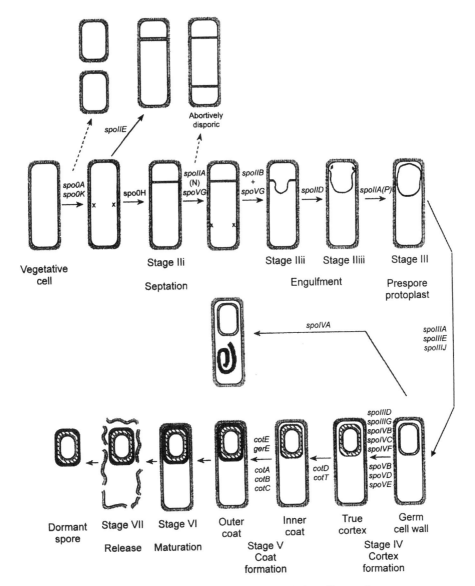

Figure 14.3. Morphological events of *B. subtilis* sporulation and the effects of mutations in various sporulation genes on morphology. The main sequence of morphological events begins with a vegetative cell and ends with the release of a mature spore. The intermediate steps that have been recognized are joined by *horizontal arrows*. Genes shown in boldface above the arrows are required for the step indicated by the arrow. Mutations in these genes give a mutant phenotype similar to the cell shown immediately before the arrow. Genes encoding proteins that are known to act at a specific point in the sequence but that do not block development are shown below the arrows. Thus, for example, the *cotT* gene encodes a protein that assembles into the inner spore coat. In a *cotT* mutant, the spores are normal in gross morphology and are apparently deficient only in the CotT protein. The locations of potential cell division events that can occur in the presence of certain mutations are indicated by *crosses*. In such cases, cells with additional septa are formed (indicated by the *dotted arrows*). Other mutations cause the organism to continue on an abnormal pathway, leading to the formation of the aberrant cells, which has been shown above the main sequence. (Adapted from Errington, J. [1993]. *Bacillus subtilis* sporulation: regulation of gene expression and control of morphogenesis. *Microbiological Reviews* 57: 1–33.)

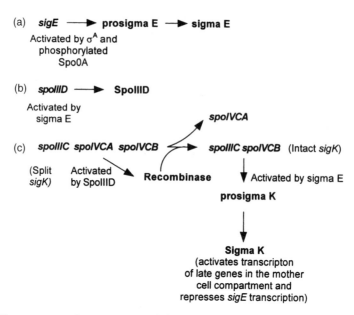

Figure 14.4. Sigma K in *B. subtilis* results from a gene rearrangement. Genes are represented by italic type. (**a**) The *sigE* gene is transcribed in the presence of the normal vegetative sigma factor σ^A and a phosphorylated protein, Spo0A. (**b**) Sigma E in its turn activates the *spoIIID* promoter. (**c**) The SpoIIID protein activates *spoIVCA*, which is inserted into the middle of the *sigK* gene. The protein product is a recombinase that removes the inserted DNA segment, leaving an intact *sigK* gene. Sigma E activates that gene to produce prosigma K protein that is eventually cleaved to give the active form.

the presence of SpoIIID protein, transcription of the *sigK* region occurs, the recombinase acts, and the σ^K concentration builds up. The σ^K protein is an inhibitor of *spoIIID* transcription. Therefore, as the σ^K concentration increases, the SpoIIID concentration decreases, and *sigK* transcription turns off. Meanwhile σ^K activates *gerE*. This gene originally appeared to affect spore germination but is now known to affect sporulation as well. SpoIIID is also an inhibitor of certain coat protein genes, including *cotD*. Owing to the decrease in SpoIIID, these genes now become active, signaling the transition from stage IV to stage V of sporulation. The GerE protein stabilizes this transition. One possible model for these events is shown in Fig. 14.5.

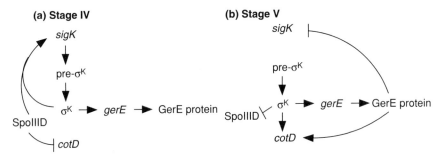

Fig. 14.5. Model for the switch from *sigK* to *cotD* transcription in the mother cell during the stage IV–V transition of sporulation. (**a**) During stage IV (cortex formation), SpoIIID stimulates *sigK* transcription by σ^K RNA polymerase and represses *cotD* transcription. σ^K RNA polymerase also transcribes *gerE*, leading to the synthesis of GerE. (**b**) The accumulation of σ^K, resulting from the processing of pro-σ^K to σ^K, causes a decrease in the level of SpoIIID, producing a switch to the stage V (coat formation) pattern of gene expression (i.e., *sigK* transcription is no longer stimulated and *cotD* transcription by σ^K RNA polymerase is no longer repressed). Continued production of GerE reinforces the switch since GerE represses *sigK* transcription and stimulates *cotD* transcription. (Redrawn from Halberg, R., Kroos, L. [1992]. Fate of the SpoIIID switch protein during *Bacillus subtilis* sporulation depends on the mother-cell sigma factor, σ^K. *Journal of Molecular Biology* 228: 840–849.)

Inteins

Chapter 6 presents information about naturally occurring RNA splicing. The previous section of this chapter discusses how genes may need to recombine to remove an insert before they can function, in other words, natural DNA splicing. The remaining possibility for splicing is protein splicing, which is the subject of this section.

Protein splicing occurs when a gene contains the carboxy and amino termini of a protein separated by a region of DNA coding for an embedded protein. The embedded protein is called an internal protein (**intein**), and it has catalytic activity. The remaining two parts of the original protein are designated **exteins** and rejoin via a peptide bond to make a single functional protein. A large intein has two domains, an endonuclease and a protein-splicing agent, while a mini-intein has only the latter.

There are a surprising number of intein-coding sequences known in eukaryotes, Bacteria, and Archaea, at least 130 in more than 48 different species.

The topic has been reviewed by Gogarten et al. (2002). Details about inteins and a current list are available on the World Wide Web site listed in Appendix 2.

The function of the endonuclease is to provide for the spread of the intein to other genes. It is an example of a **homing endonuclease**, a moderately site-specific recombination agent (see Chapter 15) that transfers parasitic DNA to an allele of its present location. Note that unlike transposons such as phage Mu, an intein does not create a mutation in a functional sense. Admittedly, the DNA sequence has changed, but the protein sequence remains intact, so there is little, if any, effect on the host cell because of the presence of the intein.

Small RNA Molecules

This class of regulators is not intuitively obvious. It consists of short nontranslated RNA molecules (**sRNA**) that, nevertheless, have a metabolic function. Most examples are individual RNA molecules, but the class does include untranslated regions on mRNA transcripts. Several members of this class have been discussed earlier (tmRNA, RNase P, and antisense RNA). Gottesman (2004) has reviewed their history and functions. Current estimates are that *E. coli* has about 4000 open reading frames and about 50–100 sRNA types.

Finding these molecules is not a simple genetic problem. Because they are not translated, gene fusions to the lac operon will not work, nor will protein array experiments. Computer searches based on the structure of known members of the group have had some success. The general criteria are presence of a promoter, about 300 nucleotides of content, and a ρ-independent transcription terminator. The content region usually shows predicted stem-and-loop structures. Definitive identification requires Northern blotting.

The effect of sRNA is not always intuitively obvious. For example, antisense RNA may destabilize the message by making it susceptible to nuclease attack. It may physically block the ribosome binding site to prevent translation. However, DsrA can improve translation of RpoS (stationary phase sigma factor) by enhancing ribosome accessibility.

Summary

There are many potential problems inherent in trying to regulate multioperon systems. Commonly used strategies include activator proteins that must be present at every promoter in every operon within the system and special sigma factors that activate the otherwise inactive promoters. It is becoming clear that small RNA molecules may also regulate either positively or negatively. As in the case of phage T4, bacteria also sometimes use cascades of sigma factors in which each new sigma factor induces synthesis of yet another sigma factor. The advantage of such a system is its ability to ensure that specific biochemical events occur in a defined sequence. An additional regulatory element that sometimes occurs both in prokaryotes and in eukaryotes is the enhancer, a discrete site to which a protein must bind for a promoter to be fully functional. Multifunctional regulatory proteins are often the subjects of intense study by geneticists because their distinct functions are associated with specific nonoverlapping domains within the molecule. Experimenters can use site-specific mutagenesis to eliminate only one function while preserving others. Examples of genes kept nonfunctional by insertion of extraneous genetic information are now known. Removal of the extra material can occur by protein splicing or recombination prior to transcription.

Questions for Review and Discussion

1. Genetically and biochemically, how would you recognize a mutation affecting (a) an enhancer; (b) an activator protein; and (c) a sigma factor?
2. Describe at least two methods that could be used to identify proteins that are members of a global regulatory network.
3. How can phosphorylation serve as a regulatory process?
4. Give some examples of proteins that serve as (a) positive autoregulators; and (b) negative autoregulators.
5. What might be the evolutionary advantage to genes with sequences inserted into the coding region?

References

Generalized

Boos, W., Shuman, H. (1998). Maltose/maltodextrin system of *Escherichia coli*: Transport, metabolism, and regulation. *Microbiology and Molecular Biology Reviews* 62: 204–229.

Dixon, R. (1998). The oxygen-responsive NIFL–NIFA complex: A novel two-component regulatory system controlling nitrogenase synthesis in γ-proteobacteria. *Archives of Microbiology* 169: 371–380.

Geiduschek, E.P. (1997). Paths to activation of transcription. *Science* 275: 1614–1616.

Gogarten, J.P., Senejani, A.G., Zhaxybayeva, O., Olendzenski, L., Hilario, E. (2002). Inteins: Structure, function, and evolution. *Annual Review of Microbiology* 56: 263–287.

Gottesman, S. (2004). The small RNA regulators of *Escherichia coli*: Roles and mechanisms. *Annual Review of Microbiology* 58: 303–328.

Gruber, T.M., Gross, C.A. (2003). Multiple sigma subunits and the partitioning of bacterial transcription space. *Annual Review of Microbiology* 57: 441–466.

Hilbert, D.W., Piggot, P.J. (2004). Compartmentalization of gene expression during *Bacillus subtilis* spore formation. *Microbiology and Molecular Biology Reviews* 68: 234–262.

Laksanalamai, P., Whitehead, T.A., Robb, F.T. (2004). Minimal protein-folding systems in hyperthermophilic archaea. *Nature Reviews: Microbiology* 2: 315–324.

Martinez-Argudo, I., Little, R., Shearer, N., Johnson, P., Dixon, R. (2004). The NifL–NifA system: A multidomain transcriptional regulatory complex that integrates environmental signals. *Journal of Bacteriology* 186: 601–610.

Narberhaus, F. (1999). Negative regulation of bacterial heat shock genes. *Molecular Microbiology* 31: 1–8.

Wösten, M.M.S.M. (1998). Eubacterial sigma-factors. *FEMS Microbiology Reviews* 22: 127–150.

Specialized

Govantes, F., Andújar, E., Santero, E. (1999). Mechanism of translational coupling in the *nifLA* operon of *Klebsiella pneumoniae*. *The EMBO Journal* 17: 2368–2377.

Horlacher, R., Xavier, K.B., Santos, H., Diruggiero, J., Kossmann, M., Boos, W. (1998). Archaeal binding protein-dependent ABC transporter: Molecular and biochemical analysis of the trehalose/maltose transport system of the hyperthermophilic archaeon *Thermococcus litoralis*. *Journal of Bacteriology* 180: 680–689.

Jack, R., De Zamaroczy, M., Merrick, M. (1999). The signal transduction protein glnK is required for NifL-dependent nitrogen control of *nif* gene expression in *Klebsiella pneumoniae. Journal of Bacteriology* 181: 1156–1162.

Joly, N., Danot, O., Schlegel, A., Boos, W., Richet, E. (2002). The Aes protein directly controls the activity of MalT, the central transcriptional activator of the *Escherichia coli* maltose regulon. *The Journal of Biological Chemistry* 277: 16606–16613.

Joly, N., Böhm, A., Boos, W., Richet, E. (2004). MalK, the ATP-binding cassette component of the *Escherichia coli* maltodextrin transporter, inhibits the transcriptional activator MalT by antagonizing inducer binding. *The Journal of Biological Chemistry* 279: 33123–33130.

Klein, G., Dartigalongue, C., Raina, S. (2003). Phosphorylation-mediated regulation of heat shock response in *Escherichia coli. Molecular Microbiology* 48: 269–285.

Mogk, A., Deuerling, E., Vorderwülbecke, S., Vierling, E., Bukau, B. (2003). Small heat shock proteins, ClpB and the DnaK system form a functional triade in reversing protein aggregation. *Molecular Microbiology* 50: 585–595.

Mujacic, M., Bader, M.W., Baneyx, F. (2004). *Escherichia coli* Hsp31 functions as a holding chaperone that cooperates with the DnaK–DnaJ–GrpE system in the management of protein misfolding under severe stress conditions. *Molecular Microbiology* 51: 849–859.

Thiel, T., Lyons, E.M., Erker, J.C. (1997). Characterization of genes for a second Mo-dependent nitrogenase in the cyanobacterium *Anabaena variabilis. Journal of Bacteriology* 179: 5222–5225.

Wu, H., Hu, Z., Liu, X.-Q. (1998). Protein *trans*-splicing by a split intein encoded in a split *dnaE* gene of *Synechocystis* sp. PCC06803. *Proceedings of the National Academy of Sciences of the USA* 95: 9226–9231.

15

Site-Specific Recombination

Site-specific recombination occurs in the temperate response of bacterio-phages, the integration of plasmids, and the movement of transposons or inteins. While elements of the basic recombination pathways discussed are in Chapter 5, there are unique aspects to each of these processes.

Major topics include:

- The way phage lambda ensures that its prophage always inserts in the same chromosomal location and in the same orientation
- Double site-specific recombinations and whether all of them proceed by the same mechanism
- Possible models for transposition

General Principles

Generalized recombination involves large regions of homologous DNA sequences, but site-specific recombination involves considerably smaller segments of DNA in which the recombination event occurs at a specific sequence (the recognition sequence). The essential components for site-specific recombination include enzyme(s) specific for the recognition sequence and two DNA duplexes, at least one of which carries that sequence. Under proper conditions, reciprocal (or nearly so) recombination occurs within or adjacent to the recognition sequence. If both DNA duplexes must carry the recognition sequence, the process is described as double site-specific recombination. The term single site-specific recombination has been used when only one DNA duplex is carrying the recognition sequence.

Double Site-Specific Recombination

Phage Lambda Integration

One especially well-studied example of double site-specific recombination is that of the integration of λ DNA to form a prophage. The λ integrase is the definitive member of a group of similar tyrosine recombinases and is the one to which the others are compared. The recognition sequence in this case is the *att* site, which was represented in Chapter 8 by the letters PP′. The corresponding region of the *Escherichia coli* genome is *att*λ, which was represented by the letters BB′. Analysis of these two regions by both genetic and heteroduplex techniques has led to the surprising conclusion that they are not homologous at all. Although the minimum size of the *att* site is about 240 bp, there is a small 7-bp sequence embedded within each *att* site that is the actual point at which the integrase enzyme acts to produce staggered cuts.

In recognition of this small, centrally located binding site, the terminology for designating the *att* sites has changed to POP′ and BOB′, where O designates the short homologous sequence (overlap region). The integration event produces prophage end points (BOP′ and POB′) that are slightly different from the original sequence. In fact, in vitro experiments have shown that integrase has difficulty binding to the right prophage end (POB′ or *att*R), and the role of the *xis* protein is to assist folding of the structure so that integrase can act. This situation would account for the genetic observations that only *int* function is necessary for integration but both *int* and *xis* are required during prophage excision unless the level of Int protein is high. In all cases, integration host factor (IHF), the product of the *himA* and *hip* genes, is necessary for normal activity. IHF also binds to cII DNA, but more weakly than to *att* DNA.

A general diagram for λ recombination at *att* is shown in Fig. 15.1. The phage *att* region is 234 bp and can be subdivided into P and P′ arms. The bacterial *att* region is only about 25 bp and is subdivided into B and B′ arms. Within

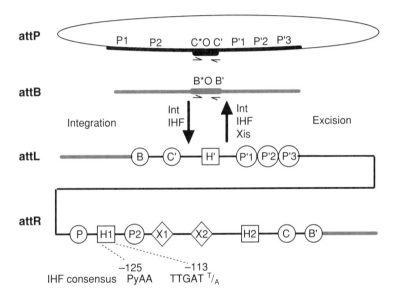

Figure 15.1. Lambda site-specific recombination. The *att* sites involved in the integration and excision reactions as well as the proteins required for each are shown. *Gray lines* are bacterial DNA, and *black lines* are phage DNA. The locations of the binding sites of the Int (O), IHF (□), and Xis (◊) have been deduced from footprinting experiments. There are two kinds of Int binding sites: those in the phage arm DNA (P) and two core sites (C). The IHF consensus sequence and coordinates relative to the center of the overlap region are shown under the sequences of the H1 sites. The *asterisks* indicate the side of the overlap region where the first DNA cuts are made. (Adapted from Thompson, J.F., Waechter-Brulla, D., Gumport, R.I., Gardner, J.F., Moitoso de Vargas, L., Landy, A. [1986]. Mutations in an integration host factor-binding site: Effect on lambda site-specific recombination and regulatory implications. *Journal of Bacteriology* 168: 1343–1351.)

each arm, footprinting experiments have identified specific sequences protected by the Int protein (identified by circles in Fig. 15.1): three in the P' arm, two in the P arm, two in the core region, and one each in the B and B' arms. IHF has three binding sites designated H in the figure, all within the phage DNA. Two of these sites flank the region of the phage DNA in which exchange occurs. Arthur Landy and his collaborators have shown that the H1 site appears to be of critical importance in regulating the integration event. Catalysis and cleavage of the core sites is a function of the carboxy-end of the integrase molecule (Tirumalai et al. 1998).

Rutkai et al. (2003) have examined the extent to which DNA pairing is important for the action of integrase. They deleted the normal bacterial *att* site and looked at the DNA sequences at which λ now managed to integrate. The leftmost portion of the overlap region was conserved in the new sites along with substantial similarity in the imperfect repeats of the flanking arms.

Kim and Landy (1992) have considered the problem of how the two ends of the prophage can find one another so that recombination can occur. The role of the additional proteins apparently is to provide appropriately bent DNA. IHF is a known DNA bending protein. If the concentration of Xis is limiting, the host Fis protein (see next section) can substitute. It too is a known DNA bending protein. The necessity for DNA bending arises from the fact that Int is a monomeric protein with two specific binding activities. The amino terminus binds with high affinity to a site in the arm, while the carboxy terminus binds with low affinity to the core site where strand exchange occurs. Therefore, when excision is to occur, each arm must be folded over a core site and the core sites juxtaposed so that exchange is possible.

Overall regulation of integration and excision is provided by several mechanisms. The end of the *xis* gene overlaps the p_1 promoter so that when Xis is being produced, Int expression is reduced. Furthermore, the weak binding of IHF to the cII region can occur only after the *att*P binding sites are all occupied. This requirement ensures that the phage DNA is ready to integrate before the temperate mode of transcription becomes predominant.

Circular Chromosome Segregation

As noted in Chapter 2, organisms with circular chromosomes must be prepared to deal with the possibility that replication or recombination might accidentally produce a single concatemeric molecule instead of two unit circles. Before chromosome segregation can occur, the chromosomes or low copy number plasmids must be monomeric (Fig. 2.10). The *E. coli* system that accomplishes this task is the *xerC* and *xerD* system. The two proteins show sequence similarities to the essential regions of the λ integrase protein. The actual site of exchange is *dif*, which is located in the terminus region between *terA* and *terC*. Mutations in *xerC*, *xerD*, or *dif* result in filamentous cells with aberrantly segregating chromosomes. The FtsK protein, a protein known to be required early in cell division, participates in the process. Li et al. (2003) bound a protein fused to the green fluorescent protein to the chromosome terminus of replication and tracked the position of the terminus using light micrography. Movement of the terminus to the center of the daughter cell occurred normally in *xerC* mutants but not in *ftsK* mutants.

Inversion Systems

Bacteria possess a variety of invertible segments whose orientation is controlled by DNA invertases (for a review, see Johnson [2002]). Invertases are enzymes that catalyze site-specific recombination between the ends of defined DNA segments. Normally, the invertible segment of DNA contains two oppositely oriented genes, only one of which is adjacent to a promoter. Inversion has the effect of turning on expression of one member of the paired genes while inactivating the other. Site-specific inversion systems are known to control flagellar proteins in *Salmonella*, fimbriae production in *E. coli*, and tail fiber production (host range determination) in phages P1 and Mu.

The Mu system is a particularly well-studied one. The invertible segment is bounded by two inverted *gix* sites, each consisting of two inversely oriented 12-bp half-sites separated by two asymmetric spacer bases. In the approximate middle of the invertible segment is a recombination enhancer element that functions independently of orientation or distance to the *gix* sites. For inversion to occur, host Fis protein (factor for inversion stimulation) must be present. Fis binds to the enhancer and causes synapsis with the two *gix* sites (Fig. 15.2). Gin protein then causes double-strand breaks at each *gix* site and catalyzes rotation of the two strands and resealing of recombined ends. The process can be processive, meaning that the reaction may occur more than once on the same DNA molecule. The result of such multiple reactions is a knotted DNA molecule.

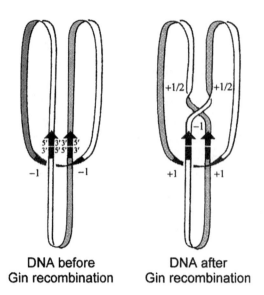

**DNA before
Gin recombination**

**DNA after
Gin recombination**

Figure 15.2. Ribbon diagrams of DNA illustrating changes caused by Gin recombination. A DNA substrate for Gin is represented schematically by a *ribbon* in which the edges are the complementary strands. One side of the ribbon is gray; the other is white. The *split arrows* indicate the inversely oriented *gix* sites bound by Gin, and the *darker box* at the bottom of the loop indicates the enhancer bound by Fis. The DNA substrates for Gin need to be (−) supercoiled. Synapsis of the two *gix* sites and the enhancer traps two (−) supercoils as shown on the *left*. The trapped crossings, or nodes, are indicated by −1. Recombination occurs via a double-strand break in each *gix* site, a 180° right-handed rotation of one pair of half-sites relative to the other and religation to generate the product diagrammed on the *right*. The rotation of the DNA creates one additional (−) node, while simultaneously overtwisting both sites by a half turn (indicated by +1/2); after deproteinization, these nodes cancel. (Reproduced with permission from Klippel, A., Kanaar, R., Kahmann, R., Cozzarelli, N.R. [1993]. Analysis of strand exchange and DNA binding of enhancer-independent Gin recombinase mutants. *The EMBO Journal* 12: 1047–1057.)

Integrons

Integrons are small pieces of modular DNA found in Bacteria. These have assumed enormous importance as it has been established that they represent a way for pathogenic organisms to share and/or exchange drug resistance genes. The basic integron codes for an integrase (*intI*) that is a member of the tyrosine recombinase superfamily whose primary member is λ integrase. Other members of the superfamily that have been discussed earlier include Cre (Phage P1) and XerCD. Like λ integrase, each *intI* gene has associated with it an *attI* site that is specific to the integrase (Collis et al. 2002), and this difference serves to distinguish different classes of integrons. The final element in an integron is a strong promoter near the *att* site. In order for the system to operate, there needs to be the equivalent of the *attB* site (Fig. 15.1), which in this case is the 59-be (59 base element) site found on gene cassettes. A gene cassette is one or more promoterless genes with an associated 59-be. As in the case of phage λ, the actual point of recombination is a core element within the *att* sites. In some cases, 59-be sites can recombine directly. The process is reminiscent of conjugative transposition because a circular intermediate is formed during capture of the mobile cassette.

While a single integron is a simple structure (Fig. 15.3), more complexity is possible. *Vibrio cholera*, for example, contains a super integron that is 126 kb long and contains 179 cassettes (Rowe-Magnus et al. 1999), most of which are inactive. Holmes et al. (2003) argue that the integrons represent a major element in genome evolution. They used PCR to sample the population of cassettes available to organisms and showed that there were both protein-encoding and noncoding cassettes to serve as the raw material for evolution (see also Chapter 17).

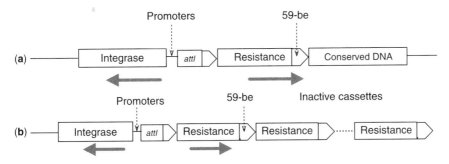

Figure 15.3. General structure of an integron. (a) An integron includes a gene for an integrase, an *att* site, and two divergent promoters. One promoter transcribes integrase while the other transcribes the cassette. The *gray arrows* show the directions of transcription. (b) In a super integron additional cassettes are present beyond the actively transcribed one.

Transposons: Single Site-Specific Recombination

Transposons are genetic elements that maintain their own integrity (i.e., their site of recombination is preserved) while integrating into a variety of sites on the target DNA (the target DNA site is not preserved). Included here are phenomena such as the highly promiscuous integration of phage Mu DNA, the integration of the R100 plasmid to form an Hfr, and the movement of various pure transposons (e.g., Tn*10*). However, all of these phenomena reduce to single site-specific recombination events catalyzed by the insertion elements bounding various transposons. For example, when R100 integrates, it usually loses its Tn*10* transposon, which means that the entire transfer region of the R plasmid as well as the antibiotic resistance genes is acting like a large transposon and hopping from one DNA molecule (the R100 plasmid) to another (the *E. coli* chromosome). To emphasize the size differences, the process has been called inverse transposition (i.e., Tn*10* stays where it is, and the rest of the DNA moves).

The molecular mechanisms proposed for transposition basically come down to two. Each involves a transposon located on a donor DNA molecule and a target site that may be located on the same or different DNA molecule. In one model, there is DNA replication involved in the process, and an intermediate cointegrate molecule is formed if the target site is on a separate molecule. A cointegrate structure is one in which two DNA molecules are fused into one. The other model requires no DNA replication as an intrinsic part of the transposition process. This mode of transposition was first suggested by Douglas Berg and is characteristic of several transposons, including Tn*5* and Tn*10*.

THINKING AHEAD

If a transposon moves by a nonreplicative mechanism, how do you get more transposons in the world?

Transposon Tn*10*

Tn*10* is the 9.3-kb transposon encoding tetracycline resistance that is found within the R100 plasmid. It carries inverted IS*10* elements at its boundaries. As detailed in Chapter 13, it has moved into and out of the DNA of a variety of bacteria and their phages and therefore must use an extremely versatile recombination system. A close examination of Tn*10* has revealed several important aspects of its behavior. As shown in Fig. 15.4, it can excise itself from a molecule precisely or imprecisely. It can also invert a region of DNA or delete a region of DNA. Prior to the discovery of transposons, these recombination events would have been classified as examples of illegitimate (nonhomologous) recombination. Lack of replication in this process was demonstrated by preparing λ phages

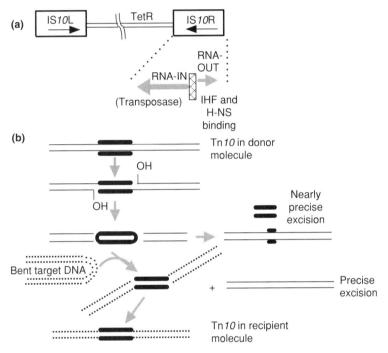

Figure 15.4. Conservative transposition of Tn*10*. (a) Tn*10* is composed of a central region coding for resistance to tetracycline with inverted repeat sequences at its ends. Each repeat is an IS*10* element, but only IS*10*R has functional transposase. The major IS*10* transcript is RNA-IN, which is named for its direction of synthesis. A minor transcript from the other DNA strand is RNA-OUT. The transcripts overlap at their 5′-ends. IHF is integration host factor. (b) When transposase is synthesized, it binds to the outer edges of IS*10* and makes a single nick. The hydroxyl group thus produced engages in a nucleophyllic attack on the opposite DNA strand catalyzed by transposase. The product is a hairpin loop. Transposase then cleaves the hairpin and uses the hydroxyl group to attack the recipient molecule. If the donor molecule is reassembled correctly, the process was precise excision. If portions of the transposon remain, the process was nearly precise excision.

carrying slightly different Tn*10* moieties. Strand separation and reannealing were used to make heteroduplexes differing at specific bases. After packaging and infection, when the products of transposition were examined, the transposon was still a heteroduplex. This could happen only if there were no replication as a part of transposition. Despite this observation, transpositions have been observed in which Tn*10* apparently remained where it was and also appeared in a new site. During transposition double-strand cuts occur at the ends of the transposon (the outer edge of the IS*10* elements) catalyzed by a *cis*-acting transposase enzyme. The IS*10* elements at the transposon ends are necessary and sufficient for transposition, implying that they code for the transposase. However, the two IS*10* elements are not identical. The left-hand element in the conventional genetic map is vestigial, as all but 13 bp at the tip can be deleted and still give transposition. Similar deletions of the right-hand element cut the frequency of transpositions by 90%.

The departing transposon leaves a double-strand gap in the donor DNA. The observed transposon duplication may be due to the nature of the repair process that acts on this gap. If there is no repair, the gap destroys the integrity of the donor DNA and the molecule cannot replicate. Proteins such as RecA are known to have the capability of joining the ends of broken DNA to repair damage such as that induced by x-rays. A similar phenomenon might occur in this case, and the result would be precise excision of the transposon. A final alternative would be recombinational repair in which case the missing transposon DNA would be replaced by transposon DNA looped from another DNA molecule and then copied. Note carefully that in this instance the replication is part of the recombination process and not a part of transposition. The net result would be the appearance of an additional copy of the transposon within the cell.

The transposition frequency for Tn*10* is about 10^{-4} per cell per generation when the cells are growing on a minimal medium. At a frequency of 10^{-5} or less, Tn*10* also promotes deletions or inversions and deletions. These rearrangements occur preferentially near the transposon itself, whereas transposition targets are located more or less randomly. The normal frequency of Tn*10* transposition is of the order of 10^{-6} to 10^{-7}. Regulation of this rate is obtained by several mechanisms. The major system involves the DNA adenine methylase (*dam*) DNA modification pathway. Transposase primarily acts on the ends of IS*10* when they are hemimethylated (i.e., immediately after replication).

A second regulatory mechanism for transposition involves synthesis of complementary RNA in a manner similar to that used to regulate copy number in R plasmids. The start of the transposase coding region in IS*10* is overlapped by a small RNA (RNA-OUT, Fig. 15.4) transcribed from an outwardly directed promoter called p-OUT. Inactivation of that promoter allows extra translation of the mRNA transcribed from p-IN (RNA-IN) and increases the transposition frequency. Transcription of the transposase gene from outside the transposon is prevented by a double-strand RNA region that sequesters the AUG codon of the transposase so that translation initiation would be difficult if not impossible. RNA-OUT by itself forms a stem-loop structure. However, the pairing between RNA-IN and RNA-OUT has more hydrogen bonds and is more stable than the intramolecular loop. Therefore, given the opportunity, RNA-IN will pair with RNA-OUT. Coexistence of multiple copies of Tn*10* within a single cell is difficult owing to a *trans*-acting repressor of transposase whose gene lies within IS*10*.

The only transposon-specific protein necessary for transposition is the transposase. However, cellular integration host factor (IHF) plays an important regulatory role, both positive and negative, with respect to transposition. Binding of IHF and transposase to the IS*10* element sharply bends the DNA, producing a **transpososome** loop—a stable complex of DNA and protein. The nucleoid DNA-binding protein H-NS is also necessary for normal transposition. Mutant cells can excise the transposon but cannot form the circular intermediate, so Swingle et al. (2004) suggest that its function is to stabilize the bent DNA.

The effect of IHF binding depends on the supercoiling state of the local DNA. If supercoiling is lacking, the effect is inhibitory, presumably on the grounds that there is already a problem with the DNA that transposition would only make worse. Kennedy et al. (1998) have shown that the transposase catalyzes four sequential reactions at each end of the transposon (Fig. 15.3): hydrolysis, transesterification by the hydroxyl group created in the first reaction, hydrolysis, and transesterification of the hydroxyl group. In the process, the enzyme creates a hairpin structure at each end of the transposon. The hairpin breaks as part of the reaction joining the end of the transposon to the new target DNA.

Transposon Tn*10* is, comparatively speaking, a specific transposon. An extensive analysis of the DNA sequence at its insertion sites has shown that there is some specificity involved. Approximately 85% of all insertions are found at the sequence NGCTNAGCN, where N represents any base. The sequence is symmetric, so there is no preferred orientation. Pribil et al. (2004) have shown that changes that make DNA bending easier (e.g., a nick) can compensate for alterations in the target sequence. After insertion of Tn*10*, there is duplication of the 9 bp forming the target site. This duplication presumably indicates that the enzyme making the incision in the target site does so with offset cuts in the manner of a type II restriction enzyme. The transposon ends are ligated to the offset cuts and the gaps filled in by DNA repair, generating the duplications (Fig. 15.4).

Phage Mu Transposition

Phage Mu can carry out transposition in either a replicative or a nonreplicative mode. The latter reaction is called the simple insertion mode and occurs during a new infection of a cell. It is a special case because the donor DNA is linear, whereas other types of transposition involve circular, supercoiled DNA molecules. Mu replication is via replicative transposition.

Consider first the replicative transposition mechanism. Initially the Mu DNA must be bent into the appropriate configuration. Tetramers of the transposase, product of gene A, bind cooperatively to three sites at each end of the phage DNA to generate a transposome. There is an enhancer sequence approximately 1 kb in from the left-hand end to which IHF binds, and a sharp bend in the DNA results. The net effect is to bring together the ends of the prophage. By itself, protein A is only 1% effective at transposition. It requires the presence of protein B for maximum effect.

The B protein is a DNA binding protein and ATPase that polymerizes onto DNA (Green and Mizuuchi 2002). In that state it assists in the selection of the target DNA. If it binds to donor DNA in the vicinity of protein A, the presence of protein A causes ATP hydrolysis and release of the B protein. Therefore, phage Mu does not transpose close to its original location. What protein B does accomplish is to bring together the target DNA and the donor DNA (Fig. 15.5).

Figure 15.5. Replicative transposition of phage Mu. (a) Overview of molecular movements. The Mu prophage (*thick gray line*), located in one DNA molecule, produces two proteins, A and B. The transposase A and host IHF bind to and bend the donor DNA. Meanwhile, the B proteins assemble on a target DNA and participate in binding of the target to the transposase. ATP hydrolysis occurs during the process, and then the B proteins are released. The transposase nicks both target and donor DNA. (b) Magnified view of the strand exchange. *Gray lines* are donor DNA and Mu prophage, *dark lines* are target DNA. The transposase makes offset cuts in both DNA molecules. The donor DNA separates and binds to the nicked ends of the prophage. Replication occurs primed by the free ends of the target DNA. The results are two copies of the prophage, duplication of target DNA sequences at the ends of the prophage, and formation of a cointegrate molecule (concatemer). Normal concatemer separating processes will resolve the monomeric DNA molecules.

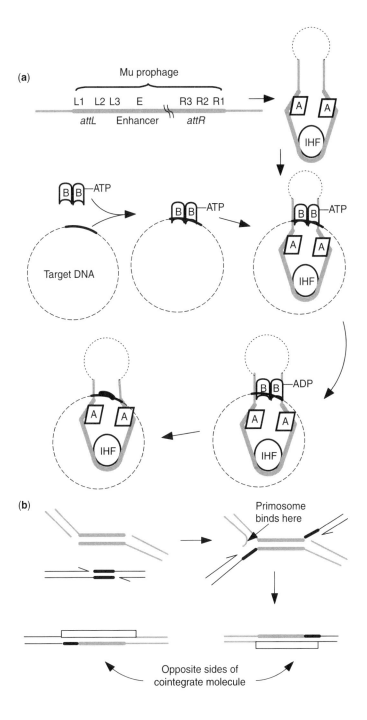

(a)

Mu prophage

L1 L2 L3   E   R3 R2 R1

attL   Enhancer   attR

Target DNA

(b)

Primosome
binds here

Opposite sides of
cointegrate molecule

In Fig. 15.5a, single-strand nicks are made at the ends of the transposon and five base offset nicks are made in the target region. As in the case of Tn*10*, it is the 3'-OH ends of the transposon that are first linked to the target. The strand transfer reaction requires one host protein, HU. Proper folding of the helices generates the strand transfer complex (STC, X-shaped structure in Fig. 15.5b) that is held together by ligating the transposon ends to the offset nicks in the target DNA. After ATP hydrolysis and exit of protein B, the ClpX chaperonin arrives to remodel the STC and disassemble the transpososomes.

The gaps remaining in the STC can serve as primers for DNA replication to yield the two structures shown in Fig. 15.5b (see the review by Nakai et al. [2001]). As the gaps of the structure are filled in, the typical 5-base repeat of the target DNA is generated. Note that if the original DNA molecules were circular, the left and right sides of each molecule were linked. Those linkages are not affected by the transposition, and so the final structure is then a single circle (cointegrate) containing two copies of Mu. A recombination event between the two copies of Mu then resolves the cointegrate into single circles each containing a recombinant Mu prophage.

To derive a simple insertion, only a slight modification must be made in the model. Instead of filling in the gaps in Fig. 15.5b, the remaining links of the transposon DNA to the donor molecule (links to gray lines) are degraded. A cointegrate structure then cannot develop. Instead, the donor DNA is left with a gap, as in the case of Tn*10*, and the target DNA (black lines) is left with the transposon plus one gap at each end of the transposon. A simple gap-filling reaction restores molecular integrity of the target.

Summary

Site-specific recombination concerns DNA duplexes that have little or no homology. If the recombination event always tends to occur at the same site on both duplexes, as in the case of λ Int or the *gin* inversion system, it is classified as double site-specific. If, on the other hand, it tends to occur at a specific site on one DNA duplex, but randomly on the other, as in the case of phage Mu or Tn*10*, it is said to be single site-specific. Each type of site-specific recombination has an appropriate protein or proteins associated with it. In the case of transposons, replication may or may not be an integral part of the transposition process. Integrons represent a way of expressing modular genes by recombining them next to an active promoter.

Questions for Review and Discussion

1. How would you determine whether a recombination process is generalized, double site-specific, or single site-specific?
2. All recombination processes begin with some sort of nicking event. Describe this initial event for each of the above processes.
3. How are each of the recombination processes completed?
4. What elements regulate the frequency with which each recombination process occurs?

References

General

Hallett, B., Sherratt, D.J. (1997). Transposition and site-specific recombination: Adapting DNA cut-and-paste mechanisms to a variety of genetic rearrangements. *FEMS Microbiology Reviews* 21: 157–178.

Johnson, R.C. (2002). Bacterial site-specific DNA inversion systems. pp. 230–271. In: Craig, N.L., Craigie, R., Gellert, M., Lambowitz, A.M. (eds.), *Mobile DNA II*. Washington, DC: American Society for Microbiology Press.

Nakai, H., Doseeva, V., Jones, J.M. (2001). Handoff from recombinase to replisome: Insights from transposition. *Proceedings of the National Academy of Sciences of the USA* 98: 8247–8254. (How phage Mu replicates its DNA.)

Rowe-Magnus, D.A., Guérout, A.-M., Mazel, D. (1999). Super-integrons. *Research in Microbiology* 150: 641-651.

Specialized

Chalmers, R., Guhathakurta, A., Benjamin, H., Kleckner, N. (1998). IHF modulation of Tn*10* transposition: Sensory transduction of supercoiling status via a proposed protein/DNA molecular spring. *Cell* 93: 897–908.

Collis, C.M., Kim, M.J., Stokes, H.W., Hall, R.M. (2002). Integron-encoded IntI integrases preferentially recognize the adjacent cognate *attI* site in recombination with a 59-be site. *Molecular Microbiology* 46: 1415–1427.

Deufel, A., Hermann, T., Kahmann, R., Muskhelishvili, G. (1997). Stimulation of DNA inversion by FIS: Evidence for enhance-independent contacts with the Gin–*gix* complex. *Nucleic Acids Research* 25: 3832–3839.

Gravel, A., Messier, N., Roy, P.H. (1998). Point mutations in the integron integrase Intl1 that affect recombination and/or substrate recognition. *Journal of Bacteriology* 180: 5437–5442.

Green, E.C., Mizuuchi, K. (2002). Dynamics of a protein polymer: The assembly and disassembly pathways of the MuB transposition target complex. *The EMBO Journal* 21: 1477–1486.

Holmes, A.J., Gillings, M.R., Nield, B.S., Mabbutt, B.C., Nevalainen, K.M.H., Stokes, H.W. (2003). The gene cassette metagenome is a basic resource for bacterial genome evolution. *Environmental Microbiology* 5: 383–394.

Jain, C. (1997). Models for pairing of IS*10* encoded antisense RNAs in vivo. *Journal of Theoretical Biology* 186: 431–439.

Kennedy, A.K., Guhathakurta, A., Kleckner, N., Haniford, D.B. (1998). Tn*10* transposition via a DNA hairpin intermediate. *Cell* 95: 125–134.

Li, Y., Youngren, B., Sergueev, K., Austin, S. (2003). Segregation of the *Escherichia coli* chromosome terminus. *Molecular Microbiology* 50: 825–834.

Pribil, P.A., Wardle, S.J., Haniford, D.B. (2004). Enhancement and rescue of target capture in Tn*10* transposition by site-specific modifications in target DNA. *Molecular Microbiology* 52: 1173–1186.

Rutkai, E., Dorgai, L., Sirot, R., Yagil, E., Weisberg, R.A. (2003). Analysis of insertion into secondary attachment sites by phage λ and by *int* mutants with altered recombination specificity. *Journal of Molecular Biology* 329: 983–996.

Tirumalai, R.S., Kwon, H.J., Cardente, E.H., Ellenberger, T., Landy, A. (1998). Recognition of core-type DNA sites by lambda integrase. *Journal of Molecular Biology* 279: 513–527.

Yamauchi, M., Baker, T.A. (1998). An ATP–ADP switch in MuB controls progression of the Mu transposition pathway. *The EMBO Journal* 17: 5509–5518.

16

Applied Bacterial Genetics

Bacterial genetics is passing through an interesting phase in its history. The new methodologies and concomitant understanding of molecular biology have given scientists new insights that are so far-reaching that it is impossible to estimate their ultimate impact. Bacterial genetics is becoming a tool for studying genetic systems other than those of bacteria and for developing processes applicable to specific industrial, ecological, pharmaceutical, or other problems (e.g., fermentation technologists and biochemists). This chapter presents some of the current trends in bacterial genetic research and some of their implications. Old and new technologies are included, for basic research in bacterial genetics does not cease but continues to reveal phenomena important to the understanding of life processes. The topics of this chapter should be viewed as summaries of subjects presently under investigation.

Major topics include:

- Special features that can be incorporated into cloning vectors
- Safety issues raised by DNA cloning experiments
- Some examples of applied DNA cloning

More Information About Recombinant DNA Technology

Chapter 2 presents a bare-bones outline of how to link disparate DNA molecules. The process is a fundamentally simple one, but as with so many other areas of science, the actual practice is more complicated than the theory. The following material provides additional information on the techniques used to join DNA molecules and some problems that can arise during the process. The following section considers how to design vectors to eliminate or ameliorate the problems.

Restriction Mapping

Restriction enzymes play a central role in linking DNA and can provide information that contributes to the construction of physical maps of DNA molecules. Restriction maps show the physical positions of specific recognition sequences on the DNA molecule of interest based on computer analysis of DNA sequence. A simple restriction map for phage λ is shown in Fig. 16.1. A more complex example of a restriction map can be found in Fig. 12.2. Restriction mapping is an excellent method for characterizing small DNA molecules quickly and accurately. If two types of DNA yield same-sized fragments for all enzymes tested, they may be identical, whereas if they yield different-sized fragments, they definitely have some DNA sequence heterogeneity. Presently, a number of laboratories are carrying out macrorestriction experiments in which entire bacterial genomes are cut with restriction enzymes whose recognition sites are comparatively rare to yield some 20–80 fragments that can be separated on pulsed field gel electrophoresis. These fragments can be thought of as a DNA fingerprint, and epidemiologists often use the technique to determine whether multiple outbreaks of a disease have a common source.

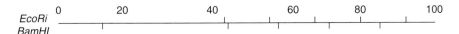

Figure 16.1. Restriction map for phage λ DNA. The *horizontal line* represents the phage DNA molecule. The numeric scale at the top represents distance as a percentage of the total genome. The short *vertical lines* represent the points of cleavage by either *Eco*RI or *Bam*HI. Note that the cleavages do not overlap. It should be remembered that the cohesive ends of λ may produce a circular molecule that would result in the two fragments at the ends of the DNA molecule merging into a single, larger fragment.

Sources of DNA for Linking

The discussion in Chapter 2 focused on linking DNA using simple restriction fragments. Several problems can arise. There may not be a restriction enzyme that cuts to yield a fragment carrying the entire region of interest. If the DNA is from a eukaryotic organism, it usually has one or more introns within it. Such DNA cannot yield a correctly processed RNA when cloned into a prokaryotic organism because the only RNA splicing known to occur in prokaryotes is self-splicing. Sometimes, the genetic map position for the desired DNA sequence has not yet been determined, and thus it is not possible to predict the nature of the fragment required for the cloning. Each of these problems has a reasonably straightforward solution, although in some cases considerably more work is involved.

When the map position of the appropriate DNA sequence is not known, a common solution is to produce a library of DNA clones. A library is prepared by taking the entire genome of the organism and breaking it up into more or less standard-sized pieces. It can be done mechanically by hydrodynamic shear or enzymatically by incomplete digestion with a restriction enzyme that has many recognition sites on the DNA (e.g., *Sau*3AI, whose recognition site is 5′GATC; Redenbach et al. [1996]). Both techniques generate a series of overlapping DNA fragments that can be sorted by size using various physical methods. One common method is to use a sucrose gradient, a centrifuge tube in which increasing concentrations of sucrose are present from top to bottom. DNA layered on the top of the gradient and subjected to centrifugation moves in the opposite manner as on an agarose gel (i.e., the largest fragments migrate the farthest). By choosing DNA from different regions of the sucrose gradient, it is possible to adjust the average size of the DNA fragments to any desired value.

The overhangs resulting from *Sau*3AI digestion (GATC) are perfectly complementary to *Bam*HI overhangs (see Table 2.2). Therefore, a vector can be cut with *Bam*HI, and *Sau*3AI-derived fragments ligated into it. One potential problem that can occur is self-closure of the vector in which the *Bam*HI ends are resealed without the addition of any *Sau*3AI fragment. Application of a phosphatase to the ends of the cut vector leaves it unable to ligate unless a DNA insert provides the missing phosphate groups.

It is possible to obtain linkers, small oligonucleotides that contain within their sequences a particular restriction site. These linkers can be blunt-end ligated to DNA fragments. In blunt-end ligation, high concentrations of DNA and ligase are used so that if the ends of DNA molecules happen to touch, they are immediately ligated. There is no formal mechanism for bringing the fragments into apposition. Once ligated, the linkers can be cut with the appropriate restriction enzyme to generate single-strand ends.

When the DNA sequence to be cloned contains introns, it is often simpler to clone a DNA sequence that corresponds to the processed mRNA molecule rather than to the actual coding region for the protein. The retroviral enzyme reverse transcriptase possesses two enzymatic activities: an RNA-dependent DNA polymerase that makes DNA from an RNA template and a ribonuclease H activity that can degrade RNA:DNA hybrids, mRNA, and poly A tails (Fig. 16.2). The product is cDNA, a double-strand DNA molecule containing one strand that is complementary to the mRNA. This DNA can then be used directly in cloning experiments if linkers are attached as above. The major problem with the technique is that the reverse transcriptase does not do well in vitro with long transcripts. It is often difficult to obtain cDNA molecules longer than 600 bp in length. One variation on this technique is to use RT-PCR in which the initial reaction is catalyzed by reverse transcriptase and then normal PCR amplifies the product.

PCR technology is also useful for a type of reverse genetics. There are cases where researchers have a protein of known function, but unknown gene location. They can sequence a bit of the protein and, from the amino acid sequence, predict what the DNA code must be. The prediction is not 100% accurate, of course, because of the degeneracy of the code. Careful choice of amino acid sequences for study will minimize this problem. Suitably degenerate primers can be made and used in PCR reactions. They will cause amplification of DNA that could code for the known protein. These sequences can be cloned and transcribed and translated in vitro to see if they produce the desired protein. In this way, a gene can be cloned without the knowledge of its actual physical

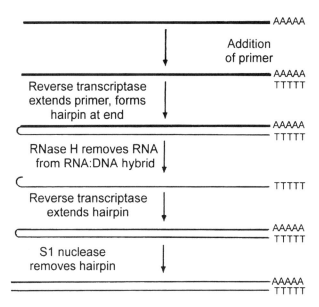

Figure 16.2. Production of a cDNA copy of an mRNA. In this example, the mRNA (*thick line*) is assumed to be a eukaryotic mRNA molecule that has a poly-A tail at its 3′-end. To begin DNA synthesis, a short primer of polydeoxythymidine hydrogen bonds to the poly-A tail. Reverse transcriptase then uses this primer to initiate and carry out the synthesis of a DNA strand (*thin line*) complementary to the mRNA base sequence. It is a characteristic of this enzyme that as it reaches the end of the mRNA molecule, it loops back on itself to form a continuous hairpin-like structure. The RNA that comprises one leg of the hairpin is removed by the RNase H activity, and the reverse transcriptase can complete synthesis of a double-strand DNA molecule. The loop of the hairpin is opened by the action of S1 nuclease to leave a double-strand DNA molecule that contains all of the sequences present in the original RNA molecule, including the poly-A tail.

location on the chromosome. Southern blotting of restriction fragments from the chromosome using the clone as a probe will identify an approximate physical position for the gene.

THINKING AHEAD

What are the potential risks in creating new combinations of genes?

Safety Considerations

The techniques of artificially linking genes together present interesting scientific and ethical problems because they require unusual standards of experimental control. Any DNA can be joined to any other DNA, regardless of species barriers or any other genetic impediments. Moreover, the linked DNA can be made compatible with essentially any organism by an appropriate choice of vector plasmid. In theory, then, any organism can be endowed with any genetic attribute. The researchers involved in the early work on linking genes were aware of the implications of this analysis, and they convened several meetings to discuss the issue. Finally, after an extensive review, the principal scientists involved in this type of research agreed to a compendium of safety guidelines that are now binding on anyone receiving research grants from the U.S. government.

The regulations address two broad categories of risk. The first potential risk is intrinsic in the host–vector combination. The presence of a modified plasmid might confer new and surprising properties such as unusual toxin production on a cell. Such cells would be dangerous both to laboratory workers and to the surrounding population. Therefore, the problem posed by this type of potential hazard is one of physical containment—keeping the plasmid inside the laboratory and in appropriate culture vessels.

The second type of potential biohazard is possible escape of the cloned DNA from its original host into some sort of laboratory contaminant. The plasmid might be perfectly harmless in the bacterium in which it was originally cloned, but might confer unwanted antibiotic resistance, pathogenicity, and so forth, on other bacteria. In this case, the problem posed is one of biological containment—keeping the plasmid in the appropriate host organism.

At present, the Recombinant DNA Advisory Committee (RAC) advises the Director of the National Institutes of Health about necessary changes in the Guidelines. The general tendency has been for the Guidelines to be relaxed as it becomes clearer that the potential hazards of recombinant DNA have been overrated. Despite these changes, the basic philosophy of the Guidelines has remained unaltered. Physical containment is rated on a scale of 1–4, with a biosafety level 1 (BSL1) facility being the normal microbiology laboratory, and a BSL4 facility being a laboratory suitable for work with the most dangerous pathogens. In such a laboratory, all air, liquids, and solids leaving the facility are sterilized, and all personnel wear airtight suits with separate air supply. After removing the suits, they shower and change clothes before leaving the facility. Biological containment is rated on a scale of 1–3, with the viability of the host cell and the safety of the plasmid considered together.

An HV1 system provides moderate containment using standard laboratory strains of bacteria and suitable plasmids. An HV2 system provides high-level containment due to weakening of the host strain by the introduction of appropriate mutations that greatly reduce its ability to grow under normally encountered environmental conditions. There is also a deliberate alteration of the vector so that so that it does not replicate well in other host bacteria, in case it escapes biologic confinement. The general survival rate for either host or vector outside the laboratory must be shown to be less than 10^{-8}. No HV3 systems have been constructed. Table 16.1 illustrates the application of the regulations to specific cases. Guidelines for large-scale (i.e., industrial) production of organisms carrying cloned DNA also exist.

For the specific case of *Escherichia coli*, the equivalent of an HV1 system is EK1, and the equivalent of an HV2 system is EK2. Roy Curtiss and coworkers constructed a suitable host strain for an EK2 system and designated it χ1776.

Table 16.1. Examples of the types of containment required for recombinant DNA experiments under NIH guidelines.

Type of Donor DNA	Type of Physical Containment[a]
Escherichia. coli	Exempt (BSL1)[b]
S. enterica serovar Typhimurium	Exempt (BSL1)
B. subtilis	Exempt (BSL1)
Drosophila	Exempt (BSL1)
Hepatitis B virus	BSL2
Marburg virus	
Complete genome	BSL4
Noninfectious segment	BSL2

Note: In each case the host bacterium is assumed to be *E. coli*.
[a]BSL1 containment represents the implementation of normal, safe microbiologic techniques; BSL2 containment requires that, in addition to BSL1 techniques, all experiments be carried out in a special biohazard laboratory with restricted access; BSL3 containment requires all of the above plus installation of double doors on the facility and the use of special containment hoods; BSL4 containment requires the use of a special facility equipped with negative air pressure, equipment to prevent direct contact of workers with the material, and sterilization of all materials before removal from the facility.
[b]This type of experiment is not regulated but good laboratory practice would be to work at the BSL1 level.

Careful tests have shown that this strain does not colonize the intestinal tract of test animals even if given in massive doses. An HV2 system is also available in *Bacillus subtilis*.

Jechlinger et al. (1998) have proposed an interesting solution to the problem of accidental release of genetically engineered organisms into the environment by preparing a cascade of repressors. They have taken the φX174 gene E (causes host cell lysis) and cloned it into a vector carrying the phage λ promoter and a temperature-sensitive repressor. The phage λ promoter controls synthesis of a second repressor, either phage 434 cI or *E. coli lacI*. Gene E is under the control of the second repressor. Under normal laboratory conditions (high temperature), the lambda cI gene product is inactive, meaning that the second repressor is synthesized and keeps gene E turned off. In temperate climates, the ambient temperature is sufficiently low that the λ repressor becomes active, shutting off synthesis of the second repressor and allowing gene E expression. The escaped cells will then lyse.

Torres et al. (2003) have taken a similar approach to the problem of accidental genetic transfer from the normal host of the cloned DNA to an accidental host. They have developed a host–vector system that incorporates both the colicin E3 and *Eco*RI genes. The immunity proteins are located on the chromosome, but the toxins are located on the plasmid flanking the cloned DNA. If the plasmid were to accidentally transfer, the toxins would promptly kill the recipient cell. The system reduces transfer by 8 orders of magnitude.

The presumed barrier to genetic transfer between prokaryotic and eukaryotic organisms is not absolute. For example, there is the Ti plasmid discussed in Chapter 12 that transfers T-DNA from bacterium to plant. The seeming identity of introns from phage T4 with those from eukaryotic mitochondria also suggests a connection between the two kingdoms. Furthermore, Carlson and Chelm (1986) reported that glutamine synthetase II from *Bradyrhizobium japonicum*, a bacterium commonly found in association with legumes, has no introns, but otherwise corresponds well to the eukaryotic enzymes with respect to amino acid sequence homology and lack of posttranscriptional modification. Brownlee (1986) suggested that such a situation could arise if a eukaryotic pseudogene (a nonfunctional region of DNA with extensive homology to an expressed DNA) were somehow transferred into the bacterium, implying that eukaryotic cells do not have a monopoly on the production of certain types of proteins. Supporting this view is an old observation that a particular strain of *E. coli* produced a protein with insulinlike activity. On the other hand, Nielsen et al. (1998) conclude that while there is some evidence that gene transfer from plant to bacterium has occurred, there is no solid evidence that the transfer occurs more often than once every million years or so.

Much of the present controversy about genetically engineered organisms revolves around the deliberate release of such organisms into the environment. The U.S. Environmental Protection Agency attempts to define suitable precautions for such activities, and genetically engineered plants, animals, and bacteria are now part of the landscape. Significant discussions are under way in many countries, especially the European Union, over labeling requirements for farm products from genetically engineered sources.

Another area of controversy concerns the type of alteration in the organism proposed for release. How much of an alteration is necessary to make a particular organism hazardous in the environment? In the specific case of *Pseudomonas syringae*, what is wanted is not a new function but, rather, a loss of function. *P. syringae* codes for a protein that causes water to freeze more readily (at a higher temperature than normal). Wild-type bacteria have been killed and mixed with snow making solutions used on ski slopes to provide a more solid product. However, the protein is also responsible for extensive crop damage when the bacterium grows on leaves. There is a spontaneous deletion mutant that lacks the ability to produce the ice protein and thus does not damage crops. Considerable discussion ensued over potential hazards to the environment before initial field trials were authorized to see if the mutated bacterium would displace the wild-type bacterium from crop plants on which it was sprayed. Rabino (1998) has surveyed a group of scientists involved in commercial genetic engineering and has summarized their concerns about various regulatory initiatives.

One way to satisfy the various controversies is to attempt to track genetically engineered organisms in the environment. This is obviously easy with sheep and corn plants, but less so with bacteria. There is a thriving cottage industry developing genetically marked strains that can be used to track the fate of microorganisms after their release from the laboratory. The trick here is to find markers that are not based on therapeutically useful antibiotics as is usually done in the research lab.

Difficulties with the Expression of Cloned DNA

One reason it has been possible to relax the guidelines is that it is much more difficult to express foreign DNA sequences in a bacterial host than was expected. There are no particular problems when DNA is transferred from one species to the same species and often none when DNA is transferred among members of the same genus, barring the presence of distinct restriction and modification systems. However, greater taxonomic distance often leads to difficulties that are not necessarily symmetric. For example, DNA from *B. subtilis* can be readily cloned and expressed in *E. coli*, but DNA from *E. coli* is not well expressed in *Bacillus*. The problem lies in the nature of the *Bacillus* promoter. Whereas *E. coli* does not require a specific base sequence outside of the −10 and −35 regions, *B. subtilis* has a requirement for a specific 8- or 9-base sequence upstream of the −10 site. Oddly enough, a particular fragment of phage T5 DNA carrying a promoter is always expressed in *B. subtilis*, but is expressed in *E. coli* only during late infection (i.e., after modification of the host RNA polymerase). Along the same lines, bacterial DNA carried by the Ti plasmid is not well expressed in plants unless an appropriate plant promoter is fused to it.

The introns discussed earlier provide a major block to expression of eukaryotic DNA in prokaryotes. It is also now known that the genetic code is not universal. In yeast, there is widespread use of UGA to code for tryptophan. The ciliated protozoans have been shown to use the codons UAA and UAG to code for glutamine. Obviously, cloned DNA carrying any translated terminator codons could be properly expressed only in bacterial cells carrying an appropriate suppressor mutation.

Another potential problem with expression is the opposite of those discussed earlier. In many cases, the eukaryotic protein is detrimental, perhaps even lethal, to the producing bacterial cell. This sort of effect provides a strong selection against maintenance of the desired clone. As discussed later, the obvious solution is to provide a regulatory mechanism so that the protein encoded by the clone is not produced until just before the experimenter is ready to harvest the cells from the culture.

Cloning Vectors

Cloning vectors must certainly provide the essential replication functions for the cloned DNA, but they can be used to fulfill other functions as well. They can be constructed so as to provide an additional safety margin for the containment of DNA constructs, and they can be used to express the cloned DNA. Vectors can be derived from simple plasmids or from phages. The minimal requirements for a vector are a replication origin, a partition function, and a selectable trait so that the genetic transformants can be readily identified regardless of the expression of the cloned DNA. Usable insertion sites are defined by those restriction enzyme recognition sequences that are present in single copy or, if double, remove only a small, nonessential portion of the plasmid. Each plasmid is identified by a lowercase "p" followed by two or more capital letters indicating the laboratory in which it was prepared and a unique number.

Cloning vectors derived from phages offer the investigator an automatic delivery system in which the DNA can be isolated already packaged into a phage coat and thereby protected from environmental influences. They circumvent the difficulties that often arise when trying to transform cells with plasmids (see Chapter 10). Two of the most commonly used phages in this regard are λ and M13. One of the major constraints on λ cloning systems is the size of the insert. Like most phages, λ does not have a variable capsid size and thus must sacrifice some phage DNA to accommodate other DNA. Clearly, if plaque-forming viruses are to be assembled, the cloned DNA must be kept small. M13 is somewhat less critical in this regard, because as a rod-shaped virus, its length is determined solely by the length of the DNA to be encapsidated. Practically speaking, DNA that is more than double the normal M13 length is undesirable and tends to shorten spontaneously by deletion. Often, the phage is modified to give it a small DNA segment coding for a selectable trait to allow for rapid detection of lysogens.

Phage λ Vectors

There are numerous cloning vectors derived from λ. Fred Blattner and coworkers prepared a series of Charon phages (named for the ferryman of the river Styx) that carry deletions from the *b*2 region through *cI* to allow space for cloned DNA. They also carry the *lacZ* gene to provide a selectable marker. Other variations on this theme are the EMBL3 and EMBL4 phages prepared by Noreen Murray and coworkers in the European Molecular Biology Laboratory. Like Charon phages, these phages are replacement vectors in which a "stuffer fragment" of DNA is used to make the phage DNA large enough to package. Removal of the fragment then makes any self-closed vectors nonviable. Moreover, the fragment carries *red* and *gam*, thereby conferring the *spi*+ phenotype (sensitivity to phage P2 inhibition). If the cloning host is a P2 lysogen, unmodified vectors cannot grow and are lost again. These two properties make it highly probable that any phage that grows on a P2-lysogenic host contains inserted DNA.

Another type of vector is the cosmid or cohesive end-containing vector. It is a standard plasmid that includes the λ cos site as a part of its DNA. After appropriate cutting and ligating experiments have been carried out in vitro, the DNA can be packaged into λ virions by cell extracts prepared from infected cells that contain the appropriate enzymes plus empty heads and tails to yield phage particles that are infectious but contain no λ DNA except for the *cos* site. The virions serve as an efficient delivery system and avoid the complications inherent in the transformation procedure for *E. coli*.

Phage M13 Vectors

Joachim Messing and collaborators developed a series of M13mp phages (more than 19 are available) that are particularly advantageous for cloning DNA to yield either a double-strand or a single-strand product. The phages are prepared in pairs (e.g., M13mp18 and M13mp19) that contain identical loci (Fig. 16.3). A portion of the amino-terminus of β-galactosidase is included in each phage for identification purposes. When mixed with a host cell protein derived from a particular deletion of the amino-terminus of β-galactosidase, the combination of the peptide from the phage (the α peptide) and the peptide from the bacterium yields a low level of β-galactosidase activity (α-complementation). When such phages are plated on the appropriate indicator strain in the presence of X-gal (5-bromo, 4-chloro, 3-indoyl-β-D-galactoside), they hydrolyze

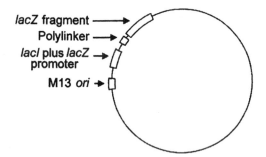

Figure 16.3. General design of the M13mp series phages. Note the presence of the α peptide region of *lacZ* and the location of the primer-complementary region just upstream of the polylinker. The unlabeled region contains the information for the phage proteins. One member of this group, M13mp18, is 7250-bp long with a 54-bp polylinker.

the indicator to yield the intensely blue indole derivative and hence, a blue plaque. Located between the promoter for the β-galactosidase fragment and the coding region is a polylinker that contains multiple unique restriction sites. DNA cloned into any one of these sites prevents α-complementation and results in a colorless plaque. Each member of a pair of phages contains the polylinker in opposite orientation.

During infection of the host cell by the phage and its associated insert DNA, the DNA replication process described in Chapter 7 occurs. This process yields single-strand DNA for packaging or extrusion into the medium.

In terms of the critical component for the M13 cloning vectors, it is the intergenic region (see Fig. 7.8) that contains the origin of replication for the rolling circle. Messing has also prepared a series of pUC plasmids that are combinations of the M13 phage origin and the *lacZ* region, together with portions of pBR322, one of the earliest cloning vectors. The plasmid can serve as a standard cloning vector; but when single-strand DNA is needed, the M13 origin can be used to provide it. Rolling circle replication is triggered by an M13 infection of the cell carrying the plasmid.

Expression Vectors

In expression vectors, the cloning site is located just downstream from a promoter and ribosome-binding site; therefore, the cloned DNA need not include either structure. The promoter may be any one of the standard promoters already discussed or a composite. For example, if an experimenter wants the cloned DNA to be regulated to prevent inhibition of host cell growth, the *trp*, *lac*, or λ promoters might do well. Of course, the appropriate repressor molecule must be supplied as a regulator. Its gene can be located on the same DNA molecule, on another plasmid, or on the bacterial chromosome itself.

However, unless a low copy number plasmid is used for the cloning vector, a chromosomal repressor locus usually cannot synthesize enough product to fully repress the plasmid transcription. High levels of cloned DNA transcription can be obtained by mutating whatever promoter is used to bring it closer to the ideal sequence. The *tac* promoter, a combination of the −35 region of the trp promoter and the −10 region of the *lac* promoter, has a high efficiency of expression, as do many conventional phage promoters. The phage T5 promoter is capable of outcompeting all *E. coli* promoters in the host cell, thereby yielding large amounts of product.

One contemporary plasmid vector, pMAL™-p2, is depicted in Fig. 16.4. This vector is based on a ColE1 plasmid and was constructed by a commercial firm. The replication functions are located in the lower left quadrant of the map in Fig. 16.4. There are two replication origins, one from ColE1 and one from phage M13. Therefore, the plasmid has the property from the ColE1 plasmid of being amplifiable by adding chloramphenicol to the medium, and single-strand DNA will result if the M13 origin is stimulated by an appropriate second plasmid. Plasmid presence is detectable either by the ampicillin resistance marker it carries or by α-complementation from the *lacZ* fragment that is fused to the MalE protein. Expression of the fusion protein depends on the *tac* promoter that is regulated by *lacI^q*, an overproducing repressor mutation. Many restriction sites are indicated on the map, but normally cloning is done into the polylinker site shown in magnified view underneath the map. Although the plasmid is mobilizable by the F plasmid, it is nonconjugative in isolation and thus provides additional containment for the cloned DNA.

This vector is intended for use in expressing cloned genes and obtaining purified proteins. If a gene is cloned into the *XmnI* site, a fusion protein results that combines MalE with the cloned gene. The terminator from the cloned gene prevents synthesis of the *lac* gene fragment. Therefore colonies of cells carrying just plasmid will have some β-galactosidase activity and turn blue in the presence of X-gal. Cells carrying cloned DNA will retain the normal color (cream). The *tac* promoter is so strong that when IPTG is added to the medium, up to 30% of cellular protein will be the MalE fusion protein. Prior to addition of the inducer, the host cell is protected from any detrimental effects of the cloned gene because it is not expressed.

To purify the cloned protein, an experimenter first isolates fusion proteins by affinity chromatography, passing a cell lysate over a column containing amylose to which MalE specifically binds. Adding maltose (which binds better to MalE than does amylose) to the column releases the now purified fusion protein. Superimposed on the *XmnI* cloning site is a protease Xa site. Treatment of the fusion protein with this protease releases the pure cloned protein from MalE.

Particular attention has been paid to the phages that, like T7, code for their own RNA polymerases. They include T3, T7, and the *Salmonella* phage SP6. Each of the polymerases recognizes its own particular promoters and ignores those of other phages or the host bacterium. The differences in the promoters are primarily at base -11. This specificity can be turned into an advantage as in the case of a clone prepared by Tabor and Richardson (1985). In one case, pBR322 was used to carry the T7 polymerase gene under the control of the $\lambda\, p_L$ promoter. The cI repressor function was provided in *trans* and prevented syn-

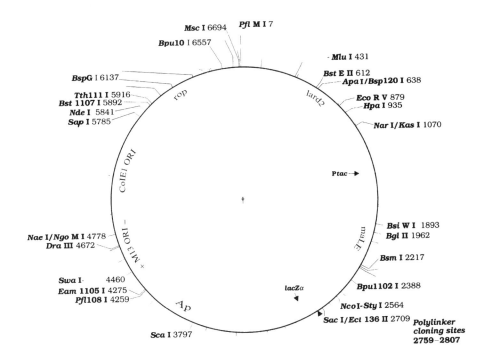

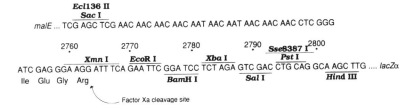

Figure 16.4. pMAL™–p2 cloning vector. This *E. coli* cloning vector is designed to create fusions between a cloned gene and the *E. coli malE* gene, which codes for malt- ose binding protein (MBP). The MBP fusion can then be expressed and purified, tak- ing advantage of the properties of MBP. The vector contains the inducible P$_{tac}$ promoter, positioned to transcribe a *malE–lacZα* gene fusion. The *lacI^q* gene encodes the Lac repressor that turns off transcription from P$_{tac}$ until isopropylthio-β-D-galac- topyranoside is added. The polylinker provides restriction endonuclease sites to insert the gene of interest, fusing it to the *malE* gene. A portion of the *rrnB* operon containing two terminators prevents transcription originating from P$_{tac}$ from inter- fering with plasmid functions. The gene for β-lactamase (Ap) and the origin of repli- cation are from pBR322. Nucleotide numbering starts at the beginning of the *lacI^q* fragment. The polylinker is shown beneath the map. The coordinates refer to the position of the 5′ base in each recognition sequence. The map also shows the rela- tive positions of the coding sequences and the origin of replication.

thesis of the T7 polymerase until the repressor was inactivated. At that time the T7 RNA polymerase was synthesized and caused the specific expression of DNA cloned downstream from a T7 promoter. When it is not certain which strand of cloned DNA needs to be transcribed, a combined T7/SP6 vector is often useful. The cloning site is flanked by inwardly directed promoters, T7 at one end and SP6 at the other. The DNA strand that is transcribed is controlled by the type of phage RNA polymerase that is added to the system.

Sometimes a particular mRNA transcript made from cloned DNA is poorly translated in the host cell. This lack of efficiency may be due to an increased susceptibility to degradative enzymes or to the use of rare codons in the mRNA. The only solution for the latter problem is either to change the host cells to those that have a more suitable pattern of codon usage, or else to use site-specific mutagenesis to change the codons themselves. In the case of the former problem, changing the structure of the mRNA molecule itself can often prevent excessive degradation. In particular, mRNA stability can be enhanced by adding sequences onto the end of the cloned DNA so that a suitable hairpin loop is formed. The loop prevents access of the exonuclease to the end of the mRNA. In *E. coli* such sequences are called REP.

Broad Host Range Vectors

As discussed in Chapter 10, certain plasmids are capable of existing in a variety of hosts. These plasmids make advantageous vectors for DNA that is to be tested in several hosts to observe its effects on metabolism. For best results, promoters must be supplied that function in all cells that are to be tested. A related entity is the shuttle vector, which carries replication origins for two different plasmids on the same vector. It allows the experimenter to carry out preliminary manipulations in a well-characterized system such as *E. coli* and then transfer the resultant clone to the organism of interest. If the clone is to be expressed in both the hosts, it may be necessary to provide two promoters as well.

Runaway Replication Vectors

Cloning vectors are under the same plasmid controls for copy number and partitioning that were discussed in Chapter 13. It is sometimes advantageous to let gene amplification take place so as to augment production of a particular product or provide large amounts of the cloned DNA from a relatively small culture volume. Amplification can be accomplished by suitably mutating the copy number control system so that a conditional mutation, usually temperature-sensitive, is in place. When the culture temperature is raised, all copy number control is lost, and the plasmid can replicate to high levels, perhaps 1000 copies or more per cell. The effect is to kill the host cell, but that is not important because the next step in the procedure is to extract the product from the cells in any case.

Artificial Chromosomes

In this era of whole genome sequencing on a massive scale, there is a real demand for cloning vectors that can handle very large inserts, of the order of 100–300 kb. Multicopy vectors would not be suitable because of the enormous drain on cell resources caused by the huge DNA inserts. Artificial chromosome vectors have been constructed based on plasmids from yeast, phage P1, or *E. coli*. The general pattern of design is similar in all cases and therefore only one vector will be considered in detail.

Al-Hasani et al. (2003) have prepared a new bacterial artificial chromosome whose design is intended to simplify cloning and expression of DNA from eukaryotes. The basic design of the vector is shown in Fig. 16.5. The vector gets

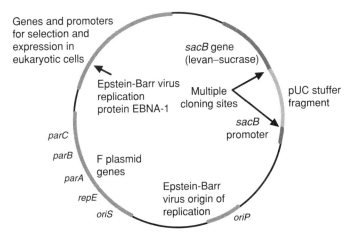

Figure 16.5. Stylized view of the pEBAC190G bacterial artificial chromosome vector of Al-Hasani et al. (2003). This vector can replicate either in *E. coli* using the F plasmid genes or in eukaryotic cells using the Epstein-Barr virus genes. If the levan–sucrase gene activates, it is lethal to *E. coli* cells growing on sucrose.

its replication and partition functions from the F plasmid and thus its very low copy number.

The selection for inserts is based on the levan–sucrase gene *sacB*. If this gene is expressed when *E. coli* is growing on sucrose, the SacB protein converts the sucrose to levan, which is toxic for the cell. The gene is normally held to a low level of expression by the pUC19 stuffer fragment inserted between the *sacB* promoter and its gene. When the vector is used for cloning, this fragment is cut multiple times by a restriction endonuclease and is no longer available. If the vector self-closes, the host cell will be rapidly poisoned because of the high level of *sacB* expression. If DNA is inserted into the multiple cloning site, the host cell will be safe.

The vector also provides the Epstein-Barr virus origin of replication and replication protein. These two components allow the vector to replicate in eukaryotic cells. Under those conditions, there are promoters and genes such as thymidine kinase for selection of phenotypes in eukaryotic cells.

Applications of DNA Cloning

Peptide Display Systems

Immunology researchers have long sought convenient, cheap sources of purified antibodies. The difficulties lie not only in the purification of the individual, relatively similar proteins, but also in making sure that all possible antibodies are present and available for purification. Some creative thinking about the Ff group of phages has led to a remarkable technique for genetic engineering of antibodies.

Instead of working with monoclonal antibody preparations, mRNA from white blood cells was used to prepare cDNA libraries that would include portions of the variable and constant regions of light and heavy chain antibodies.

The PCR technique allowed experimenters to amplify appropriate portions of the library by choosing the necessary primers. It is possible to actually join heavy and light chain variable regions by modified PCR reactions and then clone them to make fusion proteins with gene product 3 of fd phage. This protein occurs at the tip of the rod-shaped phage, and thus the cloned variable region is readily accessible to the antigen. Phage stocks represent combinatorial libraries of antibodies, the combinations developed during the random PCR reactions. Enrichment of phages expressing appropriate variable regions is accomplished by passing the phage stock over a short affinity column containing the antigen of interest, a process called **biopanning**.

There is a very interesting alternative method for a more directed preparation of combinatorial antibody libraries. It involves taking advantage of the *loxP*–Cre site-specific recombination system and creating two cloning vectors. One is an fd phage vector and the other is a plasmid vector. Both vectors carry unique antibiotic resistance genes as well as a normal *loxP* site and one mutated in such a way that the two sites are incapable of recombination within a given vector. However, they can recombine with their corresponding site on the other vector when a phage infection provides Cre protein to their host cell. The recombination events generate cointegrate structures (fusions of the two vectors). Because the normal and mutated *loxP* sites flank the DNA coding for the antibody heavy chains, new combinations of heavy and light antibody chains arise from the recombination. Reassorted clones are identifiable by new combinations of antibiotic resistance markers.

Ren and Black (1998) have extended the technique even further by showing that they could create a similar system in phage T4. They created one phage carrying an IgG anti-egg white lysozyme antibody and another displaying the HIV-1 CD4 receptor. From these observations, they concluded that the proteins were folded into their native conformation.

In an interesting reversal of the process, Etz et al. (2001) have shown that bacterial phage receptor proteins can be used to display peptides on the bacterial cell surface. The peptides must be inserted into the receptor protein at a point where the protein loops to return through the membrane. Depending on the receptor protein used, insertions of 40–86 amino acids were possible without loss of function.

Laboratory Reagents

The enzymes involved in the cloning process are specialized and were originally expensive. In many cases, their cost has been substantially reduced by the simple expedient of cloning the appropriate DNA sequences. The expression vectors described earlier can provide large quantities of the protein products of cloned genes. In this way, the availability of the Klenow fragment of DNA polymerase I (used for DNA synthesis when error correction is unwanted), the RecA protein, and numerous restriction enzymes has been greatly increased. Proteins used in industry are also good candidates for cloning. The enzyme renin, used in the cheese industry, has been successfully cloned. The manufacturers hope that this new source of supply will free them from the highly variable present market in which the enzyme must be extracted from calf stomachs.

APPLICATIONS BOX

How can you use bacteriophages to effectively treat human diseases?

Phage as Therapeutic Agents

The idea of using lytic phage as therapeutic agents is an old one and was immortalized in the novel *Arrowsmith* in 1925. While they never had much use in the United States owing to the successful development of antibiotics, phages have been used for decades in Eastern Europe as a treatment for a variety of diseases (see http://www.mansfield.ohio-state.edu/~sabedon/ for an extensive bibliography). The concern, however, has always been that mutations to phage resistance will undermine the treatment. Westwater et al. (2003) have developed a new system using the leaky phage M13, which has therapeutic potential. They took the M13 phagemid and added genes coding for addiction proteins (Gef and ChpBK). The normal host, of course, carries the immunity genes, but the pathogens should not and should therefore be killed by phage infection. Note that the usefulness of any phage therapy is limited by the host range of the virus in question.

Other Applications of Bacterial Genetics

DNA cloning is not the only possible use for bacterial genetics. Important as cloning is, it is only one component of a broader picture. The discussion that follows presents some of the ways in which the information from the preceding chapters can be applied to solve specific problems.

Mutagenicity Testing

An interesting new role for bacteria was pioneered by Bruce Ames and coworkers. They were concerned with the problem of detecting carcinogenic (cancer-causing) substances within the environment and screening chemicals for potential carcinogenic effects prior to commercial use. The traditional method has been to test suspected carcinogens in animals. Massive doses of chemicals are used, long periods of time are required, and maintaining the animals is both expensive and unpopular with a substantial segment of the population. For these reasons, animal testing is not well suited for routine screening, although it is essential as a confirmatory test. What is needed is an inexpensive, quick test that can distinguish the chemicals that have no effect from those that may have an effect. This need is fulfilled by mutagenicity testing, or Ames testing.

The theoretical basis for the test lies on the observation that nearly all proved carcinogens that act directly (i.e., attack the DNA, in contrast to indirect hormonal action) are also mutagens. It should, however, be noted that the converse has not been proved, and not all mutagens are known carcinogens. Therefore, when considering how to screen for carcinogens, Ames proposed that initial mutagenicity testing might be helpful. The chemicals not shown to be mutagenic would be considered noncarcinogenic (unless suspected of hormonal activity), and those shown to be mutagenic would be subjected to further testing.

Ames and coworkers set out to construct a bacterial system suitable for rapidly screening chemicals for mutagenic properties. The most convenient bacterial genetic systems involve positive selection for the desired event, so the type of mutation they chose to study was reversion, the reacquisition of a genetically controlled trait. Specifically, they chose to study the reversion of histidine mutations in *Salmonella enterica serovar* Typhimurium. No single strain would be suitable for detecting all types of mutations, but by assembling a collection of mutant strains that rarely revert spontaneously, they were able to detect frameshifts, amino acid substitutions, and reversion of nonsense mutations. A similar collection of tryptophan mutants has been prepared with *E. coli*. Experiments have also been performed using induction of a λ lysogen as an indicator of DNA damage. In all cases, increased sensitivity can be obtained by working with DNA repair-deficient strains. Various tissue culture cell lines can be used to test the effect of chemicals on eukaryotic cells. Regardless of the type of system used, the basic technique is the same.

It is easy to demonstrate the effect of a known mutagen. For the *Salmonella* strain, a plate containing all required nutrients except histidine is spread with a lawn of bacteria. A small amount of the mutagen is placed on the center of the plate and allowed to diffuse. Generally, the high concentration of the chemical kills all the bacteria in its immediate vicinity; but as the chemical is diluted by diffusion, cells can grow. The mutagen allows more colonies than expected to grow in this zone, and the increase in the number of colonies observed is a rough indication of the mutagenicity (and hence potential carcinogenicity) of the tested chemical.

In the formal mutagenicity test, the cells are treated with carefully measured doses of chemicals and then plated on selective medium or tested for induction of the prophage. Many chemicals are not themselves mutagenic but can be converted to proximate mutagens in the course of detoxification in the liver. Therefore, the standard screening for mutagenicity also involves treatment of the chemical with a liver extract prior to adding the cells. Any proximate mutagens that may be formed are then further modified by bacterial enzymes, such as transacetylases, to become direct mutagens. If the investigator suspects that the chemical to be tested cannot enter the bacterial cell, a transfection system using DNA from specialized transducing phages carrying the same set of mutations can be used, and the same selection for reversion is possible.

Based on tests with known carcinogens, the Ames mutagenicity testing system shows excellent predictive capabilities. An extensive body of literature exists demonstrating that the *Salmonella* system is about 84% sensitive (i.e., it finds 84% of known mutagens), and the *E. coli* system is about 91% sensitive. The speed, simplicity, low cost, and ability to quantitate the risk of the bioassays for mutagenicity have made them a worldwide standard for preliminary experimentation. Numerous commercial laboratories will perform the test, and specialized equipment is available to automate the counting process.

Recombineering

The techniques of site-directed mutagenesis are relatively labor intensive and not well suited to the replacement of lengthy sequences. Costantino and Court (2003) have developed a DNA replacement system that is based on the phage λ Red system. The Red system consists of three genes, *gam*, *exo*, and *beta*. Recombination between duplex molecules requires all three proteins, but recombination with single-strand DNA does not. In that case, only the Beta protein is required. The recombination event is not the one described earlier but instead occurs at the replication fork, apparently by substituting for an Okazaki fragment.

The recombineering system uses oligonucleotides (short, single-strand DNA) that are easily prepared by DNA synthesis machines such as that described in Chapter 2. Of course, the heteroduplex created by the substitution of the oligonucleotides is subject to mismatch repair. Costantino and Court report that in mismatch repair mutants, up to 25% of all cells incorporate the oligonucleotide that was introduced by electroporation.

Genetic Transfer in Nature

One obvious question that needs to be considered is to what extent all of the genetic transfer processes discussed in the earlier chapters actually occur in nature. Robert Miller and coworkers have shown that gene transfer among *Pseudomonas aeruginosa* members certainly occurs in their natural habitat. They found that there is a plentiful population of bacteriophages in fresh water and transduction of single or closely linked genes occurs at roughly the same frequency as in laboratory experiments (10^{-6} to 10^{-5} per colony forming unit). They also observed conjugal transfer of specific plasmids within test chambers suspended in a freshwater lake. However, simulations of the transfer under laboratory conditions gave transconjugants that frequently carried gene rearrangements or deletions. As noted in Chapter 10, there is substantial evidence in favor of naturally occurring genetic transformation. The movement of antibiotic markers via R plasmids is further evidence for spontaneous genetic transfer.

Transduction has always been a neglected area in natural population. Miller (2001) has reviewed the evidence that phage populations are actually quite large in natural waters, both fresh and marine. The observed frequencies of generalized transduction are in line with those observed in the laboratory, and marine bacteriophages can infect freshwater organisms like *E. coli* and cause transduction. Therefore, Miller concludes that natural transduction is an important source of bacterial variation.

Based on these studies, it seems clear that genetic transfer can occur under natural conditions. The existence of broad host range plasmids and bacteriophages strongly suggests that genes can move between disparate organisms. Transposons can mediate gene transfer between DNA replicons within the same organism and thereby provide new genes to the plasmids. It is difficult, however, to demonstrate the extent of such transfer because of lack of selectable markers and difficulty in identifying specific recipient strains in a natural environment.

Ecogenomics

The new technologies in molecular genetics are having an impact on the field of microbial ecology. As scientists try to determine which organisms are present in the environment, they have turned to tools such as PCR to identify the organisms present. Unexpectedly, it appears that there are many environmental organisms that are present in significant numbers (based on DNA analysis) but cannot be cultured. This is particularly true in ocean environments where Béjà (2004) reports estimates that more than 99% of the organisms have not been cultured. The new science of ecogenomics proposes to use techniques such as bacterial artificial chromosomes to "archive" the DNA of these unculturable organisms. The archives will allow scientists to sequence the DNA and look for relationships, even if the organism to which the DNA belongs is unknown.

Summary

The applications of bacterial genetics are many and varied, and they lean heavily on the great diversity within the bacterial kingdom. Genetic engineering still holds center stage with respect to the major achievements of bacterial genetics. Many useful proteins have been prepared by cloning the appropriate DNA into *E. coli* or some other bacterium. Scientists can point to numerous enzymes and other proteins whose price has been greatly reduced by successful cloning. Numerous specialized cloning vectors have been prepared to facilitate these enterprises along with techniques that allow the preparation of cDNA from purified mRNA.

There are important safety concerns about genetically engineered bacteria. Those concerning laboratory safety have been well studied, and suitable precautions have been taken. At present, however, the pressing issue is the extent to which it is appropriate to release organisms of this type into the environment. Careful ecological studies are required to prevent accidental damage to the ecosystem. These studies will require the ability to track organisms within a natural environment.

Questions for Review and Discussion

1. How would you try to assess the possibility for an artificial genetic construct to use one of the standard genetic transfer mechanisms to escape from an organism under: (a) laboratory conditions; (b) environmental conditions?
2. Suppose that you could create a truly designer cloning vector (you could insert any traits that you wished at any position). What traits would you include? Why?

3. Suppose that cloning vector size was a real problem. What would be the absolute minimum traits that ought to be included in a useful cloning vector?
4. What are the potential barriers working against both genetic engineers who are trying to clone genes into new organisms and natural genetic exchange?

References

General

Béjà, O. (2004). To BAC or not to BAC: Marine ecogenomics. *Current Opinion in Biotechnology* 15: 187–190.

Miller, R.V. (2001). Environmental bacteriophage–host interactions: Factors contribution to natural transduction. *Antonie van Leeuwenhoek* 79: 141–147.

Nielsen, K.M., Bones, A.M., Smalla, K., van Elsas, J.D. (1998). Horizontal gene transfer from transgenic plants to terrestrial bacteria—A rare event? *FEMS Microbiology Reviews* 22: 79–103.

Rabino, I. (1998). Societal and commercial issues affecting the future of biotechnology in the United States: A survey of researchers' perceptions. *Naturwissenschaften* 85: 109–116.

U.S. Environmental Protection Agency. *Microbial Products of Biotechnology: Final Regulations Under the Toxic Substances Control Act Summary* (Fact sheet). http://www.epa.gov/opptintr/biotech/fs-001.htm

Specialized

Al-Hasani, K., Simpfendorfer, K., Wardan, H., Vadolas, J., Zaibak, F., Villain, R., Ioannou, P.A. (2003). Development of a novel bacterial artificial chromosome cloning system for functional studies. *Plasmid* 49:184–187.

Chin, J.W., Cropp, T.A., Anderson, J.C., Mukherji, M., Zhang, Z., Schultz, P.G. (2003). An expanded eukaryotic genetic code. *Science* 301: 964–967.

Costantino, N., Court, D.L. (2003). Enhanced levels of λ Red-mediated recombinants in mismatch repair mutants. *Proceedings of the National Academy of Sciences of the USA* 100: 15748–15753.

Etz, H., Minh, D.B., Schellack, C., Nagy, E., Meinke, A. (2001). Bacterial phage receptors, versatile tools for display of polypeptides on the cell surface. *Journal of Bacteriology* 183: 6924–6935.

Jechlinger, W., Szostak, M.P., Lubitz, W. (1998). Cold-sensitive E-lysis systems. *Gene* 218: 1–7.

Ren, Z.J., Black, L.W. (1998). Phage T4 SOC and HOC display of biologically active, full-length proteins on the viral capsid. *Gene* 215: 439–444.

Torres, B., Jaenecke, S., Timmis, K.N., García, J.L., Díaz, E. (2003). A dual lethal system to enhance containment of recombinant micro-organisms. *Microbiology* 149: 3595–3601.

Westwater, C., Kasman, L.M., Schofield, D.A., Werner, P.A., Dolan, J.W., Schmidt, M.G., Norris, J.S. (2003). Use of genetically engineered phage to deliver antimicrobial agents to bacteria: An alternative therapy for treatment of bacterial infections. *Antimicrobial Agents and Chemotherapy* 47: 1301–1307.

17

Bacterial and Bacteriophage Evolution

One of the exciting changes that has occurred in the field of bacterial genetics in the past decade is the development of methods and information permitting the study of real evolutionary genetics with respect to bacteria and their genomes. Sufficient data are now available to permit researchers to address problems similar to those that have held the interest of geneticists studying eukaryotic organisms for many years. Although the bacterial evolutionary situation is not totally understood, our present comprehension has shed new light on classic problems in genetics and offers the promise of even more interesting results in the years to come.

Major topics include:

- Some of the possible measures of evolutionary change
- Uncertainties in assessing relatedness of organisms
- Changes in genomes with time; the emergence of new functions

- Identifying the function of unknown proteins
- The organization of genes on chromosomes
- One theory about how present-day chromosomes arose

What Is Evolution?

In the most literal sense, evolution is change occurring over time. For a geneticist, the change in question is, of course, base changes in the nucleic acid comprising the genome of a cell or a virus—in other words, a mutation of the sort discussed in Chapter 3. Most people's understanding of evolution is that it is a gradual process. However, that cannot be absolutely true. Mutations occur in an all-or-none, discontinuous manner. Either a genetic locus is mutant or it is not. Whether that mutation has any effect on the cell phenotype is an issue discussed in Chapter 3.

Implications of Strain Differences

The classic view of evolution is the one that has been proposed by Charles Darwin—the "survival of the fittest." This concept implies that as mutations accumulate in a population of individuals, some individuals will be better able to survive and to reproduce than others. The survivors are, therefore, better fitted to the current set of environmental conditions. An important point that is often forgotten, however, is that better fitted is not the same thing as best fitted. The end product of evolution is not necessarily the best-adapted organism for a particular ecological niche that could possibly be developed; it is merely better adapted than any of its competitors.

In many cases, experiments have shown that a particular cellular process is optimized rather than maximized. A good example of optimization is seen in the case of bacterial ribosomes. Charles Kurland and coworkers have demonstrated that ribosomes in the standard laboratory strain of *Escherichia coli* can mutate to have greater translational fidelity or greater speed, but they cannot maximize both the traits at the same time. Mikkola and Kurland tested seven natural isolates of *E. coli* and showed that their growth rates were initially relatively variable and generally slower than laboratory strains, although the mass of cells in a stationary phase culture was roughly constant for all strains tested. This observation indicates that the efficiency of glucose utilization is essentially maximized in all strains. Mikkola and Kurland tested their strains using a chemostat, a continuous culture device in which fresh medium is continuously added and old medium removed. After 280 generations, the growth rates for the natural isolates converged on the value seen in laboratory strains, although the cell mass per culture volume remained unchanged. In vitro tests demonstrated that the ribosomes had mutated to better optimize mRNA translation, but the nature of the mutations was not determined. The conclusion to be drawn from these results is that current growth conditions apply selection to cultures, and thus extensive laboratory studies do not necessarily reflect what is happening in the natural environment.

Another point to be considered is the founder effect. When a population arises from a few individuals, the distribution of alleles within the founding members may not reflect the distribution of alleles within the worldwide population. Thus, some unusual genes may be present just by chance. These unusual genes may or may not contribute to the overall fitness of the final population, but their rarity in the world at large guarantees that geneticists examining the isolated population will notice them. Thus, the mere presence of a gene in a successful population does not necessarily indicate a beneficial effect from that gene.

THINKING AHEAD

What is the advantage to a cell in having DNA that cannot be expressed? If there is no advantage, why do cells keep untranscribed DNA in their genomes instead of reducing the genome size so that they can replicate faster?

Cryptic Genes

It is also important not to substitute human judgment for observation of natural processes. For example, it seems intuitively obvious that **cryptic genes**, genes that are never expressed, should be undesirable. However, cryptic genes are commonly found in bacteria. *Lactobacillus, Bacillus, Escherichia, Salmonella,* and many other genera contain inactive genes encoding functions that prove useful under particular circumstances. These genes may code for amino acid biosynthesis, sugar degradation, or more unusual metabolic functions. They can be activated by mutation and appropriately selected, but in some cases when the selection is removed the genes soon become cryptic again.

The best-studied examples occur in *E. coli* and involve genes for transporting and metabolizing rare sugars. They are the *bgl, cel,* and *asc* operons. The *bgl* operon codes for three proteins (Fig. 17.1). The first is an antiterminator protein (BglG) that can be phosphorylated. In its dephosphorylated form, it is an antiterminator of transcription. The second is a member of the phosphoenolpyruvate-dependent sugar transport system (BglF), an example of PTS enzyme II. This particular enzyme transports salicin or arbutin, and phosphorylates the sugar at the same time. When inducing substrates are absent, it phosphorylates BglG. The third Bgl protein is a phospho-β-glucosidase (BglB) that degrades the sugars. Standard laboratory strains of *E. coli* do not express the *bgl* operon because of a defective promoter. Insertion of IS*1* or IS*5*, 78–125 bp upstream of the promoter or base substitutions in the CRP-cAMP binding site of the promoter, can activate the operon. While it might seem that activation by insertion would be a useless characteristic, in fact it is not particularly rare. Manna et al. (2001) have shown that the relevant area is a hot spot for insertion of phage Mu. Tn*1* and Tn*5* also preferentially insert here. Transcription always begins in the same location, so the mechanism of activation is not simple creation of a new promoter. There are two transactivators of the *bgl* operon known, *bglJ* and *leuO*. Neither is normally expressed.

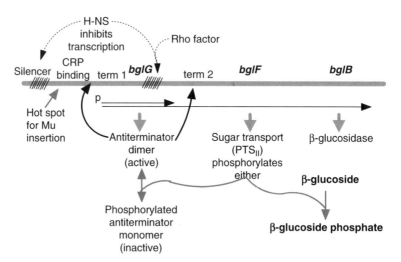

Figure 17.1. The *bgl* operon of *E. coli*. This operon is normally cryptic because the promoter is inactive. Insertions near the CRP binding site can activate the promoter as can certain proteins synthesized elsewhere on the genome. The operon includes a gene for sugar transport, degradation, and regulation.

Barry Hall and coworkers detected the *asc* operon during a series of experiments on directed evolution. They began with a strain deleted for the *bgl* operon and a related operon called *cel*. Sequentially, they selected for strains that could first use arbutin, then arbutin and salicin, and finally arbutin, salicin, and cellobiose. Each selected event could be either the return of function to a pre-existing cryptic gene or an alteration in substrate specificity of the protein product of an already expressed gene. Hall and Xu (1992) have shown that there are strong sequence homologies between the Asc active proteins and those of the *bgl* operon. Interestingly, the Bgl regulatory protein is a positive regulator, while the Asc regulatory protein is a repressor with strong homology to *galR*.

The questions that obviously arise concerning cryptic genes are how they originate and why are they maintained in the population. Cryptic genes that resemble other genes in the same genome are taken to be **paralogous**, meaning that they result from DNA duplication followed by independent evolution of the two sets of genes. In the present discussion, the *bgl* and *asc* operons would be considered paralogous, given the sequence similarities demonstrated by Hall and Xu. Their simultaneous maintenance in the population raises some additional issues.

Directly repeated sequences are recombinationally unstable, tending to form loops that are excised like a λ prophage. This instability can be counter-acted by sequence divergence within the paralogous genes and by physical distance between the duplications. Because the intervening DNA would be lost during recombination, recombinants formed from widely separated duplicated genes are likely to be inviable. In the case of *bgl* and *asc*, they are separated by about 26 minutes of DNA, including genes for all the ribosomal proteins. Therefore, recombinational loss is unlikely.

Note that all the cryptic genes just discussed code for normal proteins that can be immediately functional. The defects in the operon are all at the level of transcription, not in the protein-coding regions. Mutations do occur in the coding regions (the sequences are not identical), but harmful mutations must be disadvantageous. The only way that could happen would be if the cryptic genes are occasionally expressed. Khan and Isaacson (1998) reported that just because an operon is not expressed in the lab does not mean that it is never expressed. They used a reporter gene fused to the *bgl* promoter to show that even though the *bgl* operon is not transcribed in laboratory cultures, it is expressed during infection of mouse liver. One possible explanation would be that something in the mouse liver induces production of LeuO protein that subsequently activates *bgl* transcription.

There is, however, evidence that for everyday metabolism, it is critical that the *bgl* operon should NOT be expressed. In addition to the inactive promoter and the requirement for inducer and antitermination by BglG, the histonelike protein H-NS binds to two silencer sites (Fig. 17.1). The one upstream prevents promoter activity, and the one downstream facilitates an interaction with Rho protein that causes transcription termination (Dole et al. 2004). Nevertheless, there is a report that a sigma-factor-like protein from *Streptomyces* can activate the *bgl* operon by itself (Baglioni et al. 2003).

Arthur Koch (2004) has argued that most mutations that occur to activated cryptic genes are of an easily revertible type. If so, that would constitute another indication of the importance of keeping cryptic genes turned off except in times of significant need. Koch also considered how that effect may interact with the observations of directed mutation (see Chapter 3).

Another example of paralogous genes occurs in *Borrelia*. The organism has a linear chromosome, 12 linear plasmids, and 9 circular plasmids. Many of the plasmid genes are either paralogs of chromosomal genes or **pseudogenes**, nonfunctional genetic structures that have a similar base sequence to other chromosomal genes.

Expression of Evolutionary Relationships

Discussions on evolution invariably result in the necessity for graphical representations of relationships between organisms. The most usual type of diagram is the evolutionary tree (Fig. 17.2). In such a tree, distances along lines represent evolutionary time. Closely related organisms occur as physically close branches of the tree. Distantly related organisms are more widely separated. If the tree

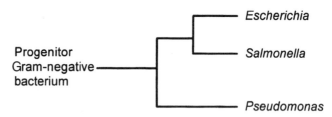

Figure 17.2. An evolutionary tree. This sample tree suggests that there was a single bacterium that gave rise to three different genera. *Pseudomonas* diverged from others earlier than *Escherichia* diverged from *Salmonella*. Divergence of the latter two was not a recent event because the branches have significant length.

includes an organism as the progenitor of all others as in Fig. 17.2, the tree is rooted. The toothlike arrangement of some trees has led scientists to refer to them as dendrograms.

Preparation of an evolutionary tree requires data that can be used to measure relatedness of organisms. Originally, these data would include such variables as physical size, shape, habitat, and so forth. However, these variables are unsatisfactory because of the limited number of possible types compared to the number of known bacteria. For example, there are only four or five fundamental cell shapes among bacteria. With the advent of macromolecular sequencing techniques, evolutionary biologists now speak of sequence similarities and use them as the measure of relatedness. The more precise method is to use nucleic acid sequence data because these will show all the changes that have occurred. However, the amino acid sequence is also informative because conserved sequence suggests functional importance. These sequencing techniques for classification are more satisfactory than the earlier methods but they still have potential pitfalls for the unwary.

Gene sequences can be similar for several reasons. One is that the genes are **orthologous**, descended from an unknown, common ancestor. Alternatively, one member of the set is the progenitor for all other members, meaning that they are **homologous**. Orthologous genes can potentially provide information about evolutionary distance. Another possibility, however, is that there is basically only one way to enzymatically perform a certain task. Therefore, if an organism makes an enzyme with that property, the shape of the enzyme will be essentially similar to all other such enzymes regardless of evolutionary lines of descent, making the genes **homoplastic**. This is an example of convergent evolution.

Molecular biologists now have sequence data from a wide variety of organisms available to them. These data are most useful when the genes being compared are highly conserved so that the changes accumulate very slowly over the time interval being examined. Many studies involve the smaller ribosomal RNA molecules because they are easy to obtain in large quantities and can be readily sequenced. Carl Woese and coworkers have published 16S rRNA sequences for a variety of organisms and expressed their data as evolutionary trees. While RNA sequences can be absolute, controversy has developed over the exact (computer-driven) methodology to be used in preparing the evolutionary trees. The Woese group uses distance measurements, estimating the fraction of positions at which the sequences differ. The assumption is that larger fractions mean greater distances on the tree. James Lake has proposed alternative calculations called evolutionary parsimony that lead to different taxonomic structures for the Bacteria and the Archaea. These calculations attempt to reconstruct step-by-step changes that might have occurred to give present-day sequences.

The fundamental principle in tree building is that of parsimony—developing the simplest tree that accommodates all of the data. Although any sort of sequence data can be used, the most desirable is DNA sequence. Protein sequence data suffer from the existence of synonymous codons, meaning that some mutations are not apparent in the amino acid sequence. One of the first issues is having DNA sequence data with enough informative bases. Sequence in and of itself is not sufficient to provide evolutionary information. There must be sequence differences between molecules from different organisms that can be used to choose between trees. If all possible trees require the same number of base changes to achieve a particular sequence difference, the difference is not phylogenetically informative (Fig. 17.3).

The issue of informative bases is one of balance. If organisms have diverged relatively recently (in an evolutionary sense), only a gene that accumulates mutations quickly will offer enough informative sites. On the other hand, if two organisms diverged a long time ago, only relatively stable genes will retain enough of their original sequence to provide information about the relatedness of the organisms. Major taxonomic trees are based on rRNA sequences because they offer a good mix of properties. Overall ribosome structure is strongly constrained by the variety of tasks that must be performed for a cell to

	↓ ↓ ↓
Progenitor sequence	A A G G C C T T
Descendant 1	A T G G G C T T
Descendant 2	A T C G G C T T
Descendant 3	A T C G C C T T

Figure 17.3. The phylogenetic information content of nucleotide sequence changes is variable. *Arrows* indicate sequence differences. The first difference is not informative, however, because all descendants have it. The second two differences begin to establish lines of descent. Descendant 2 is less closely related to the progenitor than the other two descendants because it has two sequence differences while the others have only one each. The data are not sufficient to explain how descendant 2 is related to either descendant 1 or 3.

survive and be competitive; hence there are regions of rRNA that change very slowly. These relatively unchanging regions provide information about relationships corresponding to geologic time. There are, however, other regions of rRNA that are less critical and do accumulate mutations, and these regions offer data about evolutionarily recent changes.

A more subtle problem is one of sequence alignment. Most DNA sequences do not align perfectly. Each organism has accumulated its share of small deletions and insertions. Therefore, a simple sequence alignment that begins at the 5'-end is unlikely to be satisfactory (Fig. 17.4). Instead, spaces need to be added to maximize the similarity, but that addition presents its own problems. Which space should be added first? A space added near the 5'-end will shift all sequences downstream, whereas a space added near the 3'-end has a much smaller effect. If the sequence in question encodes a protein, the proper reading frame must be preserved. If the sequence is for an rRNA molecule, the proper secondary and tertiary structures must be maintained. Once again, special computer programs are available to prepare and compare possible sequence alignments. The basic strategy is to align the sequences so that only a minimum number of changes are needed to produce the diverged sequences under study. Qi et al. (2004) have presented a new computer modeling program for evolutionary tree construction using protein sequence that attempts to answer some of the earlier criticisms.

Even the base composition of the DNA molecules being compared can have an impact on tree analysis. Steel and coworkers (1993) have presented a mathematic analysis showing that if two organisms have the same $(G + C)$ content in their DNA, problems with evolutionary trees can arise. Standard methods of calculating trees assume that similar $(G + C)$ content implies relatedness, whereas in fact the similarity may be due to convergent evolution. Steel and coworkers provide equations to discount random similarities resulting from similarities in base composition. The effect is most dramatic when the $(G + C)$ content, which can range from 25% to 75%, is either very high or very low.

Organisms can and do differ in the % $(G + C)$ in their DNA, mainly due to wobble in the third base of codons. Lawrence and Ochman (1997) present data to show that the % $(G + C)$ of the third base in a codon increases faster than the increase in the overall % $(G + C)$ of different organisms, reaching as high as 95% in *Micrococcus luteus*. On the other hand, the % $(G + C)$ in the second position increases very slowly, ranging from a low of 33% to a high of 45%. The first base in the codon falls in between the other two ranges. Therefore, base changes in position 3 of the codons have little taxonomic meaning because that position

```
Ec  ATGAAAGCGT TAACGGCCAG GCAACAAGAG GTGTTTGATC TCATCCGTGA 50
St  .......... .......... .......... .......... ......G.. 50
Ecc .......TA. ...A..A... ...G..GC.. ..T.A...C. ......C.. 50
Pp  ...TTGAAAC G...C.AC.. C...GCC..A A.TC.C.CGT .....AAGCG 50
Pa  ...C.GAA.C G...C..C.. C...GCC... A.CC.CTCCT .....AAGCG 50

Ec  TCACATCAGC CAGACAGGTA TGCCGCCGAC GCGTGCGGAA ATCGCGCAGC 100
St  .......... .......... ......A... .......... .......... 100
Ecc ....T..TGCG ..A..C..A. .......A.. ...G...... .T..T..A. 100
Pp  CTG.C.GGAA G.C.AC..CT .C.....T.. C..C..C..G ..T..T...G 100
Pa  CTG.C.GGAA G.CCAC..CT .T........ C..G...... ....C...G 100

EC  GTTTGGGGTT CCGTTCCCCA AACGCGGCTG AAGAACATCT GAAGGCGCTG 150
St  .......... .......... .......G.. .......C.. T..A..... 150
Ecc AAC.A..... T..C..T..C ..T....... .......... ...A..A... 150
Pp  AGC....C.. .AAG..G..C ..T..C..C. .G..G..C.. C......C..T 150
Pa  AAC.C..C.. .AAG..G..G .....C..C. .G..G..C.. C........... 150

Ec  GCACGCAAAG GCGTTATTGA AATTGTTTCC GGCGCATCAC GCGGGATTCG 200
St  ..G....... .G..GC.... ...C...... ........G A..T..C.. 200
Ecc ..G..T.... .T..G..... ......A..G ...G..T. .T..T.... 200
Pp  ..C.....G. ...CG..C.. ...GACGC.G ...T..C... ...C..C.. 200
Pa  .......G. ...CC..C.. ...GAC.C.G ...C.T.... ...C..C.. 200

Ec  TCTG------ ---------- --TTGCAGGA AGAGGAAGAA GGGGTTGCCGC 232
St  ....------ ---------- --..A..... ...A..G..C ..A..A.... 232
Ecc ....------ ---------- --C..AT... ...A..GACG ..TA.T..T. 232
Pp  CA.CCCTGGC CTGGAAGCCA AGGCT----- -..A....CC ..CC....CA 244
Pa  CA.TCCCGGC TTCGAACCGC ATGCCGCCA. C..C..T..G ..CC.....G 250

Ec  TGGTAGGTCG TGTGGCTGCC GGTGAACCAC TTCTGGCGCA ACAGCATATT 282
St  .T..C..G.. ..C..G..G .........G. .......... G......... 282
Ecc ....T..... C.....C..A .........G. .G........ GG.A..C..C 282
Pp  .CA.C..C.. G..C..... ....CG..GA ....T..CG. ......C..C 294
Pa  ..A.C..A.. G..C..C... ..C.C..GA .C..C..CG. ......A.C..C 300

Ec  GAAGGTCATT ATCAGGTCGA TCCTTCCTTA TTCAAGCCGA ATGCTGATTT 332
St  .....C..... .C........ ........GC.G .......A..C GC....... 332
Ecc ...T...GC. .......T.. C..G.GA.G ...A..C GC........ 332
Pp  ..GCAATCC. GCA.CA..A. C...G....C ...C.C..CC .A..C..C.A 344
Pa  ..G.AATCC. GC.G.A..A. ...CG....C .....T..TC GC..C..C.A 350

Ec  CCTGCTGCGC GTCAGCGGGA TGTGATGAA AGATATCGGC ATTATGGATG 382
St  .......... ...G....T. .......... .....C....T .......... 382
Ecc TT.......G ..G.....C. .......... .A......T .......... 382
Pp  TT........ ..ACA...C. ..AGC..... G..CG..... ...CT.C..C. 394
Pa  ....T..... ..GC....C. ..AGC..... G..C...... ...C.C..C. 400

Ec  GTGACTTGCT GGCAGTGCAT AAAACTCAGG ATGTACGTAA CGGTCAGGTC 432
St  .G..T..... ...G..A... ......G... ..C..C.... T..C.....G 432
Ecc .C..T..A.. ...C...... ......AG.A ...G..A.. T......A.T 432
Pp  .C...C..... ...G.....C .CCTGC.GT. .A.CC..C.. ...C...A.. 444
Pa  .C...C..... ...C..C..C GTC..C.GC. .A.CG..C.. ...C.....G 450

Ec  GTTGTCGCAC GTATTGATGA CGAAGTTACC GTTAAGCGCC TGAAAAAACA 482
St  .......T.. .C....... T.....G..A .A..A..... .......G.. 482
Ecc ..C.....G. .......C.. T.....G..G .G........ .......G.. 482
Pp  ...G..G... .C..C.G... ......C... ...C.....T .C..GCGCG. 494
Pa  ...G.....G. .G..C.GC.. G..G..C..G ..G..A...T .C..GCGCG. 500

Ec  GGGCAATAAA GTCGAACTGT TGCCAGAAAA TAGCGAGTTT AAACCAATTG 532
St  .......... ..G..G...C ....G..... C......... ..C..G..A. 532
Ecc ........CG ..AC.C..CC .CG.T..... .GAA...... GCC..G.... 532
Pp  A....GC... ...TGG...C .TG.C..... CCC...A..C GCC..C..C. 544
Pa  A....GC..G ...TGG...C ..G.G..... CCCT.....C GCT..G..C. 550

Ec  TCGTTGACCT TCGTCAGCAG AGCTTCACCA TTGAAGGGCT GGCGGTTGGG 582
St  .G..G..T.. G..CG.A..A .....T..G. .......... ......A..C 582
Ecc .T..C..... G........A .....TT.G. .....CT. A......... 582
Pp  AA..C..... GAAAG.A... GAGC.GGTG. .C..G..CT. .AGC..C..C 594
Pa  AA..C..T.. GAAGG..... GAAC.G.T.. .C....CT. .AGC..C..C 600

Ec  GTTATTCGCA ACGGCGACTG GCTGTAACAT ATCTCTGAGA CCGCGTGCCG 632
St  ..C....... .......A.. .T..TAGTCT CTTTTTAATC TCCTTGTAAG 632
Ecc ..G....T. ..A........ .AGCTAACCA TTCGCAGAGA TGCACTGCTC 632
Pp  ..C......C G.TGA----- -------TCC AGGAGGCGTC ATGCAGCAGT 632
Pa  ..G..C..AC G.TGA----- -------CAG GAGATACCAT GCAGACCTCC 638

Ec  CCTGGCGTCG CGGTTTGTTT TTCATCTCTC TTCATCAGGC TTGTCTGCAT 682
St  CCGCCATCCG GCAATCGTGT AGCCTGATGG CGCTGCGCTT ATCAGGCCTA 682
Ecc GTCTGTCTGG GTCAGTGCAT CATGTCCTGT ATTCATCGTT TAGCGTGTTA 682
Pp  TCATTCACGC ACCCGAGCAA GCCCAGTTGC CCCTGTTCGA AGCATTCCTC 682
Pa  CACTCGCTGC CCAGCGCCCA GTTGCCACTG TTCCAGGAAG CGTTCTGGGC 688

Ec  GGCATTCCTC ACTTCATCTG ATAAAGCACT CTGGCATCTC GCCTTACCCA 732
St  CGGGAATGCA GTTCCTGAGA TGATTAATTT GTAGGCCGGA TAAGGCGTTA 732
Ecc ATCTGCTAAC CATATATATT TAGTTACATT TCGCGCGCAT TTTCTACGAT 732
Pp  GCCCAGCCCG TGCTGCCAGG CCTGAAAGCC AGGGAACCGG CGCGCAAGAG 732
Pa  CAGCAACGGC GCTCCCTTGC TCGACGATGT CATCGACAGC CCTTCCAGCG 738

Ec  TGATTTTCTC CAATATCACC GTTC 756
St  CGTCGGCCATC CGGCAATGCG CTCG 756
Ecc CCCTATTACC CTCTGTTTTT TCAC 756
Pp  CAGCCAGCCC GAGCTGTTCA GCGA 756
Pa  CCTCCATCGA GGAACCCGCT GCCT 760
```

Figure 17.4. DNA sequence comparison of the coding region of the *lexA* genes and the sequences immediately downstream from the termination codons of *E. coli* (Ec), *S. enterica serovar* Typhimurium (St), *Erwinia carotovora* (Ecc), *Pseudomonas putida* (Pp), and *Pseudomonas aeruginosa* (Pa). The nucleotide sequence of the *lexA* gene of *E. coli* is shown for comparison. The nucleotides are numbered starting from the first nucleotide of the ATG initiation codon. The termination codon is at about residue 606. Both these codons are underlined. Identical residues in the open reading frame are depicted by a *dot* and nucleotide substitutions are indicated by the appropriate letter. *Dashes* indicate insertions needed to maintain the alignment. (From Garriga, X., Calero, S., Barbe, J. [1992]. Nucleotide sequence analysis and comparison of the *lexA* genes from *Salmonella typhimurium*, *Erwinia carotovora*, *Pseudomonas aeruginosa*, and *Pseudomonas putida*. *Molecular and General Genetics* 236: 125–134.)

is constrained to conform by the overall % (G + C) in the organism, and they should not be heavily weighted in the construction of phylogenetic trees. Changes in the second base should be the most informative, if there is a method to identify them. For a review of the genetic code, see Table 17.1.

Two methods have been proposed for resolving the problem of sorting evolutionary fluctuation from self-imposed restrictions on the code. The first depends on identifying **signature sequences**, regions in the aligned proteins where sequence changes are very informative (i.e., present in all members of one group and not in members of another). Gupta (1998) argues forcefully that only comparisons of protein sequences can avoid the problems inherent in DNA composition.

However, there is another possible approach (Karlin et al. 1998). This method takes advantage of the enormous amount of DNA sequence now available. The authors have prepared a mathematical analysis showing how the relative abundance of dinucleotides in each DNA strand is characteristic of each organism, a **genome signature**. Interestingly, the dinucleotide statistics are sufficient to reflect nearly all of the nonrandomness in genomes. There is little advantage to trinucleotide or tetranucleotide analysis according to their data.

Assuming that genetic relatedness has taxonomic significance, many evolutionary biologists have developed trees that attempt to graphically illustrate both the relatedness of particular genera or species and their common ancestry. Carl Woese and coworkers have been particularly active in this regard, and a sample of their work is presented in Fig. 17.5. Based on studies of 16S rRNA, they have proposed that all living organisms should be categorized into three domains: *Archaea*, *Bacteria*, and *Eukarya*, with eukaryotes descended from archaea.

Table 17.1. Differences in codon usage in *Rhodobacter capsulatus*: Comparison of codon usage in *Rhodobacter* and *E. coli*.

Amino Acid	Codon	Fractional Codon Usage for Each Amino Acid					
		Rhodobacter	*E. coli*	Rc-Fru	R-Pho	R-Nif	R-Crt
Gly	GGG	0.22	0.02	0.31	0.14	0.18	0.29
Gly	GGA	0.03	0.00	0.01	0.03	0.02	0.04
Gly	GGU	0.09	0.59	0.07	0.12	0.10	0.08
Gly	GGC	0.66	0.38	0.61	0.70	0.71	0.59
Glu	GAG	0.52	0.22	0.61	0.44	0.52	0.56
Glu	GAA	0.48	0.78	0.39	0.56	0.48	0.44
Asp	GAU	0.38	0.33	0.48	0.25	0.43	0.41
Asp	GAC	0.62	0.67	0.52	0.75	0.57	0.59
Val	GUU	0.09	0.51	0.08	0.09	0.08	0.09
Val	GUC	0.47	0.07	0.44	0.50	0.45	0.42
Ala	GCG	0.46	0.26	0.52	0.37	0.46	0.50
Ala	GCA	0.03	0.28	0.02	0.04	0.02	0.03
Ala	GCU	0.05	0.35	0.02	0.11	0.03	0.04
Ala	GCC	0.46	0.10	0.44	0.48	0.49	0.43
Lys	AAG	0.76	0.26	0.74	0.80	0.67	0.80
Lys	AAA	0.24	0.74	0.26	0.20	0.33	0.20
Asn	AAU	0.18	0.06	0.44	0.05	0.18	0.30
Asn	AAC	0.82	0.94	0.56	0.95	0.82	0.70
Met	AUG	1.00	1.00	1.00	1.00	1.00	1.00
Ile	AUA	0.01	0.00	0.02	0.00	0.00	0.00
Ile	AUU	0.07	0.17	0.06	0.05	0.10	0.07
Ile	AUC	0.92	0.83	0.91	0.95	0.90	0.93
Thr	ACG	0.32	0.07	0.38	0.24	0.36	0.42
Thr	ACA	0.02	0.04	0.03	0.01	0.02	0.01
Thr	ACU	0.04	0.35	0.03	0.04	0.02	0.05
Thr	ACC	0.62	0.55	0.57	0.71	0.60	0.52
Trp	UGG	1.00	1.00	1.00	1.00	1.00	1.00
Cys	UGU	0.12	0.49	0.25	0.00	0.13	0.10
Cys	UGC	0.88	0.51	0.75	1.00	0.88	0.90
Tyr	UAU	0.42	0.25	0.58	0.31	0.42	0.54
Tyr	UAC	0.58	0.75	0.42	0.69	0.58	0.46

Table 17.1. (*Continued*)

Amino Acid	Codon	Fractional Codon Usage for Each Amino Acid					
		Rhodobacter	*E. coli*	Rc-Fru	R-Pho	R-Nif	R-Crt
Phe	UUU	0.14	0.24	0.33	0.06	0.21	0.16
Phe	UUC	0.86	0.76	0.67	0.94	0.79	0.84
Ser	UCG	0.56	0.04	0.57	0.59	0.45	0.55
Ser	UCA	0.01	0.02	0.00	0.01	0.02	0.01
Ser	UCU	0.04	0.34	0.03	0.03	0.02	0.05
Ser	UCC	0.17	0.37	0.20	0.15	0.20	0.13
Ser	AGU	0.02	0.03	0.02	0.02	0.01	0.03
Ser	AGC	0.21	0.20	0.18	0.19	0.30	0.23
Arg	CGG	0.28	0.00	0.27	0.14	0.32	0.38
Arg	CGA	0.02	0.01	0.00	0.01	0.02	0.03
Arg	CGU	0.11	0.74	0.07	0.20	0.08	0.13
Arg	CGC	0.55	0.25	0.63	0.63	0.54	0.44
Arg	AGG	0.02	0.00	0.03	0.01	0.02	0.01
Arg	AGA	0.01	0.00	0.00	0.00	0.01	0.01
Gln	CAG	0.80	0.86	0.79	0.85	0.84	0.77
Gln	CAA	0.20	0.14	0.21	0.15	0.16	0.23
His	CAU	0.44	0.17	0.69	0.19	0.45	0.57
His	CAC	0.56	0.83	0.31	0.81	0.55	0.43
Leu	CUG	0.64	0.83	0.76	0.54	0.66	0.61
Leu	CUA	0.00	0.00	0.01	0.00	0.00	0.00
Leu	CUU	0.13	0.04	0.10	0.09	0.13	0.19
Leu	CUC	0.19	0.07	0.11	0.35	0.16	0.12
Leu	UUG	0.04	0.03	0.03	0.02	0.05	0.08
Leu	UUA	0.00	0.02	0.00	0.00	0.00	0.00
Pro	CCG	0.61	0.77	0.60	0.76	0.55	0.58
Pro	CCA	0.01	0.15	0.00	0.00	0.03	0.01
Pro	CCU	0.03	0.08	0.01	0.02	0.04	0.04
Pro	CCC	0.35	0.00	0.39	0.23	0.38	0.37

Abbreviations: Rc, *R. capsulatus*; R, *Rhodobacter*; Fru, fructose catabolic genes; Pho, photosynthetic genes; Nif, nitrogen utilization genes; Crt, carotenoid biosynthetic genes. (From Wu, L.-F., Saier, Jr., M.H. [1991]. Differences in codon usage among genes encoding proteins of different function in *Rhodobacter capsulatus*. *Research in Microbiology* 142: 943–949.)

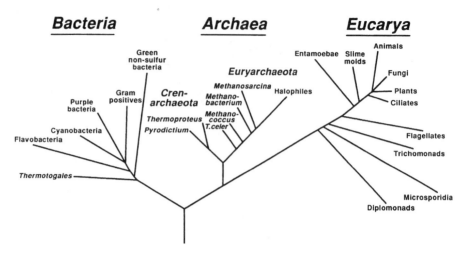

Figure. 17.5. Universal phylogenetic tree in rooted form, showing the three domains: *Archaea*, *Bacteria*, and *Eukarya*. The position of the root was determined by using the paralogous gene couple translation elongation factors EFTu and EFG. (From Woese, C.R. [1994]. There must be a prokaryote somewhere: Microbiology's search for itself. *Microbiological Reviews* 58: 1–9.)

However, Gupta (1998) argues that his analysis suggests that eukaryotes developed from a fusion of an Archaean and a Gram-negative bacterium. Fusion hypotheses have long been used to explain the origins of mitochondria and chloroplasts, so this theory merely extends those ideas. Simonson et al. (2005) have refined the fusion hypothesis and have proposed that the universal phylogenetic tree does not begin with a root but rather with a ring that represents the fusion event.

One current area of controversy is the contribution to evolution of **horizontal** (lateral) gene transfer. In a horizontal transfer, an organism receives a gene not by inheritance from its parent or by mutation, but by any of the genetic transfer processes discussed earlier. Lake and coworkers argue that the genes most likely to successfully transfer horizontally are the "operational" genes, those not involved in transcription, translation, or DNA replication. In support of their claim, they examined 312 sets of orthologous genes in six completely sequenced bacterial genomes and found evidence that horizontal transfer is more or less constant over time and did not occur at substantially higher frequencies in the remote past. Boucher et al. (2003) have reviewed the impact of horizontal transfer of genes on the taxonomy of prokaryotes. Novichkov et al. (2004) have developed a computer model that attempts to identify when the rate of evolutionary change does not match the amount of change seen in individual genes. From their data, they believe they can identify the minority of changes due to horizontal transfer and estimate that 70% of the genes studied fit a model of vertical transfer. However, some gene families show large anomalies. Ochman et al. (2005) show that horizontal transfer may actually maintain species integrity and does not prevent phylogenetic reconstruction at many levels.

Genomics and Proteomics

Bacteria

Genomics is the study of patterns in the genome of an organism, and **proteomics** is the study of the proteins it produces and the conditions under which they are produced. Genomics follows logically from the collection of DNA sequence data. Proteomics became possible for the same reason. Once the DNA sequence of an organism is known, computer programs can identify open reading frames using fairly simple assumptions. Initially, the correlation of open reading frames with real proteins requires a good genetic map, and for this reason *E. coli* is one of the best-studied organisms. However, as the data accumulate, other approaches are possible. For example, x-ray crystallography of proteins provides a three-dimensional structure to the protein that can be correlated with the DNA sequence. Knowing which part of the protein binds to a ligand like ATP means that DNA sequences coding for ATP binding can be identified. With that information in hand, now the computer programs can search known open reading frames for those that appear to bind ATP. It therefore becomes possible to predict probable functions of a protein without ever isolating it.

Two technologies have provided significant proteomic information. They are two-dimensional gel electrophoresis (separation of proteins by molecular weight and charge) and DNA microarray technology. Both these techniques allow the experimenter to determine which genes are active under a given set of conditions. In one case, the actual proteins are observed as spots on the gel, and in the other case, mRNA is assayed as a surrogate for detecting the protein itself.

Subtle variations are also possible. For example, Lee and Lee (2003) discuss the advantages of using time-of-flight mass spectrometry as an adjunct to two-dimensional gels. They looked at the effect of starvation, temperature shock, and oxidative damage on gene expression.

Despite all this effort, much remains to be done, even in *E. coli*. Matte et al. (2003) point out that only 50% of the *E. coli* open reading frames have an experimentally verified protein associated with them. Another 30% have a function attributed to them based on similarity of DNA sequence to known proteins in other organisms.

Bacteriophage

For many years the role of bacteriophages in bacterial evolution was ignored. Recently, however, more careful studies have shown that there is a significant impact. Chibani-Chennoufi et al. (2004) have reviewed the evidence that in estuarine waters in the summer the concentration of bacteria is about 10^6/ml while that of their phages is 10^7/ml. This tenfold excess of virus over host cells means that there is the enormous potential for transduction of DNA from one cell to another. One estimate is that there are roughly 10^{30} phages on the planet, making them the most numerous biological entity.

Other estimates are in accord with these ideas. One strain of *E. coli* O157 may have 18 different prophages, accounting for 16% of its genome (Canchaya et al. 2003). The question then arises, what is the selective advantage that keeps a prophage in the population? One possibility is superinfection immunity, protection for the host cell. However, there are other positive advantages. The λ *bor* gene confers serum resistance on pathogens. There are also many cases of lysogenic conversion where the prophage encodes a toxin that enhances the pathogenicity of the host.

Specific Examples of Evolution

Evolution does not occur in the abstract. Changes in DNA sequence can be reflected by changes in RNA sequence, protein amino acid sequence, and other more subtle changes. Some of these changes offer additional information to the evolutionary biologist.

THINKING AHEAD

Bacteria do not use all codons with equal frequency. Which organisms would you expect to have the most biased codon usage? Why?

Evolution of Genomes

One intriguing question about bacterial genomes is how bacteria can develop two chromosomes both carrying essential genes. One suggested answer comes from the work of Itaya and Tanaka (1997), who performed "genetic surgery" on the *Bacillus subtilis* 168 chromosome. They inserted two partial neomycin resistance cassettes into the chromosome and inserted a low copy number plasmid origin of replication in between. The neomycin cassettes contained enough sequence similarity to recombine. Successful recombination between the cassettes would excise a fragment 310 kb in length, and the recombination event would generate a neomycin resistance gene to provide a selectable marker. When they applied selection, they obtained neomycin-resistant bacteria. Pulsed field gel electrophoresis showed that the original chromosome had lost 310 kb and a new, circular subgenome was present. This type of event could presumably occur in nature, especially given the tendency of plasmids to integrate into chromosomes.

Another issue in genome evolution is what amount of variability is the result of mutations passed on to the progeny cells (vertical transmission) and how much is the result of genetic transfer processes like conjugation, transduction, or genetic transformation (horizontal transfer). Lawrence and Ochman (1997) and Ochman et al. (2005) have looked at variation in the (G + C) content of *E. coli* and *Salmonella enterica* DNA, and they estimate that 31 kb of DNA accumulate due to horizontal transfer over the course of one million years. Much of this DNA presumably comes from virus infections, either as transduction or as new genes acquired with a prophage.

Richard Lenski's research group has an interesting, ongoing experiment in evolution. They have been propagating a set of 12 populations of *E. coli* B, all derived from a common ancestor, for over 10 years (Lenski et al. 2003). Each population is subcultured daily in a glucose-limited minimal medium, so that approximately 20,000 generations have elapsed for each population. Some populations have evolved mutator mutations and some have not. DNA sequencing of small regions of the genome has allowed them to estimate that nonmutator populations have about three synonymous base substitution mutations and mutator populations have about 250 in the 20,000 generations. Given that these mutations should be neutral in their effect, they allow an estimate of mutation rate that is independent of selection. For the duration of the experiment, the fitness of the populations increased by about 70%.

Brüssow et al. (2004) review the contribution of bacteriophages to genetic evolution. Phage studies are difficult because of a lack of fossil record and concomitant molecular clock. Nevertheless, it is clear (see later) that phage modularity is a significant evolutionary tool. Phages also have a significant impact on their host bacteria in addition to the obvious effects of transduction. This is due to the extra genes that may be carried by prophages to enhance the pathogenicity of their host cells. These genes can encode toxins, proteins that change the antigenicity of the host, or enzymes that protect the host from mammalian defenses like superoxide or white blood cells.

Evolution of the Genetic Code

The genetic code itself is not invariant. Minor differences in the genetic code are discussed in Chapter 3. Each time a tRNA suppressor mutation appears, the genetic code changes slightly for that particular organism. In nearly all cases, the change does not become fixed in the population, but it could theoretically do so. Certain mitochondria (presumably derived from early bacteria) routinely translate the RNA codon UAA as tryptophan instead of termination.

The synonymous codons also offer opportunities for change. Individual organisms do not use synonymous codons with equal frequency. Instead, they exhibit characteristic patterns of codon preference (Table 17.1) that are reflected in the relative abundance of individual tRNA types. It is possible for mutations to occur that will not alter the amino acid sequence of a protein but that will contribute to a change in the (G + C) ratio of an organism's DNA.

Changes in codon usage that favor the more abundant tRNA species improve the translation of a particular mRNA molecule, and the converse is also true. The effect of these changes is to make it difficult for viruses infecting one organism to function equally well in its evolved relative. Similarly, the accumulating sequence differences can contribute to a genetic isolation of an evolving population. For example, one can predict the ability of particular DNA sequences to participate in recombination following genetic transformation based on knowledge of the sequence of the homologous gene in the recipient cell. This approach offers an additional method for attempting to define bacterial species, one based on sexual isolation.

Evolution of Proteins

There are three possible effects of mutations on proteins. The changes may be silent and not affect the amino acid sequence. The protein function may remain but the amino acid sequence may change, or both the function and the amino acid sequence may change. Each effect contributes different information to the evolutionary geneticist.

Silent changes are advantageous because they presumably result in minimal change in the selective pressure on the organism. The DNA (G + C) content may be trivially altered, but that effect should be small, at least for any single mutation. The lack of selection means that silent changes should accumulate steadily with time. Presumably, the more the silent changes seen, the greater the time since two different organisms diverged from their common ancestor.

Changes that preserve function but alter amino acid sequence are valuable to the structural biologist. They offer clues as to which regions of the protein molecule are critical and which are dispensable. They also provide an indication of which regions of the molecule must remain hydrophobic or hydrophilic. An example of diverged protein sequences is shown in Fig. 17.6. As more protein sequences become available in the computer databases, it becomes possible for scientists to identify the physiologic function of unknown proteins by looking for common modules. For example, ATP binding sites, DNA binding sites, and membrane-embedded proteins all have easily recognized structural motifs in their amino acid sequences.

Evolution of Regulatory Sequences

Chapters 4 and 14 carry extensive discussion of changes in regulatory sequences. New promoters can arise or old promoters can disappear. Similarly, enhancer sequences can be added or subtracted. Operators can mutate to be independent of repressors or repressors can mutate so that they always bind to operators and never to inducers. The general problem with evolutionary analysis of such structures is that they are of limited extent. The shorter the sequence being studied, the less likely the chance that there will be many phylogenetically informative changes. Moreover, promoter sequences are strongly constrained by functionality considerations. For these reasons, regulatory regions have not been extensively studied from the point of view of evolution.

Evolution of Mitochondria and Other Endosymbionts

Andersson et al. (1998) have reported the complete genome sequence for the obligate intracellular parasite *Rickettsia prowazekii*. It is 1100 kb and includes 834

```
                         Helix-Turn- Helix
        ascG   1 MTTMLEVAKRAGVSKATVQRVLSGNGYVSQETKDRVFQAVEESGYRPNLL  50
                 |.|: :||: |||| |||||||:.... .|:..: | ||:|. :|:||
        galR   1 MATIKDVARLAGVSVATVSRVINNSPKASEASRLAVHSAMESLSYHPNAN  50

                         .                     .                .
        ascG  51 ARNLSAKSTQTLGLVVTNTLYHGIYFSELLFHAARMAEEKGRQLLLADGK 100
                 ||.|....|:|:|||||.:.   : :|:.:: ...:| ..|. ||::::|
        galR  51 ARALAQQTTETVGLVVGDV..SDPFFGAMVKAVEQVAYHTGNFLLIGMGY  98

        ascG 101 HSAEEERQAIQYLLDLRCDAIMIYPRFLSVDEIDDIIDAHSQPIMVLNRR 150
                 |..:.|||||: |: ||.|:::.::::. .::..::. : ..:::::||
        galR  99 HNEQKERQAIEQLIRHRCAALVVHAKMIPDADLASLMK.QMPGMVLINRI 147

        ascG 151 LRKNSSHSVWCDHKQTSFNAVAELIMAGHQEIAFLTGSMDSPTSIERLAG 200
                 |. ..::: |.: ..: |. .||..|| |:::|.:. ... :||.|
        galR 148 LPGFENRCIALDDRYGAWLATRHLIQQGHTRIGYLCSNHSISDAEDRLQG 197

        ascG 201 YKDALA.SMVLRSMKNLSLTVNGRLPAGRRVEMLLERGAKFSALVASNDD 249
                 | |||| | : . : :.:. : .::. :. ||:|| .|.|:.. ||.
        galR 198 YYDALAESGIAANDRLVTFGEPDESGGEQAMTELLGRGRNFTAVACYNDS 247

        ascG 250 MAIGAMKALHERGVAVPEQVSVIGFDDIAIAPYTVPALSSVKIPVTEMIQ 299
                 || ||| .|::.|:.||::::|:||||| :..|. | |..|:.|:..|
        galR 248 MAAGAMGVLNDNGIDVPGEISLIGFDDVLVSRYVRPRLTTVRYPIVTMAT 297

        ascG 300 EIIGRLIFMLDGGDFSPKT..FSGKLIRRDSLIAPSR*........     335
                 : : : : |. .:.. | ||..|:||.|: .||
        galR 298 QAAELALALADNRPLPEITNVFSPTLVRRHSVSTPSLEASHHATSD     343
```

Figure 17.6. Alignment of the *ascG*-encoded *asc* repressor with the *galR*-encoded galactose operon repressor. The helix segments of the helix-turn-helix region (DNA-binding motif) are underlined. Perfect matches are indicated by a *vertical line*. Mismatches are assigned a similarity score. A pair of *dots* (:) indicates a score ≥0.5, while a *single dot* (.) indicates a score of 0–0.5. the individual letters indicate separate amino acids. A, alanine; B, asparagine or aspartic acid; C, cysteine; D, aspartic acid; E, glutamic acid; F, phenylalanine; G, glycine; H, histidine; I, isoleucine; K, lysine; L, leucine; M, methionine; N, asparagine; P, proline; Q, glutamine; R, arginine; S, serine; T, threonine; V, valine; W, tryptophan; Y, tyrosine; Z, glutamine or glutamic acid. (From Hall, B.G., Xu, L. [1992]. Nucleotide sequence function, activation, and evolution of the cryptic *asc* operon of *Escherichia coli* K-12. *Molecular biology and Evolution* 9: 688–706.)

protein-coding genes. None of the genes codes for anaerobic glycolysis, but there is a fully functional tricarboxylic acid cycle and electron transport. In effect, it is a mitochondrion, and there are detectable similarities between mitochondrial genes and *R. prowazekii* genes of similar function. Roughly 24% of the genome is noncoding, suggesting that these are remnants of silenced genes that may be lost without harm to the bacterium or be mutated to new functions.

More recently the same group (Canback et al. 2002) has examined the relationships among glycolytic enzymes in Bacteria, Archaea, and Eukarya. They compared the results from *Bartonella henselae*, a rickettsial relative, and found that there is little evidence of exchange among the three groups (mitochondria excluded), with the exception of some transfer from cyanobacteria to eukarya.

Another source of information about endosymbionts is the bacteria required by certain insects for normal metabolic functions. Several of these organisms have sequenced genomes, and certain features are apparent (Werne-green 2002). The genomes are very AT-rich and have accumulated significant numbers of deleterious mutations. The synonymous substitution rate in *Buchnera* is four times that in *E. coli*.

Genetic Structure of the Chromosome

Bacteria

There are several levels at which chromosome organization is visible. For example, there are clusters of genes in operons. How might such clusters arise? One model is the selfish operon model (Lawrence 2003) that argues for grouping of genes so that they can travel as a functional unit via transduction or other genetic transfer process. In addition, a group of genes that is coordinately regulated can use a cis-acting regulator, one that has a lower affinity binding constant. Boucher et al. (2003) argue that genes not so clustered will be unable to affect their new host in a positive sense because only the entire group of genes can carry out the pathway and participate in Darwinian selection.

In a related experiment, Audit and Ouzounis (2003) performed a computer analysis of the complete DNA sequence of 86 Bacteria and Archaea, looking for patterns in gene clustering and/or strand preference. They found what they describe as long-range correlations, meaning that at whatever scale greater than 2 kb they use to examine the chromosome, they find that genes tend to assemble and arrange themselves in the same orientation. The strand orientation bias is greater in organisms with analogs to PolC, suggesting that the mode of DNA replication may play a role in selecting for gene arrangements. The overall effect, regardless of scale, is operon-like.

Rocha (2004) points out that the presence of repeated elements leads to chromosome instability and selection for certain gene orders. For example, the linear chromosome of *Streptomyces coelicolor* has its essential genes clustered in the middle near the origin of replication. Repeated elements tend to occur out on the arms, which are unstable.

Bacteriophage

Some people have been declaring this decade to be the "Age of Phage" owing to the discovery not only of the important ecological roles played by bacteriophages, but also of the lessons to be learned from their genomics. As reviewed by Hendrix (2003), most bacteriophage genomes appear to be modular (Fig. 17.7), with each module a mosaic of genes for a particular viral pathway. However, the distinction between temperate and lytic phages is an important one genetically. In the prophage state, phages are subject to all the processes that affect the host chromosome, including lateral gene transfer. The lytic phages, on the other hand, lyse their host cell so rapidly that it is difficult to imagine how significant lateral gene transfer could occur. Therefore, families such as the T4 phage are likely to represent essentially pure vertical gene transmission.

Hendrix and coworkers have identified genes within modules that seem to have no corresponding gene in otherwise similar modules. They have designated these genes as "morons," units of more DNA that include a promoter and a transcription terminator. Morons often have a different $(G + C)$ content, which makes them correspondingly easier to identify. In pathogenic bacteria they are frequently associated with virulence factors and other functions not essential to phage growth.

Summary

Nucleic acid and protein sequence data are consistent with models that call for evolution by gene duplication and subsequent evolutionary divergence. Bacterial chromosomes exhibit clusters of genes regardless of the scale used in the examination. Relatedness measures allow experimenters to construct evolutionary trees to try to visually express the ways in which organisms or genes have evolved from each other. Two common methods of tree construction are distance measurements and evolutionary parsimony, although neither of them is perfectly satisfactory. Difficulties arise in locating informative base changes that are not so frequent so as to blur the evolutionary trail and in aligning the sequences

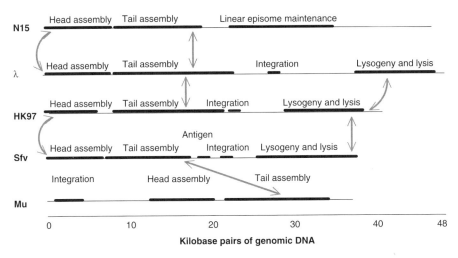

Figure 17.7. Modular design of bacteriophages. Shown in the picture are stylized genomes of five bacteriophages that infect enteric hosts. The *thin horizontal lines* represent the individual genomes. The *thicker lines* represent mosaic modules as labeled. *Gray arrows* link modules that belong to the same DNA sequence families. Note that the order of modules is often preserved. (Adapted from Lawrence, J.G., Hatfull, G.F., Hendrix, R.W. [2002]. Imbroglios of viral taxonomy: Genetic exchange and failings of phenetic approaches. *Journal of Bacteriology* 184: 4891–4905.)

so as to minimize the number of mutations needed to achieve present-day sequences. One potentially confusing aspect of bacterial evolution is that cryptic genes may be preserved intact over evolutionary time (many thousands of generations). Bacteriophages have the potential to contribute greatly to discussions on evolution because of their ubiquitous presence, ease of isolation, and modular construction.

Questions for Review and Discussion

1. How would you decide whether two bacteria are related? How would you quantify the degree of relatedness?
2. Will the possibility of convergent evolution affect your answer to question 1? Why or why not?
3. If a bacterium received a piece of DNA that coded for proteins using different codon preferences than its own, what would be the effect? Given that you

have a bacterium with pieces of DNA having different codon preferences, what might happen over evolutionary time?

4. How would you decide whether a particular gene that is presently expressed is normally cryptic?

5. What are the advantages of having a modular construction to a bacteriophage?

References

General

Boucher, Y., Douady, C.J., Papke, R.T., Walsh, D.A., Boudreau, M.E.R., Nesbø, C.L., Case, R.J., Doolittle, W.F. (2003). Lateral gene transfer and the origins of prokaryotic groups. *Annual Review of Genetics* 37: 283–328.

Brüssow, H., Canchaya, C., Hardt, W.-D. (2004). Phages and the evolution of bacterial pathogens: From genomic rearrangements to lysogenic conversion. *Microbiology and Molecular Biology Reviews* 68: 560–602.

Casjens, S. (1998). The diverse and dynamic structure of bacterial genomes. *Annual Review of Genetics* 32: 339–377.

Chibani-Chennoufi, S., Bruttin, A., Dillmann, M.-L., Brüssow, H. (2004). Phage–host interaction: An ecological perspective. *Journal of Bacteriology* 186: 3677–3686.

Dawkins, R. (1989). *The Selfish Gene*. New York: Oxford University Press. (A new edition of a classic book on evolutionary molecular biology. Most examples are not bacterial, but the point of view is an important one. For insights into how the scientific process works, don't miss the endnotes.)

Gupta, R.S. (1998). Protein phylogenies and signature sequences: A reappraisal of evolutionary relationships among archaebacteria, eubacteria, and eukaryotes. *Microbiology and Molecular Biology Reviews* 62: 1435–1491.

Hendrix, R.W. (2003). Bacteriophage genomics. *Current Opinion in Microbiology* 6: 506–511.

Karlin, S., Campbell, A.M., Mrázek, J. (1998). Comparative DNA analysis across diverse genomes. *Annual Review of Genetics* 32: 185–225.

Lake, J.A., Jain, R., Rivera, M.C. (1999). Mix and match in the tree of life. *Science* 283: 2027–2028.

Lawrence, J.G. (2003). Gene organization: Selection, selfishness, and serendipity. *Annual Review of Microbiology* 57:419–440.

Lee, P.S., Lee, K.H. (2003). *Escherichia coli*—A model system that benefits from and contributes to the evolution of proteomics. *Biotechnology and Bioengineering* 84: 801–814.

Matte, A., Sivaraman, J., Ekiel, I., Gehring, K., Jia, Z., Cygler, M. (2003). Contribution of structural genomics to understanding the biology of *Escherichia coli*. *Journal of Bacteriology* 185: 3994–4002.

Posada, D., Crandall, K.A., Holmes, E.C. (2002). Recombination in evolutionary genomics. *Annual Review of Genetics* 36: 75–97.

Riesenfeld, C.S., Schloss, P.D., Handelsman, J. (2004). Metagenomics: Genomic analysis of microbial communities. *Annual Review of Genetics* 38: 525–552.

Rocha, E.P.C. (2004). Order and disorder in bacterial genomes. *Current Opinion in Microbiology* 7: 519–527.

Wernegreen, J.J. (2002). Genome evolution in bacterial endosymbionts of insects. *Nature Reviews: Genetics* 3: 850–861.

Specialized

Andersson, S.G.E., Zomorodipour, A., Andersson, J.O., Sicheritz-Ponten, T., Alsmark, U.C.M., Podowski, R.M., Naslund, A.K., Eriksson, A.S., Winkler, H.H., Kurland, C.G. (1998). The genome sequence of *Rickettsia prowazekii* and the origin of mitochondria. *Nature* 396: 133–140.

Audit, B., Ouzounis, C.A. (2003). From genes to genomes: Universal scale-invariant properties of microbial chromosome organisation. *Journal of Molecular Biology* 332: 617–633.

Baglioni, P., Bini, L., Liberatori, S., Pallini, V., Marri, L. (2003). Proteome analysis of *Escherichia coli* W3110 expressing an heterologous sigma factor. *Proteomics* 3: 1060–1065.

Canback, B., Andersson, S.G.E., Kurland, C.G. (2002). The global phylogeny of glycolytic enzymes. *Proceedings of the National Academy of Sciences of the USA* 99: 6097–6102.

Dole, S., Nagarajavel, V., Schnetz, K. (2004). The histone-like nucleoid structuring protein H-NS represses the *Escherichia coli bgl* operon downstream of the promoter. *Molecular Microbiology* 52: 589–600.

Itaya, M., Tanaka, T. (1997). Experimental surgery to create subgenomes of *Bacillus subtilis* 168. *Proceedings of the National Academy of Sciences of the USA* 94: 5378–5382.

Khan, M.A., Isaacson, R.E. (1998). In vivo expression of the β-glucoside (*bgl*) operon of *Escherichia coli* occurs in mouse liver. *Journal of Bacteriology* 180: 4746–4749.

Lawrence, J.G., Ochman, H. (1997). Amelioration of bacterial genomes: Rates of change and exchange. *Journal of Molecular Evolution* 44:383–397.

Lenski, R.E., Winkworth, C.L., Riley, M.A. (2003). Rates of DNA sequence evolution in experimental populations of *Escherichia coli* during 20,000 generations. *Journal of Molecular Evolution* 56: 498–508.

Manna, D., Wang, X., Higgins, N.P. (2001). Mu and IS*1* transpositions exhibit strong orientation bias at the *Escherichia coli bgl* locus. *Journal of Bacteriology* 183: 3328–3335.

Novichkov, P.S., Omelchenko, M.V., Gelfand, M.S., Mironov, A.A., Wolf, Y.I., Koonin, E.V. (2004). Genome-wide molecular clock and horizontal gene transfer in bacterial evolution. *Journal of Bacteriology* 186: 6575–6585.

Ochman, H., Lerat, E., Daubin, V. (2005). Examining bacterial species under the specter of gene transfer and exchange. *Proceedings of the National Academy of Sciences of the USA* 102: 6595–6599.

Qi, J., Luo, H., Hao, B. (2004). CVTree: A phylogenetic tree reconstruction tool based on whole genomes. *Nucleic Acids Research*, 32 (Suppl. 2): W45–W47.

Simonson, A.B., Servin, J.A., Skophammer, R.G., Herbold, C.W., Rivera, M.C., Lake, J.A. (2005). Decoding the genomic tree of life. *Proceedings of the National Academy of Sciences of the USA* 102: 6608–6613.

Appendix 1

Laws of Probability and Their Application to Prokaryote Cultures

Chapter 1 includes discussions of some of the problems associated with the analysis of prokaryotic genetic systems, and Chapters 3 and 6 include some additional mathematical analysis. Another theoretical problem that must be considered is that of sampling. As noted earlier, it is not generally possible to recover all of the progeny from a cross because of the necessity of using some sort of selective technique to find a few recombinant individuals among many parental types. Therefore, when designing prokaryotic genetic experiments, it is essential to be certain that the sample is representative of the entire population of organisms and that it is of a suitable size to compensate for the random variations observed in any physical procedure. The former concern can be alleviated by good culture agitation techniques, which provide a homogeneous population for sampling. The latter concern is the subject of this appendix.

Definition of Probability

The concept of probability is, in many respects, an intuitive one. We frequently speak of "the chance" that a particular event will take place. However, what is really being discussed is the probability of the occurrence of the event. Mathematically, this probability is expressed either as a proper fraction or as a decimal fraction between 0 and 1, although in casual usage it is frequently converted to the corresponding percentage. The numbers 0 and 1 have special meanings, with 0 referring to an event that can never occur and 1 referring to an absolute certainty.

Although other systems are possible, for the purposes of this book, all probabilistic events can be characterized in one of two ways, either successes (s) or failures (f). Note that there may be many ways to succeed or fail in a particular system.

Example 1: If success means throwing a die (a cube on which each face bears a unique number from one through six) and having an even number turn up, then there are three ways to succeed and three ways to fail. If all of the numbers on the die are equally likely to turn up (i.e., the die is not "loaded"), we say that all numbers are equally probable and the chance (probability) of success is 50%, or 0.5.

Each time the die is thrown constitutes a **trial** of the system. According to the analysis in Example 1, on each trial there is a 50% chance of success, yet, as is demonstrated later, a series of five trials, all of which produce even numbers, can be expected to occur 3.1% of the time. Therefore, when considering the concept of probability, it is necessary to keep in mind that the calculated value for an outcome is really just the proportion of time a particular outcome is expected to occur. The difference between the calculated proportion and unity is the proportion of time an outcome can be expected not to occur. However, expectation does not equate with reality, as anyone who reads the weather forecast knows. Random fluctuations inherent in the physical world can be expected to affect any probabilistic event so that a calculated probability is valid only when a large number of trials is involved. This concept is expressed in the rigorous definition of **probability** by saying that, as the number of trials (t) approaches infinity, the probability of success (p) is the limit of the ratio of the number of ways to succeed (s) divided by the number of ways to succeed plus the number of ways to fail (f), or

$$p = \lim_{t \to \infty} \frac{s}{s + f} \tag{A.1}$$

assuming all outcomes occur with the same frequency. In the case of the die, $s = f$ because all numbers are equally probable; so $p = s/(s + f) = s/(s + s) = s/2s = 0.50 = 50\%$, as previously noted.

In order to have a feeling for the kinds of data that led to the development of Eq. (A.1), take a look at Table A.1 in which the results from a series of coin tosses are presented. By assuming that a coin can never land on its edge and that a head is equal to a success, these data can be made to fit the same model discussed earlier. Several points are worth emphasizing. Although it seems intuitively obvious that heads are as likely to occur as tails, the observed proportion is only 46/100 instead of 50/100. This difference between observed and predicted values is not unusual, however, because only 100 trials were involved here instead of a substantially larger number. Also note that, if only subsets of data are considered, the results can be even more skewed: in the seventh column there are 80% tails, whereas in the ninth column there are 70% heads. In the short run, wide variations in frequency are possible. Nevertheless, these frequency variations do not change the overall limit function of Eq. (A.1), and even after a run of seven consecutive tails the probability of obtaining a head on the next coin toss is still only 50%.

Dependent Versus Independent Events

Thus far, the examples used to illustrate the calculations of probability have been chosen for their intuitive clarity. In the real world, other cases are far more prevalent. It is frequently necessary to deal with the results of a series of

Table A.1. Distribution of heads and tails of a coin flipped 100 separate times.

T	T	T	T	T	T	H	T	H	T
T	T	H	H	T	H	T	H	T	T
T	T	H	T	H	H	T	H	T	H
T	H	T	H	H	H	H	T	H	T
H	H	H	H	H	T	T	T	H	T
T	H	H	T	H	T	T	H	H	H
H	T	H	T	T	H	T	H	H	H
T	T	T	H	T	T	T	T	T	T
H	H	T	T	T	H	T	T	H	H
H	H	H	T	T	H	T	T	H	T

Note: T means that a tail was observed, and an H means that a head was observed. In total, there were 46 heads and 54 tails.

samples (trials) and to try to infer the nature of the entire population from the composition of this rather limited number of samples.

When dealing with a case involving multiple samples, it is important to distinguish between independent events such as coin tosses or die throws and dependent events. For a dependent event, the probability of a successful outcome on a subsequent trial is influenced by the results of the preceding trial(s). The same is not true for an independent event. In order to illustrate this point, consider a paper bag containing 20 red marbles and 5 blue ones. It is clear that the probability of drawing out a red marble from the bag is 20/(20 + 5), or 80%. If, before the second trial, the first marble is returned to the bag and the bag shaken well, the process is called sampling with replacement. In such a case, the events are independent and the probability of success is constant at each trial. However, if the first marble is not returned to the bag, it is sampling without replacement, and the events are dependent because the probability of drawing out a red marble on the second trial is 19/(19 + 5) = 79.2% if a red marble was chosen first but 20/(20 + 4) = 83.3% if a blue marble was chosen first.

It is also necessary to distinguish between outcomes that are mutually exclusive and those that are not. The usual example of a mutually exclusive outcome is the result of drawing a single card from a complete deck of cards. The card cannot be both a seven and an eight, for example, yet it may be that drawing either a seven or an eight would be considered a success. Such a case forces the reconsideration of the method for calculating the probability of a particular outcome for the case of multiple events.

Example 1 reconsidered: Suppose a single die is to be rolled, and the result is considered a success if the number that appears on top is even. The probability of rolling a two is 1/(1 + 5) = 1/6. Similarly, the probability of rolling a four is 1/6 and that of rolling a six is also 1/6. The outcomes are mutually exclusive, but any one is acceptable. This proportion is expressed mathematically by saying that the overall probability of success is the sum of the individual probabilities of each mutually exclusive successful outcome, or 1/6 + 1/6 + 1/6 = 1/2, which is the same result as observed earlier.

Example 2: Suppose the problem given in Example 1 is reworded so that, instead of asking for any even number on one roll of the die, we want each of the even numbers in turn as the die is rolled three times (first a two, then a four, then a six). The probability of getting a two on the first roll is 1/6, the probability of getting a four on the second roll is 1/6, and the probability of getting a six on the third roll is also 1/6. Each probability is independent of the others, and the outcomes are not mutually exclusive. Therefore, one-sixth of the time when the die is rolled, the number that appears will be a two. In the case where a two does appear, only one-sixth of the time will the next number be a four. Therefore the probability of a two and then a four is $1/6 \times 1/6 = 1/36$. Moreover, if we now ask for a six, the probability becomes $1/6 \times 1/6 \times 1/6 = 1/216$. Note that the same result is obtained for the cases where the numbers appear in a different order or where the same preselected number appears three consecutive times.

Example 3: Suppose the conditions are the same as in Example 1 except that you wish to know the probability of rolling an even number five consecutive times. By the reasoning given in Examples 1 and 2, the probability of rolling an even number is $1/2$, so the probability of rolling two consecutive even numbers is $1/2 \times 1/2 = 1/4$, the probability of rolling three consecutive even numbers is $1/2 \times 1/2 \times 1/2 = 1/8$, and the probability of rolling five consecutive even numbers is $1/2 \times 1/2 \times 1/2 \times 1/2 \times 1/2 = 1/32$.

In summary, for the case of multiple trials, the overall probability is the product of the probabilities of a successful outcome at each individual trial. For the case in which the successful outcomes for a single trial are mutually exclusive, the probability of success in that individual trial is the sum of the probabilities for each possible successful outcome.

Application of Binomial Expansion to Probability Theory

Permutation and Combination

Till now, all of the examples considered have had an easily countable number of outcomes so that the ratio $s/(s + f)$ has had a readily determined numerical value, and the notion of the limit has not been necessary. For the type of probability calculation involved in bacterial genetics, this situation generally does not pertain. Instead, the population of cells, viruses, etc. is so large that exact calculations become more difficult, if not impossible. Nevertheless, it is possible to speak rigorously about certain types of probability by introducing the concept of permutation and combination.

The terms permutation and combination refer to samples taken from a population composed of individually identifiable members. For example, if a bag contains a set of marbles, each of which bears a unique number, there are conceptually two ways to sample the population. One might begin by removing marbles one at a time from the bag and placing them in a row. Suppose, first, a group of five marbles is removed from the bag and their positions in the line noted. It is clear that, if the marbles were returned to the bag and then again removed one marble at a time, each individual marble might be removed in a different sequence. The sequence of numbers in the first sample might be 12345, and the sequence in the second sample might be 54321. The composition of the two groups is identical, but the sequence in which they were obtained is different. Therefore, the two groups represent different permutations of the same combination of items.

When the term **permutation** is used, it refers not only to the composition of a sample but also to the ordering of the items within a sample. On the other hand, the term **combination** refers only to the overall composition of a sample without regard to any sort of internal order within the sample.

If the total size (N) of the population to be sampled is known and the size of the sample (n) is specified, it is possible to calculate precisely how many permutations and combinations can exist. The number of permutations (P) can be calculated from the formula:

$$P_{N,n} = \frac{N!}{(N-n)!}$$
$$= \frac{N(N-1)(N-2)\cdots(N-n+1)(N-n)(N-n-1)\cdots(3)(2)(1)}{(N-n)(N-n-1)\cdots(3)(2)(1)} \tag{A.2}$$
$$= N(N-1)(N-2)\cdots(N-n+2)(N-n+1)$$

where the sign ! means factorial. The **factorial** of a number is the product of the specified number and each integer less than itself down to and including 1. Thus, 3! $= 3 \times 2 \times 1$. The third form of the equation, which is more cumbersome to write but easier to calculate, is obtained by dividing the numerator and the denominator by the quantity $(N-n)!$ The number of possible combinations (C) that can occur is, of course, less than the number of permutations and is calculated from the formula:

$$C_{N,n} = \frac{N!}{n!(N-n)!}$$
$$= \frac{N(N-1)(N-2)\cdots(N-n+2)(N-n+1)}{n!} \tag{A.3}$$

Example 4: Consider the case of a bag of red and blue marbles, each of which has a unique number on it. There are ten red marbles and five blue marbles, and a sample of three marbles is to be chosen at random. The number of different permutations is

$$P_{N, n} = P_{15, 3} = \frac{15!}{12!} = 15 \times 14 \times 13 = 2730$$

whereas the number of different combinations is only

$$C_{N, n} = C_{15, 3} = \frac{15!}{3!12!} = \frac{15 \times 14 \times 13}{3 \times 2 \times 1} = 455.$$

Remember that a sample consisting of marbles number 15, 1, and 7 (in that order) is a different permutation from 7, 15, and 1 but represents the same combination. Roughly 26% of the time all of the marbles will be red ($10/15 \times 9/14 \times 8/13 = 0.264$), but it is still possible to distinguish among the marbles because of their numbers.

Binomial Expansion

When endeavoring to deal with a large population that is to be sampled, the theory can be greatly simplified if there are only two possible outcomes, s and f, as given earlier. Each sample can then be categorized as having a particular number of successes and a particular number of failures. In the case of the bag of red and blue marbles, for example, red might be considered a success. Then, if a sample were removed and found to contain only one red marble and two blue marbles, it would have one success and two failures. Such a distribution presumably reflects the actual proportion of red to blue marbles in the entire population. However, if the population is very large and the sample size relatively small, it is difficult to evaluate the significance of the distribution in a single sample due to random fluctuations. In order to obtain more information about the entire population, additional samples would be necessary. One way to determine the ratio of red to blue marbles would be to take a large number of samples, sum the total number of red and blue marbles, and assume that their ratio is the ratio in the larger population. This method can be accurate if enough samples are taken, but it wastes some of the information available from the sampling procedure. A better and more economic way to utilize the information from the samples is to look not only at the red/blue ratio within a sample but also at the frequency with which that sample class appears. The most frequently appearing sample class is presumably most reflective of the general population.

In order to develop a mathematic basis for the procedure just outlined, it is necessary to reverse the problem. Suppose the population of marbles in the bag is known to have a certain proportion of red marbles (a) and a certain proportion of blue marbles (b). These proportions are determined from Eq. (A.1) (e.g., a = number of red marbles/[number of red marbles + number of blue marbles]). Then, of necessity, $a + b = 1$, as no other color of marbles is possible. Furthermore, assume that the number of marbles involved is so large that removing a few samples does not significantly change a or b. Then, if a sample of five marbles is removed as before, the probability of obtaining a particular arrangement of four red marbles and one blue marble is $a \times a \times a \times a \times b$ or a^4b because the outcomes are not mutually exclusive. The probabilities for other arrangements can be calculated in a similar manner. For instance, the probability would be a^2b^3 for a sample consisting of two red and three blue marbles. However, these probabilities are actually the probabilities only for that particular permutation of marbles, whereas in fact we are interested in the probability of various combinations.

The expression can be corrected to reflect the number of possible combinations by noting that the probability for a particular sample of size n is always equal to $a^rb^{(n-r)}$, where r is the number of successes in the sample. Equation (A.3), however, shows that the number of combinations of N things taken n at a time is equal to $C_{N,n}$. Therefore, the actual probability of a sample that has a specific amount of success (r) is $C_{N,n}[a^rb^{(n-r)}]$ (where r may range in value from 0 up to n). The term $C_{N,n}[a^rb^{(n-r)}]$ is merely one term of the binomial expansion $(a + b)^n$, however, and therefore, the complete probability listing for all possible samples is $(a + b)^n$, which is the binomial expansion.

The binomial expansion thus represents a mathematically exact way of presenting the probabilities of obtaining various samples from a given population. Samples distributed in such a way that they can be described by the binomial expansion are said to be binomially distributed. The binomial distribution has several important advantages. The mean (average) amount of success obtained in a series of trials using a particular population is

$$m = np \tag{A.4}$$

where p is the probability of success in a single trial (removal of a single item) and n is the total number of trials made. Sometimes the term m is referred to as the **expectation**, as np represents the expected number of successes in a sample of size n. Random variations result in actual samples with varying amounts of success distributed about the mean value for the amount of success in the population as a whole. The degree of scatter of the amount of success in samples from the same population is usually expressed as the standard deviation (σ). The smaller the numerical value for σ, the more homogeneous are the samples. For the binomial distribution, the standard deviation is easily calculated as

$$\sigma = \sqrt{npq} \qquad\qquad (A.5)$$

where n and p are defined as in Eq. (A.4) and $q = 1 - p$ the probability of failure in a single trial.

Poisson Approximation

Although the binomial distribution is mathematically precise, it is cumbersome to evaluate. Therefore, numerous methods that approximate the binomial distribution under certain conditions have been developed. Two such methods are the **normal distribution** and the **Poisson distribution**. The normal distribution is familiar to every student as the famous "bell-shaped curve" so often used to assign grades to classes. Although its use is widespread, it is not as convenient for bacterial genetics as the Poisson distribution, as it functions best when the value for p is near 0.5, and bacterial genetics rarely studies events that are so frequent.

The Poisson approximation to the binomial distribution, however, was specifically developed to deal with rather rare events, cases in which p is considerably less than 0.5. It involves placing limitations on m as defined in Eq. (A.4). If n is very large and p is very small, m becomes a number of modest size, of the order of 0.1–5.0. Under these conditions, the probability of exactly r successes in a sample of size n is

$$P = \frac{e^{-m}m^r}{r!} \qquad\qquad (A.6)$$

The number e is an irrational transcendental number, chosen for philosophical reasons that need not concern us, which is the base of the Napierian or natural system of logarithms. Its approximate value is 2.7182828. . . . Logarithms based on the Napierian system are usually abbreviated ln to distinguish them from logarithms based on the number 10, which are abbreviated log. Some selected values of e^{-m} are given in Table A.2. In addition to the advantages of the binomial distribution, the Poisson approximation is unique because the standard deviation

Table A.2. Values of e^{-m} for use in Eq. A.6.

m	e^{-m}	m	e^{-m}	m	e^{-m}	m	e^{-m}	m	e^{-m}
0.1	0.905	0.8	0.449	1.5	0.223	2.4	0.091	3.8	0.022
0.2	0.819	0.9	0.407	1.6	0.202	2.6	0.074	4.0	0.018
0.3	0.741	1.0	0.368	1.7	0.183	2.8	0.061	4,2	0.015
0.4	0.670	1.1	0.333	1.8	0.165	3.0	0.050	4.4	0.012
0.5	0.607	1.2	0.301	1.9	0.150	3.2	0.041	4.6	0.010
0.6	0.549	1.3	0.273	2.0	0.135	3.4	0.033	4.8	0.008
0.7	0.497	1.4	0.247	2.2	0.111	3.6	0.027	5.0	0.007

Note: More extensive values can be found in standard reference works such as the *Handbook of Chemistry and Physics* published by the Chemical Rubber Publishing Company.

is equal to the square root of the mean, because as p becomes very small, q approaches a value of 1. Therefore, $\sqrt{npq}$ tends to be approximated by $\sqrt{np}$ for small p, and from Eq. (A.4) it is the same as $\sqrt{m}$.

The Poisson approximation is particularly advantageous for bacterial geneticists because in most cases the researcher is dealing with a very large population of organisms and is looking for a rare event by means of selection. An experimental design can be considerably aided by mathematical analysis.

Example 5: A researcher wishes to study the progeny produced by a single phage-infected cell. He begins with 10^8 bacteria and adds to them 10^6 phage particles. If he removes a sample of 100 bacteria from the culture, what is the probability that the sample will have no infected cells? One infected cell? Two infected cells? Three or more infected cells?

All that is necessary for the application of Eq. (A.6) is the knowledge of m and r. Assume that success is equated with the recovery of a phage-infected cell. By Eq. (A.4), $m = np$, and from the statement of the problem we know that $n = 100$ bacteria, and $p = 10^6/10^8 = 0.01$ phage per bacterium. Therefore, $m = 1$ phage particle per sample. The amount of success is represented by r, and according to the statement of the problem, taking a success to be a phage-infected cell, r is equal to 0, 1, 2, and 3 or more. Using the formula for the case in which $r = 0$, we find

$$P = \frac{e^{-m} m^r}{r!} = \frac{e^{-1} \cdot 1^0}{0!}$$

which is not difficult to evaluate if one can determine a value for 0!. By convention, both 1! and 0! are taken as being equal to 1. Because any number raised to the zero power is equal to 1, for $r = 0$, $P = e^{-1} = 0.367$ using the value of e^{-m} given in Table A.2. By similar reasoning, for $r = 1$,

$$P = \frac{e^{-1} \cdot 1^1}{1!} = e^{-1} = 0.367$$

and for $r = 2$,

$$P = \frac{e^{-1} \cdot 1^2}{2!} = \frac{e^{-1}}{2} = 0.184.$$

Each of the probabilities thus far calculated represents the probability of an exact number of phage-infected cells (successes) per sample. The last part of the problem asks for the case of three or more phage-infected cells. In almost all cases, this type of question is best answered by calculating all of the probabilities for the cases that are not included and then subtracting them from unity. For this specific case, the probability of three or more infected cells is equal to $1 - P_0 - P_1 - P_2 = 1 - 0.367 - 0.367 - 0.184 = 0.082$.

Therefore, 73.4% of the time the researcher may expect to find either zero or one phage-infected cell in the sample. If this figure is not frequent enough (i.e., if there are too many cases of multiply infected cells), by changing either n or p the probabilities can be altered to suit the experimental design. For example, if only 50 bacteria are taken per sample, $m = np = 50 \times 0.01 = 0.5$. Then

$$P_0 = \frac{e^{-0.5} \cdot 0.5^0}{0!} = 0.607$$

and

$$P_1 = \frac{e^{-0.5} \cdot 0.5^0}{1!} = 0.303$$

and the researcher has either zero or exactly one phage-infected cell 91% of the time.

When the sample size was 100, the probabilities for the $r = 0$ and $r = 1$ case were identical, but it was not the case when the sample size was 50. This situation is indicative of the way in which the Poisson distribution is skewed for small values of m. Figure A.1 shows the Poisson distribution for several values of m. Only when $m = 5$ does the distribution become symmetric. As m becomes smaller, the zero success case naturally predominates, and the width of the curve becomes smaller because the standard deviation is $\sqrt{m}$. By the time m reaches a value of 0.1, zero successes occur 90.5% of the time.

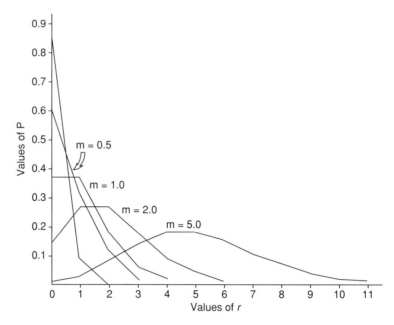

Figure A.1. Poisson distribution. Values of P have been calculated for selected values of m using Eq. (A.6).

The problem of inferring distributions within the larger population from the composition and frequency of various samples has already been mentioned. This type of analysis can be done easily with Eq. (A.6).

Example 6: The same phage researcher who was laboring in Example 5 has removed a series of samples from a culture. When the samples are tested, 74.1% of them have no phage-infected cells. What is the average number of phage-infected cells in the culture?

In this case it is necessary to solve Eq. (A.6) for m. Although it can be done for any value of r, the most convenient values are $r = 0$ or $r = 1$. Beginning with $P = e^{-m}m^r/r!$ and substituting $r = 0$, we have $P = e^{-m} \times 1/1$ so that $P = e^{-m}$ or $-m = \ln P$. However, $P = 0.741$, so $\ln P = -0.3 = -m$, so $m = 0.3$. Therefore, on the average, there is three-tenths of an infected cell per sample, or put another way, there are three infected cells per ten samples. Once m is known, it is of course possible to calculate either n or p using Eq. (A.4), provided the other value is known.

Another occasional use of the Poisson distribution is to verify that a series of samples, in fact, does reflect a random distribution of success in the population as a whole. This type of analysis is discussed in Chapter 6 in connection with some experiments by Benzer.

Summary

Probability is the likelihood that a particular event will occur. For the case in which there are only two outcomes, successes and failures, probability can be rigorously defined as

$$\lim_{n \to \infty} \frac{s}{s+f}$$

where n is the total number of trials, s is the number of ways to succeed, and f is the number of ways to fail.

The terms permutation and combination refer to samples that contain more than one item. If the order of the items within the sample is considered, the sample is a specific permutation. If the order of the items is not important, the sample is one type of combination. The number of possible permutations is $P_{N,n} = N! / (N - n)!$, whereas the number of possible combinations is $C_{N,n} = N!/n!/(N - n)!$, where N is the number of items in the population and n is the number of items in the sample.

The frequency distribution of all possible combinations of successes and failures within samples of a certain size is given by the binomial expansion $(a + b)^n$, where a is the proportion of successes in the total population, b is the proportion of failures, and n is the sample size. Because this quantity is somewhat difficult to evaluate, an approximation method is generally used. The most useful approximation method for bacterial genetics is that of Poisson, which is $P = (e^{-m}m^r)/r!$, where r is the number of successes in the sample, e is the base of the natural logarithms, and $m = np$. The explicit assumptions of the approximation are that n, the sample size, is very large and p, the proportion of successes in the population, is very small, so that m, the expectation, is a number of modest size.

Problems

Answers to the odd-numbered problems are provided at the end of this section.

1. If the cat goes into the closet one time in five when the door is open, and you forget to check for the cat one time in four before you close the door, what is the probability that you will close the cat in the closet?

2. A child's toy consists of a cone on which five rings of different diameters can be stacked. There is only one way in which the rings can be stacked so that they will all fit on the cone at the same time. If the child chooses the rings randomly, what is the probability the toy will be assembled properly?

3. In a simplified version of roulette, there are 36 numbers on a wheel. A random selector mechanism chooses one number at each trial. If you must pay one dollar for each trial but win 30 dollars if the number on which you bet is selected, what is your expected dollar loss per trial? How many times should you expect to be able to play if you begin with a stake of exactly 60 dollars?

4. If you have a tube containing 20 bacteria and you add to it two phage particles that each infect a separate cell, what is the probability that any particular bacterial cell you select is infected by a phage? If the phages are added so that both could infect the same cell, what is the probability that the bacterial cell you select is infected by both phage particles? Given that the bacterial cell you select is phage-infected, what is the probability that it is infected by both phage particles?

5. Using a standard deck of cards, what is the probability of drawing out one card and having it be either an ace or a king? What is the probability of drawing two cards and obtaining one king and one ace (assume that you replace the first card before drawing the second)? What is the probability of drawing two cards and obtaining first a king and then an ace (assume that you do not replace the first card)? What is the probability of drawing two cards without replacement and obtaining two kings?

6. A friend has a bag of candy containing 30 jelly beans of a kind you do like and 10 jelly beans of a kind you do not like. What is the probability that you will like the first two jelly beans you select (your friend will be offended if you put any of them back)? If you select four jelly beans from the bag, how many would you expect to like?

7. After a sudden flood, all the labels on the culture collection have washed off. You have a test tube rack containing 20 bacterial strains, and you know that 10 of the strains are donor cells and 10 are recipients. You have an experiment to do in which it is necessay to mix a donor with a recipient. What is the probability that the first strain you select will be a donor? What is the probability that you will select first a donor and then a recipient? What is the probability that the two strains you randomly select will be a donor and a recipient?

8. Suppose in a particular bacterial strain all mutations occur with a frequency of 10^{-6} and that there are three separately mutable sites on the bacterial genome that give rise to the same phenotype. What is the probability of observing a mutant phenotype? If you require a bacterium that is mutant for both the first phenotypic character discussed earlier and a second phenotypic character that is determined at a single separate locus, what is the probability of observing it?

9. If you have 15 billiard balls and you are going to choose a group of five, how many permutations could you get? How many combinations could you get?

10. The same flood as in Problem 7 washed the labels off six reagent bottles. Three of these bottles are necessary for a particular enzyme assay. How many combinations of three bottles can you make from the six unlabeled bottles? What is the probability of selecting the correct three for your enzyme assay? Assuming that the enzyme assay requires the addition of the reagents in a specific order and assuming that you have correctly selected the three bottles, what is the probability that you will add the reagents in the correct order? What is the probability that you will both select the correct three bottles and add the reagents in the correct order?

11. A somewhat eccentric professor has a fashion plate with a collection of 200 shirts. Unfortunately, 20 of them are grease-stained, and he is too befuddled to notice. Assuming he selects a shirt at random each morning and hangs it back in the closet each night, what is the expectation that the professor will wear a grease-stained shirt on any given day? What is the probability that he will not wear a stained shirt in 5 days? What is the probability that he will wear exactly one stained shirt in 5 days? What is the probability that he will wear two or more stained shirts in 5 days?

12. A masochistic bicyclist is taking a ride across the desert. If the probability that he will get a thorn in either of his tires is 0.1 for every 100 m he travels, what are the probabilities of a thornless journey of 100 m, 500 m, or 1000 m?

13. A culture of bacteria has accidentally been contaminated with a second type of bacterium at a level such that, of every 10^8 bacteria removed from the culture, there are 10^7 bacteria that are contaminants. If the culture is streaked on an agar plate to purify it, what is the probability that the first colony examined will be the contaminant rather than the correct bacterium? What is the probability that a sample of ten colonies will not have any contaminants? What is the probability of finding at least one contaminating colony in a sample of ten colonies?

14. If you assume that one genetic exchange event will be observed for every 10^5 cells sampled, how many cells should be removed per sample in order for the probability of the samples having at least one genetic exchange to be 50%? Suppose you wish to limit the frequency of multiple genetic exchanges (two or more) per sample to less than 10%. How should you adjust the size of the sample?

15. In the experiment described in Problem 14, you observe that, when samples of 10^5 cells are taken, 67% of the samples show no genetic exchanges. Recalculate the expectation for a genetic exchange event.

16. Culture aliquots of 10^8 cells are tested to see if they contain any cells carrying a particular mutation. It is observed that 5% of the samples contain no mutant cells. What is the average number of mutant cells per sample? What is the frequency with which the mutant cells are observed in the culture?

17. A culture of bacteria is infected with phage at the ratio of 20 bacteria for every phage particle, and aliquots of 100 bacteria are taken. How many phage-infected cells would you expect to find in a sample (assume that all phages infect a cell)? What is the probability of having no phage-infected cells in the sample? What should the sample size be so that the probability of no phage-infected cells becomes 5%?

18. A series of samples has been taken from a bacterial culture, and it has been observed that 67% of the samples have no mutant cells in them and 27% of the samples have exactly one mutant cell. What percentage of the samples would you expect to have two mutant cells? Three mutant cells? At least four mutant cells?

Answers to Odd-Numbered Problems

1. $\dfrac{1}{5} \times \dfrac{1}{4} = \dfrac{1}{20} = 0.05$

3. $\$\dfrac{1}{6}$; 360 times

5. $\dfrac{4}{52} + \dfrac{4}{52} = \dfrac{2}{13} = 0.154, 2\left(\dfrac{4}{52} \times \dfrac{4}{52}\right) = \dfrac{2}{169} = 0.012;$

   $\dfrac{4}{52} + \dfrac{4}{51} = \dfrac{4}{663} = 0.0060; \dfrac{4}{52} \times \dfrac{3}{51} = \dfrac{1}{221} = 0.0045$

7. $\dfrac{10}{20} = 0.5; \dfrac{10}{20} \times \dfrac{10}{19} = \dfrac{5}{19} = 0.263; 2\left(\dfrac{10}{20} \times \dfrac{10}{19}\right) = \dfrac{10}{19} = 0.526$

9. $P_{15,5} = \dfrac{15!}{10!} = 15 \times 14 \times 13 \times 12 \times 11 = 360,360;$

   $C_{15,5} = \dfrac{15!}{5!10!} = \dfrac{360,360}{5 \times 4 \times 3 \times 2 \times 1} = 3003$

11. $\dfrac{20}{200} = 0.1; P_{5,0} = \dfrac{e^{-0.5} \bullet 0.5^0}{0!} = e^{-0.5}$

    $= 0.607$ or $0.9 \times 0.9 \times 0.9 \times 0.9 \times 0.9 = 0.590$

    $P_{5,1} = \dfrac{e^{-0.5} \bullet 0.5^1}{1!} = 0.303$ or $5 \times 0.9^4 \times 0.1 = 0.328;$

    $1 - 0.607 - 0.303 = 0.090$ or $1 - 0.590 - 0.328 = .082$

13. $\dfrac{10^7}{10^8} = 0.1; P_{10,0} = \dfrac{e^{-1} \bullet 1^0}{0!} = 0.368; 1 - 0.368 = 0.632$

15. $0.67 = \dfrac{e^{-m} m^0}{0!}$ or $e^{-m} = 0.67;$ and $m = 0.40;$ or 4 genetic exchanges/10^6 cells

17. $\dfrac{1}{20} \times 100 = 5; P_{100,0} = \dfrac{e^{-5} \bullet 5^0}{0!} = e^{-5} = 0.007; 0.05 = e^{-m};$

    so $m = 3 = np$ and $n = \dfrac{3}{p} = \dfrac{3}{0.05} = 60$ cells/sample

Appendix 2

Useful World Wide Web Sites for Bacterial Genetics

Links to these and other sites more recently discovered are located at the web page for this book:

http://lifesciences.asu.edu/bactgen/

Bacillus Genetic Stock Center: http://www.bgsc.org/

Bacteriophage-related sites together with a large body of historical information: www.phage.org/

Canadian-sponsored program to use computers to model an *E. coli* cell: http://www.projectcybercell.ca/

Cogent has databases that list all of the sequenced genomes and their predicted proteins:

http://maine.ebi.ac.uk:8000/services/cogent/

Complete genome sequences:

http://www.ncbi.nlm.nih.gov/entrez/query.fcgi?db=Genome

Department of Energy Joint Genomics Institute provides extremely rapid sequencing of total genomes: http://www.jgi.doe.gov/

DNA Genome Atlas provides a new way to look at large-scale structure: http://www.cbs.dtu.dk/services/GenomeAtlas/

DNA Interactive allows you to experiment with tools to explore DNA structure and function: http://www.dnai.org/

E. coli Genetic Stock Center: http://cgsc.biology.yale.edu

Eco-Cyc Encyclopedia of *Escherichia coli* K12 Genes and Metabolism: http://eco-cyc.org/

German Collection of Microorganisms and Cell Cultures: http://www.dsmz.de/index.htm

Inteins: http://www.neb.com/neb/inteins.html

International Sequencing Consortium Large-Scale Sequencing Project Database. Use the database on this site to find currently active sequencing projects: www.intlgenome.org/

Kyoto Encyclopedia of Genes and Genomes is now available. It has a variety of information including genes and biochemical pathways: http://www.genome.ad.jp/kegg/

Phage Page is a compendium of topics related to phage T4: http://www.mbio.ncsu.edu/ESM/Phage/Phage.html

Pseudomonas genome database: http://www.pseudomonas.com/

Salmonella Genetic Stock Center: http://www.ucalgary.ca/~kesander

Salmonella working DNA sequence is now available online: http://www.salmonella.org/

SRS review of genomes and gene functions: http://srs.ebi.ac.uk/

Statistics and probability. Included here are examples of the binomial distribution and the Poisson distribution: http://socr.stat.ucla.edu/htmls/SOCR_Distributions.html

Streptomyces genome information: http://www.jic.bbsrc.ac.uk/SCIENCE/molmicro/Strept.html

TmRNA: http://www.indiana.edu/~tmrna/

Viral taxonomy web site with descriptions of a wide variety of viruses that infect prokaryotes: http://www.ncbi.nlm.nih.gov/ICTVdb/Ictv/fr-fst-h.htm#Archaea

Answers to Application Questions

Chapter 1

1. Use fluorescent-labeled proteins to tag specific portions of the chromosome. Then use fluorescence microscopy to visualize the origin of chromosome replication at different times during the cell cycle. Gordon, G.S., Sitnikov, D., Webb, C.D., Teleman, A., Straight, A., Losick, R., Murray, A.W., Wright, A. (1997). Chromosome and low copy plasmid segregation in *E. coli*: Visual evidence for distinct mechanisms. *Cell* 90: 1113–1121.

Chapter 2

1. Fuse the *polC* gene (codes for major subunit of DNA polymerase III) to a gene coding for a fluorescent protein. Fluorescence microscopy will show the

physical location of PolIII complexes in the bacterial cell. Migocki, M.D., Lewis, P.J., Wake, R.G., Harry, E.J. (2004). The midcell replication factory in *Bacillus subtilis* is highly mobile: Implications for coordinating chromosome replication with other cell cycle events. *Molecular Microbiology* 54: 1365–2958.

Chapter 5

1. Gene knockout analysis. Baliga, N.S., Bjork, S.J., Bonneau, R., Pan, Min, Iloanusi, C., Kottemann, M.C.H., Hood, L., DiRuggiero, J. (2004). Systems level insights into the stress response to UV radiation in the halophilic Archaeon *Halobacterium* NRC-1. *Genome Research* 14: 1025–1035.

Chapter 7

1. Guo and Chen (1997) set up an in vitro reaction in which they could substitute an inactive pRNA molecule for one of the six pRNAs associated with a precapsid. Under those conditions, complete blockage of DNA packaging occurred.

Chapter 12

1. Use pulsed field gel electrophoresis of restriction digests. Pang, X., Zhou, X., Sun, Y., Deng, Z. (2002). Physical map of the linear chromosome of *Streptomyces hygroscopicus* 10–22 deduced by analysis of overlapping large chromosomal deletions. *Journal of Bacteriology* 184: 1958–1965.

Chapter 16

1. Inject bacteriophage directly into diseased animals. Huff, W.E., Huff, G.R., Rath, N.C., Balog, J.M., Donoghue, A.M. (2003). Bacteriophage treatment of a severe *Escherichia coli* respiratory infection in broiler chickens. *Avian Diseases* 47: 1399–1405.

Glossary

A site   location on a ribosome where an incoming tRNA molecule binds so that it can match up with the mRNA anticodon. See also **E site** and **P site**.

ABC transporter   a combination of a periplasmic, high-affinity binding protein, two cytoplasmic membrane-embedded transport proteins, and an associated cyto-plasmic ATPase that performs active transport.

Abortive transductant   a cell that has received a piece of transducing DNA that fails to recombine or replicate, but does persist for a long time. The transducing DNA can express its gene(s) and alter the phenotype of the single cell in the clone that carries it.

Adaptive evolution   change in the genotype of an organism that seems to be directed by the prevailing selective conditions.

Addiction module   a pair of genes that make a long-lived toxin and a short-lived antitoxin. Loss of the antitoxin results in cell death.

Agroinfection   transmission of viral DNA into a plant cell via the T-DNA from a Ti plasmid.

Alkylating agents   chemicals that add alkyl groups such as methyl or ethyl residues to nucleotide bases.

Annotation   refers to the process of analyzing a DNA sequence to identify known genes and regions predicted to code for proteins based on computer comparisons to known sequences.

Antirepressor   a protein molecule produced by phage P22 that antagonizes the action of other temperate phage repressors.

Antisense RNA   an RNA molecule, usually small, that is complementary to a portion of an mRNA molecule. Its binding normally prevents some transcription or translation activity.

Antitermination   binding a protein or proteins to an RNA polymerase holoenzyme complex, thereby preventing it from recognizing what would normally be a transcription stop signal.

Attenuator   a portion of the leader sequence of an mRNA molecule within which premature termination of transcription can occur. The frequency of termination is empirically controlled by the rate of translation of a short peptide.

Autotroph   an organism that requires some specific biochemical compound for growth (amino acid, vitamin, etc.).

Autoradiography   exposure of an x-ray film by the decay of radioactive atoms contained in a specimen placed against the film.

Backward mutation   a mutation that restores the genotype to its previous configuration, a reversion.

Bacteriocin   a substance produced by a plasmid-carrying bacterium that kills non-producing cells (i.e., those without the plasmid).

Bacteriophage   a virus that infects bacteria.

Base analog   a chemical that resembles a nucleotide base but does not have the same base-pairing properties.

Base excision repair   restoration of normal DNA base structure by removal of a region including defective bases.

Base substitution   a mutation that occurs when one nucleotide base is substituted for another.

Blotting   transferring nucleic acids or proteins from an electrophoretic gel to a solid support by covering the gel with the supporting material and allowing the separated molecules to move from the gel to the solid surface. The movement can be the result of diffusion or an electric field perpendicular to the gel.

Branch migration   when two DNA molecules recombine there is a point of crossover (a branch), a region within which strands are exchanging. This crossover point is free to migrate (slide) along the duplexes because as it slides hydrogen bonds are made and broken at the same rate so that no energy input is required for its movement.

Budding   a type of cell division in which a daughter cell develops as a protrusion of steadily increasing size from a mother cell.

Burst size   the average number of virus particles released when an infected host cell lyses.

Catabolite (or glucose) repression   the presence of glucose in a culture medium prevents certain operons from being expressed even if their inducers are present.

Catenane   interlocked circular molecules, usually DNA.

Chi site   a sequence of eight DNA bases that, in the proper orientation, stimulates nicking activity of exonuclease V and thereby initiates recombination.

Cis–trans **test**   a genetic test that compares the phenotypes of cells carrying paired DNA molecules. In one case (*cis*), one molecule carries two mutations affecting genes coding for diffusible products and the other carries none. In the second case (*trans*), each molecule carries one of the same two mutations. If the mutations are in different genes, both cases will exhibit a wild-type phenotype, while if the mutations are in the same gene, only the *cis* case will have a wild-type phenotype.

Cistron   a region of DNA within which the *cis–trans* test is negative.

Cloning   introducing linked DNA into a cell so that it can replicate. The result is a large number of cells (e.g., a colony) all carrying the same DNA complement.

Closed complex   RNA polymerase bound to an intact DNA helix, one of the first stages in transcription initiation. See also open complex.

Codon   a triplet of bases in a DNA strand that codes for an amino acid.

Cointegrate molecule   a single, circular DNA molecule formed from two smaller circular molecules by recombination.

Colony   a macroscopically visible mound of cells on an agar plate. Well-isolated colonies are descended from individual founding cells.

Competence   a physiological state in which a bacterial cell will take up naked DNA molecules and transport them into its cytoplasm.

Competence pseudopilus   the DNA transport structure used in genetic transformation whose components are either products of *pil* genes or genes that are similar.

Complementation   when a product, or products, encoded on one DNA molecule can replace a defective function encoded on another DNA molecule in the same cell.

Concatemer   a macromolecule composed of identical subunit macromolecules joined end-to-end. An example would be a DNA molecule containing several complete viral genomes.

Conditional mutation   a change in the base sequence of a DNA molecule that results in a mutant phenotype that is expressed only in a particular situation such as high temperature.

Conjugative plasmid   a self-replicating piece of DNA that is capable of transferring itself from one cell to another while the two cells are touching.

Constitutive   a gene that is always expressed at high levels, an unregulated gene.

Context effects   a particular base sequence confers a specific property, such as the ability to bind to an mRNA codon, only when the flanking sequences include certain features.

Copy number   the ratio of the number of molecules of a plasmid in a bacterial cell to the number of chromosomes in that same cell.

Cotransduction frequency   the percentage of total transductants receiving a particular unselected marker.

Covalently closed   a macromolecule held in a circular configuration by covalent chemical bonds. Usually applied to DNA molecules.

Cryptic gene   a gene that is not normally expressed but that codes for a fully functional protein.

Cryptic plasmid   a plasmid that does not have any obvious function within a cell.

Cryptic prophage   a prophage that has mutated or suffered deletions so that it is no longer capable of reactivating and producing a virus. There may, however, be functional genes within it that can activate during an induction process.

Cutting in trans   a repair process during which the presence of damage in one DNA helix induces cutting in a homologous region within a different helix.

D-loop   a loop of single-strand DNA formed within a DNA helix as a result of invasion of a homologous, third strand of DNA.

DNA linking   using DNA ligase to join together DNA fragments in vitro.

Deletion mutation   removal of one or more bases from a DNA molecule.

Denature   to eliminate higher order structures in macromolecules. In the case of DNA, to break the hydrogen bonds holding a double-helix together.

Density gradient   a tube containing a solution whose density increases from top to bottom.

Diauxie   the phenomenon of biphasic growth discovered by Jacques Monod. *Escherichia coli* cells growing on a mixture of glucose and another sugar will exclusively use glucose until the supply is exhausted. Only then will they express the necessary genes to use the other sugar.

Diploid   a cell that carries two complete copies of each chromosome.

Directed transposition   forcing an insertion mutation at a particular spot or within a particular gene.

Distal marker   in conjugation any gene that is located in a portion of the chromosome that is transferred after the selected marker.

Double site-specific recombination   a DNA strand exchange event that occurs at a specific site located on both molecules.

E site   location on a ribosome where an exiting tRNA molecule pauses until it is displaced when a new tRNA molecule enters the A site.

Eclipse complex   during genetic transformation, the stage at which the donor DNA binds to a membrane protein and becomes incapable of retransformation if extracted from that cell.

Eclipse phase   the time during a virus infection when there are no intact virus particles within the host cell.

Electrotransformation   uptake of DNA induced by a pulse of high-voltage electricity.

Embedded gene   a gene that is located within the physical boundaries of another gene.

Endonuclease   an enzyme that cuts a phosphodiester bond within a DNA molecule.

Episome   a plasmid that can integrate itself into the bacterial chromosome.

Excinuclease (excision endonuclease)   an enzyme that makes a nick near the damaged DNA bases so that excision repair can occur.

Extein   the portion of a protein molecule that is retained after splicing has occurred.

F pili   long, thin, hairlike structures found on the surface of cells carrying an F plasmid.

F plasmid   the independently replicating DNA molecule responsible for conjugation in *Escherichia coli*.

F⁻ phenocopy   an F plasmid-carrying cell that has not synthesized any F pili and consequently behaves like an F⁻ cell in conjugation.

F-prime   an F plasmid that includes some bacterial DNA as well as its own sequences.

Fertility inhibition   the presence of one plasmid represses the conjugal functions of another plasmid within the same cell.

Filter mating   conjugation that occurs when a mixture of cells is immobilized on the surface of a microporous filter to give a very high cell density.

Fluctuation test   an experiment developed by Luria and Delbrück to show that bacteriophage mutations occur randomly, not as the result of applied selection.

Folded chromosomes   bacterial genomes as they are found in intact nucleoids. The various superhelical loops are folded on one another so that the total DNA molecule occupies comparatively little space.

Forward mutations   changes in the base sequence of a genome that alter the conventional wild-type genotype.

Founder effect   when a genetic trait in a subpopulation occurs because the first individuals in that subpopulation happened to carry that trait rather than because the trait provides a selective advantage to the members of the subpopulation.

Gene fusions   artificial DNA constructs in which the beginning of one gene is joined (fused) to another so that a composite gene product is produced.

Generalized transducing particle   a virion that carries only host DNA, no viral DNA is present.

Generation time   the time required for a microbial population to double in number.

Genetic equilibrium   a steady state in which the rate of forward mutation at a particular locus is balanced by the rate of back word mutation at the same locus so that the percentage of mutants in the population is constant.

Genetic transformation   the process of genetic transfer in which DNA from a donor cell diffuses through a medium, is taken up by the recipient cell, and arrives in the cytoplasm.

Genome   the collection of chromosomes constituting the genetic information of a cell.

Genome equivalent   the amount of DNA equal in mass to one complete cellular genome.

Genome signature   a statistical characteristic of an organism that describes the occurrence of dinucleotide pairs on a DNA strand.

Genomics   the study of patterns in the genome of an organism.

Genotype   a catalog of all genes in an organism.

Ghost   the protein shell of a bacteriophage that has lost its nucleic acid.

Global regulatory network   a group of widely scattered operons all regulated by the same repressor and/or activator.

Gratuitous inducer   a molecule with enough structural similarity to the normal inducer of an operon that the gratuitous inducer will trigger transcription of the operon even though it is not a substrate for the enzymes that will be synthesized.

Haploid   a cell that has only one copy of each chromosome.

Heat shock response   synthesis of specific proteins in response to exposure to supraoptimal temperatures.

Hershey circles   the nicked circular DNA molecules formed by phage lambda DNA after it arrives in the bacterial cytoplasm.

Heteroduplex   a double-strand helix of DNA in which the two strands were originally located in nonidentical molecules. A heteroduplex may contain regions of nonhomology but corresponding length or regions where certain sequences are present on one strand and totally absent on the other.

Heterothallic   a mating system in which the two participants are morphologically distinct.

Hets   heterozygous T4 phages resulting from unfinished recombination or dissimilar terminal redundancies.

Hfr strain   a cell with a conjugative plasmid integrated into its chromosome. The physical linkage means that large portions of the bacterial chromosome can transfer whenever the plasmid is conjugating itself into another cell.

High frequency transducing (HFT) lysate   a collection of temperate phage virions among which specialized transducing particles are numerous.

Holin   a membrane protein that creates an open channel, usually for the egress of virus particles.

Holliday structure   the x-shaped structure formed by two recombining DNA molecules.

Holoenzyme   a protein complex that includes a core enzymatic activity and various proteins that serve to modify and regulate that activity. An example is RNA polymerase.

Homing endonuclease   an agent of site-specific recombination that acts to insert parasitic DNA elements such as inteins or integrons into regions of DNA that are roughly homologous to its current location.

Homoplastic   in evolutionary genetics, the situation that prevails when there is one or a limited number of biochemical ways to accomplish a particular task. In such an instance, all individuals carrying that biochemical trait will appear to have related sequences, even if they have no common ancestor.

Horizontal gene transfer   the inheritance of genes via transduction, genetic transformation, or conjugation instead of from a parental cell.

Host range   the list of those organisms that can be infected by a particular virus or serve as conjugal recipients for a particular plasmid.

Immunity region   the portion of a temperate virus genome that contains the code for the repressor protein and the sites to which the repressor binds.

Incompatibility   the inability of two plasmids to coexist in the same cell.

Inducer   a molecule that triggers transcription of an operon.

Induction   the turning on of genes. It may represent removal of a repressor and/or addition of an activator.

Infectious center   anything that will produce a viral plaque, i.e., a virus particle or a virus-infected cell.

Insertion mutation   a mutation caused by the insertion of one or more bases into a DNA molecule.

Integrase   the enzyme that catalyzes the integration of phage λ DNA during the formation of a lysogen.

Integrative suppression   elimination of a temperature-sensitive DNA replication deficiency owing to the insertion of a normally replicating plasmid into the mutated DNA molecule.

Integron   a DNA element capable of site-specific recombination that captures genes associated with a sequence that is the substrate of its integrase.

Intein   a protein coding region embedded inside the gene of another protein that catalyzes its own removal after protein synthesis.

Intercalating agent   a chemical resembling a nucleotide base that can slip between bases in a DNA helix.

Intergenic suppressor   a mutation in a second gene that eliminates the phenotype caused by a mutation in another gene.

Interrupted mating   a conjugation experiment in which the aggregated cells are separated or DNA replication is inhibited so that DNA transfer stops at a particular time. This type of experiment allows an investigator to determine the gene order.

Interruptions   nicks that naturally occur within individual strands of phage T5 DNA.

Intervening sequence (intron)   naturally occurring pieces of DNA inserted into a gene. In order for the gene to be properly expressed, its RNA transcript must be processed to remove the intron(s).

Intragenic suppressor   a mutation in a gene that eliminates the phenotype caused by a preexisting mutation in that gene.

Inverted repeat   two identical or nearly identical DNA sequences, one of which has been rotated 180° with respect to the other. The repeated sequences may be separated by extended stretches of nonrepetitive DNA. Inverted repeats demarcate the ends of transposons.

Isoschizomers   restriction enzymes that recognize the same sequence of DNA bases but come from different organisms and may cut the DNA in different ways.

Iteron   a short sequence on plasmid DNA that is directly repeated several times and is involved in the regulation of copy number by binding a protein required for initiation of plasmid DNA replication.

Jackpot tube   in the fluctuation test, a tube that has a disproportionate number of mutant cells, indicating that its inoculum probably included at least one mutant cell.

Lacuna   a hole that develops within a bacterial lawn as a result of bacteriocin production.

Latent phase   the time interval between synchronous infection of a culture by virus particles and the beginning of cell lysis.

Leader sequence   the region of an mRNA transcript lying between its 5′-end and the coding sequence for the first major protein.

Lethal zygosis   killing of recipient cells caused by an excess of Hfr cells during conjugation.

Linkage group   a group of genes that tend to be coinherited.

Low-frequency transducing (LFT) lysate   a collection of temperate phage virions among which specialized transducing particles are rare.

Lysis from without   infecting a culture with such a high multiplicity of bacteriophage particles that all cells lyse as a result of numerous attempts to introduce viral DNA into their cytoplasm.

Lysis inhibition   a phenomenon observed in phage T4 in which reinfection of a cell during the latent phase delays the onset of cell lysis.

Lysogen   a cell that carries a prophage, a quiescent form of a virus.

mRNA class   temporal grouping of mRNA molecules based on the onset of transcription. Examples include delayed early, immediate early, late, and middle.

Male-specific phage   a virus that infects only plasmid-carrying bacteria.

Maturation   assembly of functional virus particles.

Merodiploid   a cell that carries two independent copies of only some genes.

Mismatch repair system   an enzymatic system that removes one of a pair of mismatched bases so that the correct base can be inserted.

Missense mutation   a base substitution mutation that yields a codon with a different meaning than the original.

Modification enzyme   an enzyme activity that must always be present when a restriction enzyme is synthesized. It recognizes the same DNA base sequence as a restriction enzyme but chemically modifies that sequence (e.g., by methylation) so that the restriction enzyme is unable to react with it.

Multiplicity of infection   the ratio of virus particles to potential host cells.

Mutagen   a substance or physical treatment that causes mutations.

Mutation   any change in the nucleotide sequence of a nucleic acid.

Muton   the smallest mutable unit in DNA, i.e., a single base.

Nick   a broken sugar phosphate bond in a double-strand nucleic acid.

Nonsense mutation   a missense mutation that changes a codon so that instead of coding for an amino acid it codes for a translation stop signal.

Nuclease   an enzyme that degrades a nucleic acid.

Nucleoids   aggregates of chromosomal DNA found in prokaryotes.

Okazaki fragments   short pieces of DNA that are biosynthetic intermediates during synthesis of the lagging strand of a conventional, replicating duplex DNA molecule.

Oligonucleotides   short chains of nucleic acid.

One-step growth experiment   a procedure in which virus particles infect cells in an approximately synchronous manner, and their progeny are prevented from infecting additional cells. It is a method for obtaining one cycle of viral growth.

Open complex   RNA polymerase bound to a DNA helix that has broken some of its hydrogen bonds and partially unwound, so that the active site of the polymerase complex fits within the helix.

Open reading frame   a region of an mRNA molecule (or its corresponding DNA) that should be translatable because it carries a ribosome binding site and the correct termination signal.

Operator   the site on a DNA molecule to which a repressor binds, thereby preventing transcription.

Operon   a group of genes that are all part of the same mRNA transcript and therefore coordinately regulated.

Orthologous   genes derived from a common ancestor or one member of the set is the progenitor for all other members, meaning that they are homologous.

P site   location on a ribosome to which is bound the tRNA molecule that is connected to the nascent peptide chain.

Palindrome   a sequence that reads the same, forward or backward. A DNA example is:

$$5'CGTACG3'$$
$$3'GCATGC5'$$

Paralogous   genes resulting from DNA duplication followed by independent evolution of the two sets of genes.

Penicillin selection   using penicillin or one of its derivatives to kill growing bacterial cells so that nongrowing, mutant cells will survive and be enriched in the population.

Peptide   a short chain of amino acids.

Phage   a bacteriophage, a bacterial virus.

Phagemid   a bacteriophage lysogenic form of DNA that is not integrated into the chromosome but rather is a high copy number plasmid.

Phenotype   those genetic traits of an organism that are presently expressed.

Phenotypic lag   the delay that occurs in expressing a mutant phenotype while nonmutant gene products are eliminated from the cytoplasm of a newly mutated cell.

Phenotypic mixing   the situation that occurs when the nucleic acid inside a virus particle does not code for one or more of the proteins making up the virion.

Pheromone   a substance that stimulates conjugal behavior.

Photoreactivation   cleavage of pyrimidine dimers in a DNA molecule resulting from exposure to long wave ultraviolet radiation.

Physical map   a genetic map on which distances are measured in terms of physically defined units such as the number of base pairs rather than in terms of recombinationally defined units.

Plaque morphology mutation   a change in a viral nucleic acid that alters the size, shape, or clarity of plaques caused by the phage.

Plaque-forming unit   anything that will form a plaque. One unit could be an individual virus particle or a virus-infected cell.

Plaque   an area of lysed cells or inhibited bacterial growth caused by infection with a virus particle.

Plasmid   a DNA molecule of less than chromosomal size that is capable of self-replication. It does not code for essential cell functions.

Plateau phase   the time period at the end of a one-step growth experimental viral infection when all infected cells have lysed and no new infections are possible.

Pock formation   in *Streptomyces*, the inhibition of aerial mycelia formation by a plasmid. The consequence of a cell carrying a plasmid is that its clone is readily visible in a bacterial lawn because of the absence of mycelia.

Polar effect   a mutation in one gene reduces or prevents expression of genes normally located 3' to it on the same mRNA molecule.

Polylinker   a region within a cloning vector that contains multiple restriction enzyme recognition sites, so that different DNA fragments can insert, essentially at the same spot.

Polysome   an mRNA molecule that is being translated by multiple ribosomes.

Positive regulatory control element   a molecule that must be present for transcription of a particular operon to occur.

Postsynaptic   occurring after homologous DNA molecules have associated.

Premature lysis experiment   an experiment in which the experimenter breaks open virus-infected cells before they would normally lyse in order to examine intermediate stages of viral infection.

Presynaptic   occurring before homologous DNA molecules have associated.

Promoter   an RNA polymerase binding site on a DNA molecule.

Prophage   a quiescent viral DNA molecule that has established itself as a permanent part of a bacterial cell. The only genes expressed on a prophage are those necessary to keep lytic functions turned off and those required for DNA replication, if the prophage is not integrated into the bacterial chromosome.

Proteomics   the study of the proteins of an organism and the conditions under which they are produced.

Protoplast   a cell that has lost its cell wall. In bacteria, a Gram-positive cell that has lost its cell wall.

Prototroph   a cell that will grow on a minimal salts medium. It can synthesize all biochemical compounds required for growth.

Proximal marker   in conjugation, any gene that is located in a portion of the chromosome that is transferred before the selected marker.

Pseudogene   a nonfunctional region of DNA with extensive sequence similarity to an expressed DNA region.

Pseudolysogen   a virus-infected cell in which the virus infection proceeds so slowly that the culture mimics a lysogen because progeny virus particles are slowly released over time. However, unlike a true lysogen, maintenance of the phage in the bacterial culture requires continual reinfection.

Quorum sensing protein   a protein that is activated by a substance produced in small quantities by a bacterial cell. When the culture is sufficiently dense (a quorum is present), the protein activates a particular response.

R plasmid   an independently replicating DNA molecule that encodes resistance to one or more antimicrobial agents.

R-determinant the portion of the R100 plasmid that carries the majority of the antibiotic resistance genes found on that plasmid.

Reading frame the pattern of triplet codons established when a ribosome first binds to an mRNA molecule. There are three possible reading frames for an mRNA.

Recognition sequence the base sequence to which restriction and/or modification enzymes bind.

Recombinase an enzyme that catalyzes DNA recombination.

Recombination rearrangement of the phosphodiester bonds linking bases together, so that altered DNA molecules are formed.

Recombination frequency the rate at which new combinations of particular genes arise when DNA molecules exchange information.

Recon the smallest possible unit of recombination, a single base pair.

Regulon a group of operons all dealing with the same general metabolic pathway that are coordinately regulated.

Rehybridize allow strands of denatured DNA to form hydrogen bonds with each other so that more or less normal duplexes develop.

Relaxed control starving a bacterial cell for amino acids does not immediately shut down rRNA and tRNA biosynthesis.

Release factor a protein cofactor that is necessary for release of the ribosome from the mRNA molecule after translation has completed.

Replica plating a printing process in which bacterial colonies are pressed against a piece of velvet so that cells transfer to the velvet, and then the velvet surface is used to inoculate new plates.

Replication the process of duplicating a genome or plasmid within a cell. The nucleic acid duplicated may be DNA or RNA.

Resistance transfer factor the portion of the R100 plasmid that is primarily concerned with conjugal transfer rather than antibiotic resistance.

Resolvase a site-specific recombination enzyme that converts concatemeric DNA into monomers.

Restriction enzyme an enzyme that binds to specific, unmodified DNA base sequences and cleaves both strands of the DNA molecule to which it is attached.

Reversion restoration of a cell's genotype to its original, nonmutant state.

Ribozyme an RNA molecule that has catalytic activity (i.e., behaves like an enzyme).

Rise phase the time interval during a one-step growth experiment within which all infected cells lyse.

Rolling circle replication a mode of DNA replication in which a nick is introduced into one strand of a circular double helix, and then replication is initiated at the nick. During replication, the newly synthesized strand displaces the corresponding old strand. The displaced strand can either serve as a template for synthesis of a complementary strand or provide a single strand of DNA for viral assembly or conjugation.

SOS repair the DNA repair system that operates in cases of large amounts of DNA damage.

Satellite phage   a bacteriophage such as P4 that can only reproduce itself in a cell carrying another type of virus to supply missing essential functions.

Segregation lag   the delay observed in the increase of mutant cell numbers after a mutation occurs. It is due to the necessity of eliminating nonmutant chromosomes from the mutant cell.

Selection   a treatment or environmental condition that allows only certain organisms to grow and/or survive.

Short-patch repair   excision repair that removes about 20 nucleotides.

Signature sequences   portions of aligned amino acid sequences where proteins clearly fall into two or more groups.

Single burst experiment   an experiment in which virus-infected cells are diluted and aliquotted into tubes so that each tube contains a maximum of one infected cell.

Single site-specific recombination   a DNA strand exchange event that occurs at a specific site with respect to one member of the pair, but at an essentially random site with respect to the other member of the pair.

Site-directed mutagenesis   causing a change in nucleotide sequence at a particular location on a nucleic acid.

sRNA   short, nontranslated RNA molecules that provide regulatory, enzymatic prosthetic group, or other functions.

Specialized transducing particle   a virion that carries a DNA molecule that is a mixture of host and viral DNA.

Spheroplast   a Gram-negative cell that has lost most of its cell wall due to inhibition of peptidoglycan synthesis.

Strand invasion   a single strand of DNA entering into a DNA duplex to form a D-loop.

Strand   one-half of a DNA duplex.

Stringent control   starving a bacterial cell for amino acids immediately shuts down rRNA and tRNA biosynthesis.

Superhelical coils   twists introduced into a covalently closed, circular DNA in addition to those inherent in the double-helical structure.

Superinfection   when a virus-infected cell is reinfected with the same or a different virus.

Suppressor   anything that eliminates the phenotypic effect of a mutation.

Surface exclusion   inhibition of DNA transfer between two cells carrying F plasmids. An F^- phenocopy is not subject to surface exclusion.

Syntrophy (cross-feeding)   a situation in which a pair of organisms, neither of which can grow by itself, manages to grow because each organism secretes the nutrient compound necessary for the other's growth.

Ternary complex   the combination of DNA helix and RNA polymerase that forms after the open complex. It is a more stable association and persists during RNA synthesis.

Time of entry   the elapsed time between when Hfr and F^- are mixed and when a plated sample of the mating mixture first yields a particular type of recombinant colony.

Topoisomerase   an enzyme that introduces or removes superhelical turns from DNA molecules.

Trans **translation**   a method of freeing up a stalled ribosome. A dual purpose tmRNA carrying an alanine residue binds to the ribosomal A site, adds its amino acid, then displaces the mRNA. The ribosome translate a short open reading frame and terminates normally. The extra amino acids tag the protein for destruction by a protease.

Transactivation   when genes on one DNA molecule code for diffusible gene products that can activate transcription on a different DNA molecule.

Transconjugant   recombinant cell resulting from conjugation.

Transcription   the process catalyzed by RNA polymerase in which an RNA copy of a portion of a DNA strand is made.

Transductant   recombinant cell resulting from transduction.

Transduction   the process of transporting previous host cell DNA inside a virion to a new cell.

Transfection   the process of genetic transformation using a viral DNA molecule, so that the end result is a virus-infected recipient cell.

Transfer DNA synthesis   DNA synthesis that occurs concomitantly with conjugal transfer of DNA.

Transfer delay   the elapsed time between when Hfr and F^- are mixed and when actual DNA transfer begins.

Transformant   recombinant cell resulting from genetic transformation.

Transformasome   a membrane vesicle formed around transforming DNA in *Hemophilus influenzae*.

Transition mutation   substitution of one pyrimidine for another pyrimidine or one purine for another purine in a DNA molecule.

Transitioning   the phenomenon observed in *Proteus mirabilis* whereby the number of r-determinants found on an R100 plasmid increases in the presence of antibiotic selection and decreases in its absence.

Translational coupling   a situation in which translation of one coding region of an mRNA molecule requires that a preceding region be translated first, in order to allow proper binding of a ribosome to the mRNA.

Translocation   movement of a ribosome and its associated nascent peptide along an mRNA molecule. During translocation, both bound tRNA molecules move to the next available binding site.

Transposase   the enzyme synthesized by transposons that catalyzes DNA transposition.

Transpososome   a stable complex of transposon DNA and protein that forms during a transposition event.

Transversion mutation   substitution of a pyrimidine for a purine, or a purine for a pyrimidine in a DNA molecule.

Vector   a self-replicating DNA molecule that serves as the basis for genetic engineering experiments. A vector provides replication and in some cases expression functions to cloned DNA.

Virion   a virus particle, with one or more nucleic acid molecules wrapped inside a protein coat.

W-reactivation   the phenomenon first observed by Jean Weigle in which ultraviolet-irradiated λ phage particles are more efficiently repaired by ultraviolet-irradiated host cells than by normal hosts.

Wobble hypothesis   the proposal by Francis Crick that the base in the third position of a codon does not strictly pair with the corresponding base in a tRNA anticodon, so that there can be synonymous codons.

Wrapping choice   a model for the formation of generalized transducing particles. The model states that the viral DNA packaging machinery occasionally errs and packages host DNA instead of viral DNA.

Zone of adhesion   a region of a Gram-negative bacterial cell wall within which the outer membrane and the inner membrane touch each other.

Zygotic induction   a prophage transferred during conjugation activates and begins a lytic cycle upon arrival in a recipient cell that lacks the appropriate viral repressor molecules.

Index